T0271258

# Medicinal Plants in the Asia Pacific for Zoonotic Pandemics

# Medicinal Plants in the Asia Pacific for Zoonotic Pandemics

*Series Editor:*
Christophe Wiart

This series provides an unprecedented comprehensive ethnopharmacological overview of the antibacterial, antifungal, and antiviral activities of the medicinal plants used traditionally in Asia-Pacific for the treatment of microbial infections. It discusses their actions and potentials against viruses (including COVID-19), bacteria, and fungi representing a threat of epidemic and pandemic diseases. Scientific names, botanical classifications, descriptions, medicinal uses and antimicrobial chemical constituents, commentaries, chemical structures, and selected bibliographical references are provided for each plant. This series is a reference for anyone involved in the discovery of leads for the treatment or prevention of microbial zoonotic pandemics.

Medicinal Plants in the Asia Pacific for Zoonotic Pandemics, Volume 1:
Family Amborellaceae to Vitaceae
*Christophe Wiart*

# Medicinal Plants in the Asia Pacific for Zoonotic Pandemics

Family Amborellaceae to Vitaceae

Volume 1

Christophe Wiart

## CRC Press
Taylor & Francis Group
Boca Raton London New York

CRC Press is an imprint of the
Taylor & Francis Group, an **informa** business

First edition published 2021
by CRC Press
6000 Broken Sound Parkway NW, Suite 300, Boca Raton, FL 33487-2742

and by CRC Press
2 Park Square, Milton Park, Abingdon, Oxon, OX14 4RN

ISBN: 978-1-032-00265-1 (hbk)
ISBN: 978-1-138-48204-3 (pbk)
ISBN: 978-1-351-05907-7 (ebk)

Typeset in Times
by codeMantra

*A ma grand mère, Renée Monllor*
*A ma mère, Flora Monllor*
*A mon épouse, Shirley Pita*
*A mon fils, Christophe Pita-Wiart*

# Contents

# Foreword

From the very beginning, the principle and purpose of medicine, in particular of pharmacology, was a struggle with natural selection for human life and health. Using natural sources, even ancient physicians were able to save patients suffering from infections of bacterial, viral, parasitic, or fungal origin, who would otherwise become severely ill or even die without treatment. Among these preparations, materials from ants and bees, scorpions and flies, snakes and frogs, as well as sponges, mushrooms, fishes, and mammals were used for centuries with more or less success. But of course, different plants, trees, and herbs, their seeds, fruits, roots, leaves, and flowers contained the most of ancient therapeutic remedies. The development of rational synthetic chemistry, starting from the 19th century, allowed to broaden the panel of compounds and to substitute compounds isolated from natural sources with those synthesized in laboratories. Since that time, owing to combinatory chemistry, desired structures and scaffolds, including those not occurring in nature, could be synthesized and used as potential drugs. The development and application of cell-free systems for high throughput screening of chemical libraries against specific targets resulted in a dramatic acceleration of drug-oriented researches. Finally, the modern computational methods of molecular modeling and virtual screening, including methods using artificial intelligence, allowed to test in silico virtual libraries consisting of hundreds of millions of compounds. Despite the wide possibilities the synthetic chemistry and rational drug design offer, it must be always taken into account that all potential drugs are supposed to be applied for the treatment of human beings. This imposes specific restrictions and sets certain requirements for their scaffolds. Indeed, enzymes of eukaryotic cells have specific preferences for the isomeric composition of compounds. One of such examples is that our ribosomes can handle only with L-, but not D-stereoisomers of amino acids to produce functional proteins. Many other synthetic scaffolds cannot be utilized by cellular enzymatic systems either. Not surprising are, therefore, the results of the study of Newman and Cragg (2020) having demonstrated that among all therapeutic agents approved from 1981 to 2019, only one-third of them was of fully synthetic origin. Other two-thirds were either completely natural or modified natural structures (i.e., those derived from a natural product with semisynthetic modification) as well as compounds made by total synthesis, but whose pharmacophore was from a natural product. These compounds, including fully natural ones, are now successfully used in both Western medicine and ethnical pharmacological systems, such as Ayurveda, traditional Chinese medicine, and many others. Secondary plant metabolites are, therefore, an inexhaustible source for new scaffolds having, with high probability, a potential biological activity. Of great importance and highest novelty and variability are the plants of poorly studied regions of rainforests in Asia, Africa, and South America. Among infectious organisms, viruses represent an important group of causative agents of diseases that are extremely hard to fight. Due to a specific life cycle, viral components disseminate within the cell after infection, and for some period, virus becomes a part of the cell. Therapeutic compounds, therefore, must be very selective and tightly discriminate viral and cellular components to kill a virus without affecting the cell. Moreover, on the battlefield among viruses and humans, one can see in real time the process of selection of viruses that are resistant to antiviral drugs. This problem is important for such dangerous pathogens as the human immunodeficiency virus, herpesviruses, influenza viruses, and many others. For some drugs and viruses, the genetic barrier for resistance is high, but in other cases, it is very low, so that viruses can overcome the inhibiting activity of a drug. This becomes especially actual when patients do not adhere to doses and regimens of drug use, thus providing a sub-therapeutical concentration of a drug in the body. As an example, treatment of human immunodeficiency virus infection with a single drug results in an easy selection of resistant variants and progression of disease despite the therapy. However, the introduction of the second and third drug into the scheme of treatment leads to the effective elimination of the virus, because the development of double- and triple-resistant viral mutants is much less probable.

Another serious problem of controlling viral infections is the continuous danger of the introduction of novel viruses into the human population. Novel viruses are regularly detected in humans their examples including Zika virus, avian influenza virus, severe acute respiratory syndrome (SARS), and Middle East respiratory syndrome (MERS) viruses and, of course, the main headache nowadays, SARS-CoV-2 causing coronavirus infection called COVID-19. Current pandemics covered all countries of the globe and already resulted in more than 40 million cases with over a million deaths. For these viruses, at least at the beginning of their circulation, there are no even experimental drugs, people appear unprotected, and the only hope is sanitary preventive measures. The development of therapeutic agents for such viruses, even in case of re-purposing of already approved drugs, takes months, if not years. Moreover, even the routine viruses that can be normally controlled by vaccination, like the Influenza virus, can undergo antigenic shift leading to the formation of the virus that is new for the human immune system. For these viruses, there are no population immunity, vaccine development, manufacturing, and distribution also take several months, so the virus spreads uncontrollably among the population causing pandemics. In the case of Influenza, the last pandemics occurred in 2009 ("swine flu").

All mentioned above clearly indicates that the search and development of novel antivirals exploring novel mechanisms and alternative viral (or cellular) targets are of great importance and a high priority for today. Prof. Christophe Wiart, the author of the presented book and scientist I am lucky to work with, is one of the most prominent world authorities in the field of ethnobotany and ethnopharmacology. He has collected, described, and characterized several hundreds of rare plants of Asian rainforests whose extracts contain novel antibiotics, antiviral, anti-parasite, antioxidant, or anticancer agents. It is hard to overestimate the importance of Prof. Wiart's efforts in the field of search and development of novel pharmaceutical agents against human diseases, in particular infectious ones caused by drug-resistant pathogens. He was he who said "…the last hope for the human race's survival … is in the rainforests of tropical Asia. The pharmaceutical wealth of this land is immense." In summary, the problem of the development of novel antivirals is of greatest priority in medicinal science, and in this regard, the book of Prof. Christophe Wiart is very desirable, timely, and recommended for reading by specialists in botany, plant biochemistry, virology, and medicinal chemistry.

**Dr. Vladimir Zarubaev**

## REFERENCE

Newman, D.J. and Cragg, G.M., 2020. Natural products as sources of new drugs over the nearly four decades from 01/1981 to 09/2019. *Journal of Natural Products*, 83, pp. 770–803. doi:10.1021/acs. jnatprod.9b01285.

# Preface

The recent emergence of life-threatening waves of viral zoonosis of increasing severity, culminating today with the COVID-19 pandemic, marks the beginning of an era that might be the last before the ultimate disappearance of human beings. Throughout history, plagues threatened human survival. The Roman Empire was struck from about 166 to 189 A.D. by a smallpox pandemic known as "the Antonine plague" claiming the lives of half of the population of the Empire including the emperor Marcus Aurelius and contributing to the socioeconomic decline of the Western Roman Empire until its fall in 476 A.D. The zoonotic bacterium *Yersinia pestis* accounted for two major plagues in the Middle Ages: "the Justinian plague" (6th century) and the "Black Death" (14th century) claiming the lives of millions. The Spanish flu that followed World War I due to a zoonotic Influenza A (H1N1) virus made more than 20 million victims. For the last about 50 years, novel life-threatening zoonotic viruses have emerged and spread globally including the human immunodeficiency virus, the Ebola virus, the Zika virus, the Middle East respiratory syndrome coronavirus, and today the ignominious severe acute respiratory syndrome coronavirus 2 (SARS-CoV-2) which according to the WHO has infected so far 124 million people, caused 2.72 million deaths, and paralyzes the whole world as never seen before.

Among the reasons for the emergence of zoonotic pandemics is the increasing contact of humans with animal reservoirs due to deforestation and industrial animal farming. According to Wolfe et al. (2005), the richness of microbes in a region, environmental changes, and increased human and animal contact with wildlife result in the emergence of new zoonotic pathogenic microorganisms. Thus, one could envisage the increasing waves of zoonosis to be biological "negative feedback loops" meant to compel humans to put an end to primary rainforest burning, carbon dioxide emission, global warming, as well as intensive poultry, swine, and other animal farming.

The pharmaceutical industry has managed to produce vaccines offering some levels of protection against SARS-CoV-2. However, if pressures on wildlife and nature continue, other new zoonotic pandemics will come with the possibility of the emergence of a completely untreatable fast killing virus, bacteria, or even fungi. Furthermore, between the time of appearance of a new virus and the time of production of a vaccine, people are left vulnerable.

The intention behind the writing of this book is to offer researchers, academics, and students a comprehensive and interrelated set of botanical, ethnopharmacological, and pharmacological evidence to facilitate the discovery of natural products or even herbal remedies for the treatment or prevention of viral, bacterial, or fungal zoonotic pandemics. It is also to offer an intellectual rationale or tool to understand and thus cogently approach the fascinating realm of drug discovery from medicinal plants. There is a law for the rational selection and use of plants for the treatment of diseases. This first volume covers the medicinal plants traditionally used for the treatment of microbial infection in Asia and the Pacific classified in the Clades Protomagnolids, Magnoliids, Monocots, Ranunculids, Core Eudicots, and Rosids. Within each clade, plants are presented according to their respective orders and families. Scientific and common names are givens as well as synonyms, habitat, distribution, botanical observation, pharmacology, commentaries, personally handmade botanical plates, as well as carefully selected references. We are now in lack of drugs for the treatment of COVID-19 and of lead molecules to face the upcoming pandemics. The French privateer Robert Surcouf (1773–1827) sailing across the Indian Ocean once said: "Each of us fights for what he lacks most," and it is now time for us in the face of these pandemics to join efforts to discover new antimicrobial agents from the medicinal plants in the Asia Pacific for zoonotic pandemics.

**Christophe Wiart**
*Kuala Lumpur*
*March 3, 2021*

# REFERENCE

Claas, E.C., 2000. Pandemic influenza is a zoonosis, as it requires introduction of avian-like gene segments in the human population. *Veterinary Microbiology, 74*(1–2), pp. 133–139.

Littman, R.J. and Littman, M.L., 1973. Galen and the Antonine plague. *The American Journal of Philology, 94*(3), pp. 243–255.

Morse, S.S., 2001. Factors in the emergence of infectious diseases. In Price-Smith, A.T., ed. *Plagues and Politics* (pp. 8–26). London: Palgrave Macmillan.

Sabbatani, S. and Fiorino, S., 2009. The antonine plague and the decline of the Roman Empire. *Le infezioni in medicina: rivista periodica di eziologia, epidemiologia, diagnostica, clinica e terapia delle patologie infettive, 17*(4), pp. 261–275.

Slingenbergh, J., Gilbert, M., Balogh, K.D. and Wint, W., 2004. Ecological sources of zoonotic diseases. *Revue scientifique et technique-Office international des épizooties, 23*(2), pp. 467–484.

Taylor, L.H., Latham, S.M. and Woolhouse, M.E., 2001. Risk factors for human disease emergence. *Philosophical Transactions of the Royal Society of London. Series B: Biological Sciences, 356*(1411), pp. 983–989.

Wagner, D.M., Klunk, J., Harbeck, M., Devault, A., Waglechner, N., Sahl, J.W., Enk, J., Birdsell, D.N., Kuch, M., Lumibao, C. and Poinar, D., 2014. *Yersinia pestis* and the Plague of Justinian 541–543 AD: a genomic analysis. *The Lancet Infectious Diseases, 14*(4), pp. 319–326.

Wolfe, N.D., Daszak, P., Kilpatrick, A.M. and Burke, D.S., 2005. Bushmeat hunting, deforestation, and prediction of zoonotic disease. *Emerging Infectious Diseases, 11*(12), p. 1822.

# Author

**Christophe Wiart, PharmD, PhD,** is an associate professor in the School of Pharmacy at the University of Nottingham, Malaysia Campus. His fields of expertise are Asian ethnopharmacology, chemotaxonomy, and ethnobotany. He has collected, identified, and classified several hundred species of medicinal plants from India, Southeast Asia, and China.

Dr. Wiart appeared on HBO's *Vice* (TV series) in season 3, episode 6 (episode 28 of the series) titled "The Post-Antibiotic World and Indonesia's Palm Bomb," April 17, 2015. It highlighted the need to find new treatments for infections that were previously treatable with antibiotics but are now resistant to multiple drugs. "The last hope for the human race's survival, I believe, is in the rainforests of tropical Asia," said ethnopharmacologist Christophe Wiart. "The pharmaceutical wealth of this land is immense."

# 1 The Clade Protomagnoliids

## 1.1 ORDER AMBORELLALES MELIKYAN, A.V. BOBROV & ZAYTZEVA (1999)

The order Amborellales consists of the single family Amborellaceae Pichon (1848).

### 1.1.1 FAMILY AMBORELLACEAE PICHON (1848)

The family Amborellaceae comprises a single plant: *Amborella trichopoda* Baill.

### 1.1.1.1 *Amborella trichopoda* Baill.

Habitat: Forests

Distribution: New Caledonia

Botanical observation: This shrub is endemic to New Caledonia and a living fossil. The leaves are simple, alternate, and exstipulate. The petiole is 0.6–1.2 cm long and hairy. The blade is 6–23 cm×3–5 cm with a wavy margin. The blade presents 6–19 pairs of secondary nerves. The inflorescence is axillary, cymose, and comprises up to 30 flowers. The perianth consists of spirally arranged tepals which are 0.4–0.8 cm long, white, and triangular. The androecium includes 12–21 stamens which are conspicuous, white, and disposed spirally on the receptacle. The gynoecium comprises 5 free carpels. The fruits are fleshy, drupaceous, 0.8–1 cm×0.4–0.8 cm, and red.

Medicinal uses: Apparently none

*Strong antimycobacterial polar extract*: The methanol extract of fruits inhibited *Mycobacterium tuberculosis* (strain 11–73 P2) with a minimum inhibitory concentration (MIC) between 1 and 2.5 μg/mL (Billo et al., 2005).

*Moderate antibacterial (Gram-positive) flavonol glycoside*: Kaempferol 3-*O*-rutinoside inhibited the growth of *Staphylococcus aureus* (ATCC 14053) and *Staphylococcus aureus* (ATCC 10231) with the MIC value of 125 μg/mL (Bisignano et al., 2000).

Kaempferol 3-*O*-rutinoside

**Commentaries**

i. When looking for antimicrobial agents from plants, the solvents currently used are classified into three main categories—(i) non-polar, (ii) mid-polar, and (iii) polar—which dissolve and remove compounds from vegetal tissues. Non-polar solvents like hexane or petroleum ether extract lipophilic compounds: triglycerides, fatty acids, and any other classes of secondary metabolites which are non-polar. Mid-polar solvents like chloroform, dichloromethane, ethylacetate, or diethyl ether extract amphiphilic secondary metabolites. Polar solvents like water, methanol, ethanol, butanol, or acetone extract hydrophilic secondary metabolites. Consider that amphiphilic antimicrobial compounds have higher chances of drugability.

ii. *Minimum inhibitory concentration*: The MIC is the lowest concentration of extract (as well as fractions and essential oils) or compound that completely inhibits the growth of bacteria in agar. The minimum bactericidal concentration (MBC) is determined by subculturing the test dilutions in sterile agar and incubating further for 18–24 h. The highest dilution that yields 0% bacterial growth is taken as the MBC. MIC is a quantitative technique of choice when assessing the antibacterial or antifungal properties of polar extracts or compounds. When the material to be tested is lipophilic, insolubility occurs and the turbidity generated interferes with optical density readings. It is thus preferable to use paper disc test on solid agar when assessing the antibacterial, antimycobacterial, or antifungal activities of lipophilic extracts or compounds.

iii. *Antibacterial (including antimycobacterial) and antifungal activities of extracts (including fractions and essential oils)*: Extracts are strongly antibacterial or antifungal for MIC below 100 μg/mL, moderately antibacterial or antifungal for an MIC of 100–1000 μg/mL, weakly antibacterial or antifungal for MIC from 1000 to 5000 μg/mL, and very weakly antibacterial or antifungal for MIC above 5000 μg/mL. Extracts are active in inhibiting the mycelial growth of filamentous fungi (when mixed in agar) for an MIC below 5000 μg/mL.

iv. *Antibacterial (including antimycobacterial) and antifungal activities of isolated compounds*: An isolated compound is strongly active for MIC values equal to or below 50 μg/mL (10 μM), moderately active for MIC from 50 to 100 μg/mL (20 μM), weakly active for MIC from 100 to 500 μg/mL (100 μM), very weakly active for MIC ranging from 500 to 2500 μg/mL (500 μM), and inactive for MIC values above 2500 μg/mL (500 μM). Compounds are active in inhibiting the mycelial growth of filamentous fungi (when mixed in agar) for an MIC below 500 μg/mL.

v. *Antiviral activities of extracts (including fractions and essential oils)*: Strongly active for $IC_{50}$ (or $EC_{50}$) values below 100 μg/mL, moderately active for $IC_{50}$ (or $EC_{50}$) of 100–1000 μg/mL, weakly active for $IC_{50}$ (or $EC_{50}$) of 1000–5000 μg/mL, and very weakly active for $IC_{50}$ (or $EC_{50}$) above 5000 μg/mL and above.

vi. *Antiviral activity of compounds*: Strongly active for $IC_{50}$ (or $EC_{50}$) values equal to or below 50 μg/mL (10 μM), moderately active for $IC_{50}$ (or $EC_{50}$) of 50–100 μg/mL (20 μM), weak activity for $IC_{50}$ (or $EC_{50}$) of 100–1000 μg/mL (200 μM), very weakly active for $IC_{50}$ (or $EC_{50}$) of 1000–5000 μg/mL, and inactive above 5000 μg/mL (1000 μM).

**References**

Billo, M., Cabalion, P., Waikedre, J., Fourneau, C., Bouttier, S., Hocquemiller, R. and Fournet, A., 2005. Screening of some New Caledonian and Vanuatu medicinal plants for antimycobacterial activity. *Journal of Ethnopharmacology*, 96(1), pp. 195–200.

Bisignano, G., Sanogo, R., Marino, A., Aquino, R., D'Angelo, V., Germano, M.P., De Pasquale, R., and Pizza, C., 2000. Antimicrobial activity of *Mitracarpus scaber* extract and isolated constituents. *Letters in Applied Microbiology*, 30, pp. 105–108.

Cos, P., Vlietinck, A.J., Vanden Berghe, D. and Maes, L., 2006. Anti-infective potential of natural products: how to develop a stronger in vitro 'proof-of concept'. *Journal of Ethnopharmacology*, 106, pp. 290–302.

Kuete, V., 2010. Potential of Cameroonian plants and derived products against microbial infections: a review. *Planta Medica*, 76, pp. 1479–1491.

Pichon, P., 1948. Les Monimiaceae, famille heterogene. *Bulletin du Museum d' Histoire Naturelle*, 20, 383–384.

Rios, J.L. and Recio, M.C., 2005. Medicinal plants and antimicrobial activity. *Journal of Ethnopharmacology*, 100, pp. 80–84.

Young, D.A., 1982. Leaf flavonoids of *Amborella trichopoda*. *Biochemical Systematics and Habitat*, 10(1), pp. 21–22.

## 1.2 ORDER NYMPHAEALES SALISB. EX BERCHT. ET J. PRESL (1820)

### 1.2.1 FAMILY NYMPHAEACEAE SALISB. (1805)

The family Nymphaeaceae consists of 5 genera and 75 species of aquatic herbs. The leaves are simple, spiral, fleshy, stipulate or exstipulate, floating, and cordate or peltate. The petioles are long and fleshy. The flowers are solitary and large. The calyx comprises four sepals. The corolla consists of up to 70 petals. The androecium includes from 14 to several hundreds of stamens. The gynoecium includes from 5 to 35 carpels free or partially united. The fruit is a fleshy, many-seeded, and irregularly dehiscent berry. Members of this family produce ellagic and gallic acid, gallotannins and ellagitannins, proanthocyanidins, flavonols, benzylisoquinoline, and quinozilidine alkaloids.

#### 1.2.1.1 *Euryale ferox* Salisb. ex K.D. Koenig & Sims

Common name: Euryale, Fox nut; makhna (Bangladesh); Qian chi (China); makhana (India)

Habitat: Lakes, ponds, slow rivers

Distribution: Pakistan, India, Bangladesh, Myanmar, China, Korea, and Japan

Botanical observation : This aquatic plant grows from thick rhizomes. The leaves are simple, floating, orbicular, cordate at base, 30–120 cm in diameter, dark purple below, and spiny. The flowers are solitary, showy, on a spiny peduncle, and up to about 5 cm in diameter. The calyx comprises 4 sepals, which are 2–3 cm long, green, and spiny below. The corolla includes numerous petals which are violet or whitish, oblong-lanceolate, and up to about 2.5 cm long. The androecium includes numerous stamens, which are about 5 mm long. The gynoecium includes 7–16 carpels which are fused. Stigma discoid, depressed, concave with as many rays as carpels. The fruits are dark purple, ovoid and 5–10 cm in diameter, fleshy, spiny, and contain numerous black seeds, which are about 5 mm in diameter.

Medicinal uses: Gonorrhea (China); placentitis, gonorrhea (India)

*Broad-spectrum antibacterial polar extract*: Methanol extract of seeds inhibited the growth of *Staphylococcus aureus* (ATCC 25923), *Escherichia coli* (ATCC 25922), *Pseudomonas aeruginosa* (ATCC 27853), *Proteus vulgaris*, and *Salmonella typhi* with MIC values of 64, 128, 64, 500, and 256 µg/mL, respectively (Parray et al., 2010).

*Antibacterial (Gram-positive) hydrophilic ellagitannin*: Corilagin (LogD = 0.9 at pH 7.4; molecular weight = 634.4 g/mol) inhibited the growth of *Staphylococcus aureus* with an inhibition zone diameter of 12 mm (100 µg/6 mm paper disc) (Fogliani et al., 2005).

*Antibacterial (Gram-positive) simple phenolic*: Gallic acid inhibited the growth of *Staphylococcus aureus* with an inhibition zone diameter of 11 mm (100 µg/6 mm paper disc) (Fogliani et al., 2005).

*Broad spectrum antifungal halo developed by polar extract*: Methanol extract of seeds (400 µg/disc) inhibited the growth of *Candida albicans* and *Penicillum notatum* with the inhibition zone diameters of 16 and 15 mm, respectively (Parray et al., 2011).

*Anticandidal halo developed by ellagitannin*: Corilagin inhibited the growth of *Candida albicans* with the inhibition zone diameter of 12 mm (100 µg/6 mm diameter paper disc) (Fogliani et al., 2005).

*Anticandidal halo developed by simple phenolic*: Gallic acid inhibited the growth of *Candida albicans* with the inhibition zone diameter of 7 mm (100 µg/6 mm paper disc) (Fogliani et al., 2005).

*Strong antiviral (enveloped segmented linear single stranded (−) RNA virus) hydrophilic simple phenol*: Gallic acid (LogD = −2.3 at pH 7.4; molecular weight = 170.1 g/mol) inhibited the replication of Influenza A virus (PR8 strain) with an $IC_{50}$ value of 8.1 µg/mL (Dao et al., 2019).

Gallic acid

*Strong (enveloped linear monopartite double-stranded DNA virus) hydrophilic simple phenol*: Gallic acid inhibited the replication of the Herpes simplex virus type-2 (MS strain) with an $IC_{50}$ value of 10 µg/mL (Kane et al., 1988).

*Strong antiviral (enveloped segmented linear single-stranded (−)RNA) hydrophilic ellagitannin*: Corilagin (LogD = 0.9 at pH 7.4; molecular weight = 634.4 g/mol) inhibited the replication of Influenza A virus (PR8 strain) with an $IC_{50}$ value of 31.2 µg/mL (Dao et al., 2019).

Corilagin

*Strong antiviral (enveloped monopartite linear double-stranded DNA) amphiphilic flavonol*: Quercetin inhibited the replication of the herpes simplex virus type-1 (F strain) with an $IC_{50}$ value of 19.2 µM (Wu et al., 2011).

Quercetin

*Moderate (enveloped monopartite linear single-stranded (+)RNA) simple phenolic*: Gallic acid inhibited the replication of human coronavirus (HCoV) NL63 with the $IC_{50}$ value of 71.4 µM (Weng et al., 2019).

*Viral enzyme inhibition by flavonols*: Kaempferol and quercetin at the concentration of 20 µM inhibited coronavirus ion channel 3a-mediated current by almost 100% (Schwarz et al., 2014) and inhibited Severe acute respiratory syndrome-associated coronavirus 3C-like protease with an $IC_{50}$ value of 73 µM (Chen et al., 2005).

**Commentaries**
  (i) The plant abounds with lignans of which (+)-syringaresinol and the sesquinolignans euryalins A–C (Song et al., 2011).
 (ii) Note that the methanol extract has both antibacterial and antifungal activities, suggesting activity on chromosomes. Corilagin, gallic acid, kaempferol, and quercetin have been identified in the flowers (Kizu and Tomimori, 2003; Mukherjee et al., 1986) and have antibacterial properties (Fogliani et al., 2005; Fu et al., 2016; Yuan et al., 2008).
(iii) Coronaviruses (CoV), members of the Coronaviridae family in the genus *betacoronavirus*, are enveloped viruses that contain a non-segmented, positive-stranded RNA (Perlman & Netland, 2009) responsible for waves of pandemic in the last 20 years. Recrudescence of patients presenting fever, cough, and breathing difficulties was observed in Foshan City in China by clinicians starting from about December 22, 2002 (Liu et al., 2012). The microbe responsible was identified as Severe acute respiratory syndrome-associated coronavirus (SARS-CoV) (Liu et al., 2012). By 2003, this virus accounted for an epidemic outbreak in South East Asia and Canada (Drosten et al., 2003), sickening more than 8000 people and claiming 774 souls (Liang et al., 2004). The Middle East respiratory syndrome coronavirus (MERS-CoV) appeared next in Saudi Arabia in 2012–2016, with a mortality rate of 35.4%. In December 2019, COVID-19 (coronavirus disease 2019) caused by the Severe acute respiratory virus-associated coronavirus 2 (SARS-CoV-2) appeared in Wuhan, China, imposing for the first time in human history complete global paralysis, ensickening around 25 millions and claiming around 900 souls by mid-2020 (WHO, 2020). These waves of increasing severity and threat for the survival of humans are red flags sent by Nature to bring humans to immediately stop the burning of rainforests for palm oil, industrial gas emissions, pollutions of the seas, and violation of animal rights (intensive farming). Should that human madness not be stopped, Nature will send continuous waves of zoonotic viruses, bacteria, or fungi more dangerous than COVID-19 that will eventually

remove humans from Earth. This is a simple negative feedback loop mechanism to stop human pressure on the environment.

(iv) Currently, there are no approved drugs against coronaviruses, but some potential viral targets have been proposed. Nucleocapsid (N) and nonstructural protein 3 (nsp3) interactions are important for coronavirus replication (Hurst et al., 2010). The nsp3 interacts with N protein through its EF motif site that contains a calcium-binding domain, which can mean that this interaction could depend on calcium (Ulasli et al., 2014). The Severe acute respiratory syndrome-associated coronavirus encodes for 3a protein, which forms ion channels that become incorporated into the membrane of the host cells (Hurst et al., 2010).

(v) Halos developed around paper disc or agar well containing extract or compound is a simple method of qualitative appreciation of antibacterial or antifungal effects. No clear guideline is available as to define the strength of the antibacterial and antifungal extracts using paper disc test or agar well for natural products from medicinal plants. This technique is a useful mean of spotting antibacterial or antifungal activity with the formation of a halo around the disc. Simply put, the presence of a halo indicates activity. Note that some strongly antibacterial principles may not diffuse well on agar, explaining why extracts or compounds although having strong activities in liquid broth assays develop small halos. An extract, fraction, or essential oil is defined as active when it develops a halo around a paper disc impregnated with a maximum load of 5000 µg/disc. Also an extract, fraction, or essential oil is defined as active when a halo develops around a well in agar filled with a maximum of 5000 µg/mL of extracts. Also essential oils are active when a halo develops around a paper disc impregnated with a maximum of 100 µL/disc or agar well filled with a maximum of 100 µL/mL or 10% (v/v). For pure compounds, there is activity when they develop a halo around a paper disc impregnated with a maximum load of 1000 µg/disc. Consider that the obtention of crystal-clear halos in paper disc test or agar well is a preliminary indication of bactericidal (including bacteriolytic) or fungicidal effects, whereas turbid halos denote regrowth indicative of bacteriostatic or fungistatic effects.

(vi) Consider that the mode of antibacterial, antifungal, and antiviral natural secondary metabolites from medicinal plants often depends on their concentrations. For instance, phenol at high concentration commands cytoplasmic coagulation but a low concentration targets other cellular components.

## References

Chen, C.N., Lin, C.P., Huang, K.K., Chen, W.C., Hsieh, H.P., Liang, P.H. and Hsu, J.T.A., 2005. Inhibition of SARS-CoV 3C-like protease activity by theaflavin-3, 3′-digallate (TF3). *Evidence-Based Complementary and Alternative Medicine*, 2(2), pp. 209–215.

Dao, N.T., Jang, Y., Kim, M., Nguyen, H.H., Pham, D.Q., Le Dang, Q., Van Nguyen, M., Yun, B.S., Pham, Q.M., Kim, J.C. and Hoang, V.D., 2019. Chemical constituents and anti-influenza viral activity of the leaves of Vietnamese plant *Elaeocarpus tonkinensis*. *Records of Natural Products*, 13(1), pp. 71–80.

Drosten, C., Gunther, S., Preiser, W., van der Werf, S., Brodt, H.R., Becker, S., Rabenau, H., Panning, M., Kolesnikova, L., Fouchier, R.A., Berger, A., Burguiere, A.M., Cinatl, J., Eickmann, M., Escriou, N., Grywna, K., Kramme, S., Manuguerra, J.C., Muller, S., Rickerts, V., Sturmer, M., Vieth, S., Klenk, H.D., Osterhaus, A.D., Schmitz, H. and Doerr, H.W., 2003. Identification of a novel coronavirus in patients with severe acute respiratory syndrome. *The New England Journal of Medicine*, 348, pp. 1967–1976.

Fogliani, B., Raharivelomanana, P., Bianchini, J.P., Bourai, S. and Hnawia, E., 2005. Bioactive ellagitannins from *Cunonia macrophylla*, an endemic Cunoniaceae from New Caledonia. *Phytochemistry*, 66(2), pp. 241–247.

Hurst, K.R., Ye, R., Goebel, S.J., Jayaraman, P. and Masters, P.S. 2010. An interaction between the nucleocapsid protein and a component of the replicase-transcriptase complex is crucial for the infectivity of coronavirus genomic RNA. *Journal of Virology*, *84*, 10276–10288.

Kane, C.J., Menna, J.H., Sung, C.C. and Yeh, Y.C., 1988. Methyl gallate, methyl-3, 4, 5-trihydroxybenzoate, is a potent and highly specific inhibitor of herpes simplex virusin vitro. II. Antiviral activity of methyl gallate and its derivatives. *Bioscience Reports*, *8*(1), pp. 95–102.

Kizu, H. and Tomimori, T., 2003. Phenolic constituents from the flowers of *Nymphaea stellata*. *Natural Medicines*= 生薬學雜誌, *57*(3), p. 118.

Liang, G., Chen, Q., Xu, J., Liu, Y., Lim, W., Peiris, J.S., Anderson, L.J., Ruan, L., Li, H., Kan, B., Di, B., Cheng, P., Chan, K.H., Erdman, D.D., Gu, S., Yan, X., Liang, W., Zhou, D., Haynes, L., Duan, S., Zhang, X., Zheng, H., Gao, Y., Tong, S., Li, D., Fang, L., Qin, P. and Xu, W. 2004. Laboratory diagnosis of four recent sporadic cases of community-acquired SARS, Guangdong Province, China. *Emerging Infectious Diseases*, *10*, pp. 1774–1781

Liu, X., Zhang, M., He, L. and Li, Y., 2012. Chinese herbs combined with Western medicine for severe acute respiratory syndrome (SARS). *Cochrane Database of Systematic Reviews*, *10*(10), CD004882.

Mukherjee, K.S., Bhattacharya, P., Mukheriee, R.K. and Ghosh, P.K., 1986, Chemical examination of *Nymphaea stellata* Willd. *Journal of the Indian Chemical Society*, *513*, pp. 530–531.

Parray, J.A., Kamili, A.N., Qadri, R., Hamid, R. and da Silva, J.A.T., 2010. Evaluation of anti-bacterial activity of *Euryale ferox* Salisb., a threatened aquatic plant of Kashmir Himalaya. *Medicinal and Aromatic Plant Science and Biotechnology*, 4(Special Issue 1), pp. 80–83.

Parray, A., Kamili, A.N., Hamid, R., Ganai, B.A., Mustafa, K.G. and Qadri, R.A., 2011. Phytochemical screening, antifungal and antioxidant activity to Euryale ferox-SALISB. A threatened aquatic plant of Kashmir Himalaya. *Journal of Pharmacy Research*, *4*(7), pp. 2170–2174.

Perlman, S. and Netland, J., 2009. Coronaviruses post-SARS: update on replication and pathogenesis. *Nature Reviews Microbiology*, 7, pp. 439–450.

Schwarz, S., Sauter, D., Wang, K., Zhang, R., Sun, B., Karioti, A., Bilia, A.R., Efferth, T. and Schwarz, W., 2014. Kaempferol derivatives as antiviral drugs against the 3a channel protein of coronavirus. *Planta Medica*, *80*(02/03), pp. 177–182.

Song, C.W., Wang, S.M., Zhou, L.L., Hou, F.F., Wang, K.J., Han, Q.B., Li, N. and Cheng, Y.X., 2011. Isolation and identification of compounds responsible for antioxidant capacity of *Euryale ferox* seeds. *Journal of Agricultural and Food Chemistry*, *59*(4), pp. 1199–1204.

Ulasli, M., Gurses, S.A., Bayraktar, R., Yumrutas, O., Oztuzcu, S., Igci, M., Igci, Y.Z., Cakmak, E.A. and Arslan, A., 2014. The effects of *Nigella sativa* (Ns), *Anthemis hyalina* (Ah) and *Citrus sinensis* (Cs) extracts on the replication of coronavirus and the expression of TRP genes family. *Molecular Biology Reports*, *41*(3), pp. 1703–1711.

Weng, J.R., Lin, C.S., Lai, H.C., Lin, Y.P., Wang, C.Y., Tsai, Y.C., Wu, K.C., Huang, S.H. and Lin, C.W., 2019. Antiviral activity of Sambucus FormosanaNakai ethanol extract and related phenolic acid constituents against human coronavirus NL63. *Virus Research*, *273*, p. 197767.

WHO, 2020. Coronavirus disease (COVID-19) pandemic. https://www.who.int/emergencies/diseases/novel-coronavirus-2019

Wu, N., Kong, Y., Zu, Y., Fu, Y., Liu, Z., Meng, R., Liu, X. and Efferth, T., 2011. Activity investigation of pinostrobin towards herpes simplex virus-1 as determined by atomic force microscopy. *Phytomedicine*, *18*(2–3), pp. 110–118.

Zaki, A.M., van Boheemen, S., Bestebroer, T.M., Osterhaus, A.D., Fouchier, R.A. 2012. Isolation of a novel coronavirus from a man with pneumonia in Saudi Arabia. *The New England Journal of Medicine*, *367*, 1814–1820.

### 1.2.1.2  *Nuphar japonica* DC.

Synonym: *Nuphar japonicum* DC.

   Common name: East Asian yellow water lily; Japanese pond lily; senkotsu (Japan)

   Habitat: Ponds, lakes, and streams

   Distribution: Japan

   Botanical observation: This nuphar grows from stout, 1–3 cm in diameter rhizomes. The leaves are simple, alternate. The petiole is elongated, up to 1 cm in diameter terete. The blades are oblong ovate, 12–35 cm×6–18 cm, with 18–44 lateral veins and blade glabrous to pubescent below. The flowers 2–3.5 cm in diameter. The calyx comprises 5 sepals yellow, rarely red-tinged, greenish toward base, obovate, and rounded at apex. The corolla comprises several petals which are spatulate and yellow. The androecium consists of numerous stamens with anthers of 2.5– 5 mm, long. The fruits are green, urceolate, 2–3.5 cm×1.6–2.3 cm, smooth, and with a conspicuous neck and yellow-ish stigmatic disk yellow, and contains numerous tiny seeds.

   Medicinal use: Wounds (China, Japan)

   *Strong antibacterial (Gram-negative) amphiphilic quinolizidine alkaloid*: 6,6′-Dihydroxythiobinupharidine (LogD = 3.5 at pH 7.4; molecular weight = 526.7 g/mol) from the rhizomes inhibited the growth of *Enterococcus faecalis* and *Enterococcus faecium* with MIC ranging from 2 to 4 μg/mL and methicillin-resistant *Staphylococcus aureus* with an MIC of 2 μg/mL. This alkaloid inhibited topoisomerase IV in *Staphylococcus aureus* (Okamura et al., 2015).

6,6′-Dihydroxythiobinupharidine

**Commentaries**

 (i) During bacterial division, topoisomerase IV catalyzes in the relaxation of the DNA chain (Levine et al., 1998). In Gram-positive bacteria, topoisomerase IV is inhibited by quinolone antibiotics leading to DNA damage and bacterial death (Munoz et al., 1996; Mitton-Fry et al., 2013). Medicinal plants in Asia and the Pacific produce a vast array of quinoline, tetrahydroisoquinoline, piperidine, and quinolizidine alkaloids representing an interesting reservoir of topoisomerase IV inhibitors. Members of genus *Nuphar* Sm. produce a unique type of quinolizidine alkaloids with antibacterial activity, such as thiobinupharidine.

 (ii) Rios and Recio (2005) suggest that MIC superior to 100 μg/mL for compounds indicates weak antibacterial activity and proposes a strong activity with an MIC of 10 μg/mL and below. According to Kuete (2010), the antibacterial activity of compounds is classified in three categories: strong for MIC < 10 μg/mL, moderate for MIC between 10 and 100 μg/mL, and weak for MIC above 100 μg/mL. These classifications are too restrictive and may result in discarding potential antibacterial candidates.

(iii) The plant produces ellagitannins (Ishimatsu et al., 1989) as well as quinolizidine alkaloids (Miyazawa et al., 1998) which may at least partially account for the medicinal use.

## References

Ishimatsu, M., Tanaka, T., Nonaka, G.I., Nishioka, I., Nishizawa, M. and Yamagishi, T., 1989. Tannins and related compounds. LXXV.: isolation and characterization of novel diastereoisomeric ellagitannins, nupharins A and B, and their homologues from *Nuphar japonicum* DC. *Chemical and Pharmaceutical Bulletin, 37*(1), pp. 129–134.

Levine, C., Hiasa, H. and Marians, K.J., 1998. DNA gyrase and topoisomerase IV: biochemical activities, physiological roles during chromosome replication, and drug sensitivities. *Biochimica et Biophysica Acta (BBA)-Gene Structure and Expression, 1400*(1–3), pp. 29–43.

Mitton-Fry, M.J., Brickner, S.J., Hamel, J.C., Brennan, L., Casavant, J.M., Chen, M., Chen, T., Ding, X., Driscoll, J., Hardink, J. and Hoang, T., 2013. Novel quinoline derivatives as inhibitors of bacterial DNA gyrase and topoisomerase IV. *Bioorganic & Medicinal Chemistry Letters, 23*(10), pp. 2955–2961.

Miyazawa, M., Yoshio, K., Ishikawa, Y. and Kameoka, H., 1998. Insecticidal alkaloids against *Drosophila melanogaster* from *Nuphar japonicum* DC. *Journal of Agricultural and Food Chemistry, 46*(3), pp. 1059–1063.

Munoz, R. and De La Campa, A.G., 1996. ParC subunit of DNA topoisomerase IV of *Streptococcus pneumoniae* is a primary target of fluoroquinolones and cooperates with DNA gyrase A subunit in forming resistance phenotype. *Antimicrobial Agents and Chemotherapy, 40*, 2252–2257.

Okamura, S., Nishiyama, E., Yamazaki, T., Otsuka, N., Taniguchi, S., Ogawa, W., Hatano, T., Tsuchiya, T. and Kuroda, T., 2015. Action mechanism of 6, 6'-dihydroxythiobinupharidine from *Nuphar japonicum*, which showed anti-MRSA and anti-VRE activities. *Biochimica et Biophysica Acta (BBA)-General Subjects, 1850*(6), pp. 1245–1252.

### 1.2.1.3. *Nymphaea nouchali* Burm.f.

Synonyms: *Nymphaea stellata* Willd.

Common names: Blue lotus, Water lily; mum phlong (Cambodia); tunjung (Indonesia); nilpadma (India, Bangladesh); hua bua (Laos); teratai kechil (Malaysia); bua fan (Thailand); Sung lam (Vietnam)

Habitat: Ponds, lakes, and slow rivers

Distribution: Tropical Asia and Australia

Botanical observation: This aquatic herb develops from a rhizome. The leaves are simple, alternate, and exstipulate. The petiole is long and slender. The blade is floating, orbicular, sagittate to cordate, 5–20×10–15 cm, and fleshy. The flowers are graceful, solitary, above the water surface, 5–15 cm across. The calyx comprises 4 sepals which are up to 3 cm long. The perianth comprises 10–30 petals which are light blue or yellowish purple, linear oblong, and up to about 10 cm long. The androecium comprises 10–30 stamens, which are about 5 mm long and with blue appendages. The gynoecium includes a syncarpous, many locular ovary developing 10–20 horned-shaped stigmas. The fruit is a globose berry which is about 5 cm in diameter, enclosed by persistent sepals and containing numerous minute seeds.

Medicinal uses: Skin infection, diarrhea, leucorrhea (India)

*Broad-spectrum antibacterial halo developed by polar extract*: Ethanol extract of seeds inhibited the growth of *Salmonella typhi*, *Pseudomonas aeruginosa*, *Escherichia coli*, *Vibrio cholerae*, *Brucella* sp., *Klebsiella pneumoniae*, *Shigella dysenteriae*, *Staphylococcus aureus*, *Bacillus cereus*, and *Enterococcus faecalis* with inhibition zones of 12, 25, 15, 13, 13, 13, 14, 20, 14, and 11 mm, respectively (Parimala & Shoba, 2014).

*Moderate broad-spectrum antibacterial polar extracts*: Methanol extract inhibited the growth of *Bacillus subtilis* (FO 3026), *Sarcinia lutea* (IFO 3232), *Xanthomonas campestris* (IAM 1671), *Escherichia coli* (IFO 3007), and *Klebsiella pneumoniae* (ATCC 10031) with the MIC values of 128, 256, 1024, 512, and 1024 µg/mL, respectively (Dash et al., 2013). Ethanol extract of seeds evoked the MIC values of 300, 30, 60, 70, 30, 100, 300, 100, 300, and 30 µg/mL for *Bacillus cereus*, *Klebsiella pneumoniae*, *Staphylococcus aureus*, *Enterococcus faecalis*, *Shigella dysenteriae*, *Vibrio cholerae*, *Pseudomonas aeruginosa*, Brucell, and *Escherichia coli*, respectively (Parimala & Shoba, 2014).

*Moderate broad-spectrum antifungal polar extract*: Ethanol extract of seeds inhibited the growth of *Curvularia* sp., *Aspergillus niger*, *Penicillium* sp., *Trichophyton mentagrophytes*, and *Candida albicans* with the MIC values of 1200, 1200, 1200, 300, and 300 µg/mL, respectively (Parimala & Shoba, 2014).

*Moderate antifungal (filamentous) mid-polar extract*: Ethylacetate extract of flowers inhibited the growth of *Trichophyton mentagrophytes*, *Epidermophyton floccosum*, *Trichophyton simii*, *Curvularia unata*, and *Scopulatiopsis* sp. with the MIC values of 500, 100, 500, 100, and 1000 µg/mL, respectively (Dash et al., 2013).

*Strong antiviral (enveloped monopartite linear dimeric single stranded (+)RNA) hydrophilic isoquinoline alkaloid*: The tetrahydrobenzylisoquinoline coclaurine (LogD = 0.2 at pH 7.4; molecular mass = 235.3 g/mol) inhibited the replication of the Human immunodeficiency virus with the $EC_{50}$ value of 0.8 µg/mL and a selectivity index (SI) >25 (Qing et al., 2017).

Coclaurine

### Commentaries

(1) Members of the genus *Nymphaea* L produce benzylisoquinolines (Liscombe et al., 2005) which are antimicrobial.

(2) Viruses have several types of genomes in terms of nucleic acid type. All bacteria and eucaryotes have double-strand DNA genome (two complementary strands, or chains, or threads, of DNA connected with each other with hydrogen bonds). In contrast, viruses may have either RNA- or DNA-based genome. Both types of viral genomes can be either double- or single-stranded. Double-stranded DNA is represented by two polynucleotide chains as is described in *The Double Helix* by Watson and Crick: two strands twisted around each other, like two ropes. Single-stranded genome means that there is only one chain of DNA instead of two. The same is true for RNA (+) and (−) genomes are explained as follows: to produce viral progeny, viral genome must be translated, i.e., converted into encoded viral proteins that will further associate into viral particles. This is realized by cellular protein-synthesizing machinery (ribosomes). There are two options. First, viral genome RNA is able to directly interact with ribosomes. In this case, it serves as messenger RNA and is designated as plus-strand (or positive RNA). The second option is when

viral genome is represented not by mRNA but a complementary chain. In this case, prior to translation, it must be converted by viral polymerase into a complementary chain, the latter serving as mRNA. In this case, the genome RNA is designated as minus-strand (or negative RNA). Some examples of negative genomes are Influenza virus, Parainfluenza virus, Ebola virus, and respiratory syncitial virus. Positive genomes: polyovirus, rhinoviruses, and Hepatitis C virus (Zarubaev V, personal communication).

## References

Dash, B.K., Sen, M.K., Alam, K., Hossain, K., Islam, R., Banu, N.A., Rahman, S. and Jamal, A.M., 2013. Antibacterial activity of *Nymphaea nouchali* (Burm. f) flower. *Annals of Clinical Microbiology and Antimicrobials*, *12*(1), p. 27.

Duraipandiyan, V. and Ignacimuthu, S., 2011. Antifungal activity of traditional medicinal plants from Tamil Nadu, India. *Asian Pacific Journal of Tropical Biomedicine*, *1*(2), pp. S204–S215.

Liscombe, D.K., MacLeod, B.P., Loukanina, N., Nandi, O.I. and Facchini, P.J., 2005. Evidence for the monophyletic evolution of benzylisoquinoline alkaloid biosynthesis in angiosperms. *Phytochemistry*, *66*(11), pp. 1374–1393.

Mukherjee, K.S., Bhattacharya, P., Mukherjee, R.K. and Ghosh, P.K., 1986. Chemical examination of *Nymphaea stellata* Willd. *Journal of the Indian Chemical Society*, *63*(5), pp. 530–531.

Parimala, M. and Shoba, F.G., 2014. In vitro antimicrobial activity and HPTLC analysis of hydroalcoholic seed extract of *Nymphaea nouchali* Burm. f. *BMC Complementary and Alternative Medicine*, *14*(1), p. 361.

Qing, Z.X., Yang, P., Tang, Q., Cheng, P., Liu, X.B., Zheng, Y.J., Liu, Y.S. and Zeng, J.G., 2017. Isoquinoline alkaloids and their antiviral, antibacterial, and antifungal activities and structure-activity relationship. *Current Organic Chemistry*, *21*(18), pp. 1920–1934.

### 1.2.1.4 *Nymphaea pubescens* Willd.

Synonyms: *Nymphaea lotus* var. *pubescens* (Willd.) Hook. f. & Thomson; *Nymphaea rubra* Roxb. ex Salisb.

Common names: Hairy water lily; lal shapla (Bangladesh); rou mao chi ye shui lian (China); tunas (the Philippines)

Habitat: Lakes, ponds, slow rivers

Distribution: Pakistan, India, Sri Lanka, Bangladesh, Myanmar, Vietnam, Indonesia, Philippines, and Papua New Guinea

Botanical observation: This elegant nuphar grows from erect rhizomes producing slender stolons. The leaves are alternate and simple. The petiole is elongated and can reach 2 m long. The blade is ovate, cordate at base, 15–26 cm in diameter, sometimes up to 45 cm in diameter, papery, pubescent below, with an a dentate margin. The flowers are showy, magnificent, up to 15 cm in diameter. The calyx comprises 4 sepals which are oblong, and 4–5 cm, up to 9 cm long. The corolla comprises from 12 to 30 petals, which are white, red, or pink; oblong, and 5–9 cm long. The androecium includes about 60 numerous stamens, which are yellow and conspicuous, and with 2 cm long anthers. The gynoecium comprises 11–20 carpels fusing into a plurilocular ovary. The fruits are globose and 5 cm in diameter containing minute seeds.

Medicinal uses: Gonorrhea (the Philippines); skin diseases, diarrhea, urinary tract infections (India).

*Antibacterial polar extract*: Methanol extract of leaves inhibited the growth of *Bacillus cereus* (ATCC 14875), *Staphylococcus aureus* (980), *Staphylococcus aureus* (ATCC 25923), methicillin-resistant *Staphylococcus aureus*, and *Enterococcus faecium* (Chankhamhaengdecha & Damrongphol, 2015).

**Commentaries**

(i) Gram-positive bacteria are more sensitive to extracts or compounds from medicinal plants at least partially because they do not have an outer membrane. In Gram-negative bacteria, the outer membrane acts as an impermeable barrier that protects bacteria from the deleterious effects of high molecular weight compounds from medicinal plants (Denyer & Maillard, 2002). Hydrophilic and amphiphilic compound can pass the outer membrane through water-filled transmembrane channels (porins or aquaporins), if they have a molecular size below or equal to 600 g/mol (Denyer & Maillard, 2002). This outer membrane is impermeable to macromolecules (like the antibiotic vancomycin or natural products with a molecular size above 600 g/mol) and it is, at least, for this reason that Gram-negative bacteria are relatively more resistant to biocides and antibiotics (Nikaido & Vaara, 1985). Saponins range in molecular weight often above 600 g/mol (Rahman, 2000) as well as tannins (Hättenschwiler & Vitousek, 2000). In addition, the outer membrane of Gram-negative bacteria is covered with hydrophilic and ionized lipopolysaccharides (LPS) (carrying a net negative charge), which hampers (not completely) the penetration of lipophilic compounds (Hancock, 1984; Denyer & Maillard, 2002).

(ii) Members of the genus *Nymphaea* L. elaborate hydrophilic natural products, yet to be identified, inhibiting the growth of Gram-positive bacteria. For instance, ethanol extract of stamens of the North American *Nymphea odorata* Aiton elicited zone inhibition diameter ranging from 21 to 24 mm against 12 strains of methicillin-resistant *Staphylococcus aureus* and exhibited a synergistic effect with amoxicillin (500 μg of extract, agar diffusion) (Mandal et al., 2010). In a previous study, ethanol extract of the leaves of the closely related *Nymphaea lotus* L. inhibited the growth of *Staphylococcus aureus* and *Streptococcus* sp. with inhibition zone diameters of 14.3 and 15.6 mm, respectively (agar diffusion) and was inactive from Gram-negative bacteria (Yisa, 2009). In line, Akinjogunla et al. (2010) noted that ethanolic extract of leaves of *Nymphaea lotus* L. inhibited the growth of methicillin-resistant *Staphylococcus aureus* and vancomycin-resistant *Staphylococcus aureus* (VRSA) isolated from wounds. The antibacterial constituents of *Nymphaea pubescens* Willd are apparently unknown. Tannins, which are not uncommon in the genus *Nymphaea* L., are most probably involved. Note that tannins often abound in aquatic plants serving as a natural defense against aquatic bacteria and fungi. Plants may combat pathogenic bacterial and fungi with "pre-formed" antimicrobial secondary metabolites (phytoanticipins). Such compounds are, for instance, tannins, phenolic, stilbene glycosides, cyanogenic glycosides, thioethers, saponins, and glucosinolates embedded in cell walls of plants (Matyssek et al., 2014). Antibacterial and/or antifungal compounds synthesized *de novo* after the plant tissue is exposed to microbial infection or damaged are termed "phytoalexins" (Van Etten et al., 1994). Plant alkaloids are generally responsible for imparting resistance to pathogenic fungi in plants (Fawceit & Spencer, 1969).

(iv) Methanol and ethanol are able to extract lipophilic, amphiphilic, and hydrophilic phenolics, terpenes, and alkaloids without discrimination. Hence, methanol or ethanol (and even hot water) extracts of medicinal plants always display some levels of antimicrobial effects.

**References**

Akinjogunla, O.J., Yah, C.S., Eghafona, N.O. and Ogbemudia, F.O., 2010. Antibacterial activity of leave extracts of *Nymphaea lotus* (Nymphaeaceae) on methicillin resistant Staphylococcus aureus (MRSA) and vancomycin resistant Staphylococcus aureus (VRSA) isolated from clinical samples. *Annals of Biological Research*, *1*(2), pp. 174–184.

Chankhamhaengdecha, S. and Damrongphol, P., 2015. Antibacterial activity of *Nymphaea pubescens* Willd. Leaves. 7th International Conference on Medical, Biological and Pharmaceutical Sciences (ICMBPS'2015).

Denyer, S.P. and Maillard, J.Y., 2002. Cellular impermeability and uptake of biocides and antibiotics in Gram-negative bacteria. *Journal of Applied Microbiology, 92,* pp. 35S–45S.

Fawceit, C.H. and Spencer D.M., 1969. *Natural Antifungal Compounds, p. 701 in Fungicides. An Advanced Treatise (A. Torgeson, Fxl.).* London: Academic Press.

Hättenschwiler, S. and Vitousek, P.M., 2000. The role of polyphenols in terrestrial ecosystem nutrient cycling. *Trends in Habitat & Evolution, 15*(6), pp. 238–243.

Mandal, S., DebMandal, M., Pal, N.K. and Saha, K., 2010. Synergistic anti–*Staphylococcus aureus* activity of amoxicillin in combination with *Emblica officinalis* and *Nymphae odorata* extracts. *Asian Pacific Journal of Tropical Medicine, 3*(9), pp. 711–714.

Matyssek, R., Schnyder, H., Oßwald, W., Ernst, D., Munch, J.C. and Pretzsch, H., 2014. *Growth and Defence in Plants.* Berlin: Springer.

Nikaido, H. and Vaara, M., 1985. Molecular basis of bacterial outer membrane permeability. *Microbiological Reviews, 49,* pp. 1–32.

Rahman, A.U. ed., 2000. *Bioactive Natural Products: Part E. Studies in Natural Products Chemistry.* Elsevier Science & Technology. USA.

Van Etten, H.D., Mansfield, J.W., Bailey, J.A. and Farmer, E.E., 1994. Two classes of plant antibiotics: phytoalexins versus "phytoanticipins". *Plant Cell, 6,* pp. 1191–1192.

Yisa, J., 2009. Phytochemical analysis and antimicrobial activity of *Scoparia dulcis* and *Nymphaea lotus. Australian Journal of Basic and Applied Sciences, 3*(4), pp. 3975–3979.

### 1.2.1.5 *Nymphaea tetragona* Georgi

Synonyms: *Castalia tetragona* (Georgi) G. Lawson; *Nymphaea acutiloba* DC.; *Nymphaea crassifolia* (Hand.-Mazz.) Nakai

Common names: Pygmy water lily; shiu lian (China)

Habitat: Lakes, ponds, and slow rivers

Distribution: Kazakhstan, Kashmir, India, Vietnam, China, Japan, and Korea

Botanical observation: It is a graceful nuphar. The leaves are simple, alternate, and grow from unbranched and erect rhizomes. The blades are ovate, cordate at base, round at apex, 5–12×3.5–9 cm, papery, and with entire margin. The flowers are 3–6 cm in diameter. The calyx comprises 4 sepals, which are broadly lanceolate, 2–3.5 cm long, and persistent. The perianth comprises 8–15 white, broadly lanceolate 2–2.5 cm petals. The androecium is conspicuous and yellow. The gynoecium includes several carpels. The fruits are globose, 2–2.5 cm in diameter, and comprise numerous tiny and ellipsoid seeds.

Medicinal uses: Fever, bronchial asthma (China)

*Antibacterial (Gram-negative) halo developed by hydrolysable tannin*: Kurihara and coworkers isolated from the fresh leaves the ellagitannin geraniin (LogD = −0.06 at pH 7.4; molecular mass = 952.6 g/mol) (50ss0 µg/8 mm discs), which developed a zone of inhibition of 13 mm against *Aeromonas salmonicida* and 14 mm against *Pseudomonas fluorescens* (Kurihara et al., 1993).

*Moderate antibacterial (Gram-negative) mid-polar extract*: Ethyl acetate extract (containing methyl gallate and pyrogallol) inhibited the growth of *Salmonella typhymurium* (KCTC 2515), *Salmonella typhymurium* (ST171), *Salmonella typhymurium* (ST482), *Salmonella typhymurium* (ST688), and *Salmonella typhymurium* (ST21A) (Hossain et al., 2014).

*Moderate antibacterial (Gram-negative) hydrolysable tannin*: Geraniin inhibited the growth of *Staphylococcus aureus* (ATCC 29213), *Salmonella arizonae* (ATCC 13314), *Escherichia coli* (ATCC 25922), *Vibrio parahematolyticus* (ATCC17802), and *Vibrio vulficus* (ATCC 27562) with the MIC values of 200, 800, 1600, 200, and 25 µg/mL, respectively (Taguri et al., 2004).

Geraniin

*Strong antiviral (enveloped monopartite linear double-stranded DNA) hydrolysable tannin*: Geraniin inhibited the replication of the Herpes simplex virus type-2 and type-1 with $IC_{50}$ values of 18.4 and 35 μM and selectivity indices of 2.8 and 1.5, respectively (Yang et al., 2007).

*Strong antiviral (non-enveloped monopartite linear single-stranded (+) RNA) hydrolysable tannin*: Geraniin inhibited the replication of Enterovirus 71 with the $IC_{50}$ value of 10 μg/mL and a selectivity index of 20 (Yang et al., 2012).

*Strong antiviral (enveloped circular double-stranded DNA) hydrolysable tannin*: Geraniin (200 μg/mL), inhibited the secretion of the Hepatitis B surface antigen by 85.8% and e-antigen by 63.7% (Li et al., 2008).

*Strong antiviral (enveloped monopartite linear dimeric single-stranded (+)RNA) hydrolysable tannin*: Geraniin inhibited the replication of Human immunodeficiency virus type-1 with the $EC_{50}$ of 0.4 μM and a selectivity index of 29.7 in MT4 cells, and inhibited reverse transcriptase with the $IC_{50}$ value of 1.8 μg/mL, and blocked virus uptake by cells (Notka et al., 2003).

**Commentaries**

(i) Ellagitannins, and in fact hydrolysable tannins, are broadly antimicrobials *in vitro* and inhibit *in vitro* all sorts of enzymes, but they are of no value for internal systemic infections. Ellagitannins are hydrolyzed in the human digestive system and are not absorbed. Also, they are hepatotoxic at high dose. These tannins have potential as topical antiseptics and styptics. The plant contains tannin (0.3% w/w of fresh plant material), which acts as a structural defense against aquatic bacteria, fungi, and algae. Consider that tannins at high concentration coagulate proteins, hence the use of tanniferous medicinal plants to treat diarrhea and dysentery.

(ii) MBC/MIC ratio of >4 and <32 indicates that the extract, essential oil, fraction, or the compound tested is bacteriostatic (i.e., bacteria are still alive but are not able to divide). If the ratio is ≥ 32, the bacteria tested is tolerant to the extract, essential oil, fraction, or the compound tested. An MBC/MIC ratio of ≤4 indicates that the extract, essential oil, fraction, or the compound tested is bactericidal (kill the bacteria). Likewise, an MFC (minimum fungicidal concentration)/MIC ratio of >4 and <32 indicates that the extract tested or the natural product tested is fungistatic (Agyare et al., 2015). An MFC/MIC ratio of ≤4 indicates that the extract, essential oil, fraction, or the compound tested or the natural product tested is fungicidal (kill the fungi) (Agyare et al., 2015).

(iii) Herpes simplex virus type-1 and type-2 belong to the family Herpesviridae. These are enveloped DNA viruses. Herpes labialis and herpes genitalis have been associated with herpes simplex virus type-1 and herpes simplex virus type-2, respectively (Chisholm & Lopez, 2011). Herpes simplex virus remains latent and reactivated by stimulus to raise painful blisters. Although there are effective drugs to treat herpes, the emergence of drug-resistant mutants has been observed.

## References

Chisholm, C. and Lopez, L., 2011. Cutaneous infections caused by herpesviridae: a review. *ArcHuman Immunodeficiency Virus Type-1 es of Pathology & Laboratory Medicine*, *135*(10), pp.1357–1362.

Clinical and Laboratory Standards Institute, 2000. Methods for dilution antimicrobial susceptibility tests for bacteria that grow aerobically. CLSI Approved Standards M7-A5, Clinical and Laboratory Standards Institute, Wayne, PA, USA.

Gohar, A.A., Lahloub, M.F. and Niwa, M., 2003. Antibacterial polyphenol from *Erodium glaucophyllum*. *Zeitschrift für Naturforschung C*, *58*(9–10), pp. 670–674.

Hossain, M.A., Park, J.Y., Kim, J.Y., Suh, J.W. and Park, S.C., 2014. Synergistic effect and antiquorum sensing activity of *Nymphaea tetragona* (water lily) extract. *BioMed Research International*, p. 562173.

Krishnan, N., Ramanathan, S., Sasidharan, S., Murugaiyah, V. and Mansor, S.M., 2010. Antimicrobial activity evaluation of *Cassia spectabilis* leaf extracts. *International Journal of Pharmacology*, *6*, pp. 510–514.

Kurihara, H., Kawabata, J. and Hatano, M., 1993. Geraniin, a hydrolyzable tannin from *Nymphaea tetragona* Georgi (Nymphaeaceae). *Bioscience, Biotechnology, and Biochemistry*, *57*(9), pp. 1570–1571.

Li, J., Huang, H., Zhou, W., Feng, M. and Zhou, P., 2008. Anti-Hepatitis B virus activities of *Geranium carolinianum* L. extracts and identification of the active components. *Biological and Pharmaceutical Bulletin*, *31*(4), pp. 743–747.

Notka, F., Meier, G.R. and Wagner, R., 2003. Inhibition of wild-type human immunodeficiency virus and reverse transcriptase inhibitor-resistant variants by *Phyllanthus amarus*. *Antiviral Research*, *58*(2), pp. 175–186.

Taguri, T., Tanaka, T. and Kouno, I., 2004. Antimicrobial activity of 10 different plant polyphenols against bacteria causing food-borne disease. *Biological and Pharmaceutical Bulletin*, *27*(12), pp. 1965–1969.

Yang, C.M., Cheng, H.Y., Lin, T.C., Chiang, L.C. and Lin, C.C., 2007. The in vitro activity of geraniin and 1, 3, 4, 6-tetra-O-galloyl-β-d-glucose isolated from *Phyllanthus urinaria* against herpes simplex virus type 1 and type 2 infection. *Journal of Ethnopharmacology*, *110*(3), pp. 555–558.

Yang, Y., Zhang, L., Fan, X., Qin, C. and Liu, J., 2012. Antiviral effect of geraniin on human enterovirus 71 in vitro and in vivo. *Bioorganic & Medicinal Chemistry Letters*, 22(6), pp. 2209–2211.

## 1.3   ORDER AUSTROBAILEYALES TAKHT. EX REVEAL (1992)

### 1.3.1   FAMILY ILLICIACEAE BERCHT. & J. PRESL (1825)

This family consists of the single genus *Illicium* L.

#### 1.3.1.1   *Illicium jiadifengpi* Chang

Common name: Jia di feng pi (China)
   Habitat: Forests
   Distribution: China
   Botanical observation: It is a handsome but poisonous medium-size tree. The leaves are simple, exstipulate, in clusters of 3–5 at distal nodes. The petioles are 1.5–3 cm long. The blades are elliptic, 7–16×2–4.5 cm, with a conspicuous midrib, present 5–8 pairs of secondary nerves, attenuate at base, and acuminate at apex. The flowers are axillary and showy. The perianth comprises 33–55, white tepals, which are linear and up to 3 cm long. The androecium is conspicuous and comprises 28–32 stamens, which are 2.7–3 mm long and white. The gynoecium includes 12–14 carpels. The fruit is star-anise-like, on a 3–4 cm long peduncle.
   Medicinal uses: None apparently for the treatment of microbial infections.
   *Moderate antiviral (enveloped circular double-stranded DNA) sesquiterpene lactone*: 2-oxo-3,4-dehydroxyneomajucin inhibited the secretion of Hepatitis B virus surface antigen and Hepatitis B virus e antigen in HepG2.2.15 cells by 28.85 and 17.53% at the concentration of 64.9 µM (Liu et al., 2015).
   *Strong antiviral (non-enveloped linear monopartite (+) RNA virus) amphiphilic diterpenes*: The abietane diterpenes 7α-hydroxycallitrisic acid (molecular weight = 316.4 g/mol) and augustanoic acid F from the roots inhibited the replication of coxsackie virus B2 with $IC_{50}$ values of 2.7 and 4.9 M, respectively (Zhang et al., 2013). The pimmarane diterpene jiadifenoic acid M isolated from the stems inhibited coxsackie virus B3 with an $IC_{50}$ value of 11.6 µM and a selectivity index of 49.3 (Zhang et al., 2014).

7α-Hydroxycallitrisic acid

### Commentaries

(i) Hepatitis B virus is a blood-borne enveloped DNA virus that belongs to the family Hepadnaviridae, which is transmitted parenterally, sexually, or from mother to child (Strauss & Strauss, 2007). This virus is hepatotropic and causes acute and chronic infections of the liver, leading to liver failure, cirrhosis, and liver cancer. According to WHO's global hepatitis report released in 2017, there were 257 million persons, or 3.5% of the population, living with chronic Hepatitis B virus infection in the world in the year 2015. A vaccine can prevent Hepatitis B virus, and the widespread use of Hepatitis B vaccine started in the 1990s and 2000s allowing a reduction in Hepatitis B fatalities (Strauss & Strauss, 2007). Once diagnosed, infected persons are placed on long-term, usually lifelong, treatment (for Hepatitis B). Lamivudine, entecavir, adefovir, telbivudine, INF-$\alpha$, and Peg-INF-$\alpha$-2a are available to treat Hepatitis B virus infection but have undesirable side effects and are costly in developing countries. Furthermore, the virus develops resistance. Therefore, one can look into natural products from plants for anti-Hepatitis B virus leads. There is a growing body of evidence suggesting that sesquiterpenes from plants are often able to inhibit the replication of Hepatitis B virus *in vitro* (Parvez et al., 2016), but sesquiterpenes are often too toxic to be used systematically. Semisynthetic derivatives could be another option.

(ii) Influenza A viruses are enveloped RNA viruses (family of Orthomyxoviridae), which are classified by the antigenic properties of two surface glycoproteins: hemagglutinin (HA) and neuraminidase (NA). Sixteen HA subtypes (H1–H16) and nine NA subtypes (N1–N9) have been defined to date. Neuraminidase cleaves the specific linkage of the sialic acid receptor, resulting in the release of the newly formed virion from the infected cell. In addition, neuraminidase may facilitate the early process of Influenza virus infection of the cells. As a result, NA has been the most important target for the development of novel anti-Influenza drugs. Consider that natural products from medicinal plants that inhibit strongly the replication of Influenza A viruses *in vitro* are mainly low to medium molecular weight polyhydroxylated and aromatic, hydrophilic, or amphiphilic substances, such as notably simple phenols and flavonols.

## References

Liu, J.F., Wang, L., Wang, Y.F., Song, X., Yang, L.J. and Zhang, Y.B., 2015. Sesquiterpenes from the fruits of *Illicium jiadifengpi* and their anti-Hepatitis B virus activities. *Fitoterapia*, *104*, pp. 41–44.

Parvez, M.K., Arbab, A.H., Al-Dosari, M.S. and Al-Rehaily, A.J., 2016. Antiviral natural products against chronic hepatitis B: recent developments. *Current Pharmaceutical Design*, 22(3), pp. 286–293.

Strauss, E.G. and Strauss, J.H., 2007. *Viruses and Human Disease*. Elsevier, USA.

Zhang, G.J., Li, Y.H., Jiang, J.D., Yu, S.S., Qu, J., Ma, S.G., Liu, Y.B. and Yu, D.Q., 2013. Anti-Coxsackie virus B diterpenes from the roots of Illicium jiadifengpi. *Tetrahedron*, 69(3), pp. 1017–1023.

Zhang, G.J., Li, Y.H., Jiang, J.D., Yu, S.S., Wang, X.J., Zhuang, P.Y., Zhang, Y., Qu, J., Ma, S.G., Li, Y. and Liu, Y.B., 2014. Diterpenes and sesquiterpenes with anti-Coxsackie virus B3 activity from the stems of *Illicium jiadifengpi*. *Tetrahedron*, 70(30), pp. 4494–4499.

### 1.3.1.2 *Illicium henryi* Diels.

Synonyms: *Illicium pseudosimonsii* Q. Lin, *Illicium silvestrii* Pavol.

Common names: Henry anise tree; hong hui xiang (China)

Habitat: Forests

Distribution: China

Botanical observation: This graceful ornamental, yet poisonous, medium-size tree or shrub grows in China. The leaves are simple, exstipulate, and arranged in clusters of 2–5. The petioles are 0.7–2 cm long. The blades are obovate-lanceolate, 6–18 × 1.2–5 cm, leathery, with conspicuous midrib, marked with 5–8 pairs of secondary nerves, acuminate at base and apex and glossy. The flowers are axillary or subterminal on 1–1.5 cm long peduncles. The corolla is light red and comprises 10–15 tepals with are elliptic and 0.7–1 × 0.4–0.8 cm. The androecium comprises 11–14 stamens. The gynoecium comprises 7–9 carpels. The fruits consist of 7–9 follicles, which are 1.2–2 long and somewhat hooked at apex.

Medicinal uses: Apparently none for microbial infections.

*Strong antiviral (enveloped circular double-stranded DNA) amphiphilic sesquiterpene*: From the stems and roots, tashironin A (molecular weight = 390.5 g/mol) showed an $IC_{50}$ value of 0.4 μM in inhibiting Hepatitis B virus surface antigen secretion with a selectivity index of 6.3 and an $IC_{50}$ value of 0.1 μM (selectivity index = 20.1) (Liu et al., 2010). From the fruits of this plant, the sesquiterpene anislactone B-inhibited the secretion of the Hepatitis B virus e-antigen by Hep G2.2.15 cell line with an $IC_{50}$ value of 70 μM and a selectivity index of 43.4 (Liu et al., 2013).

*Moderate antiviral (enveloped linear monopartite dimeric single-stranded (+)RNA) sesquiterpene*: Tanshironine A inhibited the replication of the Human immunodeficiency virus type-1 (strain IIIB) in C8166 cells with the $EC_{50}$ value of 41.8 μM and a selectivity index of 6.2 (Song et al., 2007).

*Moderate antiviral (enveloped circular double-stranded DNA) neolignan*: From the stems and roots *threo*-4,9,9′-trihydroxy-3,3′-dimethoxy-3,3′-dimethoxy-8-*O*-4′-neolignan-7-*O*-β-D-glucopyranoside and (–)-dihydroconyferyl alcohol inhibited the secretion of the Hepatitis B virus surface antigen with the $IC_{50}$ values of 0.5 mM (selectivity index=1.9) and 0.06 mM (selectivity index = 8.8) and inhibited the secretion of the Hepatitis B virus e antigen with the $IC_{50}$ value of 0.8 (selectivity index=1.4) and 0.5 mM (selectivity index = 1.1) using Hepatitis B virus-transfected Hep G2.2.15 cell line (Liu et al., 2011).

*Moderate antiviral (enveloped circular double-stranded DNA) lignan*: From the stems and roots *threo*-4,9,9′-trihydroxy-3,3′-dimethoxy-3,3′-dimethoxy-8-*O*-4′-neolignan-7-*O*-β-D-glucopyranoside inhibited the secretion of the Hepatitis B virus surface antigens with the $IC_{50}$ value of 0.5 mM (selectivity index=1.9) and the secretion of the Hepatitis B virus e antigen with the $IC_{50}$ value of 0.8 mM (selectivity index = 1.4) (Liu et al., 2011).

*Moderate antiviral (enveloped circular double-stranded DNA) phenylpropanoid*: From the stems and roots, (–)-dihydroconyferyl alcohol inhibited Hepatitis B virus surface antigen secretion with an $IC_{50}$ value of 0.06 mM (selectivity index = 8.8) and inhibited Hepatitis B virus e-antigen secretion with an $IC_{50}$ value of 0.5 mM (selectivity index =1.1) using Hepatitis B virus-transfected Hep G2.2.15 cell line (Liu et al., 2011).

*Weak antiviral (enveloped monopartite linear single-stranded (+) RNA) cyclic monoterpene epoxyde*: 1,8-cineole inhibited the replication of the Avian infectious bronchitis virus with the $IC_{50}$ of 610 μM by acting after penetration of the virus into the cell (Yang et al., 2010b).

### Commentaries

(i) The plant produces the cyclohexene phenol acid shikimic acid (Singh et al., 2020) as well as an essential oil containing mainly safrole, myristicin, and 1,8-cineole (Liu et al., 2015), which are antibacterial and antifungal.

(ii) *Illicium verum* Hook.f. (star anise) produces shikimic acid, which is used for the production of Oseltamivir (Tamiflu®) used in the treatment and prophylaxis of both Influenza virus A and Influenza virus B. It also produces the phenylpropanoid anethole, which inhibits the growth of antibiotic-resistant bacteria, including *Acinetobacter baumannii* with an MIC value of 110 μg/mL via membrane disruption (Yang et al., 2010a), *Salmonella typhi* murium, and *Staphylococcus aureus* with MIC values of 75 and 100 μg/mL, respectively (Karapinar & Aktuğ, 1987).

### References

Karapinar, M. and Aktuğ, S.E., 1987. Inhibition of foodborne pathogens by thymol, eugenol, menthol and anethole. *International Journal of Food Microbiology*, 4(2), pp. 161–166.

Liu, J.F., Jiang, Z.Y., Geng, C.A., Zhang, Q., Shi, Y., Ma, Y.B., Zhang, X.M. and Chen, J.J., 2011. Two new lignans and anti-HBV constituents from *Illicium henryi. Chemistry & Biodiversity*, *8*(4), pp. 692–698.

Liu, X.C. and Liu, Z.L., 2015. Analysis of the essential oil of *Illicium henryi* Diels root bark and its insecticidal activity against *Liposcelis bostrychophila* Badonnel. *Journal of Food Protection*, *78*(4), pp. 772–777.

Liu, J.F., Jiang, Z.Y., Zhang, Q., Shi, Y., Ma, Y.B., Xie, M.J., Zhang, X.M. and Chen, J.J., 2010. Henrylactones A–E and anti-HBV constituents from *Illicium henryi. Planta Medica*, *76*(02), pp. 152–158.

Singh, P., Gupta, E., Mishra, N. and Mishra, P., 2020. Shikimic acid as intermediary model for the production of drugs effective against influenza virus. In *Phytochemicals as Lead Compounds for New Drug Discovery* (pp. 245–256). Elsevier, USA.

Song, W.Y., Ma, Y.B., Bai, X., Zhang, X.M., Gu, Q., Zheng, Y.T., Zhou, J. and Chen, J.J., 2007. Two new compounds and anti-HIV active constituents from *Illicium verum. Planta Medica*, *73*(04), pp. 372–375.

Yang, J.F., Yang, C.H., Chang, H.W., Yang, C.S., Wang, S.M., Hsieh, M.C. and Chuang, L.Y., 2010a. Chemical composition and antibacterial activities of *Illicium verum* against antibiotic-resistant pathogens. *Journal of Medicinal Food*, *13*(5), pp. 1254–1262.

Yang, Z., Wu, N., Fu, Y., Yang, G., Wang, W., Zu, Y. and Efferth, T., 2010b. Anti-infectious bronchitis virus (IBV) activity of 1, 8-cineole: effect on nucleocapsid (N) protein. *Journal of Biomolecular Structure and Dynamics*, *28*(3), pp. 323–330.

## 1.3.2 FAMILY SCHISANDRACEAE BLUME (1830)

The family Schisandraceae consists of 2 genera and about 40 species of woody climbers. The perianth is made of tepals, the androecium includes 4–80 stamens, and the gynoecium is made of numerous carpels. The fruits are groups of apocarps on an elongated receptacles. Seeds 1–5 (or more) per apocarps. Members of this family mainly produce series of unusual sesquiterpenes and triterpenes with interesting anti-HIV properties *in vitro*.

### 1.3.2.1 *Kadsura angustifolia* A.C. Sm.

Synonym: *Kadsura guangxiensis* S.F. Lan
   Common name: Xia ye nan wu wei zi (China)
   Habitat: Forests
   Distribution: Vietnam and China
   Botanical observation: This is a beautiful climber. The leaves are simple, alternate, and exstipulate. The petioles are 1–1.7 cm long. The blade is elliptic, 9.5–14 cm×2.5–4.5 cm, thin, with 7–13 pairs of secondary nerves, cuneate at base, serrulate at margin, and acuminate at apex. The perianth comprises 9–15 tepals, which are whitish and 0.7–0.8×0.5–0.6 mm. Male flowers present 50 stamens. Female flowers are endowed with a gynoecium with up to 80 carpels. The fruit consists of an aggregate of apocarps, which are 0.9–1 cm×0.8–0.9 cm.
   Medicinal uses: Apparently none for microbial infections.
   *Strong antiviral (enveloped monopartite linear dimeric single-stranded (+)RNA) amphiphilic cycloartane triterpene*: Angustific acid A (LogD = 4.2 at pH 7.4; molecular mass = 464.6 g/mol) inhibited the replication of the Human immunodeficiency virus type-1 with an $EC_{50}$ value of 6.1 µg/mL in C8166 cells with a selectivity index of more than 32.8 (Sun et al., 2011).

Angustific acid A

*Strong antiviral (enveloped monopartite linear dimeric single-stranded (+)RNA) amphiphilic lignan*: The dibenzocyclooctadiene lignan binankadsurin A (LogD = 3.2 at pH 7.4; molecular mass = 402.4 g/mol) inhibited the Human immunodeficiency virus type-1 with an $EC_{50}$ of $3.8\,\mu M$ (Gao et al., 2008).

Binankadsurin A

### Commentary

The Human immunodeficiency virus belongs to the family Retroviridae in the genus *Lentivirus*. Type-2 is less aggressive than type-1 (Kanki et al., 1994) and mainly affecting West Africa (Poulsen et al., 1993). This enveloped single-stranded positive-sense RNA virus in humans targets CD4 T cells and CXCR4 receptors present on the surface of the cells, leading to cellular destruction, and immunodeficiency, leading to AIDS. According to the WHO, the AIDS pandemic caused 770,000 fatalities and 37.9 million people were living with HIV/AIDS worldwide in 2018, touching principally developing countries. What was in the 1990s a death sentence is now, thanks to a mammoth of research work, a chronic illness which needs the continuous intake of drugs. Reverse transcriptase inhibitors such as Zidovudine, Didanoside, Zalcitabne, Stavudine, Lamivudine, and Abacavir have been licensed and used for the lifelong treatment of AIDS (Lim et al., 1997), as well as protease inhibitors such as Saquinavir, Ritonavir, Indinavir, Nelfinavir, Amprenavir, and Lopinavir, which lock the formation of reverse transcriptase, integrase, protease, and structural proteins. However, long-term use of these reverse transcriptase and protease inhibitors leads to side effects, and viral resistance is on the rise. HIV is of zoonotic (Chimpanzees to humans) origin, like The Severe acute respiratory syndrome-associated coronavirus (Hemelaar, J., 2012). Destruction of the primary rainforest is correlated with increased exposure to zoonotic viruses.

**References**

Gao, X.M., Pu, J.X., Huang, S.X., Yang, L.M., Huang, H., Xiao, W.L., Zheng, Y.T. and Sun, H.D., 2008. Lignans from *Kadsura angustifolia*. *Journal of Natural Products*, 71(4), pp. 558–563.

Hemelaar, J., 2012. The origin and diversity of the HIV pandemic. *Trends in Molecular Medicine*, 18(3), pp. 182–192.

Kanki, P.J., Travers, K.U., Marlink, R.G., Essex, M.E., MBoup, S., Gueye-NDiaye, A., Siby, T.I.D.I.A.N.E., Thior, I.B.O.U., Sankale, J.L., Hsieh, C.C. and Hernandez-Avila, M., 1994. Slower heterosexual spread of HIV-2 than HIV-1. *The Lancet*, 343(8903), pp. 943–946.

Poulsen, A.G., Aaby, P., Gottschau, A., Kvinesdal, B.B., Dias, F., Mølbak, K. and Lauritzen, E., 1993. HIV-2 infection in Bissau, West Africa, 1987–1989: incidence, prevalences, and routes of transmission. *Journal of Acquired Immune Deficiency Syndromes*, 6(8), pp. 941–948.

Sun, R., Song, H.C., Wang, C.R., Shen, K.Z., Xu, Y.B., Gao, Y.X., Chen, Y.G. and Dong, J.Y., 2011. Compounds from *Kadsura angustifolia* with anti-HIV activity. *Bioorganic & Medicinal Chemistry Letters*, 21(3), pp. 961–965.

### 1.3.2.2 *Kadsura longipedunculata* Finet & Gagnep.

Synonyms: *Kadsura discigera* Finet & Gagnep.; *Kadsura omeiensis* S.F. Lan; *Kadsura peltigera* Rehder & E.H. Wilson

Common name: Nan wu wei zi (China)

Habitat: Forests

Distribution: China, Vietnam

Botanical observation: It is a magnificent woody climber. The leaves are simple, alternate and exstipulate. The petiole is 0.6–1.7 cm long. The blade is elliptic 5.5–12 cm×2–4.5 cm with 4–8 pairs of secondary nerves, cuneate at base, and acuminate at apex. The flowers are axillary on 1.2–16 cm peduncles. The perianth includes 10–15 tepals which are yellowish, 0.4–1.3×0.3–0.6 cm. The male flowers include 26–54 stamens. The female flowers include 20–58 carpels. The fruit consists of a globose aggregate of apocarps, which are red, 0.6–1.1×0.4–0.6 cm, and edible.

Medicinal uses: Apparently no use for microbial infections.

*Moderate antibacterial (Gram-positive) essential oil*: Essential oil of the stem bark inhibited the growth of *Bacillus subtilis* (ATCC 6051), *Staphylococcus saprophyticus* (ATCC 15305), *Staphylococcus epidermidis* (ATCC 14990), *Staphylococcus aureus* (ATCC 29213), *Streptococcus agalactiae* (ATCC 27956), *Streptococcus pyogenes* (ATCC 12344), *Enterococcus faecalis* (ATCC 29212), vancomycin-resistant *Enterococcus faecalis* (VRE), methicillin-resistant *Staphylococcus aureus* (NCTC 10442) (Mulyaningsih et al., 2010).

*Very weak broad-spectrum antibacterial monoterpenes*: Borneol from the stem bark inhibited the growth of *Bacillus subtilis* (ATCC 6051), *Staphylococcus saprophyticus* (ATCC 15305), *Staphylococcus epidermidis* (ATCC 14990), *Staphylococcus aureus* (ATCC 29213), *Streptococcus agalactiae* (ATCC 27956), and *Streptococcus pyogenes* (ATCC 12344) with MIC/MBC values of 2/4, 1/4, 1/4, 4/8, 1/2, and 1/2 mg/mL, respectively (Mulyaningsih et al., 2010). Camphene from the stem bark inhibited the growth of *Bacillus subtilis* (ATCC 6051), *Staphylococcus saprophyticus* (ATCC 15305), *Staphylococcus epidermidis* (ATCC 14990), *Streptococcus agalactiae* (ATCC 27956), and *Streptococcus pyogenes* (ATCC 12344) with MIC/MBC values of 2/4, 1/4, 2/4, 1/1, and 1/2 mg/mL, respectively (Mulyaningsih et al., 2010).

*Very weak anticandidal essential oil*: Essential oil of the stem bark of the plant inhibited the growth of *Candida albicans* (ATCC 90028), *Candida glabrata* (ATCC MYA 2950), and *Candida parasilopsis* (ATCC 22019) with MIC/MBC values of 2/4, 2/4, and 1/2 mg/mL, respectively (Mulyaningsih et al., 2010).

*Strong antiviral (enveloped monopartite linear dimeric single-stranded (+)RNA) amphiphilic furan lignans*: The lignans kadlongirin A and kadlongirin B (LogP = 2.7; molecular weight = 446.5 g/mol) inhibited the replication of the Human immunodeficiency virus type-1 with $EC_{50}$ values of 51.8 and 16 µg/mL and selectivity indices values of 1.9 and 6.7, respectively (Pu et al., 2008).

Kadlongirin B

*Strong antiviral (enveloped monopartite linear dimeric single-stranded (+)RNA) amphiphilic cadinane sesquiterpene*: 2,7-dihydroxy-11,12-dehydrocalamenene (molecular mass = 230.3 g/mol) inhibited the Human immunodeficiency virus type-1 with a $EC_{50}$ value of 12.8 µg/mL and a selectivity index of 4.8 (Pu et al., 2008).

2,7-Dihydroxy-11,12-dehydrocalamenene

*Viral enzyme inhibition by 3,4-seco-cycloartane triterpene*: schisanlactone A from the roots and stems inhibited the Human immunodeficiency virus type-1 protease with an $IC_{50}$ value of 20 µM (Sun et al., 2006).

*Viral enzyme inhibition by dibenzocyclooctadiene lignan*: longipedunin A from the roots and stems inhibited the replication of the Human immunodeficiency virus type-1 protease with an $IC_{50}$ value of 50 µM (Sun et al., 2006).

### Commentary

Kadlongirin A and kadlongirin B only differ by the presence of an epoxide moiety in kadlongirin B resulting in a significant increase in anti-HIV activity *in vitro*. Epoxide and endoperoxide moieties in natural products from medicinal plants are antiviral (also antibacterial and antifungal) pharmacophores. These groups generate reactive oxygen species or open to form covalent bonding with amino acids in peptides and proteins as well as genetic material.

**References**

Agyare, C., Apenteng, J.A., Adu, F., Kesseih, E. and Boakye, Y.D., 2015. Antimicrobial, antibiotic resistance modulation and cytotoxicity studies of different extracts of *Pupalia lappacea*. *Pharmacologia*, 6(6), pp. 244–257.

Mulyaningsih, S., Youns, M., El-Readi, M.Z., Ashour, M.L., Nibret, E., Sporer, F., Herrmann, F., Reichling, J. and Wink, M., 2010. Biological activity of the essential oil of *Kadsura longipedunculata* (Schisandraceae) and its major components. *Journal of Pharmacy and Pharmacology*, 62(8), pp. 1037–1044.

O'Neill, A. and Chopra, I., 2004. Preclinical evaluation of novel antibacterial agents by microbiological and molecular techniques. *Expert Opinion on Investigational Drugs*, 13.

Pu, J.X., Gao, X.M., Lei, C., Xiao, W.L., Wang, R.R., Yang, L.B., Zhao, Y., Li, L.M., Huang, S.X., Zheng, Y.T. and Sun, H.D., 2008. Three new compounds from *Kadsura longipedunculata*. *Chemical and Pharmaceutical Bulletin*, 56(8), pp. 1143–1146.

Sun, Q.Z., Chen, D.F., Ding, P.L., Ma, C.M., Kakuda, H., Nakamura, N. and Hattori, M., 2006. Three new lignans, longipedunins AC, from *Kadsura longipedunculata* and their inhibitory activity against HIV protease. *Chemical and Pharmaceutical Bulletin*, 54(1), pp. 129–132.

### 1.3.2.3 *Schisandra bicolor* W.C. Cheng

Synonyms: *Schisandra bicolor* var. *tuberculate* (Y. W. Law) Y. W. Law; *Schisandra tuberculata* Y. W. Law; Schisandra wilsoniana A.C. Sm.

Common name: Er se wu wei zi (China)

Habitat: Forests

Distribution: China

Botanical observation: It is a magnificent woody climber. The leaves are simple, alternate, and exstipulate. The petiole is slender and 1.3–6.7 cm long. The blade is elliptic, 7–12×3.5–8 cm, with 4–6 pairs of secondary veins 4–6, cuneate at base, margin entire, and apex acute. The flowers are axillary on 1.7–5.5 cm long peduncles. The perianth includes 6–11 tepals, which are whitish to red and up to 0.4–0.8 cm×0.3–0.6 cm. The male flowers present five stamens merged in a somewhat pentagonal mass. The female flowers include more than 50 carpels. The fruit is an aggregate of apocarps, which are red and 1–1.6×1–1.6 cm.

Medicinal use: cough (China)

*Strong antiviral (enveloped monopartite linear dimeric single stranded (+)RNA) dibenzocyclooctadiene lignan*: Marlignan L (molecular mass = 404.5 g/mol) from leaves and stems inhibited the replication of the Human immunodeficiency virus type-1 with an $EC_{50}$ of 16.4 µg/mL and a selectivity index of 16.4 (Yang et al., 2010a). Wilsonilignan C isolated from the fruits inhibited the replication of the Human immunodeficiency virus type-1 with an $EC_{50}$ of 2.8 µg/mL (Yang et al., 2010b).

*Weak antiviral (enveloped circular double-stranded DNA) dibenzocyclooctadiene lignan*: Schisantherin C from fruits inhibited Hepatitis B virus surface antigen and e-antigen secretion by 59.7 and 34.7%, respectively, at 50 µg/mL (Ma et al., 2009).

**References**

Ma, W.H., Lu, Y., Huang, H., Zhou, P. and Chen, D.F., 2009. Schisanwilsonins A–G and related anti-HBV lignans from the fruits of *Schisandra wilsoniana*. *Bioorganic & Medicinal Chemistry Letters*, 19(17), pp. 4958–4962.

Yang, G.Y., Li, Y.K., Wang, R.R., Li, X.N., Xiao, W.L., Yang, L.M., Pu, J.X., Zheng, Y.T. and Sun, H.D., 2010a. Dibenzocyclooctadiene lignans from *Schisandra wilsoniana* and their anti-HIV activities. *Journal of Natural Products*, 73(5), pp. 915–919.

Yang, G.Y., Li, Y.K., Wang, R.R., Xiao, W.L., Yang, L.M., Pu, J.X., Zheng, Y.T. and Sun, H.D., 2010b. Dibenzocyclooctadiene lignans from the fruits of *Schisandra wilsoniana* and their anti-HIV activities: original article. *Journal of Asian Natural Products Research*, 12(6), pp. 470–476.

### 1.3.2.4   *Schisandra lancifolia* (Rehder & E.H. Wilson) A.C. Sm.

Synonym: *Schisandra sphenanthera* var. *lancifolia* Rehder & E.H. Wilson

  Common name: Xia ye wu wei zi (China)

  Habitat: Forests

  Distribution: China

  Botanical observation: It is a woody climber that grows in China. The leaves are single, alternate, and exstipulate. The petiole is 0.3–1.5 cm long. The blade is elliptic, 3.5–7.5 cm × 1.5–4 cm, papery, with 4–6 pairs of secondary nerves, cuneate at base, denticulate at margin, and acute at apex. The flowers are axillary on 0.9–5.6 cm long peduncles. The perianth consists of 6–8 tepals, which are whitish or pink, up to 0.3–0.6 cm × 0.3–0.5 cm. The male flowers present 8–19 stamens. The female flowers include 14–24 carpels. The fruit is an aggregate of red, 0.4–0.9 × 0.4–0.7 cm, and edible apocarps on a 3–6.5 cm long torus.

  Medicinal use: Traumatic injuries (China)

  *Strong antiviral (enveloped monopartite linear dimeric single-stranded (+)RNA) lipophilic dibenzocyclooctadiene lignan*: Schilancifolignan A (molecular mass = 430.5 g/mol), B, and C isolated from the leaves and stems inhibited the replication of the Human immunodeficiency virus type-1 with $EC_{50}$ values of 2.3, 3.8, and 3.2 μg/mL and selectivity indices of 8.0, 12.5, and 6.9, respectively (Yang et al., 2010).

Schilancifolignan A

*Moderate antiviral (enveloped monopartite linear dimeric single-stranded (+)RNA) nortriterpenes*: Lancifodilactone G and Lancifodilactone F isolated from the leaves and stems inhibited the replication of the Human immunodeficiency virus type-1 with the $EC_{50}$ values of 95.4 and 20.6 μg/mL and selectivity indices >1.8 and >6.6, respectively (Xiao et al., 2005a, 2005b). From the stems, the nortriterpenes schilancitrilactone C inhibited the replication of the Human immunodeficiency virus type-1 with the $EC_{50}$ value of 27.5 μg/mL (Luo et al., 2012).

### References

Luo, X., Shi, Y.M., Luo, R.H., Luo, S.H., Li, X.N., Wang, R.R., Li, S.H., Zheng, Y.T., Du, X., Xiao, W.L. and Pu, J.X., 2012. Schilancitrilactones A–C: three unique nortriterpenoids from *Schisandra lancifolia*. *Organic Letters*, 14(5), pp. 1286–1289.

Xiao, W.L., Zhu, H.J., Shen, Y.H., Li, R.T., Li, S.H., Sun, H.D., Zheng, Y.T., Wang, R.R., Lu, Y., Wang, C. and Zheng, Q.T., 2005a. Lancifodilactone G: a unique nortriterpenoid isolated from *Schisandra lancifolia* and its anti-HIV activity. *Organic Letters*, 7(11), pp. 2145–2148.

Xiao, W.L., Li, R.T., Li, S.H., Li, X.L., Sun, H.D., Zheng, Y.T., Wang, R.R., Lu, Y., Wang, C. and Zheng, Q.T., 2005b. Lancifodilactone F: a novel nortriterpenoid possessing a unique skeleton from *Schisandra lancifolia* and its anti-HIV activity. *Organic Letters*, 7(7), pp. 1263–1266.

Yang, G.Y., Fan, P., Wang, R.R., Cao, J.L., Xiao, W.L., Yang, L.M., Pu, J.X., Zheng, Y.T. and Sun, H.D., 2010. Dibenzocyclooctadiene lignans from *Schisandra lancifolia* and their anti-human immunodeficiency virus type-1 activities. *Chemical and Pharmaceutical Bulletin*, 58(5), pp. 734–737.

### 1.3.2.5  *Schisandra micrantha* A.C. Smith

Synonyms: *Schisandra elongata* Baillon var. *dentata* Finet & Gagnepain; *Schisandra elongata gracilis* A. C. Smith.

Common name: Xiao hua wu wei zi (China)

Habitat: Forests

Distribution: India, Myanmar, China

Botanical observation: It is a woody climber. The leaves are simple, alternate, extipulate, and gathered at the tips of stems. The petiole is 0.5–2.2 cm long. The blade is narrowly elliptic 3–7.5 cm×2.5–4 cm, with 4–5 pairs of secondary nerves, cuneate at base, serrulate at margin, and acuminate at apex. The flowers are axillary on 1.2–6.1 cm pedicel. The perianth includes 5–9, yellowish tepals, which are up to 6 mm long. The male flower includes 7–14 stamens. The female flower comprises 14–22 carpels. The fruit is an aggregate of apocarps, which are red, about 8 mm long on 1.5–3.5 cm long torus.

Medicinal uses: None for microbial infections apparently

*Moderate (enveloped monopartite linear dimeric single-stranded (+)RNA) hydrophilic nortriterpene*: Micrandilactone B and C (LogD = 0.6 at pH 7.4; molecular mass = 534.6 g/mol) inhibited the replication of the Human immunodeficiency virus type-1 with the $EC_{50}$ values of 42.6 (selectivity index = 4.6) and 7.7 μg/mL (selectivity index > 25.9), respectively (Li et al., 2005).

### Reference

Li, R.T., Han, Q.B., Zheng, Y.T., Wang, R.R., Yang, L.M., Lu, Y., Sang, S.Q., Zheng, Q.T., Zhao, Q.S. and Sun, H.D., 2005. Structure and anti-HIV activity of micrandilactones B and C, new nortriterpenoids possessing a unique skeleton from *Schisandra micrantha*. *Chemical Communications*, 23, pp. 2936–2938.

### 1.3.2.6  *Schisandra propinqua* (Wall.) Baill.

Synonym: *Kadsura propinqua* Wall.

Common name: He rui wu wei zi (China)

Habitat: Forests

Distribution: India, Nepal, Myanmar, China, Thailand, and Indonesia

Botanical observation: It is a woody climber. The leaves are simple, alternate, and exstipulate. The petiole is 0.4–2.7 cm long. The blade is elliptic 7–11.5 cm×1–4 cm with 5 or 6 pairs of secondary nerves, cuneate at base, serrulate at apex, and acuminate at apex. The flowers are axillary on 0.2–1.7 cm long peduncles. The perianth consists of 8–15 tepals, which are whitish pink, and up to 0.4–6 cm×0.3–0.5 cm. The male flowers present 4–11 stamens. The female flowers present up to about 50 carpels. The fruits are aggregates of apocarps, which are edible, red, about 9 mm long on a 2–6.5 cm long torus.

Medicinal use: Apparently none for microbial infections

*Strong antiviral (enveloped monopartite linear dimeric single-stranded (+)RNA) dibenzocyclooctadiene lignan*: Tiegusanin G (molecular weight = 660.7 g/mol) from the aerial parts inhibited the replication of the Human immunodeficiency virus with the $EC_{50}$ value of 7.9 μM and a selectivity index > 25 (Li et al., 2009).

Tiegusanin G

**Reference**

Li, X.N., Pu, J.X., Du, X., Yang, L.M., An, H.M., Lei, C., He, F., Luo, X., Zheng, Y.T., Lu, Y. and Xiao, W.L., 2009. Lignans with anti-HIV activity from *Schisandra propinqua* var. sinensis. *Journal of Natural Products*, 72(6), pp. 1133–1141.

### 1.3.2.7 *Schisandra rubriflora* (Franch.) Rehder & E.H. Wilson

Synonym: Schisandra chinensis var. rubriflora Franch.

Common name : Hong hua wu wei zi (China)

Habitat: Forests

Distribution: India, Myanmar, China

Botanical observation: It is a woody climber. The leaves are simple, alternate, exstipulate and regrouped at node and glabrous. The petioles are 1–3 cm long. The blades are obovate, 5.5–11.5 cm × 2.5–5.5 cm, papery, with 5–7 pairs of secondary nerves, cuneate at base, serrulate at margin, and acuminate at apex. The flowers are axillary, solitary; on 1–5 cm long slender peduncles. The perianth includes 6–9 tepals, which are deep purplish red, and 7–1.6×0.7–1.3 cm. The male flowers include 34–66 stamens, which are whitish. The female flowers include about 50–70 carpels. The fruits consist of an aggregate of red, and about 1 cm long apocarps.

Medicinal uses: Apparently none for microbial infections.

*Strong antiviral (enveloped monopartite linear dimeric single-stranded (+)RNA) amphiphilic dibenzocyclooctadiene lignan*: Gomisin $M_1$ (LogD = 4.2 at pH 7.4; molecular mass = 386.4 g/mol) from the fruits inhibited the replication of the Human immunodeficiency virus type-1 with an $EC_{50}$ below 0.6 mM and a selectivity index > 68 (Chen et al., 2006). Rubrilignans A and B inhibited the replication of the Human immunodeficiency virus type-1 with the $EC_{50}$ values of 2.2 and 1.8 μg/mL and the electivity indexes of 35.5 and 18.6, respectively (Mu et al., 2011). Wulignan A2 from the fruits inhibited the replication of the Human immunodeficiency virus type-1 with an $EC_{50}$ of 9.1 μg/mL and a selectivity index of 9.4 (Xiao et al., 2010).

Wulignan A2

*Enzyme inhibition by polar extract*: Aqueous extract of fruits at the concentration of 250 µg/mL inhibited inhibited the Human immunodeficiency virus type-1 protease by 47.7% (Xu et al., 1996).

## Commentaries

(i) Flavonoids and hydrolyzable tannins are often antiviral *in vitro*, but they are in general of limited therapeutic use because they do not have good oral bioavailability due, at least partially, to bacterial degradation in the guts. Dibenzocyclooctadiene lignans might have better bioavailability and represent an interesting class of anti-HIV agents that could be examined for *in vivo* effects and possibly preclinical trials. Consider that anti-HIV natural products from medicinal plants are not uncommonly able to inhibit the replication of coronaviruses. Thus, one could endeavor to examine the anti-Severe acute respiratory syndrome-associated coronavirus *in vitro* activity of dibenzocyclooctadiene lignans.

(ii) Economical and quick *in vivo* models using mice would be useful to assess the antiviral, antibacterial, or even the antifungal activities of extracts or compounds because extracts or compounds active *in vitro* are often inactive or less active *in vivo*.

## References

Chen, M., Kilgore, N., Lee, K.H. and Chen, D.F., 2006. Rubrisandrins A and B, lignans and related anti-HIV compounds from *Schisandra rubriflora*. *Journal of Natural Products*, *69*(12), pp. 1697–1701.

Mu, H.X., Li, X.S., Fan, P., Yang, G.Y., Pu, J.X., Sun, H.D., Hu, Q.F. and Xiao, W.L., 2011. Dibenzocyclooctadiene lignans from the fruits of *Schisandra rubriflora* and their anti-HIV-1 activities. *Journal of Asian Natural Products Research*, *13*(05), pp. 393–399.

Xiao, W.L., Wang, R.R., Zhao, W., Tian, R.R., Shang, S.Z., Yang, L.M., Yang, J.H., Pu, J.X., Zheng, Y.T. and Sun, H.D., 2010. Anti-HIV activity of lignans from the fruits of *Schisandra rubriflora*. *ArcHuman Immunodeficiency Virus Type-1 es of Pharmacal Research*, *33*(5), pp. 697–701.

Xu, H.X., Wan, M., Loh, B.N., Kon, O.L., Chow, P.W. and Sim, K.Y., 1996. Screening of traditional medicines for their inhibitory activity against HIV-1 protease. *Phytotherapy Research*, *10*(3), pp. 207–210.

### 1.3.2.8  *Schisandra chinensis* (Turcz.) Baill

Synonyms: *Kadsura chinensis* Turcz.; *Maximowiczia amurensis* Rupr.; *Maximowiczia chinensis* (Turcz.) Rupr.; *Schisandra chinensis* var. *leucocarpa* P.H. Huang & L.H. Zhuo

    Common name: Wu wei zi (China)

    Habitat: Forests

    Distribution: Korea, China, and Japan

    Botanical observation: It is a beautiful woody climber. The leaves are simple, alternate, and exstipulate and regrouped at nodes The petiole are 0.9–4 cm long. The blade is obovate, 4.5–8 cm×2.5–6.5 cm, papery, with veins 4–6 pairs of secondary nerve, cuneate at base, serrate at margin, and acuminate at apex. The flowers are axillary on slender pedicels. The corolla comprises 5–9 tepals, which are white and 0.6–1.1 long. Male flowers present five stamens. The female flowers are endowed with a gynoecium of more than 13 carpels. The fruits comprise an aggregate of apocarps which are red, edible, and up to 7 mm long.

    Medicinal uses: Gonorrhea (China)

    *Strong antiviral (enveloped circular double-stranded DNA) dibenzocyclooctadiene lignans*: Schinlignan G from the fruit inhibited the secretion of Hepatitis B virus surface antigen (IC$_{50}$ = 9.7 μg/mL; SI = 111.73) and of Hepatitis B virus e-antigen (IC$_{50}$ = 11.2 μg/mL; SI = 96.78). This lignan showed potent inhibitory activities against Hepatitis B virus DNA replication with an IC$_{50}$ of 5.1 μg/mL (Xue et al., 2015).

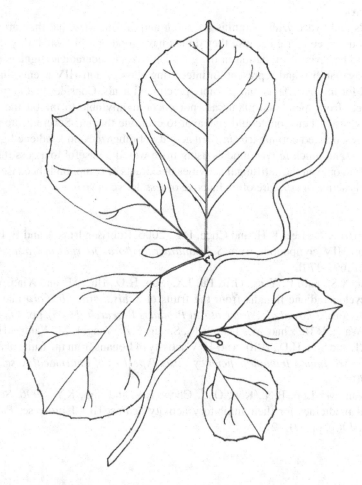

*Schisandra chinensis* (Turcz.) Baill

Schinlignan G

*Strong antiviral (enveloped monopartite linear dimeric single-stranded (+)RNA) dibenzocycloocta-diene lignans*: Nicotinoylgomisin Q from leaves and stems inhibited the replication of the Human immunodeficiency virus type-1 with an $EC_{50}$ of 17.8 µM and a selectivity index > 11.1 (Shi et al., 2014). Schizandrin B the replication of the Human immunodeficiency virus type-1 with an $EC_{50}$ of 15 µM (Xu et al., 2015).

*Viral enzyme inhibition by lignan*: Schizandrin B inhibited the Human immunodeficiency virus type-1 RNA-dependent DNA polymerase activity with the $IC_{50}$ of 29 (Xu et al., 2015).

### Commentaries

(i)   After the fusion of HIV into CD4 cells, the viral RNA is replicated into DNA by a viral enzyme called reverse transcriptase, which is an RNA-dependent RNA polymerase (Kati et al., 1992). Efavirenz (Sustiva®) is an inhibitor of RNA-dependent RNA polymerase used as part of the treatment for HIV type-1. The fruit being edible, one could have some interest to develop teas or juices that could be used to improve the well-being of AIDS patients.

(ii)  The plant is used in China for the treatment of COVID-19 in combination with oseltamivir phosphate (Luo, 2020).

### References

Kati, W.M., Johnson, K.A., Jerva, L.F. and Anderson, K.S., 1992. Mechanism and fidelity of human immunodeficiency virus reverse transcriptase. *Journal of Biological Chemistry*, 267(36), pp. 25988–25997.

Luo, A., 2020. Positive SARS-Cov-2 test in a woman with COVID-19 at 22 days after hospital discharge: a case report. *Journal of Traditional Chinese Medical Sciences*.

Shi, Y., Zhong, W., Chen, H., Wang, R., Shang, S., Liang, C., Gao, Z., Zheng, Y., Xiao, W. and Sun, H., 2014. New Lignans from the leaves and stems of *Schisandra chinensis* and their anti- HIV-1 activities. *Chinese Journal of Chemistry*, 32(8), pp. 734–740.

Xu, L., Grandi, N., Del Vecchio, C., Mandas, D., Corona, A., Piano, D., Esposito, F., Parolin, C. and Tramontano, E., 2015. From the traditional Chinese medicine plant *Schisandra chinensis* new scaffolds effective on HIV reverse transcriptase resistant to non-nucleoside inhibitors. *The Journal of Microbiology*, 53(4), p. 288.

Xue, Y., Li, X., Du, X., Li, X., Wang, W., Yang, J., Chen, J., Pu, J. and Sun, H., 2015. Isolation and anti-Hepatitis B virus activity of dibenzocyclooctadiene lignans from the fruits of *Schisandra chinensis*. *Phytochemistry, 116*, pp. 253–261.

## 1.4   ORDER CHLORANTHALES R. BROWN (1835)

### 1.4.1   FAMILY CHLORANTHACEAE R. BROWN EX SIMS (1820)

The family Chloranthaceae consists of 4 genera and 75 species of trees, shrubs, or herbs. The leaves are simple, opposite, decussate, or whorled and stipulate leaves. The flowers are minute. The perianth is reduced to a scale-like tepal or absent. The androecium includes 1–3 stamens. The gynoecium consists of a single minute carpel containing a single ovule, which is pendulous and orthotropous. Plants in this family produce antiviral sesquiterpenes and flavonoids.

#### 1.4.1.1   *Chloranthus japonicus* Siebold

Synonyms: *Chloranthus mandshuricus* Rupr., *Tricercandra japonica* (Siebold) Nakai
  *Tricercandra quadrifolia* A. Gray
  Common names: Yin xian cao (China); hitori-shizuka (Japan)
  Habitat: Wet spots in forests
  Distribution: China, Korea, and Japan
  Botanical observation: It a rhizomatous and erect herb that grows up to 50 cm tall. The leaves are simple, opposite, stipulate and clustered at the top of stem into a false whorl of 4 leaves. The petiole is 0.8–1.8 cm long. The blade is broadly elliptic, 8–14 cm × 5–8 cm, serrate, cuneate at base, acute at apex, glossy, somewhat dull dark green and fleshy, and with 6–8 pairs of secondary nerves. The inflorescence is a terminal, 3–5 cm long spike. The flowers are linear and pure white. The androecium includes 3 stamens. The ovary is ovoid. The fruit is a drupe, which is green, globose, and about 3 mm across.

*Antifungal (filamentous) halo developed by amphiphilic lindenane sesquiterpene*: Chloranthalactone A (also named dehydro-shizukanolide) isolated from the rhizomes inhibited the growth of *Fusarium solani* and *Rhizopus oryzae* with inhibition zone diameter of 11 and 36 mm, respectively (50 µg/8 mm disc) (Kawabat et al., 1981).

*Strong antiviral (enveloped linear dimeric single-stranded (+)RNA) amphiphilic lindenane sesqui- terpenes*: Shizukaol B (LogD = 1.5 at pH 7.4; molecular weight = 732.8 g/mol) inhibited the replication of the Human immunodeficiency virus (wild-type), Human immunodeficiency virus (RT-K103N), and Human immunodeficiency virus (RT-K103N) with the $EC_{50}$ values of 0.2, 0.4, and 0.5 µM, respec- tively, in C8166 cells (Fang et al., 2011). Chlorajaponilides F and G as well as sarcandrolide F, and shizukaol E inhibited the replication of the Human immunodeficiency virus type-1 (wild-type HIV) with the $EC_{50}$ values of 3, 17.1, 3.2, and 5.4 µM, respectively (Yan et al., 2016).

Shizukaol B

*Strong antiviral (enveloped monopartite linear single-stranded (+)RNA) amphiphilic lindenane sesquiterpenes*: Chlorajaponilide F (molecular mass = 716.7 g/mol), sarcandrolide F, and shizukaol E inhibited the replication of the Hepatitis C virus with the $EC_{50}$ values of 3, 9.3, and 1.6 µM, respectively (Yan et al., 2016).

Chlorajaponilide F

**Commentaries**

(i) The Hepatitis C virus (HCV) belongs to the genus *Hepacivirus* and the family Flaviviridae (Payne, 2017). This RNA virus is transmitted by blood and accounts for hepatocellular carcinoma in humans (Payne, 2017). There is no vaccine for this virus because numerous genetic types exist and it is treated with protease inhibitors, telaprevir, or boceprevir, as well as sofosbuvir, the liver phase I metabolite of which inhibits HCV NS5B polymerase, thus inhibiting the HCV-RNA synthesis by RNA chain termination (Bhatia et al., 2014). *Fusarium solani* (family Nectriaceae) is a filamentous fungus that infects plants but also an opportunistic fungus in humans (Guarro & Gene, 1995). *Rhizopus oryzae* (family Mucoraceae) is a plant pathogen filamentous fungi, which is also opportunistic in humans (Kok et al., 2007).

(ii) Consider that natural products able to strongly inhibit the replication of the Herpes simplex virus from medicinal plant in the Clades covered in this first volume have often medium to high molecular masses and are not uncommonly amphiphilic or lipophilic.

**References**

Bhatia, H.K., Singh, H., Grewal, N. and Natt, N.K., 2014. Sofosbuvir: a novel treatment option for chronic hepatitis C infection. *Journal of Pharmacology & Pharmacotherapeutics*, 5(4), p. 278.

Fang, P.L., Cao, Y.L., Yan, H., Pan, L.L., Liu, S.C., Gong, N.B., Lü, Y., Chen, C.X., Zhong, H.M., Guo, Y. and Liu, H.Y., 2011. Lindenane disesquiterpenoids with anti-HIV-1 activity from *Chloranthus japonicus*. *Journal of Natural Products*, 74(6), pp. 1408–1413.

Guarro, J. and Gene, J., 1995. Opportunistic fusarial infections in humans. *European Journal of Clinical Microbiology and Infectious Diseases*, 14(9), pp. 741–754.

Kawabata, J., Tahara, S. and Mizutani, J., 1981. Isolation and structural elucidation of four sesquiterpenes from *Chloranthus japonicus* (Chloranthaceae). *Agricultural and Biological Chemistry*, 45(6), pp. 1447–1453.

Kok, J., Gilroy, N., Halliday, C., Lee, O.C., Novakovic, D., Kevin, P. and Chen, S., 2007. Early use of posaconazole in the successful treatment of rhino-orbital mucormycosis caused by *Rhizopus oryzae. Journal of Infection*, 55(3), pp. e33–e36.

Payne, S., 2017. *Viruses: From Understanding to Investigation*. Academic Press, USA.

Yan, H., Ba, M.Y., Li, X.H., Guo, J.M., Qin, X.J., He, L., Zhang, Z.Q., Guo, Y. and Liu, H.Y., 2016. Lindenane sesquiterpenoid dimers from *Chloranthus japonicus* inhibit HIV and HCV replication. *Fitoterapia*, 115, pp. 64–68.

## 1.4.1.2   *Sarcandra glabra* (Thunb.) Nakai

Synonyms: *Ardisia glabra* (Thunb.) A. DC.; *Chloranthus glaber* (Thunb.) Makino

Common names: Glabrous sarcandra; cao shan hu (China); kayu duri duri (Indonesia); gipah, kari kari (the Philippines); soi rung (Vietnam)

Habitat: Riverbanks, roadsides, and forests

Distribution: India, Sri Lanka, Laos, Thailand, Vietnam, Malaysia, the Philippines Cambodia China, Korea, and Japan.

Botanical observation: It is a shrub that grows up to 1.5m tall and which at first glance has somewhat a look of Holly. The stems are cylindrical, and enlarged at node. The leaves are simple, decussate, and stipulate. The petiole is 0.5–2cm long. The blade is elliptic, 6–20×2–8cm, thick, acute at base, serrate at margin, acute at apex, and with 5–7 pairs of secondary nerves, which are raised. The inflorescence is a terminal spike, which is 1.5–4cm long. The flowers are tiny without perianth, and include 1 fleshy anther and a globose ovary. The fruits are globose drupes which are red, glossy, and about 4mm across.

Medicinal uses: Urinary tract infection (the Philippines)

*Antibacterial (Gram-positive) halo developed by flavonol glycoside*: Kaempferol-3-*O*-β-D-glucuronide inhibited the growth of *Staphylococcus aureus* (Yuan et al., 2008).

*Antibacterial (Gram-positive) halo developed by flavonol*: Kaempferol inhibited the growth of *Staphylococcus aureus* (Yuan et al., 2008).

*Antibacterial (Gram-positive) halo developed by coumarin glycoside*: Isofraxidin-7-*O*-beta-D-glucopyranoside inhibited the growth of *Staphylococcus* (Yuan et al., 2008).

*Strong antiviral (enveloped segmented linear single-stranded (–)RNA) hydrophilic flavonol*: Kaempferol (LogD = 0.8 at pH 7.4; molecular weight = 286.2 g/mol) inhibited the replication of the Influenza virus A/chicken/Rostock/34 (H7N1) with an $EC_{50}$ value of 18.8 µg/mL and a selectivity index of 4.9 (Pantev et al., 2006).

Kaempferol

*Strong antiviral (enveloped monopartite linear double-stranded DNA) amphiphilic flavanone*:
Pinostrobin (LogD = 3.5 at pH 7.4; molecular weight = 270.2 g/mol) (Yuan et al., 2008) inhibited
the replication of the Herpes simplex virus type-1 (F strain) with an $IC_{50}$ value of 22.7 µg/mL and
evoked viral envelope desquamation shape distortion, and destruction (Wu et al., 2011).

Pinostrobin

*In vivo antiviral (enveloped monopartite linear double-stranded DNA) amphiphilic flavanone*:
Pinostrobin given orally at the dose of 50 mg/Kg/day for 7 days to mice infected with Herpes
simplex virus evoked 100% inhibition of mortalities (Wu et al., 2011).

*Viral enzyme inhibition by hydrophilic flavonol*: Kaempferol inhibited Hepatitis C virus NS3
serine protease activity with an $IC_{50}$ value of 1.6 µM (Zuo et al., 2005).

**Commentary:**

Medium molecular weight, amphiphilic, flavonoids, stilbenes, lignans, and isoquinoline
alkaloids from medicinal plants in the Clades treated in this book are often able to strongly
inhibit the replication of Herpes simplex viruses *in vitro*.

**References**

Pantev, A., Ivancheva, S., Staneva, L. and Serkedjieva, J., 2006. Biologically active constituents of a polyphenol extract from *Geranium sanguineum* L. with anti-influenza activity.
*Zeitschrift für Naturforschung C*, *61*(7–8), pp. 508–516.

Wu, N., Kong, Y., Zu, Y., Fu, Y., Liu, Z., Meng, R., Liu, X. and Efferth, T., 2011. Activity
investigation of pinostrobin towards Herpes simplex virus-1 as determined by atomic force
microscopy. *Phytomedicine*, 18(2–3), pp. 110–118.

Yuan, K., Zhu, J.X., Si, J.P., Cai, H.K., Ding, X.D. and Pan, Y.J., 2008. Studies on chemical
constituents and antibacterial activity from n-butanol extract of *Sarcandra glabra*. *China
Journal of Chinese Materia Medica*, *33*(15), pp. 1843–1846.

Zuo, G.Y., Li, Z.Q., Chen, L.R. and Xu, X.J., 2005. In vitro anti-HCV activities of *Saxifraga
melanocentra* and its related polyphenolic compounds. *Antiviral Chemistry and
Chemotherapy*, 16(6), pp. 393–398.

# 2 The Clade Magnoliids

## 2.1 ORDER PIPERALES BERCHT. & PRESL. (1820)

### 2.1.1 FAMILY ARISTOLOCHIACEAE A.L. de Jussieu (1789)

The family Aritolochiaceae consists of 9 genera and about 600 species of discrete rhizomatous herbs, shrubs, or climbers which are often poisonous and have somewhat a monocotyledonous look. The leaves are simple, alternate, and exstipulate. The flowers are solitary or cymose and terminal or axillary. The perianth is 3-lobed and often very characteristically pipe-shaped and sometimes very large. The androecium includes 6–12 stamens. The gynoecium includes 4–6 carpels fused into a stout style with 3–6 lobes. The capsules are often ridged, dehiscent, and contain minute flat-winged seeds.

#### 2.1.1.1 *Apama corymbosa* (Bl.) O. Ktze

Synonyms: *Asiphonia piperiformis* Griff.; *Bragantia corymbosa* Griff.; *Bragantia melastomifolia* (C.Presl) Duch.; *Strakaea melastomifolia* C.Presl; *Thottea corymbosa* (Griff.) Ding Ho
  Common names: Akar julong, akar surai (Malaysia)
  Habitat: Jungle paths
  Distribution: Malaysia, Indonesia, and the Philippines
  Botanical observation: It is a shrub that grows up to 5 m tall. The stems are zigzag shaped, terete, and easy to break. The leaves are simple, alternate, and exstipulate. The blade is ovate to lanceolate, thin, 6.5–17.5 × 2.5–8.5 cm, with 3–5 pairs of secondary nerves, acute at base, and acuminate at apex. The inflorescences are terminal or axillary lax-branched corymbs, which are about 10 cm long. The perianth is minute and includes 3 lobes. The androecium comprises 7–10 stamens. The fruits are slender and linear, about up to 30 cm long, and contain numerous trigonous and rugose seeds.
  Medicinal use: Boils (Malaysia)
  *Antibacterial (Gram-positive) halo developed by polar extract:* Methanol extract of bark (1 mg/6 mm disc) inhibited the growth of *Bacillus cereus* and *Bacillus subtilis* with inhibition zone diameters of 12 and 10 mm, respectively (Wiart et al., 2004).
  *Anticandidal polar extract:* Methanol extract of bark (1 mg/6 mm disc) inhibited the growth of *Candida albicans* with an inhibition zone diameter of 13 mm (Wiart et al., 2004).

#### Reference
Wiart, C., Mogana, S., Khalifah, S., Mahan, M., Ismail, S., Buckle, M., Narayana, A.K. and Sulaiman, M., 2004. Antimicrobial screening of plants used for traditional medicine in the state of Perak, Peninsular Malaysia. *Fitoterapia, 75*(1), pp. 68–73.

#### 2.1.1.2 *Aristolochia bracteata* Retz.

Synonyms: *Aristolochia bracteolata* Lam.
  Common name: Worm killer; karalakam (India)
  Habitat: Roadside, waste lands
  Distribution: Pakistan, India, and Sri Lanka
  Botanical observation: It is a herb that grows up to 60 cm tall with an unpleasant smell. The stems are terete or angled and woody at base. The petiole is 0.3–6.5 cm long, striate, and glabrous. The leaves are simple, spiral, and exstipulate. The blade is glaucous, ovate-cordate, 3–10.5 cm × 2.5–7.5 cm, acute at apex, cordate at base, and glaucous beneath. The inflorescences are solitary and axillary.

The perianth is 2.5–4.5 cm long, irregular, tubular, greenish yellow, and with a reddish-brown to dark purple limb. The androecium includes 6 stamens, which are sessile, and united with a 6-lobed gynostegium. The ovary is oblong, hexagonal, 5 mm long, and glabrous. The capsules are 1.5–3 cm × 1.3–1.5 cm, 12-ribbed, glabrous, glaucous, and contain numerous seeds which are flat and dark brown.

Medicinal use: Putrefied wound (India)

*Broad antibacterial hydrophilic phenanthrene alkaloid:* The aristolactam-type alkaloid aristolochic acid (also known as aristolochic acid I or aristolochic acid A; LogD = 0.4 at pH 7.4; molecular mass = 341.2 g/mol) inhibited the growth of *Bacillus subtilis* and *Escherichia coli* with the inhibition zone diameters of 17 and 16 mm, respectively (500 µg/5 mm disc) (Angalaparameswari et al., 2012).

Aristolochic acid

*Moderate antibacterial polar extract:* Methanol extract inhibited the growth of *Staphylococcus aureus* (ATCC 6538), *Bacillus subtilis* (ATCC 6059), and *Micrococcus flavus* (SBUG 16) with the MIC values of 250, 500, and 250 µg/mL, respectively (Mothana et al., 2011).

*Strong antibacterial (Gram-negative) hydrophilic phenanthrene alkaloid:* Aristolochic acid inhibited the growth of *Moraxella catarrhalis* (GTC 01544) with MIC/MBC values of 25/50 µg/mL (Suliman Mohamed et al., 2014).

Commentaries: (i) *Moraxella catarrhalis* is a Gram-negative, aerobic, diplococcus responsible for otitis media and respiratory infections, including pneumonia (Verduin et al., 2002). (ii) The antibacterial mechanism of aristolochic acid is yet not fully understood but probably results from DNA damage (Pfau et al., 1990). Consider that aristolochic acid is carcinogenic and of no therapeutic use.

### References

Angalaparameswari, S., Saleem, T.M., Alagusundaram, M., Ramkanth, S., Thiruvengadarajan, V., Gnanaprakash, K., Chetty, C.M. and Pratheesh, G., 2012. Anti-microbial activity of aristolochic acid from root of *Aristolochia bracteata* Retz. *International Journal of Biological Life Sciences, 8*, pp. 2–4.

Mothana, R.A., Kriegisch, S., Harms, M., Wende, K. and Lindequist, U., 2011. Assessment of selected Yemeni medicinal plants for their in vitro antimicrobial, anticancer, and antioxidant activities. *Pharmaceutical Biology, 49*(2), pp. 200–210.

Pfau, W., Schmeiser, H.H. and Wiessler, M., 1990. Aristolochic acid binds covalently to the exocyclic amino group of purine nucleotides in DNA. *Carcinogenesis, 11*(2), pp. 313–319.

Suliman Mohamed, M., Timan Idriss, M., Khedr, A.I., Abd AlGadir, H., Takeshita, S., Shah, M.M., Ichinose, Y. and Maki, T., 2014. Activity of *Aristolochia bracteolata* against *Moraxella catarrhalis*. *International Journal of Bacteriology*.

Verduin, C.M., Hol, C., Fleer, A., van Dijk, H. and van Belkum, A., 2002. *Moraxella catarrhalis*: from emerging to established pathogen. *Clinical Microbiology Reviews, 15*(1), pp. 125–144.

### 2.1.1.3 *Aristolochia contorta* Bunge

Common name: Bei ma dou ling (China)

Habitat: Thickets and mountains

Distribution: China, Japan, and Korea

Botanical observation: It is a climber with an unpleasant smell. The stems are terete, striate, and glabrous. The leaves are simple, spiral, and exstipulate. The petiole is up to 7 cm long. The blade is deltoid-cordate, 3–13 × 3–10 cm, papery, with 2–3 pairs of secondary nerves, cordate at base, and acute or obtuse at apex. The inflorescences are axillary racemes. The perianth is yellow-green with purple veins, tubular with a 3 cm long limb. The androecium includes 6 stamens united with a 6-lobed gynostegium. The ovary is hexagonal and minute. The capsules are dehiscent, obovoid, 3–6.5 cm × 2.5–45 cm, and contain numerous seeds, which are flat and winged.

Medicinal use: Rabies (Korea)

*Strong antibacterial (Gram-positive) amphiphilic aristolactam-type alkaloid:* Aristolactam N-(6′-trans-p-coumaroyl)-β-D-glucopyranoside (LogD = 2.5 at pH 7.4; molecular mass = 601.6 g/mol) isolated from the roots inhibited the growth of *Bacillus subtilis* and *Sarcina lutea* with MIC values of 43.8 and 175 μg/mL, respectively (Lee & Han, 1992).

Aristolactam N-(6′-trans-p-coumaroyl)-β-D-glucopyranoside

Commentary: Consider that medicinal plants in Asia and Pacific are seldom used for the traditional treatment of rabies, and these are not uncommonly from the family Aristolochiaceae. Natural products with strong *in vitro* antibacterial activity against Gram-positive bacteria from the plants belonging to the Clades covered in this volume are mainly of the phenolic type and isoquinolines with medium to high molecular weight and amphiphilic.

### Reference

Lee, H.S. and Han, D.S., 1992. A new acylated N-glycosyl lactam from *Aristolochia contorta*. *Journal of Natural Products, 55*(9), pp. 1165–1169.

#### 2.1.1.4 *Aristolochia debilis* Sieb. & Zucc.

Synonyms: *Aristolochia longa* Thunb.; *Aristolochia recurvilabra* Hance; *Aristolochia sinarum* Lindl.
Common name: Na dou ling (China)
Habitat: Thickets and mountains
Distribution: China, Korea, and Japan
Botanical observation: It is a climber. The stems are terete, striate, and glabrous. The leaves are simple, spiral, and exstipulate. The petiole is 1–2 cm long. The blade is ovate or oblong-ovate to sagittate, 3–6 × 1.5–3.5 cm, papery, with 2–3 pairs of secondary nerves, the base cordate, and the apex acute or round. The inflorescences are axillary and solitary. The perianth is yellow-green, dark purple at throat, with a curved and 2–2.5 cm long tube, and a 2–3 cm long limb. The androecium includes 6 stamens united with a 6-lobed gynostegium. The ovary is hexagonal. The capsules are about 6 cm long, and contain numerous seeds, which are flat and winged.
Medicinal use: Wounds (China)
*Moderate broad-spectrum antibacterial non-polar extract:* Hexane extract (enriched in triglycerides) of roots inhibited the growth of *Escherichia coli* (ATCC 8739), *Salmonella typhimurium* (ATCC 14028), *Enterococcus faecium* (ATCC 19434), *Streptococcus agalactiae*, and *Staphylococcus aureus* (ATCC 6538) with MIC values of 1000, 1000, 125, 125, and 1000 µg/mL, respectively (Dhouioui et al., 2016).
*Broad spectrum antibacterial phenanthrene alkaloids:* Aristolochic acid and the aristolactam-type alkaloid aristolactam Ia isolated from the roots showed antibacterial activity against *Escherichia coli*, *Pseudomonas aeruginosa*, *Streptococcus faecalis*, *Staphylococcus aureus*, and *Staphylococcus epidermidis* (Hinou et al., 1990).
Commentary: A broad array of aristolochic acids and aristolactam types have been isolated from the family Aristolochiaceae (Kumar et al., 2003). Of these a very few have been assessed for their antibacterial, antifungal, and antiviral activities. One could argue that the mutagenic effects of these alkaloids could be a barrier to their therapeutic developments, but original chemical frameworks could be found as antimicrobial agents and even antibiotic potentiators.

#### References

Dhouioui, M., Boulila, A., Jemli, M., Schiets, F., Casabianca, H. and Zina, M.S., 2016. Fatty acids composition and antibacterial activity of *Aristolochia longa* L. and *Bryonia dioica* Jacq. growing wild in Tunisia. *Journal of Oleo Science, 65*(8), pp. 655–661.
Hinou, J., Demetzos, C., Harvala, C. and Roussakis, C., 1990. Cytotoxic and antimicrobial principles from the roots of *Aristolochia longa*. *International Journal of Crude Drug Research, 28*(2), pp. 149–151.
Kumar, V., Prasad, A.K. and Parmar, V.S., 2003. Naturally occurring aristolactams, aristolochic acids and dioxoaporphines and their biological activities. *Natural Product Reports, 20*(6), pp. 565–583.

#### 2.1.1.5 *Aristolochia indica* L.

Common names: Indian birthwort; isharmul (Bangladesh); isvaramuli; hukka-bel (India)
Habitat: Forest margins
Distribution: India, Bangladesh, and the Andamans
Botanical observation: It is a climber that grows to 5 m long. The stem is angled and glabrous. The leaves are simple, alternate, and exstipulate. The petiole is 1–1.5 cm long. The blade is oblong, cordate at base, acuminate at apex, 5–10 × 2–4 cm, and fleshy. The inflorescences are axillary racemes. The perianth is tubular, greenish, curved, globose at base, and about 1 cm in diameter and develops a tube, which reaches 2.5 cm in length. The androecium includes 6 stamens united with the gynostegium. The ovary is cylindrical and ribbed. The capsules are pendulous, dehiscent, 6-valved, and contain deltoid seeds.
Medicinal uses: Rabies, leucorrhea, tooth infections, diarrhea, cough, leprosy, fever (India)

*Broad spectrum antibacterial essential oil:* Essential oil of aerial part (5 µL/disc) inhibited the growth of *Pseudomonas aeruginosa*, *Bacillus subtilis*, *Staphylococcus aureus*, *Escherichia coli*, and *Salmonella typhimurium* with inhibition zone diameters of 10, 9, 8, 7, and 7 mm, respectively (Shafi et al., 2002).

*Broad spectrum antibacterial halo developed by polar extract:* Ethanol extract of the whole plant (500 µg/disc) inhibited the growth of multidrug-resistant *Escherichia coli*, *Klebsiella pneumoniae*, *Proteus mirabilis*, *Enterobacter aerogenes*, *Enterococcus faecalis*, *Vibrio cholerae*, *Staphylococcus aureus*, *Staphylococcus epidermidis*, and *Bacillus subtilis* with inhibition zone diameters ranging from 12 to 15 mm (Venkatadri et al., 2015).

*Moderate broad-spectrum antibacterial polar extract:* Methanol extract of leaves inhibited the growth of *Acinetobacter baumannii* (ATCC 17978), *Klebsiella pneumoniae* (ATCC 6059), *Pseudomonas aeruginosa* (ATCC 7221), and *Staphylococcus aureus* (ATCC 6538) with MIC/MBC values of 150/640, 90/355, 55/220, and 200/750 µg/mL, respectively (Naz et al., 2017).

Commentary: Essential oils of medicinal plants are almost always antibacterial (Gram-positive) and/or antifungal *in vitro*.

### References

Naz, R., Ayub, H., Nawaz, S., Islam, Z.U., Yasmin, T., Bano, A., Wakeel, A., Zia, S. and Roberts, T.H., 2017. Antimicrobial activity, toxicity and anti-inflammatory potential of methanolic extracts of four ethnomedicinal plant species from Punjab, Pakistan. *BMC Complementary and Alternative Medicine, 17*(1), p. 302.

Shafi, P.M., Rosamma, M.K., Jamil, K. and Reddy, P.S., 2002. Antibacterial activity of the essential oil from *Aristolochia indica*. *Fitoterapia, 73*(5), pp. 439–441.

Venkatadri, B., Arunagirinathan, N., Rameshkumar, M.R., Ramesh, L., Dhanasezhian, A. and Agastian, P., 2015. In vitro antibacterial activity of aqueous and ethanol extracts of *Aristolochia indica* and *Toddalia asiatica* against multidrug-resistant bacteria. *Indian Journal of Pharmaceutical Sciences, 77*(6), p. 788.

### 2.1.1.6 *Aristolochia tagala* Cham.

Synonym: *Aristolochia roxburghiana* Klotzsch

Common names: Er ye ma dou ling (China); kelayar (Indonesia); akar ketola hutan (Malaysia); goan-goan (the Philippines); krachao mot (Thailand)

Habitat: Mountains and forests

Distribution: India, Bhutan, Cambodia, Indonesia, Japan, Nepal, Malaysia, Myanmar, the Philippines, Thailand, and Vietnam

Botanical observation: It is a climber that grows to 10 m long. The stem is terete. The leaves are simple, alternate, and exstipulate. The petiole 2.5–4 cm long. The blade is ovate-cordate or oblong-ovate, 6–25 × 4–15 cm, papery, with 3–5 pairs of secondary nerves, cordate at base and acute or acuminate at apex. The racemes are 2–6 cm long and axillary. The perianth is tubular, curved 5–10 cm long, greenish yellow, purplish at throat, and with an oblong limb. The androecium includes 6 stamens united in a 6-lobed gynostegium. The ovary is cylindrical and ribbed. The capsules are pendulous and dehiscent, which are obovoid-globose, 3.5–5 × 2–3.5 cm, and contain numerous triangular seeds.

Medicinal use: Otitis (the Philippines)

*Broad-spectrum antibacterial polar extract:* Ethanol extract of leaves inhibited the growth of *Staphylococcus lentus*, *Bacillus cereus*, and *Serratia marcescens* with inhibition zone diameters of 23, 25, and 16.6 mm, respectively (30 µg/6 mm disc) (Hercluis & Koilpillai, 2018).

*Broad-spectrum antifungal halo developed by polar extract:* Ethanol extract of leaves inhibited the growth of *Candida albicans*, *Candida dubliniensis*, and *Cryptococcus neoformans* with inhibition zone diameters of 10.5, 11, and 10.3 mm, respectively (30 µg/6 mm disc) (Hercluis & Koilpillai, 2018).

**Reference**

Hercluis, D. and Koilpillai, Y.J., 2018. Evaluation of the antimicrobial efficacy of *Aristolochia Tagala* leaf extract against selected human pathogenic bacteria and fungi. *Int. J. Innov. Res. Tech.* 5(2);344-347.

### 2.1.1.7 *Asarum forbesii* Maxim.

Common names: Forbes's Chinese Wild Ginger; du heng (China)

  Habitat: Forests

  Distribution: China

Botanical observation: It is a rhizomatous herb. The leaves are simple. The petiole is up to 15 cm long. The blade is cordate to reniform, 3–8 × 3–8 cm, cordate at base, obtuse to rounded at apex, fleshy, variegated and Araceaeous-like. The flowers grow from the rhizome on a 2 cm long peduncle. The calyx is dark purple, cylindric to campanulate, 1.5 cm –2.5 × 1 cm, and 3-lobed, the lobes broadly lanceolate. The androecium includes 12 stamens. The gynoecium includes an ovary with 6 locules and 6 styles. The capsules are fleshy and dehiscent.

  Medicinal use: Fever (China)

  *Moderate antibacterial (Gram-negative) amphiphilic phenyl propanoid:* Methylisoeugenol (LogD = 2.7 at pH 7.4; molecular mass = 178.2 g/mol) inhibited growth of *Campylobacter jejuni* (NCTC 11168) with an MIC value of 125 µg/mL (Rossi et al., 2007).

  *Viral enzyme inhibition by polar extract:* Methanol extract at the concentration of 100 µg/mL inhibited RNA-dependent DNA polymerase activity of Human immunodeficiency virus type-1 reverse transcriptase and the Human immunodeficiency virus type-1 protease by 3.8 and 20.9%, respectively (Min et al., 2001).

  Commentaries: (i) The essential oil of this plant contains mainly methyl isoeugenol, isoelemicin, and α-asarone (Zhang et al., 2005). (ii) *Campylobacter jejuni* is a Gram-negative bacterium that accounts for enteritis and diarrhea. (iii) A member of the genus *Asarum L.* is used in China for the treatment of COVID-19 (Yang et al., 2020). (iv) Note that natural products from medicinal plants in the Clades treated in this volume able to strongly inhibit the replication of HIV are mainly medium to high molecular weight and amphiphilic or lipophilic lignans, isoquinoline alkaloids, triterpenes, sesquiterpenes, and phenolics.

**References**

Min, B.S., Kim, Y.H., Tomiyama, M., Nakamura, N., Miyashiro, H., Otake, T. and Hattori, M., 2001. Inhibitory effects of Korean plants on HIV -1 activities. *Phytotherapy Research,* *15*(6), pp. 481–486.

Rossi, P.G., Bao, L., Luciani, A., Panighi, J., Desjobert, J.M., Costa, J., Casanova, J., Bolla, J.M. and Berti, L., 2007. (E)-Methylisoeugenol and elemicin: antibacterial components of *Daucus carota* L. essential oil against *Campylobacter jejuni. Journal of Agricultural and Food Chemistry, 55*(18), pp. 7332–7336.

Yang, Q.X., Zhao, T.H., Sun, C.Z., Wu, L.M., Dai, Q., Wang, S.D. and Tian, H., 2020. New thinking in the treatment of 2019 novel coronavirus pneumonia. *Complementary Therapies in Clinical Practice, 39*, p. 101131.

Zhang, F., Xu, Q., Fu, S., Ma, X., Xiao, H. and Liang, X., 2005. Chemical constituents of the essential oil of *Asarum forbesii* Maxim (Aristolochiaceae). *Flavour and Fragrance Journal, 20*(3), pp. 318–320.

### 2.1.1.8 *Asarum sieboldii* Miq.

Synonym: *Asiasarum sieboldii* (Miq.) F. Maek.

  Common names: Korean wild ginger; han cheng xi xin (China); jok do ri (Korea)

  Habitat: Forests

  Distribution: China and Korea

Botanical observation: It is a rhizomatous herb. The leaves are simple. The petiole is up to 18 cm long. The blade is cordate to ovate-cordate, 4–11 × 4.5–13.5 cm, acuminate or acute at apex, fleshy, and glossy. The flowers grow from rhizomes on a 4 cm long peduncle. The calyx is dark purple, urceolate to campanulate, 6–8 × 1–1.5 cm, and with 3 triangular-ovate lobes, which are about 1 cm long. The androecium includes 12 stamens. The gynoecium includes an ovary with 6 locules and 6 styles. The fruit is a fleshy dehiscent capsule.

Medicinal uses: Colds (China); fever, cough (Korea)

*Antibacterial (Gram-positive) halo developed by amide alkaloid:* Pellitorine, (50 μg/disc) inhibited the growth of *Listeria monocytogenes* with an inhibition zone diameter of 9 mm (Oh et al., 2010).

*Moderate antibacterial (Gram-positive) monoterpenes:* (±)Car-3-ene-2,5-dione and (±)-asarinol A (50 μg/disc) inhibited the growth of *Listeria monocytogenes* with an inhibition zone diameter of 10 mm (Oh et al., 2010). (±)-car-3-ene-2,5-dione and (±)-asarinol A, inhibited the growth of *Listeria monocytogenes* with an MIC value of 125 μg/mL (Oh et al., 2010).

*Moderate antibacterial (Gram-positive) furofuran lignans:* (-)-Asarinin and (-)-sesamin (50 μg/disc) inhibited the growth of *Listeria monocytogenes* with inhibition zone diameters of 10 and 9 mm, respectively (Oh et al., 2010). (-)-Asarinin and (-)-sesamin, inhibited the growth of *Listeria monocytogenes* with an MIC value of 125 μg/mL (Oh et al., 2010).

*Moderate antibacterial (Gram-positive) phenylpropanoids:* γ-Asarone and methyl eugenol (50 μg/disc) inhibited the growth of *Listeria monocytogenes* with an inhibition zone diameter of 10 mm (Oh et al., 2010). γ-Asarone and methyl eugenol (50 μg/disc) inhibited the growth of *Listeria monocytogenes* with an MIC value of 125 μg/mL (Streptomycin: 42.5 μg/mL) (Oh et al., 2010).

*Very weak antibacterial amide alkaloid:* Pellitorine inhibited the growth of *Listeria monocytogenes* with an MIC value of 500 μg/mL (Oh et al., 2010).

*Strong broad-spectrum antifungal essential oil:* Essential oil from the roots inhibited the growth of *Candida albicans* (ATCC10231), *Cryptococcus neoforman* (H99), *Aspergillus niger* (ATCC8642), and *Aspergillus fumigatus* (Af237). For *Candida albicans* (ATCC10231), the MIC was 5 μL/mL (Han, 2007).

*Strong antifungal amphiphilic phenylpropanoid:* Kakuol (LogD = 1.9 at pH 7.4; molecular mass = 194.1 g/mol) from the rhizomes inhibited *Botrytis cinerea*, *Cladosporium cucumerinum*, and *Colletotrichum orbiculare* with the MIC values of 50, 30, and 10 μg/mL, respectively (Lee et al., 2005).

Kakuol

*Antiviral (enveloped segmented linear single-stranded (−) RNA virus) polar extract:* Methanol extract of roots (30 g/300 mL) reduced the count of Influenza virus A/H5N1 (A/Vietnam/1194/04 (H5N1)-NIBRG-14) by 4 –Log (Lee et al., 2010).

Commentary: (i) It appears that members of the family Aristolochiaceae have not been studied much for antiviral activity. The anti-Influenza A and B drug Tamiflu® (*oseltamivir*) was developed from shikimic acid synthesized by *Illicium verum* L., illustrating the contention that "paleoherbs" are a fascinating resource for the development of antiviral drugs.

(ii) The twentieth century has been the theater of three ignominious major pandemics due to Influenza A viruses: the "Spanish flu" of 1928 (over 40 million fatalities), the Asian flu of 1957 (over 4 million fatalities), and the "Hong Kong flu" of 1968 (1 million fatalities). According to the World Health Organization, humans acquire avian Influenza A viruses from birds, including poultry and migratory birds (Influenza virus subtypes A(H5N1), A(H7N9), and A(H9N2)) and swine (Influenza virus subtypes A(H1N1), A(H1N2), and A(H3N2)). One example is the Hong Kong outbreak of Avian Influenza A virus (H5N1) that caused death in 6 of 18 infected persons in 1997 and by July 2006, 54 countries across three continents had been affected by this virus (Ooi et al., 2014). Influenza A viruses are enveloped RNA viruses that belong to the genus Influenza virus A in the family Orthomyxoviridae. In humans, this virus evokes fever, cough, and shortness of breath, and pneumonia, which can be fatal. Note that Influenza virus type A (and type B) causes recurrent epidemics almost every year, leading to significant human morbidity and mortality (Peiris et al., 2007). The adamantanes (amantadine and rimantadine) and the neuraminidase inhibitors (oseltamivir and zanamivir) are the two currently available classes of drugs that are specifically active against Influenza viruses (Peiris et al., 2007).

(iii) Cinanserin is a synthetic inhibitor of the 3C-like proteinase of Severe acute respiratory syndrome-associated coronavirus and strongly reduces virus replication (Chen et al., 2005), which bears an amide moiety linked to benzyl groups, suggesting that amide alkaloids in plants could be of value in the fight against coronavirus infection. What about pellitorine?

### References

Chen, L., Gui, C., Luo, X., Yang, Q., Günther, S., Scandella, E., Drosten, C., Bai, D., He, X., Ludewig, B. and Chen, J., 2005. Cinanserin is an inhibitor of the 3C-like proteinase of severe acute respiratory syndrome coronavirus and strongly reduces virus replication in vitro. *Journal of Virology, 79*(11), pp. 7095–7103.

Han, K.H., 2007. Antifungal activity of essential oil from *Asarum sieboldii* against epidermal and opportunistic pathogenic fungi. *The Korean Journal of Mycology, 35*(1), pp. 58–60.

Lee, J.Y., Moon, S.S. and Hwang, B.K., 2005. Isolation and antifungal activity of kakuol, a propiophenone derivative from *Asarum sieboldii* rhizome. *Pest Management Science: Formerly Pesticide Science, 61*(8), pp. 821–825.

Lee, J.H., Van, N.D., Ma, J.Y., Kim, Y.B., Kim, S.K. and Paik, H.D., 2010. Screening of antiviral medicinal plants against avian influenza virus H1N1 for food safety. *Korean Journal for Food Science of Animal Resources, 30*(2), pp. 345–350.

Oh, J., Hwang, I.H., Kim, D.C., Kang, S.C., Jang, T.S., Lee, S.H. and Na, M., 2010. Anti-listerial compounds from Asari Radix. *Archives of Pharmacal Research, 33*(9), pp. 1339–1345.

Ooi, V.E.C., Chan, P.K.S., Chiu, L.C.M., Sun, S.S.M. and Wong, H.N.C., 2014. Antiviral activity of Chinese medicine–derived phytochemicals against avian influenza A (H5N1) virus. *Korean Journal for Food Science of Animal Resources, 20*(Supplement 4).

Peiris, J.M., De Jong, M.D. and Guan, Y., 2007. Avian influenza virus (H5N1): a threat to human health. *Clinical Microbiology Reviews, 20*(2), pp. 243–267.

### 2.1.1.9 *Thottea grandiflora* Rottb.

Common names: Purple Totthea; bunga semubut, hempedu beruang (Malaysia)

Botanical observation: It is a woody undershrub that grows to about 1 m tall. The stem is zigzag shaped, somewhat articulated, hairy, and terete. The leaves are simple, alternate, and exstipulate. The blade is oblong-ovate, 15–35 cm × 10–15 cm, glossy, round at base, and acuminate at apex. The racemes are axillary and bear magnificent flowers. The perianth is 8–10 cm long, purple, 3-lobed, the lobes oblong. The androecium includes 6 stamens free or adnate to the style column. The gynoecium includes a 4-angular and 4-celled ovary. The fruit is a capsule, which is dehiscent, about 20 cm long, linear, and containing numerous ellipsoid, 3-angular seeds.

Medicinal use: Fever (Malaysia)

Medicinal uses: Apparently none for the treatment of microbial infections.

Commentary: (i) This plant is an example (among many others (!)) of rare, unique, and precious plants that may disappear soon under the bulldozers of the palm oil corporations. Once a plant has disappeared, its active principles remain forever a mystery and a therapeutic chance may have been missed. Consider that a primary rainforest cannot be reconstituted by humans. It took millions of years of evolution for the primary rainforest of Southeast Asia to develop. This forest is a gift of the Mother Nature as a source of extraordinary molecules for the treatment of diseases. (ii) Members of the genus *Thottea* Rottb. have displayed antibacterial and antifungal effects. The antimicrobial effects of *Thottea grandiflora* Rottb., which is used for fever, need to be assessed.

### Reference

Anilkumar, E.S., Nishanth Kumar, S., Latha, P.G., Dan, M. and Kumar, D., 2014. A comparative study on the in-vitro antimicrobial activity of the roots of four thottea species.

## 2.1.2 FAMILY PIPERACEAE GISEKE (1792)

The family Piperaceae comprises 7 genera and about 1100 species of tropical climbers with a monocotelydenous-look. The stems are terete, smooth, and articulate. The leaves are simple, spiral, alternate, opposite, or verticillate, and exstipulate. The inflorescences are axillary spikes. Perianth absent. The androecium includes 2–6 stamens. The gynoecium includes 2–7 carpels. The fruits are drupaceous or baccate. Plants in this family produce amide alkaloids, isoquinolines, and styryl-lactones.

### 2.1.2.1 *Piper betle* L.

Common names: Betel, betel vine; lou ye (China); pan (Bangladesh); kun ya (Myanmar); plu (Thailand); trâu luong (Vietnam)

Habitat: Cultivated

Distribution: Native to the Indo-Malayan region cultivated in tropical Asia.

Botanical observation: It is a climber that grows to about 3 m long. The stems are terete, somewhat articulated, and rooting at nodes. The leaves are simple, spiral, and exstipulate. The petiole is about 3 cm long. The blade is cordate, glossy, fleshy, 6.5–15.5 × 4.5–11.5 cm, with 3 pairs of secondary nerves, the base cordate and apex acuminate. The inflorescences are white spikes, which are leaf-opposed and somewhat cylindrical. The androecium includes 2 stamens. The ovary is minute and develops 4 or 5 stigma. The drupes are fused into a red cylindrical mass.

Medicinal use: Boils (Malaysia)

*Moderate broad-spectrum antibacterial polar extract:* Ethanol extract of leaves inhibited the growth of *Acinetobacter baumannii*, *Bacillus cereus*, *Escherichia coli*, *Klebsiella pneumoniae*, *Listeria monocytogenes*, *Pseudomonas aeruginosa*, *Salmonella typhi*, *Salmonella typhimurium*, *Shigella flexneri*, *Staphylococcus aureus*, *Streptococcus mutans*, and *Streptococcus pyogenes* with MIC values ranging from 125 to 250 µg/mL (Limsuwan et al., 2009).

*Strong broad-spectrum antibacterial essential oil:* Essential oil of leaves 15 (µL/disc) inhibited the growth of *Salmonella typhi*, *Streptococcus enteriditis*, *Escherichia coli*, *Clostridium pefringens*, and *Campylobacter jejuni* with inhibition zone diameters of 16, 18.5, 15, 16, and 34 mm, respectively (Wannissorn et al., 2005).

*Antibiotic potentiators lignans:* The neolignans (-)-acuminatin, (-)-denudatin B, and puberulin D, displayed synergistic effects with norfloxacin towards *Staphylococcus aureus* (strain SA1199B; overexpresses the *nor*A gene encoding NorA MDR efflux pump). These compounds had no direct antibacterial effects (Sun et al., 2016).

*Antibiotic potentiators benzocyclohexenes:* Ferrudiol, ellipeiopsol B, and zeylenol displayed synergistic effect with norfloxacin towards *Staphylococcus aureus* (strain SA1199B; overexpresses the *nor*A gene encoding NorA MDR efflux pump. These compounds had no direct antibacterial effects (Sun et al., 2016). Ferrudiol, ellipeiopsol B, and zeylenol inhibited EtBr efflux in *Staphylococcus aureus* (strain SA1199B) (Sun et al., 2016).

Commentary: The multidrug transporter NorA is responsible for the efflux of fluoroquinolones out of the cytoplasm of *Staphylococcus aureus* (Blanco et al., 2016). NorA belongs to the major facilitator efflux pumps group. Inhibiting efflux pumps is a strategy to increase the sensitivity of bacteria to antibiotics.

## References

Blanco, P., Hernando-Amado, S., Reales-Calderon, J.A., Corona, F., Lira, F., Alcalde-Rico, M., Bernardini, A., Sanchez, M.B. and Martinez, J.L., 2016. Bacterial multidrug efflux pumps: much more than antibiotic resistance determinants. *Microorganism, 4*(1), p. 14.

Limsuwan, S., Subhadhirasakul, S. and Voravuthikunchai, S.P., 2009. Medicinal plants with significant activity against important pathogenic bacteria. *Pharmaceutical Biology, 47*(8), pp. 683–689.

Sun, Z.L., He, J.M., Wang, S.Y., Ma, R., Khondkar, P., Kaatz, G.W., Gibbons, S. and Mu, Q., 2016. Benzocyclohexane oxide derivatives and neolignans from Piper betle inhibit efflux-related resistance in *Staphylococcus aureus*. *RSC Advances, 6*(49), pp. 43518–43525.

Wannissorn, B., Jarikasem, S., Siriwangchai, T. and Thubthimthed, S., 2005. Antibacterial properties of essential oils from Thai medicinal plants. *Fitoterapia, 76*(2), pp. 233–236.

### 2.1.2.2 *Piper cubeba* L.f.

Synonyms: *Cubeba cubeba* (L. f.) H. Karst.; *Cubeba officinalis* Raf.; *Piper caudatum* Houtt. non Vahl.

Common names: Cubeb, tailed pepper; walga-miris (Sri Lanka); val-milaku, kankola (India); kemukus (Indonesia, Malaysia), tiêu thất (Vietnam)

Habitat: Around villages, cultivated

Distribution: Probably native to Indonesia, India, Bangladesh, and Malaysia

Botanical observation: It is a climber that grows to up to 10 m long. The stems are terete, striate, angular, fleshy, and articulated. The leaves are simple, spiral, and exstipulate. The petiole is about 2 cm long. The blade is ovate to oblong, to broadly elliptic, membranous, cordate, and asymmetrical at base, 8–15 cm × 2–9 cm, with 3–4 pairs of secondary nerves, and coriaceous, acuminate at apex. The inflorescences are erect tail-like spikes opposite to the leaves, about 4 cm × 3 mm, and yellowish. The androecium includes 2 stamens. The gynoecium includes an ovary and 3–5 stigmas. The drupes are globose, about 8 mm in diameter, yellowish red, glossy, on short and slender pedicels.

Medicinal use: Gonorrhea (Sri Lanka, Taiwan, Malaysia, Indonesia)

*Strong broad-spectrum essential oil:* Essential oil of fruits (containing mainly ß-cubebene, sabinene, and cubebol) (10 mm well; 6 µL/well) inhibited the growth of *Staphylococcus aureus* (MTCC 3103), *Bacillus subtilis* (MTCC 1790), *Escherichia coli* (MTCC 1672), and *Salmonella typhi* (MTCC 733) with inhibition zone diameters of 50.4, 72.3, 80, and 100 mm, respectively (Singh et al., 2008).

*Antiviral (enveloped monopartite linear single-stranded (+) RNA) polar extract:* Aqueous extract of fruits inhibited the replication of the Hepatitis C virus protease with an $IC_{50}$ value of 18 µg/mL (Hussein et al., 2000).

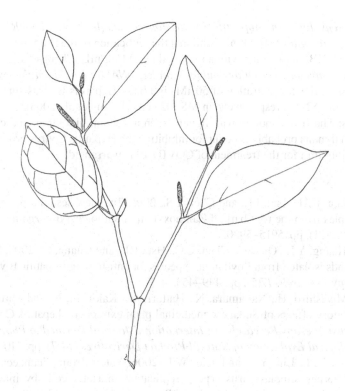

*Piper cubeba* L.f.

*Moderate antiviral (enveloped monopartite linear double-stranded DNA) dibenzylbutyrolactone lignan:* Yatein inhibited the replication of Herpes simplex virus type-1 with an $IC_{50}$ value of 30.6 μM (Kuo et al., 2006).

*Moderate antimycobacterial dibenzylbutyrolactone lignan:* Hinokinin inhibited the growth of *Mycobacterium tuberculosis*, with an MIC value equal to 62.5 μg/mL (Marcotullio et al., 2014).

*Moderate antibacterial (Gram-positive) dibenzylbutyrolactone lignan:* Hinokinin inhibited the growth of *Enterococcus faecalis*, *Streptococcus salivarius*, *Streptococcus sanguinis*, *Streptococcus mitis*, *Streptococcus mutans*, and *Streptococcus sobrinus* with MIC values of 380, 250, 250, 250, 320, and 280 μg/mL, respectively (Marcotullio et al., 2014).

*Moderate anticandidal dibenzylbutyrolactone* lignan: Hinokinin inhibited the growth of *Candida albicans* with an MIC value of 280 μg/mL (Marcotullio et al., 2014).

*Strong antiviral (enveloped linear monopartite single-stranded (+)RNA) amphiphilic dibenzylbutyrolactone lignan:* Hinokinin (LogD = 3.1 at pH 7.4; molecular mass = 354.3 g/mol) at the concentration of 20 μM inhibited plaque formation by the Severe acute respiratory syndrome-associated coronavirus in Vero cells by more than 50% (Wen et al., 2007).

Hinokinin

*Strong antiviral (enveloped monopartite linear dimeric single-stranded (+)RNA) amphiphilic dibenzylbutyrolactone lignan:* Hinokinin inhibited the replication of the Human immunodeficiency virus type-1 (strain IIIB), replication with an $IC_{50}$ value < 0.1 μg/mL (Cheng et al., 2005).

*Strong antiviral (enveloped circular double-stranded DNA) amphiphilic dibenzylbutyrolactone lignan:* Hinokinin (at the concentration of 50 μM) inhibited Hepatitis B virus surface antigen and e-antigen by 68.1 and 52.3%, respectively, in MS-G2 cells (Huang et al., 2003).

Commentaries: The fruits abound with lignans such as yatein, hinokinin, and cubebin (Ruslan et al., 2007). Such lignans probably account for inhibiting the Hepatitis C virus protease. Preclinical examination of hinokinin for the treatment of COVID-19 is warranted.

### References

Cheng, M.J., Lee, K.H., Tsai, I.L. and Chen, I.S., 2005. Two new sesquiterpenoids and anti-HIV principles from the root bark of Zanthoxylum ailanthoides. *Bioorganic & Medicinal Chemistry, 13*(21), pp. 5915–5920.

Huang, R.L., Huang, Y.L., Ou, J.C., Chen, C.C., Hsu, F.L. and Chang, C., 2003. Screening of 25 compounds isolated from Phyllanthus species for anti-human Hepatitis B virus in vitro. *Phytotherapy Research, 17*(5), pp. 449–453.

Hussein, G., Miyashiro, H., Nakamura, N., Hattori, M., Kakiuchi, N. and Shimotohno, K., 2000. Inhibitory effects of Sudanese medicinal plant extracts on hepatitis C virus (HCV) protease. *Phytotherapy Research: An International Journal Devoted to Pharmacological and Toxicological Evaluation of Natural Product Derivatives, 14*(7), pp. 510–516.

Kuo, Y.C., Kuo, Y.H., Lin, Y.L. and Tsai, W.J., 2006. Yatein from Chamaecyparis obtusa suppresses herpes simplex virus type 1 replication in HeLa cells by interruption the immediate-early gene expression. *Antiviral Research, 70*(3), pp. 112–120.

Marcotullio, M.C., Pelosi, A. and Curini, M., 2014. Hinokinin, an emerging bioactive lignan. *Molecules, 19*(9), pp. 14862–14878.

Ruslan, K., Batterman, S., Bos, R., Kayser, O., Woerdenbag, H.J. and Quax, W.J., 2007. Lignan profile of Piper cubeba, an Indonesian medicinal plant. *Biochemical Systematics and Ecology Habitat, 35*(7), pp. 397–402.

Singh, G., Kiran, S., Marimuthu, P., de Lampasona, M.P., De Heluani, C.S. and Catalán, C.A.N., 2008. Chemistry, biocidal and antioxidant activities of essential oil and oleoresins from Piper cubeba (seed). *International Journal of Essential Oil Therapeutics, 2*(2), pp. 50–59.

Wen, C.C., Kuo, Y.H., Jan, J.T., Liang, P.H., Wang, S.Y., Liu, H.G., Lee, C.K., Chang, S.T., Kuo, C.J., Lee, S.S. and Hou, C.C., 2007. Specific plant terpenoids and lignoids possess potent antiviral activities against severe acute respiratory syndrome coronavirus. *Journal of Medicinal Chemistry, 50*(17), pp. 4087–4095.

### 2.1.2.3 *Piper longum* L.

Synonyms: *Chavica roxburghii* Miq.

Common names: Indian long pepper, jaborandi pepper, long pepper; pipul (Bangladesh); dipli, prik-hang, sa-kan (Thailand); morech ansai (Cambodia); tat bat (Vietnam)

Habitat: Cultivated

Distribution: India, Sri Lanka, Nepal, Vietnam, Malaysia, and China

Botanical observation: It is a climber which grows to up to 5 m long. The stems are terete, smooth, and articulated. The leaves are simple, spiral, and exstipulate. The petiole is about 3 cm long. The blade is ovate to ovate-oblong, membranous, 3–5 cm × 7–10.5 cm, acuminate at apex cordate or oblique at base, and with 3–4 pairs of secondary nerves. The inflorescences are erect spikes opposite to the leaves, up to about 6.5 cm long. The androecium includes 2 stamens. The gynoecium includes an ovary, which is ovoid with 3 stigmas. The drupes are globose, about 2 mm in diameter, yellowish red, glossy, and sessile.

Medicinal uses: Cholera, tuberculosis, flu (Bangladesh); fever (China)

*Strong broad-spectrum antibacterial amphiphilic amide alkaloids:* Piperlonguminine (LogD = 2.7 at pH 7.4; molecular weight = 273.3 g/mol) isolated from the fruits inhibited the growth of *Bacillus sphaericus* (ATCC 14577), *Bacillus subtilis* (ATCC 6051), *Staphylococcus aureus* (ATCC 9144), *Escherichia coli* (ATCC 25922), *Pseudomonas syringae* (ATCC 13457), and *Salmonella typhimurium* (ATCC 23564) with MIC values of 20, 9, 12.5, 150, 75, and 175 µg/mL, respectively (Srinivasa Reddy et al., 2001). Piperine (LogD = 2.5 at pH 7.4; molecular mass = 285.3 g/mol) isolated from the fruits inhibited the growth of *Bacillus sphaericus* (ATCC 14577), *Bacillus subtilis* (ATCC 6051), *Staphylococcus aureus* (ATCC 9144), *Escherichia coli* (ATCC 25922), *Pseudomonas syringae* (ATCC 13457), and *Salmonella typhimurium* (ATCC 23564) with MIC values of 25, 12, 12.5, 160, 50, and 75 µg/mL, respectively (Srinivasa Reddy et al., 2001). Pellitorine (LogD = 3.8 at pH 7.4; molecular mass = 223.3 g/mol) isolated from the fruits inhibited the growth of *Bacillus sphaericus* (ATCC 14577), *Bacillus subtilis* (ATCC 6051), *Staphylococcus aureus* (ATCC 9144), *Escherichia coli* (ATCC 25922), *Pseudomonas syringae* (ATCC 13457), and *Salmonella typhimurium* (ATCC 23564) with MIC values of 25, 12.5, 20, 150, 75, and 200 µg/mL, respectively (Srinivasa Reddy et al., 2001).

Piperlonguminine

Piperine

Pellitorine

*Weak antiviral (enveloped linear double-stranded DNA) amide alkaloids:* 3β, 4α-dihydroxy-1-(3-phenylpropanoyl)-piperidine-2-one and (2E,4E,14Z)-6-hydroxyl-N-isobutyleicosa-2,4,14-trienamide from the fruits inhibited HBsAg with $IC_{50}$ values of 1.8 and 0.2 mM, respectively (Yang et al., 2013). 3β, 4α-dihydroxy-1-(3-phenylpropanoyl)-piperidine-2-one and (2E, 4E, 14Z)-6-hydroxyl-

N-isobutyleicosa-2,4,14-trienamide from the fruits inhibited the Hepatitis B e-antigen with $IC_{50}$ values of 2.2 and 2.0 mM, respectively (Yang et al., 2013). Erythro-1-[1-oxo-9(3,4-methylenedioxyphenyl)-8,9-dihydroxy-2E-nonenyl]-piperidine, threo-1-[1-oxo-9(3,4-methylenedioxyphenyl)-8,9-dihydroxy-2 E-nonenyl]-piperidine, piperine, and guineesine from the fruits inhibited the secretion of Hepatitis B virus surface-antigen and Hepatitis B virus e-antigen in Hep G 2.2.15 cell line, with $IC_{50}$ values of 0.13, 0.11, 0.15 and 0.05 mM for the Hepatitis B virus surface-antigen, and $IC_{50}$ values of 0.16, 0.11, 0.14 and 0.05 mM for Hepatitis B virus e-antigen, respectively (Jiang et al., 2013).

Commentaries: (i) The plant is used in India for the treatment of Chikungunya virus infection (Viswanathan et al., 2008). The Chikungunya virus is an enveloped RNA virus belonging to the genus Alphavirus of family *Togaviridae* transmitted by mosquitoes. In humans, symptoms of the Chikungunya virus infection include an abrupt onset of fever, skin rash, disabling joint pain, rashes, and myalgia. The Chikungunya virus was responsible for a first epidemic in 1952 in Tanzania, followed, because of worldwide proliferation of the mosquitoes (because of global warming and increasing surfaces of plastic garbage), by epidemics in Africa, the Indian subcontinent, Southeast Asia, South and Central America, and Reunion Island. For example, in 2005, a Chikungunya virus outbreak on Reunion Island caused more than 60% of Chikungunya virus patients to experience arthralgia during the 3 years that followed acute infection (Schilte et al., 2013). There is a need for the development of selective and potent inhibitors of Chikungunya virus (Corlay et al., 2014). Note that the antifilarial drug suramin, which presents a number of amide moieties, evoked some levels of protection in C57BL/6 mice against Chikungunya virus (Kuo et al., 2016). Consider that flavonoids, diarylheptanoids, and isoquinolines with medium molecular weight and amphiphilic tend to strongly inhibit the Chikungunya virus *in vitro*.

(ii) The plant is used for the treatment of COVID-19 in China (Wang et al., 2020) and Siddha healers in South India (Sasikumar et al., 2020). The use of this plant for COVID-19 in different traditional systems of medicine is a red flag for the presence of principles that protect the body against Severe acute respiratory syndrome-associated coronavirus 2, and this plant should undergo clinical trials.

(iii) The COVID-19 tragedy may impose on letting go of the current industrial/corporate therapeutic strategies of using solely Western drugs for the treatment of microbial infections and to open our minds to accept the possibility of using concomitantly Western drugs and traditional herbal medicine. As such, it should be ensured that pharmacy students get at least 4 years of serious training in medicinal plants' therapeutic properties and uses (pharmacognosy). Pressures from the big pharmas (looking for more financial profits) via accreditation boards are currently managing to eradicate the teaching of pharmacognosy from schools of pharmacy, labelling this subject as "obscure"(!). As a consequence, current pharmacy graduates from universities in Asia accredited by the West do not know anything about the clinical usefulness and toxicity of medicinal plants. These students are just trained to sell the pills of the big pharmas, are not trained to observe and think by themselves, and will be replaced soon by robots. COVID-19 is a warning sign from Mother Nature that industries and corporations are abusing life forms to dangerous levels, and if deforestation and destruction on nature continues, other viruses, bacteria, or fungi (more dangerous than COVID-19) will be produced (negative feedback loop mechanism) and most humans may disappear, (except self-suffcent uncontacted tribes).

(iv) Piperine inhibited NorA efflux pump in *Staphylococcus aureus* and efflux pumps in *Mycobacterium* species (Sharma et al., 2019).

### References

Corlay, N., Delang, L., Girard-Valenciennes, E., Neyts, J., Clerc, P., Smadja, J., Guéritte, F., Leyssen, P. and Litaudon, M., 2014. Tigliane diterpenes from Croton mauritianus as inhibitors of chikungunya virus replication. *Fitoterapia*, 97, pp. 87–91.

Jiang, Z.Y., Liu, W.F., Zhang, X.M., Luo, J., Ma, Y.B. and Chen, J.J., 2013. Anti-hepatitis B VIRUS active constituents from Piper longum. *Bioorganic & Medicinal Chemistry Letters*, 23(7), pp. 2123–2127.

Sasikumar, R., Priya, S.D. and Jeganathan, C., 2020. A case study on domestics spread of SARS-CoV-2 pandemic in India. *International Journal of Advanced Science and Technology, 29*(7), pp. 2570–2574.

Schilte, C., Staikovsky, F., Couderc, T., Madec, Y., Carpentier, F., Kassab, S., Albert, M.L., Lecuit, M. and Michault, A., 2013. Chikungunya virus-associated long-term arthralgia: a 36-month prospective longitudinal study. *PLoS Neglected Tropical Diseases, 7*(3).

Sharma, A., Gupta, V.K. and Pathania, R., 2019. Efflux pump inhibitors for bacterial pathogens: from bench to bedside. *The Indian Journal of Medical Research, 149*(2), p. 129.

Srinivasa Reddy, P., Jamil, K., Madhusudhan, P., Anjani, G. and Das, B., 2001. Antibacterial activity of isolates from *Piper longum* and Taxus baccata. *Pharmaceutical Biology, 39*(3), pp. 236–238.

Wang, S.X., Wang, Y., Lu, Y.B., Li, J.Y., Song, Y.J., Nyamgerelt, M. and Wang, X.X., 2020. Diagnosis and treatment of novel coronavirus pneumonia based on the theory of traditional Chinese medicine. *Journal of Integrative Medicine*.

Yang, J., Su, Y., Luo, J.F., Gu, W., Niu, H.M., Li, Y., Wang, Y.H. and Long, C.L., 2013. New amide alkaloids from *Piper longum* fruits. *Natural Products and Bioprospecting, 3*(6), pp. 277–281.

### 2.1.2.4 *Piper nigrum* L.

Common names: Black pepper, white pepper; hu jiao (China); lada hitam, lada puteh (Malaysia); kurumilagu (India); kalomarich (Bangladesh); burakku (Japan); phrik thai (Thailand); trieu (Vietnam)

Habitat: Cultivated

Distribution: Malaysia and Indonesia

Botanical observation: It is a climber with terete, smooth, and articulated stems rooting at nodes. The leaves are simple, spiral, and exstipulate. The petiole is about 1.5 cm long. The blade is ovate to elliptic, coriaceous, 4–6 cm × 9–11 cm, acuminate at apex, rounded to oblique at base, and marked with 3 pairs of secondary nerves. The spikes are opposite to the leaves and up to about 13 cm long. The androecium includes 2 stamens. The ovary is ovoid with 3 stigma. The drupes are globose, about 8 mm in diameter, yellowish, red, glossy, and sessile.

Medicinal use: Cholera, colds, and diarrhea (China)

*Broad-spectrum moderate antibacterial polar extract:* Methanol extract of fruits inhibited the growth of *Providencia stuartii* (ATCC 1296), *Klebsiella pneumoniae* (KP55), *Klebsiella pneumoniae* (KP63), *Klebsiella pneumoniae* (K2), and *Klebsiella pneumoniae* (K24) with MIC values ranging from 128 to 1024 µg/mL (Noumedem et al., 2013). Methanol extract of fruits inhibited the growth of *Escherichia coli* (ATCC 8739), *Escherichia coli* (ATCC 10536), *Escherichia coli* (W3110), *Escherichia coli* (MC4100), *Escherichia coli* (AG100A), *Escherichia coli* (AG100Atet), *Escherichia coli* (AG102), and *Escherichia coli* (AG100) with MIC values of 128, 256, 256, 256, 128, 256, 512, and 256 µg/mL, respectively (Noumedem et al., 2013).

*Moderate antibacterial (Gram-positive) amide alkaloids:* 8Z-*N*-isobutyleicosatrienamide, pellitorine, trachyone, pergumidiene, and isopiperolein B inhibited *Staphylococcus aureus* with MIC values of 34, 56, 30, 29, and 36 µM, respectively (Reddy et al., 2004). 8Z-*N*-isobutyleicosatrienamide, pellitorine, trachyone, pergumidiene, and isopiperolein B inhibited *Bacillus subtilis* with the MIC values of 34, 28, 30, 58 and 36 µM, respectively (Reddy et al., 2004).

*Antibiotic potentiator amide alkaloid:* Piperine decreased the MIC of ethidium bromide twofold at 32 µg/mL and fourfold at 64 µg/mL against *Myocbacterium smegmatis* (ATCC 700084). Piperine significantly enhanced accumulation and decreased the efflux of EtBr in *Mycobacterium smegmatis* (ATCC 700084), which suggests that it has the ability to inhibit mycobacterial efflux pump (Jin et al., 2011).

Commentary: (i) Note that for millions of years bacteria and fungi have coexisted, especially at the rhizosphere level, either as commensals or pathogens with flowering plants and have developed efflux pumps to survive antimicrobial defense plant products. It is also possible to infer that plants have developed inhibitors of such an efflux pump to control resistant phytopathogenic bacteria and fungi (Blanco et al., 2018). Therefore, one could see flowering plants as a fascinating source of efflux pump inhibitor (activity, assemblage, or and expression). Further, it is said that antibiotics in bacteria are produced to inhibit the growth of other bacteria but the concentration of these antibiotics in soils seems to be lower than MIC suggesting that antibiotics may have additional, yet unknown, functions between bacterial species (JFF Weber, personal communication) (ii) The plant is used in India for the treatment of Chikungunya virus infection (Viswanathan et al., 2008).

### References
Blanco, P., Sanz-García, F., Hernando-Amado, S., Martínez, J.L. and Alcalde-Rico, M., 2018. The development of efflux pump inhibitors to treat Gram-negative infections. *Expert Opinion on Drug Discovery, 13*(10), pp. 919–931.

Jin, J., Zhang, J., Guo, N., Feng, H., Li, L., Liang, J., Sun, K., Wu, X., Wang, X., Liu, M. and Deng, X., 2011. The plant alkaloid piperine as a potential inhibitor of ethidium bromide efflux in Mycobacterium smegmatis. *Journal of Medical Microbiology, 60*(2), pp. 223–229.

Noumedem, J.A., Mihasan, M., Kuiate, J.R., Stefan, M., Cojocaru, D., Dzoyem, J.P. and Kuete, V., 2013. In vitro antibacterial and antibiotic-potentiation activities of four edible plants against multidrug-resistant gram-negative species. *BMC Complementary and Alternative Medicine, 13*(1), p. 190.

Reddy, S.V., Srinivas, P.V., Praveen, B., Kishore, K.H., Raju, B.C., Murthy, U.S. and Rao, J.M., 2004. Antibacterial constituents from the berries of *Piper nigrum. Phytomedicine, 11*(7), pp. 697–700.

Viswanathan, M.V., Raja, D.K. and Khanna, S.D., 2008. Siddha way to cure Chikungunya.

Volleková, A., Košt'álová, D., Kettmann, V. and Tóth, J., 2003. Antifungal activity of Mahonia aquifolium extract and its major protoberberine alkaloids. *Phytotherapy Research, 17*(7), pp. 834–837.

### 2.1.2.5 *Piper sarmentosum* Roxb.
Synonyms: *Chavica hainana* C. DC.; *Chavica sarmentosa* (Roxb.) Miq.; *Piper albispicum* C. DC.; *Piper baronii* C. DC.; *Piper brevicaule* C. DC.; *Piper lolot* C. DC.; *Piper pierrei* C. DC.; *Piper saigonense* C. DC.

Common names: Jia ju (China); kadok, sirih tanah (Malaysia); patai-butu (the Philippines); môrech ansai (Cambodia); cha plu (Thailand); tiêu lôt (Vietnam)

Habitat: Forests, villages

Distribution: India, Cambodia, Laos, Vietnam, Malaysia, Indonesia, and the Philippines

Botanical observation: The plant grows to about 1 m long and is rooting at the nodes. The stems are terete, hairy when young, and articulated. The leaves are simple, spiral, and exstipulate. The petiole is about 2.5 cm long. The blade is ovate to lanceolate, asymmetrical, fleshy, glossy, 6–14 × 6–13 cm, cordate at base, acute at apex, and with 3–7 pairs of secondary nerves. The spikes are opposite to the leaves, up to about 1 cm long, and whitish. The androecium includes 2 stamens. The ovary is ovoid with 4 stigma. The drupes are subglobose, 4-angled, about 3 mm in diameter, and sessile on an erect infructescence, which is cylindrical and about 3 cm long.

Medicinal uses: Fever, cough (Malaysia)

*Moderate antimycobacterial polar extract:* Methanol extract inhibited the growth of *Mycobacterium tuberculosis* (H37Rv ATCC 25618) with and an MIC value of 800 µg/mL (Mohammad et al., 2011).

*Strong antimycobacterial amphiphilic pyrrolidine amide alkaloids:* N-[9-(3,4-Methylenedioxyphenyl)-2E, 4E, 8E-nonatrienoyl]pyrrolidine inhibited the growth of *Mycobacterium tuberculosis* H37Ra with the MIC value of 25 µg/mL (Tuntiwachwuttikul et al., 2006). Pellitorine (LogD = 3.8 at pH 7.4; molecular mass = 223.3 g/mol) isolated from the fruits inhibited the growth of *Mycobacterium tuberculosis* H37Ra with the MIC value of 25 µg/mL (Rukachaisirikul et al., 2004).

N-[9-(3,4-Methylenedioxyphenyl)-2E,4E,8E-nonatrienoyl]pyrrolidine

*Strong antimycobacterial lipophilic pyrrolidine alkaloid:* 1-(3,4-methylenedioxyphenyl)-1E-tetradecene isolated from the fruits inhibited the growth of *Mycobacterium tuberculosis* H37Ra with an MIC value of 25 μg/mL (Rukachaisirikul et al., 2004).

1-(3,4-methylenedioxyphenyl)-1E-tetradecene

*Strong antibacterial (Gram-positive) alkylphenol:* From the aerial parts, sarmentosumol A inhibited the growth of *Staphylococcus aureus* with the MIC value of 7 μg/mL (Yang et al., 2013).

*Strong anticandidal pyrrolidine amide alkaloids:* Brachyamide B and sarmentosine possessed antifungal activity with $IC_{50}$ values of 41.8 and 32.8 μg/mL, respectively, against a clinical isolate of *Candida albicans* (Tuntiwachwuttikul et al., 2006).

Sarmentosine

*Weak antibacterial phenylpropanoid:* 1-allyl-2,6-dimethoxy-3,4-methylenedioxybnzene isolated from the leaves inhibited the growth of *Escherichia coli* and *Bacillus subtilis* with the MIC of 100 ppm (Masuda et al., 1991).

*Strong antifungal (filamentous) amphiphilic pyrrolidine amide alkaloid:* Brachyamide B (LogD = 3.5 at pH 7.4; molecular mass = 327.4 g/mol) isolated from the aerial parts inhibited the growth of *Cryptococcus neoformans* (ATCC 90 113)$_4$, with the $IC_{50}$ of 7.1 μg/mL (Shi et al., 2017).

Brachyamide B

*Strong antiviral (enveloped monopartite linear (-) RNA) extract:* Ethanol extract inhibited the proliferation of Vesicular stomatitis virus with an $IC_{50}$ value of 20 µg/mL (Hamidi et al., 1996).

Commentary: Natural products able to strongly inhibit the replication of the Vesicular stomatitis virus from the plant belonging to the Clades listed in this volume and are amphiphilic, with low to medium molecular weight phenolics.

### References

Hamidi, J.A., Ismaili, N.H., Ahmadi, F.B. and Lajisi, N.H., 1996. Antiviral and cytotoxic activities of some plants used in Malaysian indigenous medicine. *Pertanika Journal of Tropical Agricultural Sciences, 19*(2/3), pp. 129–136.

Masuda, T., Inazumi, A., Yamada, Y., Padolina, W.G., Kikuzaki, H. and Nakatani, N., 1991. Antimicrobial phenylpropanoids from *Piper sarmentosum. Phytochemistry, 30*(10), pp. 3227–3228.

Mohamad, S., Zin, N.M., Wahab, H.A., Ibrahim, P., Sulaiman, S.F., Zahariluddin, A.S.M. and Noor, S.S.M., 2011. Antituberculosis potential of some ethnobotanically selected Malaysian plants. *Journal of Ethnopharmacology, 133*(3), pp. 1021–1026.

Rukachaisirikul, T., Siriwattanakit, P., Sukcharoenphol, K., Wongvein, C., Ruttanaweang, P., Wongwattanavuch, P. and Suksamrarn, A., 2004. Chemical constituents and bioactivity of *Piper sarmentosum. Journal of Ethnopharmacology, 93*(2–3), pp. 173–176.

Shi, Y.N., Liu, F.F., Jacob, M.R., Li, X.C., Zhu, H.T., Wang, D., Cheng, R.R., Yang, C.R., Xu, M. and Zhang, Y.J., 2017. Antifungal amide alkaloids from the aerial parts of *Piper flaviflorum* and *Piper sarmentosum. Planta Medica, 83*(01/02), pp. 143–150.

Tuntiwachwuttikul, P., Phansa, P., Pootaeng-on, Y. and Taylor, W.C., 2006. Chemical constituents of the roots of *Piper sarmentosum. Chemical and Pharmaceutical Bulletin, 54*(2), pp. 149–151.

Yang, S.X., Sun, Q.Y., Yang, F.M., Hu, G.W., Luo, J.F., Wang, Y.H. and Long, C.L., 2013. Sarmentosumols A to F, new mono-and dimeric alkenylphenols from *Piper sarmentosum. Planta Medica, 79*(08), pp. 693–696.

### 2.1.3 FAMILY PEPEROMIACEAE SMITH (1981)

The family Peperomiaceae comprises 4 genera and 1000 species of fleshy tropical herbs resembling Piperaceae. The leaves are simple, alternate, opposite, or verticillate, and exstipulate. The inflorescence is spikes. The androecium includes 2 stamens. The gynoecium includes 1–3 carpels.

#### 2.1.3.1 *Peperomia pellucida* (L.) Kunth

Synonyms: *Micropiper pellucidum* (L.) Miq.; *Peperomia concinna* (Haw.) A. Dietr.; *Peperomia ephemera* Ekman; *Peperomia knoblecheriana* Schott; *Peperomia praetenuis* Trel.; *Peperomia translucens* Trel.; *Piper concinnum* Haw.; *Piper pellucidum* L.; *Verhuellia knoblecheriana* (Schott) C. DC.

Common names: Cao hu jiao (China); ketumpangan air (Indonesia, Malaysia); ulasiman bato (the Philippines); diya Thippili (Sri Lanka); phak krasang (Thailand); rau càng cua (Vietnam)

Habitat: Moist, shaddy, and mossy drains

Distribution: Tropical Asia and Pacific

Botanical observation: It is a gracile herb which grows to about 20 cm tall. The stems are slender, fleshy, terete, and glabrous. The leaves are simple, spiral, and exstipulate. The petiole is 1–2 cm long. The blade is deltoid, 1–3.5 × 1–3.5 cm, membranous, translucent, cordate at base, acute or obtuse at apex, and with 2–3 pairs of secondary nerves. The inflorescences are terminal or leaf-opposed spikes, which are slender and up to about 6 cm long. The androecium includes 2 stamens. The gynoecium includes an ellipsoid ovary. The drupes are minute and black.

Medicinal uses: Boils, and abscesses (the Philippines)

*Weak broad-spectrum antibacterial polar extract:* Butanol extract (4 mg/disc) inhibited the growth of *Escherichia coli, Pseudomonas aeruginosa, Salmonella typhi, Klebsiella pneumonia,* and *Staphylococcus aureus* with inhibition zone diameters of 20, 20, 18, 18, and 18 mm, respectively (Khan & Omoloso, 2002).

*Strong antiviral (enveloped monopartite linear dimeric single-stranded (+)RNA) amphiphilic phenolic:* Pellucidin A (LogD = 4.1 at pH 7.4; molecular weight = 388.4 g/mol) inhibited the replication of the Human immunodeficiency virus type-1 with the $EC_{50}$ value of 6.6 µM and a selectivity index >4.3 (Thongphichai et al., 2019). At the concentration of 200 µg/mL, this compound inhibited reverse transcriptase by 15.2% (Thongphichai et al., 2019).

Pellucidin A

Commentaries: (i) The plant produces apiole (Bayma et al., 2000) and series of lignans such as sesamin (Xu et al., 2006). Sesamin inhibited Influenza type A H1N1-induced cytokine production in peripheral blood mononuclear cells (Fanhchaksai et al., 2016). (ii) One could have the curiosity to examine the anti-Severe acute respiratory syndrome-associated coronavirus 2 activities of pellucidin A, which inhibit angiotensin-converting enzyme *in vitro* (Ahmad, 2019). The Severe acute respiratory syndrome-associated coronavirus surface spike (S) protein, which is a type I membrane-bound protein projecting from virus envelope, is responsible for attaching the virus to the host cell of the receptor angiotensin-converting enzyme 2 (ACE2) (Li et al., 2003). Thus, angiotensin-converting enzyme inhibitors have the potential to inhibit the entry of the virus in the host cell.

### References

Ahmad, I., 2019. A new angiotensin-converting enzyme inhibitor from *Peperomia pellucida* (L.) Kunth, 2019. *Asian Pacific Journal of Tropical Biomedicine, 9*(6), pp. 257–262.

Bayma, J.D.C., Arruda, M.S.P., Müller, A.H., Arruda, A.C. and Canto, W.C., 2000. A dimeric ArC2 compound from *Peperomia pellucida. Phytochemistry, 55*(7), pp. 779–782.

Fanhchaksai, K., Kodchakorn, K., Pothacharoen, P. and Kongtawelert, P., 2016. Effect of sesamin against cytokine production from influenza type A H1N1-induced peripheral blood mononuclear cells: computational and experimental studies. *In Vitro Cellular & Developmental Biology-Animal, 52*(1), pp. 107–119.

Khan, M.R. and Omoloso, A.D., 2002. Antibacterial activity of *Hygrophila stricta* and *Peperomia pellucida. Fitoterapia, 73*(3), pp. 251–254.

Li, W., Moore, M.J., Vasilieva, N., Sui, J., Wong, S.K., Berne, M.A., Somasundaran, M., Sullivan, J.L., Luzuriaga, K., Greenough, T.C., Choe, H. and Farzan, M, 2003. Angiotensin-converting enzyme 2 is a functional receptor for the SARS coronavirus. *Nature, 426,* pp. 450–454.

Thongphichai, W., Tuchinda, P., Pohmakotr, M., Reutrakul, V., Akkarawongsapat, R., Napaswad, C., Limthongkul, J., Jenjittikul, T. and Saithong, S., 2019. Anti-HIV activities of constituents from the rhizomes of *Boesenbergia thorelii. Fitoterapia, 139,* p. 104388.

Xu, S., Li, N., Ning, M.M., Zhou, C.H., Yang, Q.R. and Wang, M.W., 2006. Bioactive compounds from *Peperomia pellucida. Journal of Natural Products, 69*(2), pp. 247–250.

### 2.1.3.2 *Peperomia blanda* (Jacq.) Kunth

Synonyms: *Micropiper langsdorffii* Miq.; *Peperomia arabica* Decne.; *Peperomia bequaertii* De Wild.; *Peperomia ciliata* Kunth; *Peperomia decipiens* C. DC.; *Peperomia dindygulensis* Miq.; *Peperomia dissimilis* Kunth; *Peperomia ellipticifolia* C. DC.; *Peperomia esquirolii* H. Lév.; *Peperomia fauriei* H. Lév.; *Peperomia formosana* C. DC.; *Peperomia glanduligera* Yunck.; *Peperomia japonica* Makino; *Peperomia laticaulis* C. DC.; *Peperomia leptostachya* Hook. & Arn.; *Peperomia macaroana* Trel. ex V.M. Badillo; *Peperomia murispica* Trel. ex V.M. Badillo; *Peperomia pseudodindygulensis* C. DC.; *Peperomia quitensis* Miq.; *Peperomia salvaje* C. DC.; *Peperomia sui* T.T. Lin & S.Y. Lu; *Piper blandum* Jacq.; *Piper blandum* Jacq.; *Piper ciliatum* (Kunth) Poir.; *Piper dissimile* (Kunth) Poir.; *Troxirum blandum* (Jacq.) Raf.

Common names: Arid-land peperomia; shi chan cao (China)

Habitat: Forests

Distribution: India, Sri Lanka Bangladesh, Myanmar, Cambodia, Malaysia, Thailand, Vietnam, China, and Japan

Botanical description: It is a herb that grows to about 45 cm tall. The stems are fleshy, terete, and reddish. The leaves are simple, spiral, and exstipulate. The petiole is 1–1.5 cm long. The blade is elliptic, 2–4 × 1–2 cm, membranous, tapered at base, rounded at apex, and with 2–3 pairs of secondary nerves. The spikes are terminal or axillary, slender and up to about 12 cm long. The androecium includes 2 stamens. The gynoecium includes an ovary that is obovoid. The nutlets are globose to ellipsoid and minute.

Medicinal uses: Apparently none for infectious diseases.

*Strong antiviral (enveloped monopartite linear segmented single-stranded (−) RNA virus) polar extract:* Ethanol extract (containing scutellarein, luteolin, and apigenin) increased the viability of DF-1 cells infected with the Avian Influenza virus A/Chicken/TW/0518/2011 (H6N1). This extract dose-dependently increased the viability of virus-infected cells, and the $EC_{50}$ value was 34.8 μg/mL (Yang et al., 2014). This extract inhibited the expression of the viral NP protein in DF-1 cells (Yang et al., 2014). It effectively inhibited the growth of H6N1 virus in DF-1 cells. The extract at a concentration of 120 μg/mL reduced the H6N1 neuraminidase activity to 24.3% (Yang et al., 2014).

*Strong antiviral (enveloped monopartite linear single-stranded (+)RNA) lignans:* rel-(7R, 8S, 70S, 80S)-40, 50-methylenedioxy-3,4,5,30-tetramethoxy-7,70-epoxylignan and rel-(7R, 8S, 70 S, 80 S)-4,5,40, 50-dimethylenedioxy-3,30-dimethoxy7,70-epoxylignan, and the dibenzylbutyrolactone lignin (2R, 3S)-2-methyl3-[bis(3′, 4′-methylenedioxy-5′-methoxyphenyl)methyl]butyrolactone inhibited the replication of the Hepatitis C virus with the $EC_{50}$ values of 4, 8.2, and 38.9 μM, respectively (Jardim et al., 2015).

*Moderate antiviral (enveloped monopartite linear double-stranded DNA) flavonol:* Apigenin and luteolin inhibited the replication of the Herpes simplex virus-1 (F strain) with the $IC_{50}$ values of 34.5 and 40.7 μM, respectively (Wu et al., 2011).

*Viral enzymes inhibition by flavonols:* Scutellarein inhibited the Severa acute respiratory syndrome-associated coronavirus ns13 ATPase with an $IC_{50}$ value of 0.8 μM (Yu et al., 2012). Apigenin at the concentration of 300 μM inhibited Aminopeptidase N by 42% and angiotensin-converting enzyme by 18% (Bauvois & Dauzonne, 2006).

Commentary: Aminopeptidase N is a transmembrane protease present in a wide variety of human tissues and cell types (endothelial, epithelial, fibroblast, leukocyte) via which coronavirus enter alveolar cells and establish an upper respiratory tract infection (Delmas et al., 1992, 1993; Yeager et al., 1992).

### References
Bauvois, B. and Dauzonne, D., 2006. Aminopeptidase-N/CD13 (EC 3.4. 11.2) inhibitors: chemistry, biological evaluations, and therapeutic prospects. *Medicinal Research Reviews, 26*(1), pp. 88–130.

Delmas, B., Gelfi, J., L'Haridon, R., Vogel, L.K., Sjostrom, H., Noren, O. and Laude, H., 1992. Aminopeptidase N is a major receptor for the entero-pathogenic coronavirus TGEV. *Nature, 357*(6377), pp. 417–420.

Delmas, B., Gelfi, J., Sjostrom, H., Noren, O. and Laude, H., 1993. Further characterization of aminopeptidase-N as a receptor for coronaviruses. *Advances in Experimental Medicine and Biology, 342*, pp. 293–298.

Jardim, A.C.G., Igloi, Z., Shimizu, J.F., Santos, V.A.F.F.M., Felippe, L.G., Mazzeu, B.F., Amako, Y., Furlan, M., Harris, M. and Rahal, P., 2015. Natural compounds isolated from Brazilian plants are potent inhibitors of hepatitis C virus replication in vitro. *Antiviral Research, 115*, pp. 39–47.

Wu, N., Kong, Y., Zu, Y., Fu, Y., Liu, Z., Meng, R., Liu, X. and Efferth, T., 2011. Activity investigation of pinostrobin towards herpes simplex virus-1 as determined by atomic force microscopy. *Phytomedicine, 18*(2–3), pp. 110–118.

Yang, C.H., Tan, D.H., Hsu, W.L., Jong, T.T., Wen, C.L., Hsu, S.L. and Chang, P.C., 2014. Anti-influenza virus activity of the ethanolic extract from *Peperomia sui. Journal of Ethnopharmacology, 155*(1), pp. 320–325.

Yeager, C.L., Ashmun, R.A., Williams, R.K., Cardellichio, C.B., Shapiro, L.H., Look, A.T. and Holmes, K.V., 1992. Human aminopeptidase N is a receptor for human coronavirus 229E. *Nature, 357*(6377), pp. 420–422.

Yu, M.S., Lee, J., Lee, J.M., Kim, Y., Chin, Y.W., Jee, J.G., Keum, Y.S. and Jeong, Y.J., 2012. Identification of myricetin and scutellarein as novel chemical inhibitors of the SARS coronavirus helicase, nsP13. *Bioorganic & Medicinal Chemistry Letters, 22*(12), pp. 4049–4054.

### 2.1.4 FAMILY SAURURACEAE MARTYNOV (1820)

The family Saururaceae consists of 4 genera and 7 species of rhizomatous, aromatic, and often stoloniferous herbs, which have somewhat monocotyledoneous looks. The leaves are simple, alternate, and stipulate. The stipules form a sheath. The inflorescences are spikes or racemes. The androecium includes 3, 6, or 8 stamens, which longitudinally dehiscent anthers. The gynoecium includes 2–4 carpels. The fruits are schizocarps or apically dehiscent capsules containing 1 to many seeds. Members of this family elaborate antimicrobial lignans and flavonoids.

#### 2.1.4.1 *Houttuynia cordata* Thunb.

Synonyms: *Polypara cochinchinensis* Lour.; *Polypara cordata* Kuntze
   Common names: Chinese lizard tail; dokudami (Japan); ji cai (China)
   Habitat: Roadsides, riverbanks, and forests
   Distribution: India, Nepal, Bhutan, Myanmar, China, Korea, Indonesia, Thailand, and Japan
   Botanical observation: It is a beautiful, stoloniferous, and rhizomatous herb that grows to about 80cm long. The stems are longitudinally ridged. The leaves are simple, alternate, and stipulate. The stipule is membranous. The petiole is fleshy. The blade is fleshy, cordate at base, acute at apex, and with 3–6 pairs of secondary nerves. The inflorescence is terminal or leaf-opposed yellow spikes, which are 2–3cm long, with 4 pure white, petal-like bracts. The androecium includes 3 stamens. The ovary is made of 3 carpels partly connate; with 3 free styles. The fruits are dehiscent capsules.

Medicinal uses: Sores, carbuncles (China); pneumonia, bronchitis (Korea); infections (Japan)

*Very weak antibacterial (Gram-positive) polar extract:* Aqueous extract inhibited the growth of *Streptococcus mutans* (MT 5091) and *Streptococcus mutans* (OMZ 176) with the MIC values of 2500 and 2500 µg/mL, respectively (Chen et al., 1989).

*Moderate antiviral (non-enveloped monopartitle linear single-stranded (+) RNA) polar extract:* Aqueous extract inhibited the replication of the Enterovirus 71 (BrCr strain) in Vero cells with the $IC_{50}$ value of 125.9 µg/mL and a selectivity index of 101.6 via a mechanism involving protein synthesis inhibition (Lin et al., 2009).

*Strong antiviral (enveloped monopartite linear double-stranded DNA) polar extract:* Aqueous extract inhibited the replication of the Herpes simplex type-2 infection in Vero cells with the $IC_{50}$ of 50 µg/mL (Chen et al., 2011).

*Strong antiviral (enveloped monopartite linear single stranded (+)RNA) aqueous extract:* Aqueous extract at 1000 µg/mL inhibited Severe acute respiratory syndrome-associated coronavirus chymotrypsin-like protease by about 40% (Lau et al., 2008). At 800 µg/mL, the extract inhibited Severe acute respiratory syndrome-associated coronavirus RNA-dependent RNA polymerase by 26% (Lau et al., 2008). Ethylacetate fraction of leaves inhibited mouse the replication of the mouse Hepatitis virus with $IC_{50}$ value of 0.9 µg/mL (selectivity index >4) (). From this extract, quercetin inhibited mouse Hepatitis virus with an $IC_{50}$ value of 125 µg/mL (selectivity index: 0.9) (Chiow et al., 2016).

*Moderate antiviral (enveloped monopartite linear single-stranded (+) RNA) amphiphilic flavonols:* Ethylacetate fraction of leaves inhibited the replication of the Dengue type-2 (New Guinea C strain) with an $IC_{50}$ value of 7.5 µg/mL (selectivity index > 22.2) (Chiow et al., 2016). From this extract, quercetin and quercetrin inhibited the replication of the Dengue type-2 (New Guinea C strain) with the $IC_{50}$ values of 176.7 and 467.2.5 µg/mL, respectively (Chiow et al., 2016).

*Weak antiviral (enveloped monopartite linear single-stranded (+) RNA) essential oil:* Essential oil inhibited the replication of the Avian infectious bronchitis virus (strain Beaudette) with the $IC_{50}$ value of 970 µg/mL (Yin et al., 2011).

*In vivo (enveloped monopartite linear single-stranded (+) RNA) antiviral essential oil:* Essential oil given orally to chicken (62.5 mg/mL in drinking water for 5 days) reduced Avian infectious bronchitis virus-induced mortality to 0% (Yin et al., 2011).

*Strong antiviral (enveloped monopartite linear dimeric single-stranded (+) RNA) amphiphilic lignans:* The furan lignans manassantin A and manassantin B (LogD = 4.8 at pH 7.4; molecular weight = 716.8 g/mol) from the rhizomes inhibited the Human immunodeficiency virus type-1-induced cytopathic effects with the $IC_{100}$ values of 1 µM and the selectivity index of 8 and 63.8, (Lee et al., 2010). Manassantin A inhibited the Human immunodeficiency virus type-1 protease with the $IC_{50}$ value of 38.9 µM (Lee et al., 2010).

Manassantin B

*Strong antiviral (enveloped monopartite linear double-stranded DNA):* 4″-O-Demethylmanassantin A, manassantin A, and manassantin B, isolated from the roots inhibited the replication of the Epstein-Barr virus with $EC_{50}$ of 7.5, 3.4, and 1.7 μM, respectively (Cui et al., 2013). The selectivity index for manassantin B was 116.4 (Cui et al., 2013).

*Strong antiviral (enveloped monopartite linear double-stranded DNA) amphiphilic flavonol glycosides:* Houttuynoid M and houttuynoid A isolated from the aerial parts inhibited the Herpes simplex virus type-1 strain F with the $IC_{50}$ values of 17.7 and 12.4 μM, respectively and selectivity indexes above 10 (Li et al., 2017). The long alkylated flavonoid glycoside houttuynoid E (molecular mass = 632.7 g/mol) inhibited Herpes simplex virus type-2 with an $IC_{50}$ of 10.2 μM and a selectivity index of 3.2 (Chen et al., 2012). Houttuynoid G and H inhibited Herpes simplex virus type-2 with the $IC_{50}$ values of 38.4 μM (selectivity index: 2.9) and 14.1 μM (selectivity index: 3.1), respectively (Chen et al., 2013).

Houttuynoid E

*Strong antiviral (enveloped monopartite linear single-stranded (+)RNA) hydrophilic flavonol glycoside:* Quercetin 7-rhamnoside (LogD = −0.5 at pH 7.4; molecular mass = 448.3 g/mol) at the concentration of 10 μg/mL inhibited the replication of the porcine Epidemic diarrhea virus (EDV CV 777) (Song et al., 2011).

Quercetin 7-rhamnoside

*Weak antiviral (non-enveloped monopartite linear single-stranded (+) RNA) hydroxycinnamic acid:* Chlorogenic acid inhibited the replication of Enterovirus 71 with an $IC_{50}$ value of 102.5 µg/mL (Lin et al., 2009).

Commentaries: (i) Ethylacetate fraction of leaves inhibited the murine coronavirus mouse Hepatitis virus with an $IC_{50}$ value much lower than quercetin suggesting the presence of other anti-coronavirus constituents. (ii) Epstein-Barr virus is a DNA virus enclosed within an icosahedral capsid, protein tegument, and lipid envelope, which belong to the family Herpesviridae. This virus causes infectious mononucleosis and is associated with several benign and malignant conditions, including Burkitt lymphoma, nasopharyngeal carcinoma, post-transplant lymphoproliferative disorders, Kikuchi histiocytic necrotizing lymphadenitis, Gianotti-Crosti syndrome, and oral hairy leukoplakia (Chisholm & Lopez, 2011). (iii) The Avian infectious bronchitis virus (IBV), a coronavirus, causes infectious bronchitis, leading to pandemics in poultry (Yin et al., 2011). Flavonols and flavanols, when hydrophilic are not able to penetrate the envelop of viruses but at high concentration may evoke agglutination, hence virucidal effects. Flavonoids often block the entry of viruses in host cells. Why?

(iv) Quercetin inhibited the Mmr (Rv3065) efflux pump in *Mycobacterium* species (Sharma et al., 2019).

(v) Natural products from medicinal plants able to strongly inhibit the replication of Dengue viruses *in vitro* in the Clades covered in this volume are mainly isoquinolines, lignans, and phenolics with medium to high molecular masses.

## References

Chen, S.D., Gao, H., Zhu, Q.C., Wang, Y.Q., Li, T., Mu, Z.Q., Wu, H.L., Peng, T. and Yao, X.S., 2012. Houttuynoids A–E, anti-herpes simplex virus active flavonoids with novel skeletons from *Houttuynia cordata. Organic Letters, 14*(7), pp. 1772–1775.

Chen, S.D., Li, T., Gao, H., Zhu, Q.C., Lu, C.J., Wu, H.L., Peng, T. and Yao, X.S., 2013. Anti HSV-1 flavonoid derivatives tethered with houttuynin from *Houttuynia cordata. Planta Medica, 79*(18), pp. 1742–1748.

Chen, C.P., Lin, C.C. and Tsuneo, N., 1989. Screening of Taiwanese crude drugs for antibacterial activity against *Streptococcus* mutans. *Journal of Ethnopharmacology, 27*(3), pp. 285–295.

Chen, X., Wang, Z., Yang, Z., Wang, J., Xu, Y., Tan, R.X. and Li, E., 2011. *Houttuynia cordata* blocks HSV infection through inhibition of NF-κB activation. *Antiviral Research, 92*(2), pp. 341–345.

Chiow, K.H., Phoon, M.C., Putti, T., Tan, B.K. and Chow, V.T., 2016. Evaluation of antiviral activities of *Houttuynia cordata* Thunb. extract, quercetin, quercetrin and cinanserin on murine coronavirus and dengue virus infection. *Asian Pacific Journal of Tropical Medicine, 9*(1), pp. 1–7.

Chisholm, C. and Lopez, L., 2011. Cutaneous infections caused by Herpesviridae: a review. *Archives of Pathology & Laboratory Medicine, 135*(10), pp. 1357–1362.

Cui, H., Xu, B., Wu, T., Xu, J., Yuan, Y. and Gu, Q., 2013. Potential antiviral lignans from the roots of *Saururus chinensis* with activity against Epstein–Barr virus lytic replication. Journal of *Natural Products, 77*(1), pp. 100–110.

Lau, K.M., Lee, K.M., Koon, C.M., Cheung, C.S.F., Lau, C.P., Ho, H.M., Lee, M.Y.H., Au, S.W.N., Cheng, C.H.K., Bik-San Lau, C. and Tsui, S.K.W., 2008. Immunomodulatory and anti-SARS activities of *Houttuynia cordata*. *Journal of Ethnopharmacology, 118*(1), pp. 79–85.

Lee, J., Huh, M.S., Kim, Y.C., Hattori, M. and Otake, T., 2010. Lignan, sesquilignans and dilignans, novel HIV protease and cytopathic effect inhibitors purified from the rhizomes of *Saururus chinensis*. *Antiviral Research, 85*(2), pp. 425–428.

Li, J.J., Chen, G.D., Fan, H.X., Hu, D., Zhou, Z.Q., Lan, K.H., Zhang, H.P., Maeda, H., Yao, X.S. and Gao, H., 2017. Houttuynoid M, an anti-HSV Active Houttuynoid from *Houttuynia cordata* featuring a Bis-houttuynin chain tethered to a flavonoid core. *Journal of Natural Products, 80*(11), pp. 3010–3013.

Lin, T.Y., Liu, Y.C., Jheng, J.R., Tsai, H.P., Jan, J.T., Wong, W.R. and Horng, J.T., 2009. Anti-enterovirus 71 activity screening of Chinese herbs with anti-infection and inflammation activities. *The American Journal of Chinese Medicine, 37*(01), pp. 143–158.

Sharma, A., Gupta, V.K. and Pathania, R., 2019. Efflux pump inhibitors for bacterial pathogens: from bench to bedside. *The Indian Journal of Medical Research, 149*(2), p. 129.

Song, J.H., Shim, J.K. and Choi, H.J., 2011. Quercetin 7-rhamnoside reduces porcine epidemic diarrhea virus replication via independent pathway of viral induced reactive oxygen species. *Virology Journal, 8*(1), p. 460.

Yin, J., Li, G., Li, J., Yang, Q. and Ren, X., 2011. In vitro and in vivo effects of *Houttuynia cordata* on infectious bronchitis virus. *Avian Pathology, 40*(5), pp. 491–498.

## 2.1.4.2 *Saururus chinensis* (Lour.) Baill.

Synonyms: *Saururopsis chinensis* (Lour.) turcz.; *Saururopsis cumingii* C. DC.; *Saururus cernuus* Thunb.; *Saururus cumingii* C. DC.; *Saururus loureiri* Decne.; *Spathium chinense* Lour.

Common names: San bai cao, li bai liang bai (China)

Habitat: Wet lands

Distribution: China, Japan, Korea, The Philippines, Taiwan, and Vietnam

Botanical observation: It is a water loving, rhizomatous, fleshy herb that grows to a length of about 1 m long. The leaves are simple, spiral, and stipulate. The stipules form sheath which are about 1 cm long. The petiole is up to about 3 cm long. The blade is ovate-lanceolate, 4–20 cm × 2–10 cm, membranous, cordate at base, acuminate at apex, with 5–7 pairs of secondary nerves, and whitish when young. The inflorescences are tail-like, slender, curved, and white spikes, which are up to about 20 cm long. The androecium includes 6 stamens. The gynoecium comprises 4 carpels. The schizocarps present 4 tuberculate mericarps, which are 3 mm long.

Medicinal uses: Boils, abscesses (China)

*Strong antiviral (non-enveloped monopartite linear single-stranded (+) RNA) polar extract:* Aqueous extract inhibited the replication of the Enterovirus 71 with the $IC_{50}$ value 31.5 μg/mL (Wang et al., 2015).

*Strong antiviral (enveloped monopartite linear dimeric single stranded (+)RNA) furan lignans:* Manassantin A, manassantin B, and saucerneol B from the rhizomes inhibited Human immunodeficiency type-1 (strain IIIB)-induced cytopathic effects (Lee et al., 2010). Manassantin A inhibited Human immunodeficiency type-1 protease with an $IC_{50}$ value of 38.9 μM (Lee et al., 2010).

*Strong antiviral (enveloped monopartite linear double-stranded DNA) furan lignans:* Saucerneol methyl ether, saucerneol D, and manassantin B from the rhizomes inhibited

Epstein-Barr viruslytic DNA replication in P3HR-1 cells with $EC_{50}$ values of 1.7, 1.0, and 1.7 µM, respectively (Cui et al., 2013).

Commentary: The Enterovirus 71 (EV-71) virus belongs to the family Picornaviridae. It is a non-enveloped, RNA neurotropic virus which accounts for encephalitis and meningitis for which no treatment exist so far (Tapparel et al., 2013). Amphiphilic planar natural products with medium molecular weight including isoquinolines and anthraquinones from the medicinal plants belonging to the Clades in this volume are often able to strongly inhibit the replication of Enterovirus 71.

### References

Cui, H., Xu, B., Wu, T., Xu, J., Yuan, Y. and Gu, Q., 2013. Potential antiviral lignans from the roots of *Saururus chinensis* with activity against Epstein–Barr virus lytic replication. *Journal of Natural Products, 77*(1), pp. 100–110.

Lee, J., Huh, M.S., Kim, Y.C., Hattori, M. and Otake, T., 2010. Lignan, sesquilignans and dilignans, novel HIV protease and cytopathic effect inhibitors purified from the rhizomes of *Saururus chinensis*. *Antiviral Research, 85*(2), pp. 425–428.

Tapparel, C., Siegrist, F., Petty, T.J. and Kaiser, L., 2013. Picornavirus and enterovirus diversity with associated human diseases. *Infection, Genetics and Evolution, 14*, pp. 282–293.

Wang, C., Wang, P., Chen, X., Wang, W. and Jin, Y., 2015. *Saururus chinensis* (Lour.) Baill blocks enterovirus 71 infection by hijacking MEK1–ERK signaling pathway. *Antiviral Research, 119*, pp. 47–56.

## 2.2 ORDER LAURALES JUSS. EX BERCHT. & J.PRESL (1820)

### 2.2.1 FAMILY HERNANDIACEAE BLUME (1826)

The family Hernandiaceae comprises 4 genera and 58 species of trees, shrubs, or climbers. The leaves are simple or palmately compound, alternate, and exstipulate. The inflorescences are axillary or terminal corymbs or cymose panicles. The perianth comprises 3–5 outer tepals and 3–5 inner tepals. The androecium includes 3–5 stamens. The ovary is 1-loculed, 1-ovuled, the ovule pendulous. The fruits are 2–4-winged or not and contain 1 seed.

#### 2.2.1.1 *Hernandia nymphaeifolia* (C. Presl) Kubitzki

Synonyms: *Biasolettia nymphaeifolia* C. Presl; *Hernandia peltata* Meisn.

Common names: Jack in a box; lian ye tong (China); kampak (Indonesia); baru laut (Malaysia); evuevu (Fiji); koron koron (the Philippines); kong fa mao (Thailand); Tung (Vietnam)

Habitat: Sea shores and beaches

Distribution: Tropical Asia and Pacific

Botanical observation: It is a tree that grows to about 10 m tall with somewhat a sinister aura. The leaves are simple, alternate, and exstipulate. The petiole is slender and up to 25 cm long. The blade is peltate, broadly lanceolate, 20–40 × 15–30 cm, glossy, marked with 2 to 3 pairs of showy secondary nerves emerging from the petiole, rounded at base, acute at apex and somewhat Menispermaceous. The inflorescence is an axillary panicle of cymes, which is up to 20 cm long and bearing minute flowers. The perianth includes 6–8 white tepals. The androecium is made of 3 stamens. The ovary develops a minute style and a dilated stigma. The drupes are ovoid, up to about 3 cm long, black and enclosed in an inflated fleshy cupule, whole structure looking like an eye or a lantern.

Medicinal uses: Urinary tract infection (Fiji)

*Strong antiviral (enveloped monopartite linear double-stranded DNA) amphiphilic dibenzyl butyrolactone lignan:* Deoxypodophyllotoxin (LogD = 2.7 at pH 2.7; molecular mass = 398.4 g/mol) inhibited the replication of the Herpes simplex virus type-1 (KOS strain) and Herpes simplex virus type-1 (G strain) with the $IC_{50}$ values of 0.004 and 0.01 µM, respectively and selectivity indices >2500 and >909, respectively (Sudo et al., 1998).

Deoxypodophyllotoxin

Commentary: The plant produces oxoaporphine alkaloids (2-*O*-methyl-7-oxolaetine, hernandonine, oxonorisocorydine) and lignans (deoxypodophyllotoxin, dehydropodophyllotoxin, and yatein) (Chen et al., 1996, 1997; Wei et al., 2018), and as discussed previously such compounds have antibacterial and or antifungal properties. Other potential antimicrobial compounds could be tetrahydrofuran lignans (Chen et al., 1996, 1997).

### References

Chen, J.J., Tsai, I.L. and Chen, I.S., 1996. New oxoaporphine alkaloids from *Hernandia nymphaeifolia. Journal of Natural Products, 59*(2), pp. 156–158.

Chen, I.S., Chen, J.J., Duh, C.Y. and Tsai, I.L., 1997. Cytotoxic lignans from formosan *Hernandia nymphaeifolia. Phytochemistry, 45*(5), pp. 991–996.

Sudo, K., Konno, K., Shigeta, S. and Yokota, T., 1998. Inhibitory effects of podophyllotoxin derivatives on herpes simplex virus replication. *Antiviral Chemistry and Chemotherapy, 9*(3), pp. 263–267.

Wei, C.Y., Wang, S.W., Ye, J.W., Hwang, T.L., Cheng, M.J., Sung, P.J., Chang, T.H. and Chen, J.J., 2018. New anti-inflammatory aporphine and lignan derivatives from the root wood of *Hernandia nymphaeifolia. Molecules, 23*(9), p. 2286.

### 2.2.1.2 *Illigera appendiculata* Bl.

Common name: Maralipit (Malaysia)

Habitat: Forest

Distribution: Malaysia, Indonesia, and the Philippines

Botanical observation: It is a climber. The leaves are alternate, 3-foliolate, and exstipulate. The petiole is slender and up to about 10cm long. The blade is elliptic-lanceolate, somewhat asymmetrical, 6–14 cm × 3–9cm, cordate at base, rounded, acuminate at apex, and with 3–6 pairs of secondary nerves. The inflorescences are axillary cymes of small flowers. The perianth consists of 5 tepals, which are about 1cm long. The androecium comprises 5 stamens, which are 4mm long. The nuts are red, 2.5–3cm long, and 2–4 winged.

Medicinal use: Boils (Malaysia)

Commentaries: This plant has apparently not been studied for its antimicrobial effects. Members of the genus *Illigera* Bl. produce cytotoxic aporphines (Ge et al., 2018), which are probably antibacterial and antifungal. Bulbocapnine and the oxoaporphines liriodenine and dicentrinone are not uncommon in this genus (Chen et al., 1997).

**References**

Chen, K.S., Wu, Y.C., Teng, C.M., Ko, F.N. and Wu, T.S., 1997. Bioactive alkaloids from *Illigera luzonensis. Journal of Natural Products, 60*(6), pp. 645–647.

Ge, Y.C., Zhang, H.J., Wang, K.W. and Fan, X.F., 2018. Aporphine alkaloids from *Illigera aromatica* from Guangxi Province, China. *Phytochemistry, 154*, pp. 73–76.

## 2.2.2 FAMILY LAURACEAE A.L. de Jussieu (1789)

The family Lauraceae consists of 54 genera and about 3500 species of mostly elegant trees or shrubs. The bark is often somewhat thick and aromatic. The leaves are simple, alternate or opposite, and exstipulate. The inflorescences are mostly axillary cymes or panicles of minute flowers. The perianth comprises 6 tepals. The androecium includes 12 stamens with anthers opening with very characteristic valves. The gynoecium includes a single carpel sheltering a single ovary, which is anatropous and pendulous. The fruits are drupes, which are often glossy, olive-shaped, and seated in persistent calyx. Members of this family produce essential oils (phenylpropanoids, monoterpenes), sesquiterpenes, lignans, furanones, fatty alcohols, flavonoids, and isoquinoline alkaloids.

### 2.2.2.1. *Cinnamomum bejolghota* (Buch.-Ham.) Sweet

Synonyms: *Cinnamomum obtusifolium* (Roxb.) Nees; *Laurus bejolghota* Buch.-Ham.; *Laurus obtusifolia* Roxb.

    Common names: Dun ye gui (China); ram tejpat, naga-dal-chini (Bangladesh)

    Habitat: Forest

    Distribution: India, Bhutan, Nepal, Bangladesh, Myanmar, Laos, Thailand, and Vietnam

    Botanical observation: It is a tree that grows to about 20 m tall. The inner bark is fragrant and used as spice. The leaves are simple, sub-opposite, and exstipulate. The petiole is up to about 1.5 cm long. The blade is elliptic-oblong, 12–30 × 4–9 cm, coriaceous, somewhat glaucous below, trinerved, base attenuate, and acute at apex. The inflorescence is a showy axillary panicles of numerous minute flowers. The perianth comprises 6 oblong lobes which are about 5 mm. The androecium includes 9 stamens. The ovary is minute and develops a slender style and a discoid stigma. The drupes are olive-shaped, about 1 cm long, green and glossy on a perianth cup.

    Medicinal uses: Wounds, diarrhea, fever (Bangladesh)

    *Broad-spectrum antibacterial halo developed by essential oil:* Essential oil of bark inhibited the growth of *Salmonella typhi, Streptococcus enteriditis, Escherichia coli, Clostridium pefringens,* and *Campylobacter jejuni* with inhibition zone diameters of 21.5, 16.6, 19.3, 59.5, and 37.5 mm, respectively, 15 µL of essential oil/disc 6mm) (Wannissorn et al., 2005).

    *Strong antibacterial essential oil:* Essential oil of bark inhibited the growth of *Bacillus cereus, Bacillus subtilis, Staphylococcus aureus, Escherichia coli, Pseudomonas aeruginosa,* and *Salmonella typhimurium* with the MIC values of 62.5, 31.2. 31.2, 31.2, 31.2, and 62.5 µg/mL, respectively (Atiphasaworn et al., 2017).

    *Moderate antifungal (filamentous) essential oil:* Essential oil of bark inhibited the growth of *Colletotrichum asianum, Colletotricum fruticola, Colletotrichum magna,* and *Colletotrichum tropica* with the MIC values of 125, 250, 500, and 250 µg/mL, respectively (Atiphasaworn et al., 2017).

    *Broad-spectrum antibacterial amphiphilic cyclic monoterpenes:* α-Terpineol (LogD = 3.0 at pH 7.4; molecular mass = 154.2 g/mol) inhibited the growth of *Escherichia coli* (ATCC 25922), *Salmonella enteritidis* (CMCC (B), and *Staphylococcus aureus* (ATCC 25923) with the MIC/MBC values of 0.7/0.7, 3.1/3.1, and 1.5/3.1 µL/mL (Li et al., 2014b). Time kill-curve showed a dose-dependent effect (Li et al., 2014). 1,8-Cineole (also known as eucalyptol, LogD = 2.8 at pH 7.4, molecular mass = 154.2 g/mol) inhibited the growth of *Escherichia coli* (ATCC 25922), *Salmonella enteritidis* (CMCC (B), /0041), and *Staphylococcus aureus* (ATCC 25923) with MIC/MBC values of 3.1/3,1, 6.2/6.2, and 6.2/6.2 µL/mL (Li et al., 2014). 1,8-Cineole evoked the rupture of cell wall and cell membrane as well as cytoplasmic.

α-Terpineol

1,8-Cineole

*Strong broad-spectrum antibacterial linear amphiphilic monoterpene:* Linalool (LogD = 3.2 at pH 7.4; molecular weight = 154.2 g/mol) inhibited the growth of *Escherichia coli* (95), *Klebsiella pneumoniae* (ATCC 10031), *Pseudomonas aeruginosa* (ATCC 9027), and *Staphylococcus aureus* (ATCC 65238), with the MIC values of 0.07, 0.1, 0.1, and 0.1 µg/mL, respectively (Radulović et al., 2007).

Linalool

*Strong anticandidal monoterpene:* Linalool inhibited the growth of *Candida albicans* (ATCC 10231) with an MIC value of 0.03 µg/mL (Radulović et al., 2007).

Commentary: Essential oil of leaves contains a majority of linalool and 1,8-cineole whereas the main constituent of essential oil of bark is α-terpineol (Baruah et al., 1997). In general, essential oils have always degrees of antibacterial and antifungal effects probably because they are phyto-anticipins or in other words, antimicrobial secondary metabolites present in plant tissues prior to bacterial or fungal infections. Pure extracted essential oils are mostly toxic when ingested orally and of no systemic antibacterial or antifungal therapeutic value. They can be used as antiseptics if non-toxic externally or desinfectants. The vapors of essential oils might be of usefulness to disinfect the air in times of pandemia. In this context, one could remember that the German mystic St Hildegarde of Bingen (1098–1179) prescribed *Thymus vulgaris* L. (a plant notorious for its antimicrobial essential oil) for plague and it was later included in the posies carried by judges and kings to protect them from disease in public.

**References**

Atiphasaworn, P., Monggoot, S. and Pripdeevech, P., 2017. Chemical composition, antibacterial and antifungal activities of *Cinamomum bejolghota* bark oil from Thailand. *Journal of Applied Pharmaceutical Science, 7*(04), pp. 069–073.

Baruah, A., Nath, S.C., Hazarika, A.K. and Sarma, T.C., 1997. Essential oils of the leaf, stem bark and panicle of *Cinnamomum bejolghota* (Buch.-Ham.) sweet. *Journal of Essential Oil Research, 9*(2), pp. 243–245.

Li, L., Li, Z.W., Yin, Z.Q., Wei, Q., Jia, R.Y., Zhou, L.J., Xu, J., Song, X., Zhou, Y., Du, Y.H. and Peng, L.C., 2014a. Antibacterial activity of leaf essential oil and its constituents from *Cinnamomum longepaniculatum. International Journal of Clinical and Experimental Medicine, 7*(7), p. 1721.

Li, L., Shi, C., Yin, Z., Jia, R., Peng, L., Kang, S. and Li, Z., 2014b. Antibacterial activity of α-terpineol may induce morphostructural alterations in *Escherichia coli. Brazilian Journal of Microbiology, 45*(4), pp. 1409–1413.

Radulović, N., Mišić, M., Aleksić, J., Đoković, D., Palić, R. and Stojanović, G., 2007. Antimicrobial synergism and antagonism of salicylaldehyde in *Filipendula vulgaris* essential oil. *Fitoterapia, 78*(7–8), pp. 565–570.

Wannissorn, B., Jarikasem, S., Siriwangchai, T. and Thubthimthed, S., 2005. Antibacterial properties of essential oils from Thai medicinal plants. *Fitoterapia, 76*(2), pp. 233–236.

## 2.2.2.2 *Cinnamomum burmannii* (Nees & T. Nees) Blume

Synonyms: *Cinnamomum chinense* Bl.; *Cinnamomum dulce* (Roxb.) Sweet; *Cinnamomum hainanense* Nakai; *Cinnamomum kiamis* Nees; *Cinnamomum miaoshanense* S.K. Lee & F.N. Wei; *Laurus burmannii* Nees & T. Nees

Common names: Malaysian Cinnamon, padang cassia; yin ziang (China); kayu manis (Indonesia, Malaysia); kami (the Philippines); suramarit (Thailand)

Habitat: Cultivated

Distribution: Native to Vietnam and China and found in tropical Asia and Pacific

Botanical observation: It is a tree that grows to about 15 m tall. The inner bark is fragrant. The leaves are simple, alternate, and exstipulate. The petiole is up to about 1.2 cm long. The blade is oblong-lanceolate, 5.5–10.5 × 2–5 cm, coriaceous, somewhat glaucous below, trinerved, attenuate at base and acute at apex acute. The panicles are axillary lax, and bear numerous minute flowers. The 6 lobes of the perianth are oblong. The androecium includes 9 stamens. The ovary is minute and develops a slender style and a discoid stigma. The drupes are olive-shaped, about 50 mm long, green and glossy on a perianth cup.

Medicinal use: Apparently none for infectious diseases.

*Broad-spectrum antibacterial halo developed by polar extract:* Methanol extract of bark (6 mg well) inhibited the growth of *Bacillus cereus, Listeria monocytogenes, Staphylococcus aureus, Escherichia coli,* and *Salmonella anatum* with the inhibition zone diameters of 15.4, 11.5, 15.7, 8.7, and 12.1 mm, respectively (Shan et al., 2007a).

*Weak broad-spectrum antibacterial phenylpropanoid:* Cinnamaldehyde from the bark inhibited the growth of *Bacillus cereus, Listeria monocytogenes, Staphylococcus aureus,* and *Escherichia coli* with the MIC/MBC values of 312.5/2500, 500/500, 156.3/625, and 156.3/312.5 µg/mL, respectively (Shan et al., 2007b).

*Strong antibacterial (Gram-positive) amphiphilic proanthocyanidin:* Procyanidin B2 (LogD = 0.9 at pH 7.4; molecular weight = 578.5 g/mol) isolated from the bark inhibited the growth of *Bacillus cereus, Listeria. monocytogenes, Staphylococcus aureus,* and *Escherichia coli* with the MIC values of 31.9/78.1, 625/625, 625/1250, and 1250/1250 µg/mL, respectively (Shan et al., 2007).

*Moderate anticandidal phenylpropanoid:* Cinnamaldehyde inhibited *Candida albicans* with an MIC value of 50 ppm (Tampieri et al., 2005).

Commentaries: (i) According to Tampieri et al. (2005), strong antifungal activity is defined for pure compounds for MIC values equal or below 50 ppm (or 50 µg/g), and for essential oil, activity is defined at MIC of 500 ppm (or 500 µg/g) or below. Here, a compound is defined as having a strong antifungal (or antibacterial) activity for MIC below 50 ppm (or 50 µg/g), moderate activity for MIC from 50 to 100 ppm (or from 50 to 100 µg/g), weak activity for MIC from 100 to 1500 ppm (or from 100 to 1500 µg/g) and inactivity above 1500 ppm (1500 µg/g). For essential oils, strong antifungal (or antibacterial) activity is defined here at an MIC of 500 ppm (or 500 µg/g) or below.

(ii) 1,8-Cineole, borneol, camphor, terpinen-4-ol, and α-terpineol are the main constituents of essential oil in this plant (Ji et al., 1991).

(iii) In Indonesia and Malaysia the plant is used as spice.

### References
Ji, X.D., Pu, Q.L., Garraffo, H.M. and Pannell, L.K., 1991. Essential oils of the leaf, bark and branch of *Cinnamomum buramannii* Blume. *Journal of Essential Oil Research, 3*(5), pp. 373–375.

Shan, B., Cai, Y.Z., Brooks, J.D. and Corke, H., 2007a. Antibacterial properties and major bioactive components of cinnamon stick (*Cinnamomum burmannii*): activity against foodborne pathogenic bacteria. *Journal of Agricultural and Food Chemistry, 55*(14), pp. 5484–5490.

Shan, B., Cai, Y.Z., Brooks, J.D. and Corke, H., 2007b. The in vitro antibacterial activity of dietary spice and medicinal herb extracts. *International Journal of Food Microbiology, 117*(1), pp. 112–119.

Tampieri, M.P., Galuppi, R., Macchioni, F., Carelle, M.S., Falcioni, L., Cioni, P.L. and Morelli, I., 2005. The inhibition of *Candida albicans* by selected essential oils and their major components. *Mycopathologia, 159*(3), pp. 339–345.

## 2.2.2.3 *Cinnamomum camphora* (L.) J. Presl

Synonyms: *Camphora officinarum* Nees; *Cinnamomum camphoroides* Hayata; *Cinnamomum nominale* (Hayata) Hayata; *Cinnamomum simondii* Lecomte; *Cinnamomum taquetii* H. Lév.; *Laurus camphora* L.; *Persea camphora* (L.) Spreng.

Common names: Camphor laurel; zhang (China); kapur (India); paruk (Myanmar)

Habitat: Cultivated

Distribution: Vietnam, China, Korea, and Japan

Botanical description: It is a tree which grows to about 25 m tall. The bark and leaves are fragrant. The leaves are simple, alternate, and exstipulate. The petiole is up to about 3 cm long. The blade is ovate-elliptic, 5.5–12 × 2–5.5 cm, somewhat coriaceous and wavy, dark green, glossy, glaucous below, with 1 pair of secondary nerves, cuneate at base, and acute at apex. The panicles are axillary, lax, and bear numerous minute flowers. The perianth is white and comprises 6 broadly lanceolate lobes, which are oblong and minute. The androecium includes 9 stamens. The ovary is ovoid and develops a short style. The drupes are globose, about 8 mm in diameter, black, and glossy on a perianth cup.

Medicinal uses: Bronchitis (Nepal); abscesses, sores (China)

*Broad-spectrum antifungal essential oil:* Essential oil of leaves inhibited the growth of *Candida albicans* with the inhibition zone diameter of 10.6 mm (3.7 µL absorbed by 5 mm disc) (Dutta et al., 2007). Essential oil was found to possess a mycostatic effect against *Aspergillus flavus* at 4000 ppm (Mishra et al., 1991) and inhibited *Candida albicans* with an MIC of 500 ppm (Tampieri et al., 2005).

*Strong broad spectrum antibacterial amphiphilic furofuran lignan:* (+)-Pinoresinol (LogD = 1.9 at pH 7.4; molecular weight = 358.3 g/mol) isolated from the leaves inhibited the growth of *Escherichia coli, Staphylococcus aureus, Pseudomonas aeruginosa, Salmonella enteritidis,* and *Bacillus subtilis* with the MIC/MBC values of 31.2/62.5, 15.6/31.2, 7.8/15.6, 31.2/62.5, and 3.9/7.8 µg/mL, respectively, via cytoplasmic membrane and cell wall destruction (Zhou et al., 2017).

(+)-Pinoresinol

*Strong broad spectrum antifungal furofuran lignan:* (+)-Pinoresinol inhibited the growth of *Candida albicans*, *Malassezia furfur*, and *Trichosporon beigelli* with the MIC values of 12.5, 25, and 25 μg/mL, respectively, via, in the case of *Candida albicans*, perturbation of the cytoplasmic membrane (Hwang et al., 2010).

*Antifungal cyclic monoterpene ketone:* Camphor inhibited the mycelial growth and affected the morphology of hyphae of *Choanephora cucurbitarum* due to cytoplasm coagulation and hyphal lysis (Pragadheesh et al., 2013).

Commentaries: (i) Essential oil in the leaves contains mainly camphor, cineole, and linalool (Wanyang et al., 1989). (ii) *Choanephora cucurbitarum* (family Choanephoraceae; order Mucorales) is a filamentous fungi, which is a pathogen for plants in the family Cucurbitaceae. (iii) Camphor is used in China for the treatment of COVID-19 (Luo, 2020; Wang et al., 2020).

### References

Dutta, B.K., Karmakar, S., Naglot, A., Aich, J.C. and Begam, M., 2007. Anticandidial activity of some essential oils of a mega biodiversity hotspot in India. *Mycoses, 50*(2), pp. 121–124.

Hwang, B., Lee, J., Liu, Q.H., Woo, E.R. and Lee, D.G., 2010. Antifungal effect of (+)-pinoresinol isolated from *Sambucus williamsii*. *Molecules, 15*(5), pp. 3507–3516.

Luo, A., 2020. Positive SARS-Cov-2 test in a woman with COVID-19 at 22 days after hospital discharge: A case report. *Journal of Traditional Chinese Medical Sciences, 7*(4): 413–417.

Mishra, A.K., Dwivedi, S.K., Kishore, N. and Dubey, N.K., 1991. Fungistatic properties of essential oil of *Cinnamomum camphora*. *International Journal of Pharmacognosy, 29*(4), pp. 259–262.

Pragadheesh, V.S., Saroj, A., Yadav, A., Chanotiya, C.S., Alam, M. and Samad, A., 2013. Chemical characterization and antifungal activity of *Cinnamomum camphora* essential oil. *Industrial crops and products, 49*, pp. 628–633.

Tampieri, M.P., Galuppi, R., Macchioni, F., Carelle, M.S., Falcioni, L., Cioni, P.L. and Morelli, I., 2005. The inhibition of *Candida albicans* by selected essential oils and their major components. *Mycopathologia, 159*(3), pp. 339–345.

Wang, S.X., Wang, Y., Lu, Y.B., Li, J.Y., Song, Y.J., Nyamgerelt, M. and Wang, X.X., 2020. Diagnosis and treatment of novel coronavirus pneumonia based on the theory of traditional Chinese medicine. *Journal of Integrative Medicine*.

Wanyang, S., Wei, H. and Guangyu, W., 1989. Study on chemical constituents of the essential oil and classification of types from *Cinnamomum camphora*. *Acta Botanica Sinica (China)*.

Zhou, H., Ren, J. and Li, Z., 2017. Antibacterial activity and mechanism of pinoresinol from *Cinnamomum Camphora* leaves against food-related bacteria. *Food Control, 79*, pp. 192–199.

### 2.2.2.4 *Cinnamomum cassia* (L.) J. Presl

Synonyms: *Cinnamomum aromaticum* Nees; *Laurus cassia* Nees & T. Nees

Common names: Chinese cinnamon; tejpat (Bangladesh), rou gi (China); kayu manis (Malaysia); thitchubo (Myanmar)

Habitat: Cultivated

Distribution: India, Bangladesh, Myanmar, Laos, Thailand, Vietnam, China, Malaysia, and Indonesia

Botanical description: It is a tree that grows to about 10 m tall. The bark is thick and aromatic. The leaves are simple, alternate or sub-opposite, and exstipulate. The petiole is up to about 3 cm long. The blade is narrowly elliptic, 8–16 × 4.5–5.5 cm, somewhat coriaceous, wavy, dark green, glossy, with 1 pair of secondary nerves, and acute at base and apex. The inflorescence is an axillary panicle, which is about 15 cm long and bears numerous minute flowers. The perianth is yellow and comprises 6 broadly lanceolate lobes, which are oblong and minute. The androecium includes 9 stamens. The gynoecium includes an ovary, which is minute, ovoid, and develops a short style. The fruits are ovoid, about 1 cm in diameter, black, and glossy on a perianth cup.

Medicinal uses: Diarrhea (Cambodia, Laos, Vietnam); fever (China); cough (Malaysia)

*Moderate broad-spectrum antibacterial halo developed by polar extract:* Methanol extract of bark (6 mg agar-well) inhibited the growth of *Bacillus cereus*, *Listeria monocytogenes*, *Staphylococcus aureus*, *Escherichia coli*, and *Salmonella anatum* with inhibition zone diameters of 10.3, 8.9, 12.1, 5, and 7 mm, respectively (Shan et al., 2007). Methanol extract of bark inhibited the growth of *Clostridium perfringens* (ATCC 13124) and *Bacteroides fragilis* (from human feces) (5 mg/8 mm disc) with inhibition zone diameters of 21 and 30 mm, respectively (Lee, 2002).

*Moderate broad-spectrum antibacterial essential oil:* Essential oil of bark inhibited the growth of *Staphylococcus aureus* (ATCC6538), *Staphylococcus epidermidis* (ATCC12228), *Streptococcus pyogenes* (ATCC19615), *Pseudomonas aeruginosa* (ATCC15442), and *Escherichia coli* (ATCC11303) with MIC values of about 500 μg/mL (Firmino et al., 2018).

*Broad-spectrum antibacterial halo developed by phenylpropanoid:* Cinnamaldehyde inhibited the growth of *Bifidobacterium longum* (ATCC 15707), *Clostridium perfringens* (ATCC 13124), and *Bacteroides fragilis* (from human feces) (1 mg/8 mm disc) with inhibition zone diameters of 10–15, 21–30, and above 30 mm, respectively (Lee, 2002).

*Broad-spectrum antibacterial halo developed by eudesmane sesquiterpene:* 1β, 6α-Dihydroxyeudesm-4(15)-ene from flower buds (30 μL, 10 mg/mL/6 mm wells) inhibited the growth of *Escherichia coli* and *Staphylococcus aureus* with inhibition zone diameters of 8.5 and 11 mm, respectively (Guoruoluo et al., 2017).

*Weak antibacterial phenylpropanoid:* Cinnamaldehyde inhibited the growth of *Staphylococcus aureus* (ATCC6538), *Staphylococcus epidermidis* (ATCC12228), *Streptococcus pyogenes* (ATCC19615), *Pseudomonas aeruginosa* (ATCC15442), and *Escherichia coli* (ATCC11303) with MIC values of 200, 200, 500, 200 and 200 μg/mL, respectively (Firmino et al., 2018).

*Antibiotic potentiator phenylpropanoid:* Cinnamaldehyde evoked at 4 × MIC in 1 h an irreversible decrease of methicillin-resistant *Staphylococcus aureus* count Log10 (CFU/mL) from 6 to 0, and was synergistic with vancomycin for methicillin-resistant *Staphylococcus aureus* with a fractional inhibitory concentration index of 0.3 (Hossan et al., 2018).

*Moderate antifungal (filamentous) essential oil:* Essential oil inhibited the growth of *Aspergillus flavus* (strains 3.2758 and 3.4408) and *Aspergillus oryzae* with MIC values of 125, 125, and 250 ppm, respectively (Kocevski et al., 2013).

*Anticandidal halo developed by eudesmane sesquiterpene:* 1β, 6α-dihydroxyeudesm-4(15)-ene from flower buds (30 μL, 10 mg/mL/6 mm well) inhibited the growth of *Candida albicans* with an inhibition zone diameter of 11 mm (Guoruoluo et al., 2017).

*Strong antiviral (enveloped monopartite linear dimeric single-stranded (+) RNA) polar extracts:* Ethanol extract of bark inhibited the replication of the Human immunodeficiency virus type-1 in MT-4 cells (Premanathan et al., 2000).

*Strong antiviral (enveloped monopartite linear single-stranded (-) RNA) polar extract:* Aqueous extract of bark inhibited the replication of the Human respiratory syncytial virus with an $IC_{50}$ of 7.2 μg/mL in HEp-2 cells and 6.8 μg/mL in A549 cells (Yeh et al., 2013). Aqueous extract of bark inhibited Human respiratory syncytial virus-induced plaque formation and viral internalization in both cell lines (Yeh et al., 2013). In A549 cells, this extract inhibited F-protein expression (Yeh et al., 2013).

*Moderate antiviral (enveloped monopartite linear single-stranded (+) RNA) proanthocyanidins:* Procyanidin A2, procyanidin B1, and cinnamtannin B1 (LogD = 1.7 at pH 7.4; molecular weight = 864.7 g/mol) inhibited plaque formation by the Severe acute respiratory syndrome-associated coronavirus with the $IC_{50}$ values of 29.9, 41.3, and 32.9 μM and selectivity indices of 37.3, 15.6, and 5.6, respectively (Zhuang et al., 2009).

Commentaries: (i) The bark contains (-)-epicatechin, as well as proanthocyanidins (Morimoto et al., 1986), which probably account for the antiviral and antibacterial properties of polar extracts. (ii) The plant is used in China for the treatment of COVID-19 (Wang et al., 2020; Yang et al., 2020). (iii) Natural products from the medicinal plants from the Clades in this book with strong anti-Severe acute respiratory syndrome-associated coronavirus activity *in vitro* are (apart from those described above) few, mostly medium molecular weight and amphiphilic isoquinolines and chalcones.

### References

Firmino, D.F., Cavalcante, T.T., Gomes, G.A., Firmino, N., Rosa, L.D., de Carvalho, M.G. and Catunda Jr, F.E., 2018. Antibacterial and antibiofilm activities of *Cinnamomum* Sp. essential oil and cinnamaldehyde: antimicrobial activities. *The Scientific World Journal.*

Guoruoluo, Y., Zhou, H., Zhou, J., Zhao, H., Aisa, H.A. and Yao, G., 2017. Isolation and characterization of sesquiterpenoids from cassia buds and their antimicrobial activities. *Journal of Agricultural and Food Chemistry, 65*(28), pp. 5614–5619.

Hossan, M.S., Jindal, H., Maisha, S., Samudi Raju, C., Devi Sekaran, S., Nissapatorn, V., Kaharudin, F., Su Yi, L., Khoo, T.J., Rahmatullah, M. and Wiart, C., 2018. Antibacterial effects of 18 medicinal plants used by the Khyang tribe in Bangladesh. *Pharmaceutical Biology, 56*(1), pp. 201–208.

Kocevski, D., Du, M., Kan, J., Jing, C., Lačanin, I. and Pavlović, H., 2013. Antifungal effect of *Allium tuberosum, Cinnamomum cassia*, and *Pogostemon cablin* essential oils and their components against population of *Aspergillus* species. *Journal of Food Science, 78*(5), pp. M731–M737.

Lee, H.S., 2002. Inhibitory activity of *Cinnamomum cassia* bark-derived component against rat lens aldose reductase. *Journal of Pharmacy and Pharmaceutical Sciences, 5*(3), pp. 226–230.

Morimoto, S., Nonaka, G.I. and Nishioka, I., 1986. Tannins and related compounds. XXXVIII.: isolation and characterization of flavan-3-ol glucosides and procyanidin oligomers from Cassia bark: *Cinnamomum cassia* BLUME. *Chemical and Pharmaceutical Bulletin, 34*(2), pp. 633–642.

Premanathan, M., Rajendran, S., Ramanathan, T. and Kathiresan, K., 2000. A survey of some Indian medicinal plants for anti-human immunodeficiency virus (HIV) activity. *Indian Journal of Medical Research, 112*, p. 73.

Shan, B., Cai, Y.Z., Brooks, J.D. and Corke, H., 2007. The in vitro antibacterial activity of dietary spice and medicinal herb extracts. *International Journal of Food Microbiology, 117*(1), pp. 112–119.

Wang, S.X., Wang, Y., Lu, Y.B., Li, J.Y., Song, Y.J., Nyamgerelt, M. and Wang, X.X., 2020. Diagnosis and treatment of novel coronavirus pneumonia based on the theory of traditional Chinese medicine. *Journal of Integrative Medicine, 18*(4): 275–283.

Yang, Q.X., Zhao, T.H., Sun, C.Z., Wu, L.M., Dai, Q., Wang, S.D. and Tian, H., 2020. New thinking in the treatment of 2019 novel coronavirus pneumonia. *Complementary Therapies in Clinical Practice, 39*, p. 101131.

Yeh, C.F., San Chang, J., Wang, K.C., Shieh, D.E. and Chiang, L.C., 2013. Water extract of *Cinnamomum cassia* Blume inhibited human respiratory syncytial virus by preventing viral attachment, internalization, and syncytium formation. *Journal of Ethnopharmacology, 147*(2), pp. 321–326.

Zhuang, M., Jiang, H., Suzuki, Y., Li, X., Xiao, P., Tanaka, T., Ling, H., Yang, B., Saitoh, H., Zhang, L. and Qin, C., 2009. Procyanidins and butanol extract of *Cinnamomi Cortex* inhibit SARS-CoV infection. *Antiviral Research, 82*(1), pp. 73–81.

### 2.2.2.5 *Cinnamomum longepaniculatum* (Gamble) N. Chao ex H.W. Li

Synonyms: *Cinnamomum inunctum* var. *longepaniculatum* Gamble
 Common names: Ka-ra-way-thee (Myanmar); zhang shu (China)
 Habitat: Forests
 Distribution: China
 Botanical observation: It is a tree that grows to about 15 m tall. The bark is gray and smooth. The leaves are simple, alternate, and exstipulate. The petiole is reddish and up to about 3.5 cm long. The blade is ovate, 8–12 × 3.5–6.5 cm, somewhat coriaceous, hairy below, with 4–5 pairs of secondary nerves, cuneate at base, and acuminate at apex. The panicles are slender, lax, about 20 cm long, and bear numerous minute flowers. The 6 lobes of the perianth are hairy, whitish, and broadly lanceolate. The androecium includes 9 stamens. The ovary is minute, ovoid, and develop a short style. The drupes are globose, about 1 cm in diameter, and glossy on a perianth cup.
 Medicinal uses: Fever, diarrhea (Myanmar)
 *Broad-spectrum antibacterial essential oil:* Essential oil of leaves inhibited the growth of *Escherichia coli* (ATCC 25922), *Salmonella enteritidis* (CMCC (B) 50041), and *Staphylococcus aureus* (ATCC 25923) with the MIC/MBC values of 3.1/3.1, 6.2/6.2, and 6.2/6.2 µL/mL, respectively (Li et al., 2014). *Escherichia coli, Salmonella enteritidis*, and *Staphylococcus aureus* were killed at 1 × MIC in 4, 8, 12 h and 2 × MIC in 2, 4, 8 h respectively. The rate of killing increased by increasing the concentration of essential oil. Time-kill curves essential oil showed a concentration-dependent effect (Li et al., 2014)
 *Strong broad-spectrum antibacterial amphiphilic monoterpenes:* 1,8-Cineole (also known as eucalyptol; LogD = 2.8 at pH 7.4; molecular weight = 154.1 g/mol) inhibited the growth of *Escherichia coli* (ATCC 25922), *Salmonella enteritidis* (CMCC (B) 50041), and *Staphylococcus aureus* (ATCC 25923) with MIC/MBC values of 3.1/3.1, 6.2/6.2, and 6.2/6.2 µL/mL, respectively (Li et al., 2014). α-Terpineol (LogD = 3.0 at pH 7.4; molecular mass = 154.2 g/mol) inhibited the growth of *Escherichia coli* (ATCC 25922), *Salmonella enteritidis* (CMCC (B) 50041), and *Staphylococcus aureus* (ATCC 25923) with MIC/MBC values of 0.7/0.7, 3.1/3.1, and 1.5/3.1 µL/mL, respectively (Li et al., 2014). Terpinene-4-ol inhibited the growth of *Escherichia coli* (ATCC 25922), *Salmonella enteritidis* (CMCC (B) 50041), and *Staphylococcus aureus* (ATCC 25923) with MIC/MBC values of 1.5/1.5, 3.1/3/1, and 1.5/1.5 µL/mL, respectively (Li et al., 2014). γ-Terpinene (LogD = 4.1; molecular weight = 136.1 g/mol) inhibited the growth of *Salmonella enteritidis* (CMCC (B) 50041) with MIC/MBC values of 3.1/3.1 µL/mL (Li et al., 2014).
 *Strong broad-spectrum antibacterial amphiphilic phenylpropanoid:* Safrole (LogD = 2.7 at pH 7.4; molecular weight = 162.1 g/mol) from essential oil of leaves inhibited the growth of *Escherichia coli* (ATCC 25922), *Salmonella enteritidis* (CMCC (B), /0041), and *Staphylococcus aureus* (ATCC 25923) (Li et al., 2014).
 Commentary: Essential oil of leaves comprises mainly 1,8-cineole and α-terpineol (Li et al., 2014).

Safrole

**References**

Li, L., Li, Z.W., Yin, Z.Q., Wei, Q., Jia, R.Y., Zhou, L.J., Xu, J., Song, X., Zhou, Y., Du, Y.H. and Peng, L.C., 2014. Antibacterial activity of leaf essential oil and its constituents from *Cinnamomum longepaniculatum*. *International Journal of Clinical and Experimental Medicine*, 7(7), p. 1721.

Li, L., Shi, C., Yin, Z., Jia, R., Peng, L., Kang, S. and Li, Z., 2014. Antibacterial activity of α-terpineol may induce morphostructural alterations in *Escherichia coli*. *Brazilian Journal of Microbiology*, 45(4), pp. 1409–1413.

### 2.2.2.6 *Cinnamomum osmophloeum* Kaneh.

Common name: Tu rou gui (China)

Habitat: Forests

Distribution: China and Taiwan

Botanical observation: It is a treelet with fragrant bark. The leaves are simple, alternate, sub-opposite, and exstipulate. The petiole is about 1 cm long. The blade is ovate-lanceolate, 8–12 × 2.5–5.5 cm, somewhat coriaceous, glaucous below, with 1–3 pair of secondary nerves, obtuse at base, and acuminate at apex. The inflorescence is a lax panicle. The perianth is minute, campanulate, hairy, and 6-lobed. The androecium includes 9 stamens. The ovary is ovoid and develops a short style and a discoid stigma. The drupes are ovoid, about 1 cm in diameter, and glossy on a perianth cup.

Medicinal uses: Apparently none for microbial infections.

*Moderate antibacterial essential oil:* Essential oil of leaves completely inhibited *Legionella pneumophila* serogroup 1 at 1000 μg/mL (Chang et al., 2008).

*Weak broad-spectrum antibacterial phenylpropanoids:* Cinnamaldehyde from the leaves inhibited the growth of *Escherichia coli*, *Pseudomonas aeruginosa*, *Enterococcus fecalis*, and *Staphylococcus aureus* with MIC of 500, 1000, 250, and 250 μg/mL, respectively (Chang et al., 2001). Cinnamaldehyde was also active against *Streptococcus epidermidis*, methicillin-resistant *Staphylococcus aureus*, *Klebsiella pneumonia*, *Salmonella* sp., and *Vibrio parahemolyticus* with MIC values of 250, 250, 1000, 500, and 250 μg/mL, respectively (Chang et al., 2001). From the same leaves eugenol was only active towards *Escherichia coli* at 1000 μg/mL whereas MIC was superior to 1000 μg/mL for other bacteria (Chang et al., 2001).

*Strong antifungal (filamentous) amphiphilic phenylpropanoids:* Cinnamaldehyde (LogD = 1.7 at pH 7.4; molecular weight = 132.1 g/mol) and eugenol (LogD = 2.4 at pH 7.4; molecular weight = 164.2 g/mol) totally inhibited the growth of *Laetiporus sulphureus* at 100 ppm (Wang et al., 2005). At 100 ppm, eugenol inhibited the growth of *Coriolus versicolor* by 71.1% and cinnamaldehyde by 100% (Wang et al., 2005).

Eugenol

Commentaries: Essential oil of leaves comprises mainly cinnamaldehyde, 1,8-cineole, and neral (Chang et al., 2001). Note that essential oil constituents within plants in the same species vary with the location and time of the year the sample was collected. Consider that essential oils are often anticandidal *in vitro*. As for filamentous fungi, strong inhibition *in vitro* is observed with amphiphilic medium molecular weight phenolics, including lignans and phenylpropanoids. The mode of antibacterial action of phenylpropanoids from essential oils includes, at least partially, membrane insults (Hossan et al., 2018).

## References

Chang, S.T., Chen, P.F. and Chang, S.C., 2001. Antibacterial activity of leaf essential oils and their constituents from *Cinnamomum osmophloeum*. *Journal of Ethnopharmacology*, 77(1), pp. 123–127.

Chang, C.W., Chang, W.L., Chang, S.T. and Cheng, S.S., 2008. Antibacterial activities of plant essential oils against Legionella pneumophila. *Water research*, 42(1–2), pp. 278–286.

Hossan, M.S., Jindal, H., Maisha, S., Samudi Raju, C., Devi Sekaran, S., Nissapatorn, V., Kaharudin, F., Su Yi, L., Khoo, T.J., Rahmatullah, M. and Wiart, C., 2018. Antibacterial effects of 18 medicinal plants used by the Khyang tribe in Bangladesh. *Pharmaceutical biology*, 56(1), pp. 201–208.

Wang, S.Y., Chen, P.F. and Chang, S.T., 2005. Antifungal activities of essential oils and their constituents from indigenous cinnamon (*Cinnamomum osmophloeum*) leaves against wood decay fungi. *Bioresource Technology*, 96(7), pp. 813–818.

## 2.2.2.7 *Cinnamomum tamala* (Buch.-Ham.) T. Nees & Nees

Synonyms: *Laurus cassia* Nees & T. Nees; *Laurus tamala* Buch.-Ham.

Common names: Indian cassia lignea; tejpat (Bangladesh); chai gui (China); tejapatra (India)

Habitat: Forests, riverbanks

Distribution: India, Bhutan, Nepal, Bangladesh, and China

Botanical observation: It is a tree that grows to about 15 m tall. The bark is gray and somewhat aromatic. The stems are quadrangular at apex. The leaves are simple, alternate, sub-opposite, and exstipulate. The petiole is up to about 1.3 cm long. The blade is ovate-oblong or lanceolate, 7.5–15.5 × 3.5–5.5 cm, somewhat coriaceous, with 1 pair of secondary nerves, base acute, and apex acuminate. The inflorescence is an axillary or terminal panicle, which is about 10 cm long and bears minute flowers. The perianth is hairy, whitish, and develops 6 lobes. The androecium includes 9 stamens. The ovary is minute, ovoid, and develop a short style. The drupes are ellipsoid, about 1 cm long, and glossy on perianth cups.

Medicinal uses: Diarrhea, chicken pox, intestinal worms, and tuberculosis (Bangladesh)

*Weak antibacterial (Gram-positive) polar extract:* Methanol extract of leaves inhibited the growth of multidrug-resistant strains of *Staphylococcus aureus* with MIC/MBC values of 1500/3400 µg/mL (Rath & Padhy, 2014).

*Antibacterial (Gram-positive) essential oil:* Essential oil of partially dried leaves inhibited the growth of methicillin-resistant *Staphylococcus aureus* (ATCC 33591) with an MIC value of 4% (v/v) (Rubini et al., 2018).

*Moderate broad-spectrum antifungal essential oil:* Essential oil of dried leaves inhibited the growth of *Candida albicans* (ATCC-90028), *Candida glabrata* (MTCC 6507), and *Candida tropicalis* (MTCC 310) with the MIC values of 0.6, 0.4, and 0.6% (v/v) (Banu et al., 2018). Essential oil of leaves inhibited *Aspergillus niger*, *Aspergillus fumigatus*, *Rhizopus stolonifera*, *Penicillium* sp., and *Candida albicans* (Banu et al., 2018).

Commentary: The essential oil of leaves contain β-caryophyllene, linalool, caryophyllene oxide (Ahmed et al., 2000), cinnamyl acetate, eugenol, and methyl eugenol (Kumar et al., 2012). Eugenol and methyl eugenol have the ability to penetrate the membranes of bacteria and to destabilize them (Saha & Verma, 2018). Consider that extracting fresh leaves, barks, roots, or fruits will result in the extraction of glycosides because plant parts that are removed and let to dry release enzymes that cleave glycosides to release aglycones in an attempt to fight microbes, insects, and other predators.

## References

Ahmed, A., Choudhary, M.I., Farooq, A., Demirci, B., Demirci, F. and Can Başer, K.H., 2000. Essential oil constituents of the spice *Cinnamomum tamala* (Ham.) Nees & Eberm. *Flavour and Fragrance Journal*, 15(6), pp. 388–390.

Banu, S.F., Rubini, D., Shanmugavelan, P., Murugan, R., Gowrishankar, S., Pandian, S.K. and Nithyanand, P., 2018. Effects of patchouli and cinnamon essential oils on biofilm and hyphae formation by Candida species. *Journal de mycologie medicale, 28*(2), pp. 332–339.

Kumar, S., Sharma, S. and Vasudeva, N., 2012. Chemical compositions of *Cinnamomum tamala* oil from two different regions of India. *Asian Pacific Journal of Tropical Disease, 2*, pp. S761–S764.

Rath, S. and Padhy, R.N., 2014. Monitoring in vitro antibacterial efficacy of 26 Indian spices against multidrug resistant urinary tract infecting bacteria. *Integrative Medicine Research, 3*(3), pp. 133–141.

Rubini, D., Banu, S.F., Nisha, P., Murugan, R., Thamotharan, S., Percino, M.J., Subramani, P. and Nithyanand, P., 2018. Essential oils from unexplored aromatic plants quench biofilm formation and virulence of methicillin resistant *Staphylococcus aureus. Microbial Pathogenesis*.

Saha, S. and Verma, R.J., 2018. Molecular interactions of active constituents of essential oils in zwitterionic lipid bilayers. *Chemistry and Physics of Lipids, 213*, pp. 76–87.

### 2.2.2.8 *Cinnamomum kotoense* Kaneh. & Sasaki

Common name: lan yu rou gui (China)

Habitat: Forests

Distribution: Taiwan

Botanical observation: It is a tree that grows to about 15 m tall. The leaves are simple, alternate, sub-opposite, and exstipulate. The petiole is reddish, up to about 1.5 cm long. The blade is ovate-oblong, 7.5–11.5 × 3.5–5.5 cm, coriaceous, with 1 pair of secondary nerves, rounded at base and acute at apex. The inflorescence is 5 cm long panicle. The drupes are ovoid, about 1.4 cm long, and glossy on perianth cups.

Medicinal uses: Apparently none

*Strong Antimycobacterial amphiphilic long chain butyrolactones:* Kotolactone A and lincomolide B (LogD = 3.7 at pH 7.4; molecular weight = 278.3 g/mol) from stem wood inhibited the growth of *Mycobacterium tuberculosis* (90–221387) with the MIC values of 40 and 10.1 µM, respectively (Chen et al., 2005). Isoobtusilactone A from stem wood inhibited *Mycobacterium tuberculosis* (90–221387) with the MIC value of 22.4 µM (Chen et al., 2005).

Lincomolide B

*Strong antimycobacterial fatty acids fraction:* A mixture of palmitic acid, margaric acid, and stearic acid inhibited the growth of *Mycobacterium tuberculosis* (90–221387) with the MIC value of 25 μM (Chen et al., 2005).

Commentaries: (i) When column chromatographies are made with large amounts of extract, minor constituents are isolated with rare chemical structures, and these can be of symbiotic bacterial, fungal, or lichen origin. Such compounds are often antimicrobials. Some research teams isolate and grow microorganisms on rainforest plants in order to extract antibiotics peptides and other substances, and all of these are not covered here. (ii) The intrinsic resistance of mycobacteria is due, at least in part, to the lipophilicity of the cell wall (Liu et al., 1996). The cell wall of mycobacteria contains an arabinogalactan linked to the wall peptidoglycan and esterified with mycolic acids (mycolic acids are long chain β-hydroxyl α-branched fatty acids) as well as trehalose-containing glycolipids, phenolic glycolipids, or glycopeptidolipid (Liu et al., 1996). Fatty acids, fatty alcohols and liposoluble and amphiphilic compounds penetrate the cell wall through its lipid domain and are able to elicit antimycobacterial effects. It can be observed that lipophilic natural products from medicinal plants are not uncommonly antimycobacterial.

### References

Chen, F.C., Peng, C.F., Tsai, I.L. and Chen, I.S., 2005. Antitubercular constituents from the stem wood of *Cinnamomum kotoense*. *Journal of Natural Products, 68*(9), pp. 1318–1323.

Liu, J., Barry, C.E., Besra, G.S. and Nikaido, H., 1996. Mycolic acid structure determines the fluidity of the mycobacterial cell wall. *Journal of Biological Chemistry, 271*(47), pp. 29545–29551.

### 2.2.2.9 *Cinnamomum zeylanicum* Blume

Synonym: *Cinnamomum verum* J. Presl.; *Laurus cinnamomum* L.

Common names: Ceylon cinnamon, true cinnamon; chek tum phka loeng (Cambodia); xi lan rou gui (China); ilayangam, tamalapatra (India); kayu manis (Indonesia, Malaysia); thit gya boe (Myanmar); kanela (the Philippines); que hoi (Vietnam)

Habitat: Cultivated

Distribution: India, Sri Lanka, Myanmar, and Indonesia

Botanical observation: It is a tree that grows to about 10 m tall. The bark is dark brown and aromatic. The stems are quadrangular at apex. The leaves are simple, sub-opposite, and exstipulate. The petiole is up to about 2 cm long. The blade is ovate-lanceolate, glossy, 10.5–16.5 × 4.5–5.5 cm, coriaceous, with 1 pair of secondary nerves, acute at base, and acuminate at apex. The panicles are axillary or terminal, up to about 20 cm long lax, yellowish green, and carrying minute flowers. The perianth is yellow, hairy, and develops 6 lobes. The androecium includes 9 stamens. The ovary is minute, ovoid, and develop a short style and a discoid stigma. The drupes are ovoid, up to about 1.5 cm long, black, and glossy on perianth cups.

Medicinal uses: Fever, diarrhea, small pox (Myanmar)

*Moderate broad-spectrum antibacterial essential oil:* Essential oil of bark inhibited the growth of *Staphylococcus aureus* (ATCC 29213), *Streptococcus pyogenes* (ATCC 19615), *Enterococcus faecalis* (ATCC 29212), *Enterococcus faecium* (ATCC 6057), *Bacillus cereus* (ATCC 11778), *Acinetobacter lwoffii* (ATCC 19002), *Enterobacter aerogenes* (ATCC 13043), *Escherichia coli* (ATCC 25922), *Klebsiella pneumoniae* (ATCC 13883), *Proteus mirabilis* (ATCC 7002), *Pseudomonas aeruginosa* (ATCC 27853), *Salmonella typhimurium* (ATCC 14028), *Clostridium perfringens* (KUKENS-Turkey), *Listeria monocytogenes* (F 1483), and *Listeria ivanovii* (F 4084) with MIC values Ranging from 100 to 1100 μg/mL (Unlu et al., 2010). The essential oil inhibited *Streptococcus pneumoniae* (ATCC 49619) and *Acinetobacter lwoffii* (ATCC 19002) with MIC below 40 μg/mL (Unlu et al., 2010). Essential oil inhibited the growth of *Escherichia coli, Staphylococcus aureus, Streptococcus mutans*, and *Lactophilus acidophilus* with MIC values of 0.8, 1.0, 0.3, and 1.4 μL/mL, respectively (Miller et al., 2015).

*Strong antimycobacterial essential oil:* Essential oil of bark inhibited *Mycobacterium smegmatis* (CMM 2067) with an MIC value of 70 μg/mL (Unlu et al., 2010).

*Strong antifungal (filamentous) polar extract:* Methanol extract of bark inhibited dermatophytes *Trichophyton interdigitale* (ATCC 200099), *Trichophyton mentagrophytes* (ATCC 9533), *Microsporium canis* (ATCC 32903), *Microsporium gypseum* (ATCC 14683), and *Trichophyton mentagrophytes* (MYA-4439) with MIC/MFC values of 0.1/0.1, 0.07/0.1, 0.06/0.06, 0.06/1.3, and 0.06/1.6 mg/mL, respectively (Ayatollahi Mousavi & Kazemi, 2015).

*Strong anticandidal essential oil:* Essential oil of bark inhibited the growth of *Candida albicans* (ATCC 10231), *Candida albicans* (ATCC 90028), *Candida parapsilosis* (ATCC 90018), and *Candida krusei* (ATCC 6258) with MIC values of 0.07, 1.1, <0.04, and <0.04 mg/mL, respectively (Unlu et al., 2010).

*Moderate broad-spectrum antibacterial phenylpropanoid:* Cinnamaldehyde inhibited the growth of *Staphylococcus aureus, Bacillus cereus, Escherichia coli, Proteus mirabilis, Klebsiella pneumoniae,* and *Pseudomonas aeruginosa* with MIC values of 62.5, 31.2, 62.5, 125, 62.5, and 125 μg/mL, respectively (Al-Bayati & Mohammed, 2009).

Cinnamaldehyde

*Antibiotic potentiator phenylpropanoid:* 20 μg/mL of cinnamaldehyde decreased the MIC of clindamycin for *Clostridium difficile* 16-fold, from 4.0 to 0.25 μg/mL (Shahverdi et al., 2007).

*Moderate antiviral (enveloped monopartite linear single-stranded (+) RNA virus) phenylpropanoid:* Cinnamaldehyde at 80 μM inhibited the replication of the influenza PR8 virus in Madin-Darby canine kidney cells when given 3 hours after viral loading as evidenced by a decrease of about 80% after 10 hour of viral yield (Hayashi et al., 2007). Given intranasally (250 μg/mouse/day) prolonged the lifespan of mice with virus-induced pneumonia compared to untreated animals (Hayashi et al., 2007). This treatment decreased virus yield by almost 1 log in bronchoalveolar lavage fluid of treated mice (Hayashi et al., 2007). At the concentration of 315 μg/mL, cinnamaldehyde inhibited the replication of Human adenovirus type 3 (ADV3) by 58.6% (Liu et al., 2009).

Commentary: The lucrative trade of cinnamon bark accounted to the ruthless colonization of Sri Lanka by Europeans (Schrikker, A., 2007) and islamization of the Malay Archipelago (Hidayat & Hidayturrahman, 2019). Medieval European doctors recommended the Theriaque of Galen (which included Cinnamon) to fight plagues (Fabbri, 2007). One can read in the Antidotarium Nicolai (thirteen century):

"Triaca mangna [sic] la quale puose Galieni e detta 'donna dele medecine'. Vale molto alle gravissime infermitar del corpo humano, agli epilentici e catalentici e poplentici e scotomaci, cefalargici, e all magrana fa grandissimo prode; alla fio caggine dela boce ed al constringnimento del petto; e ottima ali artetici a agli asmatici emottoici, iterici, idropici, epiplemonicis. Recipe trociscorum sciliti corum lb. ii; piperis longi lb. i et s.; trociscorum tiri, trociscorum diacoralli, ana lb. i; xilobalsimo on. vii; oppei tebauci, agarici, oppi, rosarum, yrei, scordeon, seminis rape, salvie, cinamomi, opopobalsami, ana on" (Fabrri, 2007).

## References

Al-Bayati, F.A. and Mohammed, M.J., 2009. Isolation, identification, and purification of cinnamaldehyde from *Cinnamomum zeylanicum* bark oil. An antibacterial study. *Pharmaceutical Biology, 47*(1), pp. 61–66.

Ayatollahi Mousavi, S.A. and Kazemi, A., 2015. In vitro and in vivo antidermatophytic activities of some Iranian medicinal plants. *Medical Mycology, 53*(8), pp. 852–859.

Fabbri, C.N., 2007. Treating medieval plague: the wonderful virtues of theriac. *Early science and medicine, 12*(3), pp. 247–283.

Hidayat, A.T. and Hidayturrahman, M., 2019, February. Spice Route and Islamization on the West Coast of Sumatra in 17th-18th Century. In *2nd International Conference on Culture and Language in Southeast Asia (ICCLAS 2018)* (pp. 48–50). Atlantis Press.

Hayashi, K., Imanishi, N., Kashiwayama, Y., Kawano, A., Terasawa, K., Shimada, Y. and Ochiai, H., 2007. Inhibitory effect of cinnamaldehyde, derived from *Cinnamomi cortex*, on the growth of influenza A/PR/8 virus in vitro and in vivo. *Antiviral Research, 74*(1), pp. 1–8.

Liu, L., Wei, F.X., Qu, Z.Y., Wang, S.Q., Chen, G., Gao, H., Zhang, H.Y., Shang, L., Yuan, X.H. and Wang, Y.C., 2009. The antiadenovirus activities of cinnamaldehyde in vitro. *Laboratory Medicine, 40*(11), pp. 669–674.

Miller, A.B., Cates, R.G., Lawrence, M., Soria, J.A.F., Espinoza, L.V., Martinez, J.V. and Arbizú, D.A., 2015. The antibacterial and antifungal activity of essential oils extracted from Guatemalan medicinal plants. *Pharmaceutical Biology, 53*(4), pp. 548–554.

Shahverdi, A.R., Monsef-Esfahani, H.R., Tavasoli, F., Zaheri, A. and Mirjani, R., 2007. Trans-cinnamaldehyde from *Cinnamomum zeylanicum* bark essential oil reduces the clindamycin resistance of *Clostridium difficile* in vitro. *Journal of Food Science, 72*(1), pp. S055–S058.

Schrikker, A., 2007. *Dutch and British colonial intervention in Sri Lanka, 1780-1815: expansion and reform* (Vol. 7). Brill, Leiden.

Unlu, M., Ergene, E., Unlu, G.V., Zeytinoglu, H.S. and Vural, N., 2010. Composition, antimicrobial activity and in vitro cytotoxicity of essential oil from *Cinnamomum zeylanicum* Blume (Lauraceae). *Food and Chemical Toxicology, 48*(11), pp. 3274–3280.

### 2.2.2.10 *Cryptocarya chinensis* (Hance) Hemsl.

Synonym: *Beilschmiedia chinensis* Hance

Common name: Hou ke gui (China)

Habitat: Forests

Distribution: China and Taiwan

Botanical observation: It is a tree that grows to about 15 m tall. The stems are terete and hairy at apex. The leaves are simple, alternate or opposite, and exstipulate. The petiole is about 1 cm long. The blade is glossy, elliptic, 6.5–11.5 × 3.5–5.5 cm, coriaceous, with 1–3 pairs of secondary nerves, cuneate at base, and acuminate at apex. The panicles are axillary, terminal about 5 cm long and bear minute yellowish flowers. The perianth is tubular and develops 6 lobes. The androecium comprises 9 stamens. The ovary is clavate and develops a slender style. The fruit is globose, angled, about 1 cm long, and purplish-black.

Medicinal use: Apparently none for microbial infections.

*Antibacterial (Gram-positive) halo developed by flavanone:* Pinocembrin inhibited the growth of *Bacillus cereus* and *Staphylococcus aureus* (Fukui et al., 1988).

*Strong antimycobacterial amphiphilic flavanone:* Pinocembrin (LogD = 2.9 at pH 7.4; molecular weight = 256.2 g/mol) isolated from the leaves inhibited the growth of *Mycobacterium tuberculosis* with the MIC value of 3.5 μM (ethambutol 6.2 μM) (Chou et al., 2011).

Pinocembrin

*Moderate antimycobacterial flavanone:* The dihydrochalcone cryptocaryone from the leaves inhibited the growth of *Mycobacterium tuberculosis* with an MIC value of 25 μM (ethambutol 6.2) (Chou et al., 2011).

*Anticandidal halo developed by flavanone:* Pinocembrin of inhibited the growth of *Candida albicans* with an minimum inhibiting mass of 32 μg/discs (Fukui et al., 1988)

*Strong antiviral (enveloped monopartite linear dimeric single stranded (+) RNA) phenanthroindolizidine:* Dehydroantofine isolated from the wood inhibited the the proliferation of the Human immunodeficiency virus type-1 in H9 cells with the $EC_{50}$ value of 1.8 μg/mL (Wu et al., 2012).

Dehydroantofine

*Strong antiviral (enveloped monopartite linear single-stranded (+) RNA) flavanone:* Pinocembrin inhibited the replication of the Zika virus (PRVABC59 Puerto Rico strain) with an $IC_{50}$ value of 17.4 μM via RNA synthesis inhibition(Le Lee et al., 2019).

*Strong antiviral (enveloped monopartite linear single-stranded (+) RNA) flavanone:* Pinocembrin at a concentration of 39 μM evoked a 0.71 $\log_{10}$ decrease in Dengue virus type-2 (New Guinea C strain) in Huh7 cells and at a concentration of 21.5 μM induced a 1.05 $\log_{10}$ decrease Chikungunya virus (strain SGEHICHD122508) in HeLa cells (Le Lee et al., 2019).

Commentary: (i) The Zika virus from the genus Flavivirus in the family Flaviviridae is a zoonotic single-stranded RNA-enveloped virus that originates from rhesus macaque in the Zika forest of Uganda (Dick et al., 1952). Zika virus infections remained sporadic until 2007, when a massive outbreak was reported in Micronesia and nearly 75% of its population was infected (Duffy et al., 2009). Subsequently, another major Zika virus epidemic occurred from 2013 to 2014 in French Polynesia (Song et al., 2017). From 2014 to 2016, the virus spread in Central and South America and the Caribbean, Pacific Islands, and Southeast Asia where Guillain-Barré syndrome, meningoencephalitis, and microcephaly cases forced the World Health Organization (WHO) to declare Zika virus a global health emergency in 2016 (Le Lee et al., 2019). To date no vaccine or drug exist to prevent or cure this dangerous zoonosis. Zoonosis will continue to emerge by waves as a consequence of industrial pressure on forests (cutting primary rainforest timbers), seas, and animals (poultries). Global warming and, as direct consequence, melting of polar ice caps pose the nightmarish threat of releasing prehistoric viruses from melted ice (Smith et al., 2004). COVID-19 is one of the first of these waves and if the environment is not respected we must expect much more aggressive zoonosis in the next 50 years. In brief, Nature has her finger on the trigger. If humans carry on with their abuses on nature (which is in all logic the most probable scenario), they will go extinct by a completely untreatable and fast-killing bullet virus, bacteria, or fungi. Dinausors may have been extinct owed to a global pandemia. Humans are next.

(ii) Natural products able to strongly inhibit the replication of the Zika virus *in vitro* from the plants belonging to the Clades in this volume are amphiphilic medium molecular weight flavonoids, lignans, and isoquinolines.

### References

Chou, T.H., Chen, J.J., Peng, C.F., Cheng, M.J. and Chen, I.S., 2011. New flavanones from the leaves of *Cryptocarya chinensis* and their antituberculosis activity. *Chemistry & Biodiversity*, 8(11), pp. 2015–2024.

Dick, G.W.A., Kitchen, S.F. and Haddow, A.J., 1952. Zika virus (I). Isolations and serological specificity. *Transactions of the Royal Society of Tropical Medicine and Hygiene*, 46(5), pp. 509–520.

Duffy, M.R., Chen, T.H., Hancock, W.T., Powers, A.M., Kool, J.L., Lanciotti, R.S., Pretrick, M., Marfel, M., Holzbauer, S., Dubray, C. and Guillaumot, L., 2009. Zika virus outbreak on Yap Island, federated states of Micronesia. *New England Journal of Medicine*, 360(24), pp. 2536–2543.

Fukui, H., Goto, K. and Tabata, M., 1988. Two antimicrobial flavanones from the leaves of *Glycyrrhiza glabra*. *Chemical and Pharmaceutical Bulletin*, 36(10), pp. 4174–4176.

Le Lee, J., Loe, M.W.C., Lee, R.C.H. and Chu, J.J.H., 2019. Antiviral activity of pinocembrin against Zika virus replication. *Antiviral Research, 167*, pp. 13–24.

Smith, A.W., Skilling, D.E., Castello, J.D. and Rogers, S.O., 2004. Ice as a reservoir for pathogenic human viruses: specifically, caliciviruses, influenza viruses, and enteroviruses. *Medical Hypotheses, 63*(4), pp. 560–566.

Song, B.H., Yun, S.I., Woolley, M. and Lee, Y.M., 2017. Zika virus: history, epidemiology, transmission, and clinical presentation. *Journal of Neuroimmunology, 308*, pp. 50–64.

Wu, T.S., Su, C.R. and Lee, K.H., 2012. Cytotoxic and anti-HIV phenanthroindolizidine alkaloids from Cryptocarya chinensis. *Natural Product Communications, 7*(6), p. 725.

### 2.2.2.11 *Laurus nobilis* L.

Common names: Bay leaves, laurel; yue gui (China); paminta dahon (the Philippines); defne (Turkey)
    Habitat: Cultivated
    Distribution: Turkey and Iran
    Botanical observation: It is a treelet (common in Southern Europe), which grows to 5 m tall with a smooth dark brown bark. The leaves are simple, alternate, and exstipulate. The petiole is about

1 cm long. The blade is elliptic, glossy, coriaceous, aromatic, 5.5–13 × 1.8–3.2 cm, coriaceous, with 10–12 pairs of secondary nerves, acute at the base wavy, apex acute. The inflorescences are axillary racemes of numerous minute flowers. The perianth is minute, yellowish, and develops 4 lobes. The androecium includes 12 stamens. The ovary is ovoid with a short style and a dilated stigma. The berries are glossy, green, and about 1 cm long.

Medicinal uses: Antiseptic (Turkey)

*Broad-spectrum antibacterial essential oil:* Essential oil of leaves inhibited the growth of *Staphylococcus aureus* (CEC 976), *Listeria monocytogenes* (EGD-e), *Enterococcus faecium* (CEC 4932), and *Bacillus subtilis* (CECT 4071) with the MIC/MBC values of 4/21, 0.5/14, 14/28, and 4/14 µL/L, respectively (Cherrat et al., 2014). This essential oil inhibited the growth of *Escherichia coli* (O157:H7) and *Yersinia enterolitica* (CECT 4315) with MIC/MBC values of 4/21 and 8/14 µL/L, respectively (Cherrat et al., 2014). Essential oil of leaves (containing mainly 1,8-cineole) inhibited the growth of *Escherichia coli* O157:H7, *Listeria monocytogenes*, *Salmonella typhimurium*, and *Staphylococcus aureus* (Dadalioğlu & Evrendilek, 2004).

*Moderate anticandidal essential oil:* Essential of leaves inhibited the growth of *Candida albicans* (CBS 562), *Candida albicans* (ATCC 60193), *Candida krusei* (CBS 573), *Candida krusei* (ATCC 3413), *Candida tropicalis* (CBS 94), *Candida tropicalis* (ATCC 750), and *Candida glabrata* (IZ 07) with the MIC/MFC values of 250/500, 250/250, 500/500, 500/500, 500/500, 250/250, and 500/500 µg/mL, respectively, by altering membrane permeability (Peixoto et al., 2017).

*Strong antibacterial (Gram-positive) flavonol glycoside:* Kaempferol-3-$O$-$\alpha$-L-(2″, 4″-$E$-p-coumaroyl)-rhamnoside isolated from the leaves inhibited the growth of methicillin-resistant *Staphylococcus aureus* strains OM481, OM505, OM584, OM623, COL, N315 and 209P, as well as *Enterococcus faecium* (FN-1) and *Enterococcus faecalis* (NCTC12201) with MIC values of 1, 1, 2, 2, 1, 1, 0.5, 8, and 4 µg/L, respectively (Otsuka et al., 2008). Kampferol-3-$O$-$\alpha$-L-(2″-$E$-p-coumaroyl-4″-$Z$-p-coumaroyl)-rhamnoside isolated from the leaves inhibited the growth of methicillin-resistant *Staphylococcus aureus* strains OM481, OM505, OM584, OM623, COL, N315, and 209P, as well as *Enterococcus faecium* (FN-1) and *Enterococcus faecalis* (NCTC12201) with the MIC values of 1, 2, 2, 2, 1, 1, 0.5, 8, and 4 µg/L, respectively (Otsuka et al., 2008).

*Antibiotic potentiator flavonol glycosides:* Kaempferol-3-$O$-$\alpha$-L-(2″, 4″-$E$-p-coumaroyl)-rhamnoside and kampferol-3-$O$-$\alpha$-L-(2″-$E$-p-coumaroyl-4″-$Z$-p-coumaroyl)-rhamnoside from the leaves inhibited the growth of methicillin-resistant *Staphylococcus aureus* and was synergistic with norfloxacin and ciprofloxacin (Liu et al., 2009).

*Antibacterial germacrane sesquiterpene lactone:* Deacetyl laurenobiolide inhibited the growth of bacteria (Fukuyama et al., 2011).

*Moderate antiviral (enveloped monopartite linear single-stranded (+) RNA) essential oil:* Essential oil inhibited the Severe acute respiratory syndrome-associated coronavirus with an $IC_{50}$ value of 120 µg/mL (Loizzo et al., 2008).

*Moderate antiviral (enveloped monopartite linear double-stranded DNA) essential oil:* Essential oil inhibited Herpes simplex virus type-1 with an $IC_{50}$ value of 120 µg/mL (Loizzo et al., 2008).

Commentaries: (i) Essential oil of leaves contains predominantly 1.8-cineole, α-terpinyl acetate, and sabinene (Sangun et al., 2007) and one could suggest that being active of both DNA and RNA viruses enveloped viruses, the essential oil might target the envelop. The essential oil inhibits Severe acute respiratory syndrome-associated coronavirus suggesting that it could be used to purify the atmosphere in public places. In this context, it worth mentioning that linen clothes had to be washed with special herbs of which *Laurus nobilis* L. to prevent *Yersinia pestis* plague in 17 century Italy. The *Reggimento contra peste* of the Spanish physician Pietro Castagno recommended at that time rubbing the body with an oil consisting of numerous ingredients of which *mirra* (*Commiphora myrrha* (*T. Nees*) Engl.), *reubarbaro* (*Rheum* spp.), *zedoaria* (zedoary, *Curcuma zedoaria* (Christm.) Roscoe), *croco* (*Crocus sativus* L.), and *aristologra* (*Aristolochia* spp.) (Vicentini et al., 2020).

## References

Cherrat, L., Espina, L., Bakkali, M., García-Gonzalo, D., Pagán, R. and Laglaoui, A., 2014. Chemical composition and antioxidant properties of *Laurus nobilis* L. and *Myrtus communis* L. essential oils from Morocco and evaluation of their antimicrobial activity acting alone or in combined processes for food preservation. *Journal of the Science of Food and Agriculture, 94*(6), pp. 1197–1204.

Dadalioğlu, I. and Evrendilek, G.A., 2004. Chemical compositions and antibacterial effects of essential oils of Turkish oregano (*Origanum minutiflorum*), bay laurel (*Laurus nobilis*), Spanish lavender (*Lavandula stoechas* L.), and fennel (*Foeniculum vulgare*) on common foodborne pathogens. *Journal of Agricultural and Food Chemistry, 52*(26), pp. 8255–8260.

Fukuyama, N., Ino, C., Suzuki, Y., Kobayashi, N., Hamamoto, H., Sekimizu, K. and Orihara, Y., 2011. Antimicrobial sesquiterpenoids from *Laurus nobilis* L. *Natural Product Research, 25*(14), pp. 1295–1303.

Liu, M.H., Otsuka, N., Noyori, K., Shiota, S., Ogawa, W., Kuroda, T., Hatano, T. and Tsuchiya, T., 2009. Synergistic effect of kaempferol glycosides purified from *Laurus nobilis* and fluoroquinolones on methicillin-resistant *Staphylococcus aureus*. *Biological and Pharmaceutical Bulletin, 32*(3), pp. 489–492.

Loizzo, M.R., Saab, A.M., Tundis, R., Statti, G.A., Menichini, F., Lampronti, I., Gambari, R., Cinatl, J. and Doerr, H.W., 2008. Phytochemical analysis and in vitro antiviral activities of the essential oils of seven Lebanon species. *Chemistry & Biodiversity, 5*(3), pp. 461–470.

Otsuka, N., Liu, M.H., Shiota, S., Ogawa, W., Kuroda, T., Hatano, T. and Tsuchiya, T., 2008. Anti-methicillin resistant *Staphylococcus aureus* (MRSA) compounds isolated from *Laurus nobilis*. *Biological and Pharmaceutical Bulletin, 31*(9), pp. 1794–1797.

Peixoto, L.R., Rosalen, P.L., Ferreira, G.L.S., Freires, I.A., de Carvalho, F.G., Castellano, L.R. and de Castro, R.D., 2017. Antifungal activity, mode of action and anti-biofilm effects of *Laurus nobilis* Linnaeus essential oil against Candida spp. *Archives of Oral Biology, 73*, pp. 179–185.

Sangun, M.K., Aydin, E., Timur, M., Karadeniz, H., Caliskan, M. and Ozkan, A., 2007. Comparison of chemical composition of the essential oil of *Laurus nobilis* L. leaves and fruits from different regions of Hatay, Turkey. *Journal of Environmental Biology, 28*(4), pp. 731–733.

Vicentini, C.B., Manfredini, S., Mares, D., Bonacci, T., Scapoli, C., Chicca, M. and Pezzi, M., 2020. Empirical "integrated disease management" in Ferrara during the Italian plague (1629–1631). *Parasitology international, 75*, p.102046.

### 2.2.2.12 *Lindera akoensis* Hayata

Synonym: *Benzoin akoense* (Hayata) Kamik.
Common names: Taiwan spicebush; tai wan xiang ye shu (China)
Habitat: Thickets
Distribution: Taiwan
Botanical observation: It is a shrub with a dark brown bark. The stems are hairy at apex. The leaves are simple, alternate, and exstipulate. The petiole is about 1.2 cm long and hairy. The blade is ovate, 3–5 × 2–3.5 cm, coriaceous, hairy, with 4 or 5 pairs of secondary nerves, cuneate at base, and acute at apex. The inflorescences are axillary fascicles of 5 or 6 flowers. The perianth is minute and includes 6 whitish to membranous tepals. The androecium includes 9 stamens. The ovary is ovate, minute, and develops a disciform stigma. The drupes are ovoid, red, glossy, and about 5 mm long.
Medicinal uses: Apparently none for microbial infections.
*Strong antimycobacterial lipophilic long chain butyrolactones:* Litseakolide A (molecular weight = 294.4 g/mol), litsenolide B2, and litsenolide C2 from the roots inhibited the growth of *Mycobacterium tuberculosis* with $IC_{50}$ values of 15, 20, and 20 μM, respectively (ethambutol 6.2 μM) (Chang et al., 2008).

Litseakolide A

### Reference

Chang, S.Y., Cheng, M.J., Peng, C.F., Chang, H.S. and Chen, I.S., 2008. Antimycobacterial butanolides from the root of *Lindera akoensis*. *Chemistry & Biodiversity,* 5(12), pp. 2690–2698.

### 2.2.2.13 *Lindera erythrocarpa* Makino

Synonyms: *Benzoin erythrocarpum* (Makino) Rehder; *Lindera erythrocarpa* var. *longipes* S.B. Liang; *Lindera funiushanensis* C.S. Zhu; *Lindera henanensis* H.B. Cui

Common names: Spice bush; hong guo shan hu jao (China); kanaguki-no-ki (Japan)

Habitat: Forests

Distribution: China, Taiwan, Japan, and Korea

Botanical observation: It is a shrub that grows to about 5 m tall. The bark is gray-brown. The stems are rough. The leaves are simple, spiral, and exstipulate. The petiole is about 1 cm long. The blade is oblanceolate, 9–12 × 4–6 cm, with 4 or 5 pairs of secondary nerves, decurrent at base, and acuminate at apex. The inflorescences are axillary, many flowered fascicles. The perianth includes 6 yellowish green, elliptic tepals. The androecium includes 9 stamens. The ovary is ellipsoid with a thick style and discoid stigma. The drupes are globose, about 1 cm in diameter, red, and and glossy.

Medicinal use: Apparently none for microbial infections.

*Fungal enzyme inhibition by cyclopenta-2,4-dienones:* Methyllinderone and linderone isolated from the stems inhibited chitin synthase 2 with $IC_{50}$ values below 20 µg/mL, respectively (Hwang et al., 2007).

*Fungal enzyme inhibition chalcone:* Kanakugiol isolated from the stems inhibited chitin synthase 2 with an $IC_{50}$ value of 23.8 µg/mL (Hwang et al., 2007).

*Strong hydrophilic antiviral (enveloped monopartite linear single-stranded (+) RNA)) hydrophilic cyclopenta-2,4-dienone genus hepacivirus:* Lucidone (LogD = −1.1 at pH 7.4; molecular weight = 256.2 g/mol) reduced Hepatitis C virus RNA levels in Ava5 cells with an $EC_{50}$ of 15 μM (Chen et al., 2013). This compound activated Nrf-2, oxidase-1 expression, and subsequent increase of biliverdin from which inhibition of HCV NS3/4A protease and stimulation of interferon response (Chen et al., 2013).

*Viral enzyme inhibition by polar extract:* Methanol extract of leaves inhibited the Human immunodeficiency virus type-1 protease (Park, 2003).

Commentaries: (i) The Hepatitis C virus is an enveloped RNA virus belonging to the family Flaviviridae and genus *Hepacivirus*, which is transmitted from blood to blood between humans. Hepatitis C may favor the development cirrhosis, hepatic failure, or hepatocellular carcinoma (Pawlotsky, 2004; Chevaliez & Pawlotsky, 2006). The disease is treated by NS5B polymerase inhibitors such as sofosbuvir, NS5A inhibitors such as ledipasvir, NS3/4A protease inhibitors such as simeprevir as well as the inosine-5-monophosphate dehydrogenase inhibitor ribavirin (Konerman & Lok 2016). (ii) Chitin is a polysaccharide that consists of unbranched chains of β-(1,4)-linked 2-acetamido-2-deoxy-D-glucose units, which is the major constituent of the wall of filamentous fungi (Farkas, 1979). The synthesis of chitin requires chitin synthetase, which is a target for antifungal agents (Georgopapadakou, 1998). (iii) Ethanol extract of roots of *Lindera aggregata* (Sims) Kosterm. (local name: *wu yao* in China) inhibited the replication of Severe acute respiratory syndrome-associated coronavirus BJ-001 and BJ-006 with $IC_{50}$ values of 88.2 and 80.6 μg/mL (Li et al., 2005). (iv) Methanol extract of leaves and stems of *Lindera obtusifolia* Bl. at a concentration of 100 μg/mL inhibited RNA-dependent DNA polymerase activity and reverse transcriptase of the Human immunodeficiency virus by 85 and 27.1%, respectively (Min et al., 2001). (v) Natural products from medicinal plants able to strongly inhibit the replication of the Hepatitis C virus decribed in this volume are often amphiphilic, with medium molecular weight, and are structurally variable.

### References

Chen, W.C., Wang, S.Y., Chiu, C.C., Tseng, C.K., Lin, C.K., Wang, H.C. and Lee, J.C., 2013. Lucidone suppresses hepatitis C virus replication by Nrf2-mediated heme oxygenase-1 induction. *Antimicrobial Agents and Chemotherapy, 57*(3), pp. 1180–1191.

Chevaliez, S. and Pawlotsky, J.M., 2006. Hepatitis C viruses genome and life cycle. *Hepatitis C viruses: genomes and molecular biology.*

Farkas, V., 1979. Biosynthesis of cell walls of fungi. *Microbiological Reviews, 43*(2), p. 117.

Georgopapadakou, N.H., 1998. Antifungals: mechanism of action and resistance, established and novel drugs. *Current Opinion in Microbiology, 1*(5), pp. 547–557.

Hwang, E.I., Lee, Y.M., Lee, S.M., Yeo, W.H., Moon, J.S., Kang, T.H., Park, K.D. and Kim, S.U., 2007. Inhibition of chitin synthase 2 and antifungal activity of lignans from the stem bark of *Lindera erythrocarpa*. *Planta Medica, 73*(07), pp. 679–682.

Li, S.Y., Chen, C., Zhang, H.Q., Guo, H.Y., Wang, H., Wang, L., Zhang, X., Hua, S.N., Yu, J., Xiao, P.G. and Li, R.S., 2005. Identification of natural compounds with antiviral activities against SARS-associated coronavirus. *Antiviral Research, 67*(1), pp. 18–23.

Konerman, M.A. and Lok, A.S., 2016. Hepatitis C treatment and barriers to eradication. *Clinical and translational gastroenterology, 7*(9), p. e193.

Min, B.S., Kim, Y.H., Tomiyama, M., Nakamura, N., Miyashiro, H., Otake, T. and Hattori, M., 2001. Inhibitory effects of Korean plants on HIV-1 activities. *Phytotherapy Research, 15*(6), pp. 481–486.

Park, J.C., 2003. Inhibitory effects of Korean plant resources on human immunodeficiency virus type 1 protease activity. *Oriental Pharmacy and Experimental Medicine, 3*(1), pp. 1–7.

Pawlotsky, J.M., 2004. Pathophysiology of hepatitis C virus infection and related liver disease. *Trends in Microbiology, 12*(2), pp. 96–102.

### 2.2.2.14 *Litsea elliptica* Blume

Synonyms: *Litsea odorifera* Valeton; *Litsea petiolata* Hook.f.; Tetranthera elliptica (Blume) Nees

Common names: Trawas (Indonesia); medang perawas (Malaysia); batikuling-surutan (the Philippines)

Habitat: Forests

Distribution: Malaysia, Indonesia, and the Philippines

Botanical observation: It is a buttressed tree which grows to 30 m tall. The bark is greyish brown and the inner bark pinkish and aromatic. The leaves are simple, spiral, and exstipulate. The blade is elliptic, 6–16 × 2–7 cm, glaucous below, with 4–8 pairs of secondary nerves, cuneate at base, and hortly acuminate to obtuse at apex. The inflorescences are axillary fascicles. The drupes are dark purple, glossy, and seated on shallow cupules.

Medicinal uses: Post-partum (Malaysia)

*Antibacterial essential oil:* Essential oil of leaves (15 μL/6 mm disc) inhibited the growth of *Salmonella typhi*, *Clostridium pefringens*, and *Campylobacter jejuni* with the inhibition zone diameters of 11, 31.5, and 14 mm, respectively (Wannissorn et al., 2005).

*Strong broad-spectrum antifungal amphiphilic linear alkanone:* Undecan-2-one (LogD = 4.1 at pH 7.4; molecular mass = 170.2 g/mol) inhibited the growth of *Candida mycoderma* and *Aspergillus niger* with the MIC/MBC values of 20/30 and 1/20 μg/mL, respectively (Kunicka-Styczyńska and Gibka, 2010).

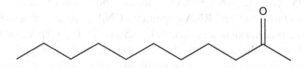

Undecan-2-one

Commentary: Essential oil of this plant contains mainly undecan-2-one, undec-10-en-2-one, and secondary alcohols, (-)-undecan-2-ol and undec-10-en-2-ol (Forney & Markovetz, 1971). Undecan-2-one evoked some levels of antibacterial activities (Gibka et al., 2009).

### References

Forney, F.W. and Markovetz, A.J., 1971. The biology of methyl ketones. *Journal of Lipid Research, 12*(4), pp. 383–395.

Gibka, J., Kunicka-Styczyńska, A. and Gliński, M., 2009. Antimicrobial activity of Undecan-2-one, Undecan-2-ol and their derivatives. *Journal of Essential Oil Bearing Plants, 12*(5), pp. 605–614.

Kunicka-Styczyńska, A.L.I.N.A. and Gibka, J., 2010. Antimicrobial Activity of Undecan-x-ones (x = 2–4). *Polskie Towarzystwo Mikrobiologów Polish Society of Microbiologists, 59*(4), pp. 301–306.

Wannissorn, B., Jarikasem, S., Siriwangchai, T. and Thubthimthed, S., 2005. Antibacterial properties of essential oils from Thai medicinal plants. *Fitoterapia, 76*(2), pp. 233–236.

### 2.2.2.15 *Litsea glutinosa* (Lour.) C.B. Rob.

Synonyms: *Litsea chinensis* Lam., *Litsea laurifolia* (Jacq.) Cordem.; *Tetranthera laurifolia* Jacq., *Sebifera glutinosa* Lour.

Common names: Common tallow laurel, Indian laurel; kukur chita, pipul-jongi (Bangadesh); chan gao mu jiang zi, chuanshu (China)

Habitat: Forests, riverbanks

Distribution: India, Bhutan, Nepal, Bangladesh, Myanmar, Thailand, Vietnam, and the Philippines

Botanical observation: It is a tree that grows to 10 m tall. The stems are hairy. The leaves simple, spiral, and exstipulate. The petiole is 1–2.5 cm long and hairy. The blade is elliptic-lanceolate, obovate, 3.5–10 × 1.5–11 cm, hairy below, with 5–12 pairs of secondary nerves, cuneate at base, obtuse at apex, glossy, and coriaceous. The inflorescences are umbels on 2–4 cm peduncles. The perianth includes 4 spoon shaped and about 5 mm long tepals. The androecium comprises about 15 stamens. The ovary is ovoid and develops a slender style. The drupes are globose and about 8 mm in diameter.

Medicinal uses: Gonorrhea (Bangladesh); ringworm, cuts, wounds (India); furuncles (China)

*Broad-spectrum antibacterial halo developed by polar extract:* Methanol extract of bark (200 µg/mL, agar diffusion) inhibited the growth of *Staphylococcus aureus* (ML 267), *Bacillus subtilis, Klebsiella pneumoniae, Shigella dysenteriae, Escherichia coli, Salmonella typhimurium,* and *Vibrio cholera* with inhibition zone diameters of 9, 10, 9.5, 10, 10, 9.5, and 11 mm, respectively (Mandal et al., 2000).

Commentary: The plant produces lignans (Wu et al., 2017). Essential oil of the fruits contains principally the linear monoterpene (E)-β-ocimene (Choudhury et al., 1996).

### References

Choudhury, S.N., Singh, R.S., Ghosh, A.C. and Leclercq, P.A., 1996. *Litsea glutinosa* (Lour.) CB Rob., a new source of essential oil from northeast India. *Journal of Essential Oil Research, 8*(5), pp. 553–556.

Mandal, S.C., Kumar, C.A., Majumder, A., Majumder, R. and Maity, B.C., 2000. Antibacterial activity of *Litsea glutinosa* bark. *Fitoterapia, 71*(4), pp. 439–441.

Wu, Y., Jin, Y., Dong, L., Li, Y., Zhang, C., Gui, M. and Zhang, X., 2017. New lignan glycosides from the root barks of *Litsea glutinosa. Phytochemistry Letters, 20*, pp. 259–262.

### 2.2.2.16 *Litsea verticillata* Hance

Synonyms: *Litsea brevipetiolata* Lecomte; *Litsea multiumbellata* Lecomte

Common names: lun ye mu jiang zi, niulali (China); bời lời cuống ngắn (Vietnam)

Habitat: Riverbanks, thickets

Distribution: Cambodia, Thailand, Taiwan, and Vietnam

Botanical observation: It is a shrub growing up to about 3 m tall. The stems are covered with yellow silky hair. The leaves are simple, 4–6-verticillate, and exstipulate. The petiole is about 5 mm long. The blade is narrow, elliptic, 6.5–20 × 2–6.5 cm, somewhat asymmetrical, with 12–14 pairs of secondary nerves, acute at base and acuminate at apex. The inflorescences are terminal clusters of 5–8 flowers. The perianth is yellowish and comprises 6 segments. The androecium presents 9 stamens. The drupes are ovoid, up to about 1.5 cm long, and seated on a discoid cupule.

Medicinal use: Traumatic injuries (China)

*Strong lipophilic antiviral (enveloped monopartite linear dimeric single stranded (+) RNA) amphiphilic eudesmane sesquiterpene:* Verticillatol (LogD = 3.1 at pH 7.4; molecular weight = 238.3 g/mol) from the leaves and twigs inhibited the replication of the Human immunodeficiency virus type-1 with the $IC_{50}$ of 34.5 µg/mL (Hoang et al., 2002).

Verticillatol

*Strong antiviral (enveloped monopartite linear dimeric single stranded (+)RNA) furofuran lignan:* (+)-Demethoxyepiexcelsin from the leaves and twigs inhibited the replication of the Human immunodeficiency virus with an $IC_{50}$ of 16.4 µg/mL (Hoang et al., 2002).

*Strong antiviral (enveloped monopartite linear dimeric single stranded (+) RNA) amphiphilic long chain butyrolactones:* 3-epilitsenolide $D_2$ and litseabutanolide from the leaves and twigs inhibited the Human immunodeficiency virus type-1 (strain IIIB) with the $IC_{50}$ of 9.9 and 40.3 µM, respectively (Zhang et al., 2005). Litseaverticillol A (LogD = 3.4 at pH 7.4; molecular weight = 234.3 g/mol), B, C, D, and E from the leaves inhibited the replication of the Human immunodeficiency virus type-1 (strain IIIB) with the $IC_{50}$ values of 5, 3, 7.1, 14.4, and 4 µg/mL, respectively (Zhang et al., 2003).

Litseaverticillol A

*Strong antiviral (enveloped monopartite linear dimeric single stranded (+) RNA) amphiphilic long chain cyclobutanone:* Isolitseane B (LogD = 3.0; molecular weight = 236.3 g/mol) from the leaves and twigs inhibited Human immunodeficiency virus type-1 (strain IIIB) with $IC_{50}$ of 38.1 µM (Zhang et al., 2005).

Isolitseane B

Commentary: The plant produces compounds belonging to different classes but yet all active against HIV. These compounds have probably some levels of activity against coronaviruses.

**References**

Hoang, V.D., Tan, G.T., Zhang, H.J., Tamez, P.A., Van Hung, N., Cuong, N.M., Soejarto, D.D., Fong, H.H. and Pezzuto, J.M., 2002. Natural anti-HIV-1 agents—part I:(+)-demethoxyepiexcelsin and verticillatol from *Litsea verticillata*. *Phytochemistry*, *59*(3), pp. 325–329.

Zhang, H.J., Tan, G.T., Hoang, V.D., Van Hung, N., Cuong, N.M., Soejarto, D.D., Pezzuto, J.M. and Fong, H.H., 2003. Natural anti-HIV-1 agents. Part 3: Litseaverticillols A–H, novel sesquiterpenes from *Litsea verticillata*. *Tetrahedron*, *59*(2), pp. 141–148.

Zhang, H.J., Van Hung, N., Cuong, N.M., Soejarto, D.D., Pezzuto, J.M., Fong, H.H. and Tan, G.T., 2005. Sesquiterpenes and butenolides, natural anti-HIV1 constituents from *Litsea verticillata*. *Planta Medica*, *71*(05), pp. 452–457.

### 2.2.2.17 *Persea americana* Mill.

Synonyms: *Laurus persea* L.; *Persea drymifolia* Schltdl. & Cham.; *Persea edulis* Raf.; *Persea floccosa* Mez; *Persea gigantea* L.O. Williams; *Persea gratissima* C.F. Gaertn.; *Persea leiogyna* S.F. Blake; *Persea nubigena* L.O. Williams; *Persea paucitriplinervia* Lundell; *Persea persea* (L.) Cockerell; *Persea steyermarkii* C.K. Allen

Common names: Avocado; avoka (Cambodia); e li (China); avokado (Indonesia, Malaysia); awokado (Thailand); bo (Vietnam)

Habitat: Cultivated, villages

Botanical observation: It is a handsome tree probably native to Central America that grows to about 10 tall. The bark is greyish green. The leaves are simple, spiral, and exstipulate. The petiole is up to about 5 cm long. The blade is glaucous below, glossy above, elliptic, ovate, or obovate, 7.5–15 × 5–7 cm, coriaceous, with 5–7 pairs of secondary nerves, cuneate or acute at base, and acute at apex. The inflorescences are cymose panicles, which are about 10 cm long and bearing minute yellowish flowers. The perianth is about 5 mm long, hairy, and includes 6 lanceolate lobes. The androecium comprises 9 stamens. The ovary is ovoid, hairy, and develops a discoid stigma. The drupe is green turning purplish, glossy, somewhat pear shaped, about 10 cm long, and containing a massive seed. Avocados from Cebu are particularly delicious.

Medicinal uses: Antiseptic (the Philippines)

*Moderate broad-spectrum antibacterial polar extract:* Ethanol extract of seeds inhibited the growth of *Listeria monocytogenes* (ATCC 7644), *Staphylococcus aureus* (ATCC 25923), *Enterococcus faecalis* (ATCC 29212), *Salmonella Enteritidis* (ATCC 13076), *Citrobacter freundii* (ATCC 8090), *Pseudomonas aeruginosa* (ATCC 27853), *Salmonella typhimurium* (ATCC 13311), and *Enterobacter aerogenes* (ATCC 13048) with the MIC values of 166.7, 416.7, 250, 208.3, 145.8, 166.7, 250, and 125 μg/mL, respectively (Raymond Chia & Dykes, 2010)

*Moderate broad-spectrum antifungal non-polar extract:* Hexane extract of seeds inhibited the growth of *Candida parapsilosis* (ATCC 22019), *Candida tropicalis*, *Candida albicans*, *Candida krusei* (ATCC 6528), *Cryptococcus neoformans*, and *Malassezia pachydermatis* with MIC values ranging from 156 to 625 μg/mL (Leite et al., 2009).

*Moderate antifungal (yeast) polar extract:* Ethanol extract of seeds inhibited the growth of *Zygosaccharomyces bailii* with an MIC value of 416.7 μg/mL (Raymond Chia & Dykes, 2010).

*Strong antimycobacterial fatty alcohols*: Avocadenol A and (2R, 4R)-1,2,4-trihydroxynonadecane from unripe fruits inhibited the growth of *Mycobacterium tuberculosis* with MIC values of 24 and 24.9 μg/mL, respectively (ethambutol: MIC: 6.2 μg/mL) (Lu et al., 2012).

Avocadenol A

*Strong antiviral (enveloped monopartite linear dimeric single-stranded (+)RNA) polar extract:* Aqueous extract of leaves inhibited the replication of the Human immunodeficiency virus type-1 (stain RF) with an $IC_{50}$ value of 18.7 µg/mL (Wigg et al., 1996).

*Strong antiviral (enveloped monopartite linear double-stranded DNA) hydrophilic flavonol glycosides:* Afzelin (also known as kaempferol 3-*O*-rhamnoside; LogD = −0.2; molecular weight = 432.3 g/mol), quercetin 3-*O*-α-D-arabinopyranoside, and quercitrin isolated from the leaves at a concentration of 50 µg/mL inhibited the replication of Herpes simplex virus type-1 by 72.5, 81.4, and 84.5%, respectively (De Almeida et al., 1998).

Afzelin

*Weak antiviral (enveloped linear double-stranded DNA) hydroxycinnamic acid:* Chlorogenic acid (3-O-caffeoyl-quinic acid) isolated from the leaves at a concentration of 200 µg/mL inhibited the replication of the Herpes simplex virus type-1 by 88.2% (De Almeida et al., 1998).

Commentaries: (i) Kaempferol-3-*O*-a-L-(2,4-bis-E-p-coumaroyl)rhamnoside isolated from the leaves of a member of the genus *Persea* Mill. inhibited ethidium bromide efflux in NorA over-expressing *Staphylococcus aureus* (SA1199B) with an $IC_{50}$ value of 2 µM ($IC_{50}$ value of 9 µM for reserpine) (Holler et al., 2012).

(ii) Note that 4′, 5′-*O*-dicaffeoylquinic acid inhibited NorA efflux pump in *Staphylococcus aureus* (Sharma et al., 2019).

(iii) Aqueous extract of leaves inhibited the replication of Aujesky disease virus (ADV) (Strain Nova Prata) and Bovine Virus Diarrhea (BVD) (strain Singer) and evoked virucidal effects (Koseki et al., 1990).

(iv) One could examine the effect of aqueous extract of leaves on the Severe acute respiratory syndrome-associated coronavirus.

### References

De Almeida, A.P., Miranda, M.M.F.S., Simoni, I.C., Wigg, M.D., Lagrota, M.H.C. and Costa, S.S., 1998. Flavonol monoglycosides isolated from the antiviral fractions of *Persea americana* (Lauraceae) leaf infusion. *Phytotherapy Research: An International Journal Devoted to Pharmacological and Toxicological Evaluation of Natural Product Derivatives, 12*(8), pp. 562–567.

Holler, J.G., Christensen, S.B., Slotved, H.C., Rasmussen, H.B., Gúzman, A., Olsen, C.E., Petersen, B. and Mølgaard, P., 2012. Novel inhibitory activity of the *Staphylococcus aureus* NorA efflux pump by a kaempferol rhamnoside isolated from Persea lingue Nees. *Journal of Antimicrobial Chemotherapy, 67*(5), pp. 1138–1144.

Koseki, I., Simoni, I.C., Nakamura, I.T., Noronha, A.B. and Costa, S.S. 1990. Antiviral activity of plant extracts against aphtovirus, pseudorabies virus and pestivirus in cell cultures. *Microbios Letters, 44*, 19–30.

Lu, Y.C., Chang, H.S., Peng, C.F., Lin, C.H. and Chen, I.S., 2012. Secondary metabolites from the unripe pulp of *Persea americana* and their antimycobacterial activities. *Food Chemistry, 135*(4), pp. 2904–2909.

Raymond Chia, T.W. and Dykes, G.A., 2010. Antimicrobial activity of crude epicarp and seed extracts from mature avocado fruit (*Persea americana*) of three cultivars. *Pharmaceutical Biology, 48*(7), pp. 753–756.

Sharma, A., Gupta, V.K. and Pathania, R., 2019. Efflux pump inhibitors for bacterial pathogens: from bench to bedside. *The Indian Journal of Medical Research, 149*(2), p. 129.

Wigg, M.D., Al-Jabri, A.A., Costa, S.S., Race, E., Bodo, B. and Oxford, J.S., 1996. In-vitro virucidal and virustatic anti HIV effects of extracts from *Persea Americana* Mill, (Avocado) leaves. *Antiviral Chemistry and Chemotherapy, 7*(4), pp. 179–183.

### 2.2.2.18 *Phoebe lanceolata* (Nees) Nees

Synonyms: *Laurus lanceolaria* Roxb.; *Ocotea lanceolata* Nees; *Ocotea ligustrina* Nees

Common names: Changpichla (Bangladesh); pi zhen ye nan (China); nake naw ruk; pheer (Nepal); chandra (India)

Habitat: Forests

Distribution: Pakistan, India, Himalayas, Nepal, Bhutan, Bangladesh, Cambodia, Laos, Vietnam, Malaysia, Thailand, China, and Indonesia

Botanical observation: It is a tree that grows to about 10 m tall. The leaves are simple, spiral, and exstipulate. The petiole is 1–2.5 cm long. The blade is elliptic-lanceolate, 13–22 × 3–5.5 cm, coriaceous, hairy below, with 9–13 pairs of secondary nerves, attenuate at base, and apex acuminate at apex. The panicles are terminal or axillary and lax. The 6 lobes of the perianth are pale green or yellowish green and minute. The androecium includes 9 stamens. The ovary is glabrous. The drupes are ovoid, about 1.2 cm long, glossy, and black.

Medicinal use: Leucorrhea (Bangladesh)

*Antibacterial essential* oil: Essential oil of leaves displayed antibacterial effects (Joshi et al., 2010).

*Strong antibacterial (Gram-positive) polar extract:* Ethanol extract of stem bark inhibited the growth of *Staphylococcus aureus*, *Streptococcus mutans*, and *Staphylococcus epidermidis* with the MIC values of 100, 100, and 50 µg/mL, respectively (Semwal et al., 2009).

*Strong antimycobacterial amphiphilic aporphine alkaloids:* Nordicentrine (LogD = 3.0 at pH 7.4; molecular weight = 325.3 g/mol) inhibited the growth of *Mycobacterium tuberculosis* with the MIC of 12.5 µg/mL (Lekphrom et al., 2009). Dicentrinone inhibited *Mycobacterium tuberculosis* (H37Rv) with MIC of 50 µg/mL (Camacho-Corona et al., 2009).

Nordicentrine

Commentary: The plant produces aporphine alkaloids including nordicentrine and dicentrinone (Arbain & Sargent, 1987). Aporphine alkaloids are antibacterial. For instance, lysicamine from *Phoebe grandis* (Nees) Merr. inhibited the growth of *Staphylococcus epidermidis, Staphylococcus aureus,* and *Bacillus subtilis* (Omar et al., 2013). Aporphine alkaloid are planar and intercalate DNA and inhibit topoisomerases (Woo et al., 1999) and are thus very often antimicrobial (and also cytotoxic). One further example is ocoteine, which inhibited the growth of *Mycobacterium tuberculosis* (H37Rv) with the MIC value of 50 µg/mL (Albarracin et al., 2017).

### References

Albarracin, L.T., Patino, O.J., Guzman, J.D., Begum, N., McHugh, T.D., Cuca, L.E. and Ávila, M.C., 2017. Aporphine alkaloids with antitubercular activity isolated from Ocotea discolor Kunth (Lauraceae). *Revista Colombiana de Química, 46*(3), pp. 22–27.

Arbain, D. and Sargent, M.V., 1987. A preliminary investigation of the alkaloid of *Phoebe Lanceolata. ASEAN Journal on Science and Technology for Development, 4*(1), pp. 41–44.

Camacho-Corona, M.D.R., Favela-Hernández, J.M.D.J., González-Santiago, O., Garza-González, E., Molina-Salinas, G.M., Said-Fernández, S., Delgado, G. and Luna-Herrera, J., 2009. Evaluation of some plant-derived secondary metabolites against sensitive and multidrug-resistant *Mycobacterium tuberculosis. Journal of the Mexican Chemical Society, 53*(2), pp. 71–75.

Joshi, S.C., Verma, A.R. and Mathela, C.S., 2010. Antioxidant and antibacterial activities of the leaf essential oils of *Himalayan Lauraceae* species. *Food and Chemical Toxicology, 48*(1), pp. 37–40.

Lekphrom R., Kanokmedhakul S., Kanokmedhakul K., 2009. Bioactive styryllactones and alkaloid from flowers of Goniothalamus laoticus. *J Ethnopharmacol*, 125(1):47–50.

Omar, H., Hashim, N.M., Zajmi, A., Nordin, N., Abdelwahab, S.I., Azizan, A.H.S., Hadi, A.H.A. and Ali, H.M., 2013. Aporphine alkaloids from the leaves of *Phoebe grandis* (Nees) Mer.(Lauraceae) and their cytotoxic and antibacterial activities. *Molecules, 18*(8), pp. 8994–9009.

Semwal, D.K., Rawat, U., Bamola, A. and Semwal, R., 2009. Antimicrobial activity of *Phoebe lanceolata* and *Stephania glabra*; preliminary screening studies. *Journal of Scientific Research, 1*(3), pp. 662–666.

Woo, S.H., Sun, N.J., Cassady, J.M. and Snapka, R.M., 1999. Topoisomerase II inhibition by aporphine alkaloids. *Biochemical Pharmacology, 57*(10), pp. 1141–1145.

## 2.3  ORDER MAGNOLIALES JUSS. EX BERCHT. & J. PRESL (1820)

### 2.3.1  FAMILY ANNONACEAE A.L. de Jussieu (1789)

The Family Annonaceae consists of about 128 genera and 2400 species of tropical trees, shrubs, or woody climber. The inner bark has often a delicate fragrance. The leaves are simple, alternate, and exstipulate. The inflorescences are mostly terminal or axillary and solitary flowers with somewhat a monocot look. The calyx includes 3 sepals. The corolla comprises 3–6 petals that are not uncommonly coriaceous. The androecium consists numerous short and packed stamens. The gynoecium includes a few to many carpels containing 1 or 2 ovules on parietal placentas. Members of this family produce notably antimicrobial diterpenes, lignans, acetogenins, and isoquinolines.

#### 2.3.1.1  *Annona reticulata* L.

Synonyms: *Annona excelsa* Kunth; *Annona humboldtiana* Kunth; *Annona humboldtii* Dun.; *Annona laevis* Kunth; *Annona longifolia* Sessé & Moc.; *Annona primigenia* Standl. & Steyerm.; *Annona reticulata* var. *primigenia* (Standl. & Steyerm.) Lundell; *Annona riparia* Kunth.

Common names: Custard apple; mo bat (Cambodia); niu xin fan li zhi (China); uto ni mbulu-makau (Fiji); nona (India); buah nona (Indonesia, Malaysia); khan tua lot (Laos); thinbaw-awza (Myanmar); shree ram phal (Nepal); ramphal (Pakistan); beretetutu (Solomon Islands); manong (Thailand); qua na (Vietnam).

Habitat: Cultivated, villages

Distribution: Tropical Asia and Pacific

Botanical observation: It is a treelet native to tropical America that grows to 6 m tall. The leaves are simple, alternate, and exstipulate. The petiole is stout and up to 1.5 cm long. The blade is lanceo-late, coriaceous, 10–17.5 × 2.5–5 cm, cuneate at base, and acuminate at apex. The inflorescence is axillary. The calyx includes 3 sepals. The corolla consists in 2 whorls of 3 petals, which are coria-ceous, yellowish green, broadly lanceolate, and 1.5–2 cm × 0.8–1.5 cm. The androecium comprises numerous stamens, which are and minute. The gynoecium includes numerous carpels. The fruit is somewhat heart shaped, light brownish pink, 10–15 cm × 5.5–12.5 cm, reticulate, edible, and con-tain numerous glossy black seeds.

Medicinal uses: Fever (Bangladesh; the Philippines)

*Broad-spectrum antibacterial halo developed by polar extract:* Methanol extract of leaves inhibited the growth of *Staphylococcus aureus, Salmonella typhi, Klebsiella pneumonia, Proteus vulgaris,* and *Pseudomonas aeruginosa* with inhibition zone diameters of 11, 10, 14, 18, and 16 mm, respectively (100 µg/mL/5 mm disc) (Chandra, 2013).

*Very weak broad-spectrum antibacterial polar extract:* Methanol extract of leaves inhibited the growth of *Bacillus subtilis, Staphylococcus aureus, Vibrio alginolyticus,* and *Vibrio cholerae* with MIC/MBC values of 5/<10, 2.5/<10, 5/<10, and 5/<10 mg/mL, respectively (Padhi et al., 2011).

*Broad-spectrum antibacterial lipophilic acetogenin:* Squamocin (also known as annonin I; LogD = 7.1 at pH 7.4; molecular weight = 622.9 g/mol) inhibited the growth of *Bacillus subti-lis, Bacillus cereus, Bacillus megaterium, Staphylococcus aureus, Sarcina lutea, Escherichia coli, Shigella dysenteriae, Shigella shiga, Shigella flexneri, Shigella sonnei, Salmonella typhi, Pseudomonas aeruginosa,* and *Klebsiella* sp. with inhibition zone diameters ranging from 12 to 21 mm (200 µg/disc) (Rahman et al., 2005).

Squamocin

*Strong antiviral (non-enveloped segmented linear double-stranded RNA) polar extract:* Aqueous extract of leaves at a concentration of 20 µg/mL inhibited the replication of Human rotavirus (HCR3) by 43.8% (Gonçalves et al., 2005).

Commentaries: (i) The seeds of the plant contain cytotoxic acetogenins, including annoreticuin, annoreticuin-9-one, bullatacin, and squamocin (Chang et al., 1998). Consider that lipophilic natural products tend to diffuse easily from paper discs onto solid agar.

(ii) Bacteria do not have mitochondria. The bacterial respiratory chain, which is located at the cytoplasmic level, consists of three complexes with quinones and reduced nicotinamide adenine dinucleotide (NADH) acting as the carriers that shuttle electrons and protons between large protein complexes. In complex I, three inner membrane respiratory enzymes (of the NADH oxidase family) are present: proton-translocating NADH-quinone (Q) oxidoreductase (NDH-1), NADH-Q oxido-reductase, which lacks an energy-coupling site (NDH-2), and the sodium-translocating NADH-Q oxidoreductase (Friedrich, 2014). Acetogenins are "mitochondrial poisons" that inhibit NADH ubi-quinone oxido-reductase (complex I) in mitochondrial electron transport systems, and inhibition of

NADH oxidase in the plasma membranes of human cells (Chang et al., 1998). Note that mitochondria may have originated from bacteria (Gray, 2012).

(iii) The plant produces a series of aporphine and benzylisoquinoline alkaloids (Matsushige et al., 2012).

### References

Chandra, M., 2013. Antimicrobial activity of medicinal plants against human pathogenic bacteria. *International Journal of Biotechnology and Bioengineering Research, 4*(7), pp. 653–658.

Chang, F.R., Chen, J.L., Chiu, H.F., Wu, M.J. and Wu, Y.C., 1998. Acetogenins from seeds of *Annona reticulata*. *Phytochemistry, 47*(6), pp. 1057–1061.

Friedrich, T., 2014. On the mechanism of respiratory complex I. *Journal of Bioenergetics and Biomembranes, 46*(4), pp. 255–268.

Gonçalves, J.L.S., Lopes, R.C., Oliveira, D.B., Costa, S.S., Miranda, M.M.F.S., Romanos, M.T.V., Santos, N.S.O. and Wigg, M.D., 2005. In vitro anti-rotavirus activity of some medicinal plants used in Brazil against diarrhea. *Journal of Ethnopharmacology, 99*(3), pp. 403–407.

Gray, M.W., 2012. Mitochondrial evolution. *Cold Spring Harbor Perspectives in Biology, 4*(9), p. a011403.

Matsushige, A., Kotake, Y., Matsunami, K., Otsuka, H., Ohta, S. and Takeda, Y., 2012. Annonamine, a new aporphine alkaloid from the leaves of *Annona muricata*. *Chemical and Pharmaceutical Bulletin, 60*(2), pp. 257–259.

Padhi, L.P., Panda, S.K., Satapathy, S.N. and Dutta, S.K., 2011. In vitro evaluation of antibacterial potential of *Annona squamosa* L. and *Annona reticulata* L. from Similipal Biosphere Reserve, Orissa, India. *Journal of Agricultural Technology, 7*(1), pp. 133–142.

Rahman, M.M., Parvin, S., Haque, M.E., Islam, M.E. and Mosaddik, M.A., 2005. Antimicrobial and cytotoxic constituents from the seeds of *Annona squamosa*. *Fitoterapia, 76*(5), pp. 484–489.

### 2.3.1.2 *Annona squamosa* L.

Synonyms: *Annona asiatica* L.; *Annona cinerea* Dunal; *Annona forskahlii* DC.; *Annona glabra* Forssk.; *Guanabanus squamosus* (L.) M. Gómez; *Xylopia frutescens* Sieb. ex Presl; *Xylopia glabra* L.

Common names: Sugar apple; sweet-sop; ata (Bangladesh); tiep srok (Cambodia); fan li zhi (China); sitaphal (India; Pakistan); sirkaja (Indonesia); apeli (Fiji); atis (Guam); banreishi (Japan); buah nona (Malaysia); ates (the Philippines); makkhiap (Thailand); mang câu ta (Vietnam)

Habitat: Cultivated, villages

Distribution: Tropical Asia and Pacific

Botanical observation: It is a tree that grows up to 6 m tall. The bark is smooth and slightly aromatic. The leaves are simple, alternate, and exstipulate. The petiole is 0.4–1.5 cm long. The blade is elliptic, thinly leathery, 5–17.5 cm × 2–7.5 cm, obtuse at base, margin entire, acute at apex, and presents 8–15 pairs of secondary nerves. The flowers are clustered and cauliflorous and 2–3 cm. The calyx comprises 3 triangular sepals. The corolla includes 3 greenish, oblong-lanceolate, 1.5–3 × 0.5–0.8 cm, fleshy, petals are purplish at base. The androecium includes numerous tiny stamens. The gynoecium includes several carpels. The fruit ovoid, 5–10 cm in diameter, green to purplish, edible, and squamose. The seeds are black, 1.4 cm long, and embedded in a white pulp.

Medicinal uses: Fever in infants (the Philippines)

*Antibacterial (Gram-negative) halo developed by polar extract:* Acetone extract of seeds inhibited the growth of *Shigella flexneri* (MTCC 1457) and *Vibrio cholerae* (MTCC 3906) with inhibition zone diameters of 8 and 18 mm, respectively (10 µg/6 mm paper disc) (Kothari et al., 2010).

*Moderate broad-spectrum antibacterial polar extracts*: Ethanol extract of leaves inhibited the growth of *Staphylococcus aureus* (ATCC 11778), *Listeria monocytogenes* (ATCC 19111), *Staphylococcus aureus* (ATCC 6538), and *Campylobacter jejuni* (ATCC 29428) with MIC values of 250, 125, 62.5, and 250 µg/mL (suggesting increase in outer membrane permeability). MBC for *Staphylococcus aureus* (ATCC 11778) was 500 µg/mL (Dholvitayakhun et al., 2012). Methanol extract of fruit inhibited *Staphylococcus aureus*, *Bacillus cereus*, *Staphylococcus epidermidis*, *Escherichia coli*, *Salmonella typhimurium*, *Shigella flexneri*, and *Pseudomonas aeruginosa* with MIC/MBC of 1250/2500, 1250/1250, 312.5/2500, 2500/5000, 2500/>5000, 2500/5000, and 2500/5000 mg/mL, respectively (Fu et al., 2016). Methanolic extract of leaves (containing the aporphine alkaloids anonaine, asimilobine, liriodenine, nornuciferine, the protoberbeine alkaloid corypalmine, and the benzylisoquinoline alkaloid reticuline) inhibited the growth of *Staphylococcus aureus* (ATCC 6538), *Enterobacter aerogenes* (ATCC 13048), *Enterobacter cloacae* (ATCC 23355), *Klebsiella pneumoniae* (ATCC 4552), *Pseudomonas aeruginosa* (ATCC 9027), *Bacillus cereus* (ATCC 14579), *Enterococcus faecalis* (ATCC 19433), *Shigella dysenteriae* (ATCC 13313), *Escherichia coli* (ATCC 10536), *Salmonella typhimurium* (ATCC 13311), and *Salmonella choleraesuis* (ATCC 10708) with MIC/MBC values of 78/78, 1250/1250, 625/625, 78/156, 1250/5000,1250/1250, 39/78, 625/625, 1250/1250, 1250/2500, and 1250/500 µg/mL, respectively. This extract was synergistic with erythromycin against *Klebsiella pneumoniae* and *Enterococcus faecalis* (Pinto et al., 2017).

*Antibacterial (Gram-positive) essential oil:* The essential oil in the leaves (containing germacrene D, bisabolene, caryophyllene oxide, bisabolene epoxide, and kaur-16-ene) inhibited the growth of *Bacillus subtilis* and *Staphylococcus aureus* (Chavan et al., 2006).

*Moderate broad-spectrum antifungal polar extract*: Methanol extract of leaves inhibited the growth of *Alternaria alternata*, *Candida albicans*, *Fusarium solani*, *Microsporum canis*, and *Aspergillus niger* with MIC values of 800, 600, 600, 400, and 400 µg/mL, respectively (broth dilution method; $10^5$ dilution; nutrient broth) (Kalidindi et al., 2015).

*Strong broad-spectrum antibacterial amphiphilic benzylisoquinoline alkaloid:* The tetrahydro benzyl isoquinole reticuline (LogD = 1.8 at pH 7.4; molecular weight = 329.3 g/mol) inhibited the growth of *Escherichia coli*, *Pseudomonas aeruginosa*, *Proteus mirabilis*, *Klebsiella pneumonia*, *Acinetobacter baumannii*, *Staphylococcus aureus*, and *Bacillus subtilis* with MIC values of 32, 32, 32, 32, 32, 64, and 64 µg/mL, respectively (Orhan et al., 2007).

Reticuline

*Broad-spectrum antibacterial acetogenin:* Squamocin and a mixture of annotemoyin-1 and annotemoyin-2, which displayed a broad-spectrum antibacterial effect (200 µg/disc) (Rahman et al., 2005).

*Antifungal (filamentous) halo developed by acetogenins:* Annotemoyin-1 and annotemoyin-2 isolated from the seeds, (when combined), inhibited the growth of *Aspergillus flavus*, *Aspergillus niger*, and *Aspergillus fumigatus* with inhibition zone diameters of 13, 18, and 10 mm (200 µg/disc), respectively (Rahman et al., 2005).

*Strong antiviral (enveloped monopartite linear dimeric single stranded (+) RNA) kaurane diterpene*: 16β-17-dihydroxy-ent-kauran-19-oic acid inhibited the replication of the Human immunodeficiency virus in H9 lymphocyte cells with an $EC_{50}$ value of 0.8 μ/mL (selectivity index > 5) (Wu et al., 1996).

*Strong antiviral (enveloped monopartite linear double-stranded DNA) amphiphilic isoquinoline alkaloid:* Reticuline (LogD = 1.8 at pH 7.4; molecular weight = 329.3 g/mol) inhibited Herpes simplex virus replication in Madin-Darby canine kidney cells with a maximum nontoxic concentration of 32 μg/mL (Orhan et al., 2007).

*Strong antiviral (enveloped monopartite linear single-stranded (−) RNA) isoquinoline alkaloid:* Reticuline inhibited Parainfluenza-3 virus in Vero cells with a maximum nontoxic concentration of 64 μg/mL, respectively (Orhan et al., 2007).

Commentaries: (i) Consider that natural products from plants with antiviral activity have often some levels of cytotoxic effects. Hence, a good antiviral property can be observed *in vitro* and at the same time have cytotoxicity for host cells. The generally accepted threshold of selectivity index in the area of antivirals development is 10, i.e., if a compound demonstrates SI 10 or higher it is considered as prospective (Vladimir Zarubaev personal communication).

(ii) Note that natural products from medicinal plants in the Clades treated in this volume able to strongly inhibit the replication of the Parainfluenza-3 viruses are mainly medium molecular weight and amphiphilic isoquinoline alkaloids.

### References

Chavan, M.J., Shinde, D.B. and Nirmal, S.A., 2006. Major volatile constituents of *Annona squamosa* L. bark. *Natural Product Research, 20*(8), pp. 754–757.

Dholvitayakhun, A., Cushnie, T.T. and Trachoo, N., 2012. Antibacterial activity of three medicinal Thai plants against *Campylobacter jejuni* and other foodborne pathogens. *Natural Product Research, 26*(4), pp. 356–363.

Fu, L., Lu, W. and Zhou, X., 2016. Phenolic compounds and in vitro antibacterial and antioxidant activities of three tropic fruits: persimmon, guava, and sweetsop. *BioMed Research International.*

Kalidindi, N., Thimmaiah, N.V., Jagadeesh, N.V., Nandeep, R., Swetha, S. and Kalidindi, B., 2015. Antifungal and antioxidant activities of organic and aqueous extracts of *Annona squamosa* Linn. leaves. *Journal of Food and Drug Analysis, 23*(4), pp. 795–802.

Kothari, V. and Seshadri, S., 2010. In vitro antibacterial activity in seed extracts of *Manilkara zapota, Anona squamosa*, and *Tamarindus indica. Biological Research, 43*(2), pp. 165–168.

Orhan, I., Özçelik, B., Karaoğlu, T. and Şener, B., 2007. Antiviral and antimicrobial profiles of selected isoquinoline alkaloids from Fumaria and *Corydalis* species. *Zeitschrift für Naturforschung C, 62*(1–2), pp. 19–26.

Pinto, N.C., Silva, J.B., Menegati, L.M., Guedes, M.C.M., Marques, L.B., Silva, T.P., Melo, R.C.D., Souza-Fagundes, E.M., Salvador, M.J., Scio, E. and Fabri, R.L., 2017. Cytotoxicity and bacterial membrane destabilization induced by *Annona squamosa* L. extracts. *Anais da Academia Brasileira de Ciências, 89*(3), pp. 2053–2073.

Rahman, M.M., Parvin, S., Haque, M.E., Islam, M.E. and Mosaddik, M.A., 2005. Antimicrobial and cytotoxic constituents from the seeds of *Annona squamosa. Fitoterapia, 76*(5), pp. 484–489.

Wu, Y.C., Hung, Y.C., Chang, F.R., Cosentino, M., Wang, H.K. and Lee, K.H., 1996. Identification of ent-16β, 17-dihydroxykauran-19-oic acid as an anti-HIV-1 principle and isolation of the new diterpenoids annosquamosins A and B from *Annona squamosa. Journal of Natural Products, 59*(6), pp. 635–637.

### 2.3.1.3 *Artabotrys suaveolens* (Blume) Blume

Synonym: *Unona suaveolens* Blume

Common names: Xiang ying zhao (China); akar chenana (Malaysia); nga pye yin (Myanmar); manaranchitam (India); susong kalabao (the Philippines); aka kai (Thailand)

Habitat: Cultivated, forests

Distribution: India, Myanmar, Thailand, Malaysia, Indonesia, and the Philippines.

Botanical observation: It is a woody climber that grows to a length of 15 m. The inner bark has a subtle fragrance. The leaves are simple, alternate, and exstipulate. The blade is elliptic, glossy, acute at base, acuminate at apex, and 5–9 cm × 2.5–3.4 cm. The flowers are fragrant, clustered on a hooklike peduncle that is persistent. The calyx consists of 3 sepals. The corolla consists of 6 petals, which are whitish and linear. The androecium comprises many stamens. The gynoecium is made of 4 carpels. The fruits are ripe carpels, which are about 0.7 cm in diameter, sessile, and globose.

Medicinal uses: Cholera (Indonesia)

*Strong antibacterial broad-spectrum amphiphilic aporphine alkaloids:* Xylopine (LogD = 2.8 at pH 7.4; molecular mass = 295.3 g/mol) inhibited the growth of *Bacillus cereus, Micrococcus* sp., and *Staphylococcus aureus* (Tsai et al., 1989) as well as *Vibrio cholerae* with an MIC value of 50 µg/mL and of *Escherichia coli* and *Shigella dysenteriae* type 5 with an MIC value of 75 µg/mL (Hossain et al., 1993). Tan et al. (2015) observed that artabotrine (also known as isocorydine; LogD = 3.1 at pH 7.4; molecular weight = 341.4 g/mol) strongly inhibited the growth of a broad array of bacteria with MIC values ranging from 1.2 to 5 µg/mL. Artabotrine was bactericidal against Gram-negative extended-spectrum β-lactamase-producing *Klebsiella pneumoniae* with an MIC value equal to 2.5 µg/mL and an MBC value equal to 2.5 µg/mL (Tan et al. 2015).

Xylopine

Commentaries: (i) Cholera is an extremely virulent, acute diarrheal infection caused by the ingestion of food or water contaminated with the Gram-negative bacterium *Vibrio cholerae* which has developed resistance to antibiotics (Das et al., 2020). This bacterium can kill within hours if untreated. During the nineteenth century, following the Tambora volcanic eruption of 1815, cholera spread across the world from its original reservoir in the Ganges delta in India in time of the humanitarian crisis killing millions of people across the globe (Hamlin, 2009). Lack of water, sanitation, and the displacement of populations during wars boost the spread of cholera, the annual burden of which has been estimated at 1.3–4.0 million cases and 21,000–143,000 deaths worldwide in 2017 according to the World Health Organization. Oral cholera vaccines (OCV) such as Dukoral®, Shanchol™, and Euvichol® prevent the disease.

(ii) *Artabotrys suaveolens* (Blume) Blume is known to elaborate the aporphine alkaloid alkaloids, suaveoline, and xylopine (Barger & Sargent, 1939). Bisabolene-type sesquiterpene endoperoxides are produced *Artabotrys hexapetalus* (L. f.) Bhandari, which is used in China to make tea (local name: *ying zhua hua*) (Zhang et al., 1988). Bisabolene-type endoperoxides are interesting candidates against *Mycobacterium tuberculosis* because 2,10-bisaboladiene-1,4-endoperoxide isolated from *Rudbeckia laciniata* L. (family Asteraceae) inhibited the growth of *Mycobacterium tuberculosis* (H37Rv) with an MIC value of about 32 μg/mL (Pauli et al., 2005). Another example of endoperoxide with strong antimycobacterial activity is the sterol ergosterol-5,8-endoperoxide isolated from *Ajuga remota* Wall. ex Benth. (family Lamiaceae), which inhibited the growth of *Mycobacterium tuberculosis* (H37Rv) with an MIC as low as 1 μg/mL (Cantrell et al., 1999).

### References

Barger, G. and Sargent, L.J., 1939. 211. The alkaloids of *Artabotrys suaveolens. Journal of the Chemical Society (Resumed)*, pp. 991–999

Cantrell, C.L., Rajab, M.S., Franzblau, S.G., Fronczek, F.R. and Fischer, N.H., 1999. Antimycobacterial ergosterol-5, 8-endoperoxide from *Ajuga remota. Planta Medica,* 65(08), pp. 732–734.

Das, B., Verma, J., Kumar, P., Ghosh, A. and Ramamurthy, T., 2020. Antibiotic resistance in Vibrio cholerae: understanding the habitat of resistance genes and mechanisms. *Vaccine, 38*, pp. A83–A92.

Hamlin, C., 2009. *Cholera: The Biography.* Oxford: Oxford University Press.

Hossain, M.S., Ferdous, A.J. and Hasan, C.M., 1993. In vitro antimicrobial activities of alkaloids from the stem bark of *Desmos longiflorus* (Roxb.) [Bangladesh]. *Bangladesh Journal of Botany (Bangladesh).*

Pauli, G.F., Case, R.J., Inui, T., Wang, Y., Cho, S., Fischer, N.H. and Franzblau, S.G., 2005. New perspectives on natural products in TB drug research. *Life Sciences, 78*(5), pp. 485–494.

Tan, K.K., Khoo, T.J., Rajagopal, M. and Wiart, C., 2015. Antibacterial alkaloids from *Artabotrys crassifolius* Hook. f. & Thomson. *Natural Product Research, 29*(24), pp. 2346–2349.

Tsai, I.L., Liou, Y.F. and Lu, S.T., 1989. Screening of isoquinoline alkaloids and their derivatives for antibacterial and antifungal activities. *Gaoxiong yi xue ke xue za zhi = The Kaohsiung Journal of Medical Sciences, 5*(3), pp. 132–145.

Zhang, L., Zhou, W.S. and Xu, X.X., 1988. A new sesquiterpene peroxide (yingzhaosu C) and sesquiterpenol (yingzhaosu D) from Artabotrys unciatus (L.) Meer. *Journal of the Chemical Society, Chemical Communications*, (8), pp. 523–524.

WHO. http://www.who.int/news-room/fact-sheets/detail/cholera

### 2.3.1.4 *Cananga odorata* (Lam.) Hook. f. & Thomson

Synonyms: *Cananga odoratum* (Lam.) Baill. ex King; *Canangium odoratum* (Lam.) king; *Unona odorata* (Lam.) Baill.; *Unona odorata* (Lam.) Dunal; *Uvaria odorata* Lam.

Common names: Ylang-ylang; chhkè srèng (Cambodia); yi lan (China); Kananga (Malay/Indonesian); kadatngan (Myanmar); Ilang-ilang (the Philippines); kradang-nga thai (Thailand); hoàng lan (Vietnam)

Habitat: Cultivated

Distribution: Southeast Asia and Pacific

Botanical observation: It is a treelet that grows to about 5 m tall. The bark is dark brown. The leaves are simple, alternate, and exstipulate. The petiole is up to 2 cm long. The blade is broadly elliptic, 9–20 × 4–15 cm, papery, somewhat wavy, dull green with 7–15 pairs of secondary nerves, base rounded, and apex acute to acuminate. The inflorescences are axillary cymes of showy and fragrant flowers on 1–5 cm long peduncles. The calyx comprises 3 sepals, which are ovate and about 5 mm long. The corolla includes 6 petals, which are yellow-green, linear, 5–8 × 0.5–1.8 cm. The androecium includes numerous minute stamens. The gynoecium comprises 10–12 carpels, which are about 5 mm long. The fruits are ripe carpels, which are black, ovoid, globose, or oblong, up to about 2.5 cm long, and containing numerous seeds.

Medicinal uses: Boils (Solomon Islands)

*Broad-spectrum antibacterial halo developed by aporphine alkaloids:* The oxoaporphines *O*-methylmoschatoline and liriodenine from the bark inhibited (200 µg/disc) the growth of both Gram-positive and Gram-negative bacteria in disc diffusion essay (Rahman et al., 2005). *O*-Methylmoschatoline (200 µg/disc) inhibited the growth of *Staphylococcus aureus, Escherichia coli, Pseudomonas aeruginosa, Salmonella typhi,* and *Klebsiella pneumonia* with zone diameters of 20, 12, 12, 20, and 22 mm, respectively (Rahman et al., 2005).

*Broad-spectrum antibacterial halo developed by simple phenolic:* 3,4-dihydroxybenzoic acid from the bark inhibited (200 µg/disc) the growth of both Gram-positive and Gram-negative bacteria in disc diffusion essay (Rahman et al., 2005).

*Moderate broad-spectrum antifungal essential oil:* Essential oil (containing mainly benzyl benzoate, benzyl salicylate, and linalool) inhibited the growth of *Candida albicans* (ATCC 48274), *Rhodotorula glutinis* (ATCC 16740), *Saccharomyces cerevisiae* (ATCC 2365), *Schizosaccharomyces pombe* (ATCC 60232), and *Yarrowia lypolitica* (ATCC 16617) with MIC values of 100, 200, 500, 200, and 30 µg/mL, respectively (Sacchetti et al., 2005).

*Strong broad-spectrum antifungal amphiphilic azaoxoaporphine alkaloid:* Sampangine (LogD = 2.7 at pH 7.4; molecular weight 232.2 g/mol) inhibited the growth of *Candida albicans* (ATCC 90028), *Candida glabrata* (ATCC 90030), *Candida kruseii* (ATCC 6258), *Aspergillus fumigatus* (ATCC 90906), and *Cryptococcus neoformans* (ATCC 90113) with MIC values of 3.1, 3.1, 6.2, 6.2, and 0.05 µg/mL, respectively by inhibiting the biosynthesis or metabolism of heme (Agarwal et al., 2008; Rao et al., 1986).

Sampangine

Commentaries: (i) Essential oil of flowers contains principally linalool and the germacrane-type sesquiterpene germacrene D (Stashenko et al., 1996). Note that the germacrane 1,10-epoxy-4-germacrene-12,8;15,6-diolide from a member of the genus *Mikania micrantha* Kunth (family Asteraceae) inhibited the replication of Respiratory syncytial virus and Parainfluenza type 3 virus with an $IC_{50}$ value of 37.4 µM and a selectivity index of 16 (But et al., 2009).

(ii) Simple phenolics are in general antibacterial on account of their abilities to, at least partially, alter cytoplasmic membrane permeability and leakage of cytoplasmic contents, lysis, inhibition of enzymes, and, at high concentrations, cytoplasmic coagulation (Fogg & Lodge, 1945; Hugo, 1967).

### References

Agarwal, A.K., Xu, T., Jacob, M.R., Feng, Q., Lorenz, M.C., Walker, L.A. and Clark, A.M., 2008. Role of heme in the antifungal activity of the azaoxoaporphine alkaloid sampangine. *Eukaryotic Cell,* 7(2), pp. 387–400.

But, P.P.H., He, Z.D., Ma, S.C., Chan, Y.M., Shaw, P.C., Ye, W.C. and Jiang, R.W., 2009. Antiviral constituents against respiratory viruses from *Mikania micrantha. Journal of Natural Products,* 72(5), pp. 925–928.

Fogg, A.H. and Lodge, R.M., 1945. The mode of antibacterial action of phenols in relation to drug-fastness. *Transactions of the Faraday Society, 41*, pp. 359–365.

Hugo, W.B., 1967. The mode of action of antibacterial agents. *Journal of Applied Bacteriology, 30*(1), pp. 17–50.

Rahman, M.M., Lopa, S.S., Sadik, G., Islam, R., Khondkar, P., Alam, A.K. and Rashid, M.A., 2005. Antibacterial and cytotoxic compounds from the bark of *Cananga odorata*. *Fitoterapia, 76*(7), pp. 758–761.

Rao, J.U.M., Giri, G.S., Hanumaiah, T. and Rao, K.V.J., 1986. Sampangine, a new alkaloid from *Cananga odorata*. *Journal of Natural Products, 49*(2), pp. 346–347.

Sacchetti, G., Maietti, S., Muzzoli, M., Scaglianti, M., Manfredini, S., Radice, M. and Bruni, R., 2005. Comparative evaluation of 11 essential oils of different origin as functional antioxidants, antiradicals and antimicrobials in foods. *Food Chemistry, 91*(4), pp. 621–632.

Stashenko, E.E., Prada, N.Q. and Martínez, J.R., 1996. HRGC/FID/NPD and HRGGC/MSD study of Colombian ylang-ylang (*Cananga odorata*) oils obtained by different extraction techniques. *Journal of High Resolution Chromatography, 19*(6), pp. 353–358.

### 2.3.1.5 *Goniothalamus malayanus* Hook. f. & Thomson

Synonym: *Goniothalamus puncticulatus* Boerl. & Koord.

Common names: sugi lado itam (Indonesia), pisang pisang (Malaysia)

Habitat: Forests

Distribution: Andaman and Nicobar Islands, Thailand, Malaysia, and Indonesia

Botanical observation: It is an endangered tree that grows, up to 10 m tall. The bark is fibrous and used to make strong ropes by locals. The inner bark has a slight delicate fragrance. The leaves are alternate, simple, and exstipulate. The petiole is stout and 0.5–1.5 cm. The blade is obovate, 12.5–28 cm × 3.5–9 cm, dark green, glossy, round at base, acuminate at apex, and with 14–16 pairs of secondary nerves. The flowers are solitary, axillary, and cauliflorous. The calyx includes 3 sepals, which are valvate, triangular 0.2–0.8 cm × 0.3–0.7 cm, and green. The perianth includes 3 small inner petals, which are up to 1.5 cm long, and 3 large outer petals, which are coriaceous, wavy, and yellowish, and up to 6 cm long. The androecium includes 80–250 tiny stamens. The gynoecium includes 8–20 carpels. The fruits are ripe carpels, which are ovoid, 1.5–3.5 cm long, 1–3 seeded, and green.

Medicinal uses: Apparently none for microbial infections.

*Moderate antimycobacterial styryl-lactone*: An extract (mainly containing altholactone) inhibited the growth of *Mycobacterium tuberculosis* (H37Rv) with an MIC value of 128 µg/mL (Macabeo et al., 2014).

*Strong broad-spectrum antibacterial amphiphilic styryl-lactone:* Altholactone (LogD = 1.2 at pH 7.4; molecular weight = 232.2 g/mol) inhibited the growth of *Staphylococcus aureus* (ATCC 25392), *Staphylococcus aureus* (ATCC 25923), *Enterococcus faecalis* (ATCC 29212), *Salmonella typhi* (ATCC 14023), and *Escherichia coli* (ATCC 35218) with MIC values of 0.6, 0.6, 0.6, 1.2, and 1.2 µg/mL, respectively (Al Momani et al., 2011).

Altholactone

*Strong antifungal (yeast) amphiphilic styryl-lactone:* Altholactone inhibited the growth of *Candida albicans* with an MIC value of 2.5 μg/mL (Al Momani et al., 2011). Altholactone exhibited antifungal activity with a high MIC value of 128 μg/mL against *Cryptococcus neoformans* and *Saccharomyces cerevisiae* (Euanorasetr et al., 2016).

Commentary: Members of the genus *Goniothalamus* (Blume) Hook. f. & Thomson synthesiz styryl-lactones, which are antibacterial and antifungal (Lekphrom et al., 2009). The antibacterial and antifungal mechanisms are yet unknown. Note that natural products that are cytotoxic for mammalian cancer cells are not uncommonly antifungal for yeast *in vitro*, and this is the case with altholactone (Inayat-Hussain et al., 2002).

### References

Al Momani, F., Alkofahi, A.S. and Mhaidat, N.M., 2011. Altholactone displays promising antimicrobial activity. *Molecules, 16*(6), pp. 4560–4566.

Euanorasetr, J., Junhom, M., Tantimavanich, S., Vorasin, O., Munyoo, B., Tuchinda, P. and Panbangred, W., 2016. Halogenated benzoate derivatives of altholactone with improved anti-fungal activity. *Journal of Asian Natural Products Research, 18*(5), pp. 462–474.

Inayat-Hussain, S.H., Osman, A.B., Din, L.B. and Taniguchi, N., 2002. Altholactone, a novel styryl-lactone induces apoptosis via oxidative stress in human HL-60 leukemia cells. *Toxicology Letters, 131*(3), pp. 153–159.

Lekphrom, R., Kanokmedhakul, S. and Kanokmedhakul, K., 2009. Bioactive styryllactones and alkaloid from flowers of *Goniothalamus laoticus*. *Journal of Ethnopharmacology, 125*(1), pp. 47–50.

Macabeo, A.P.G., Lopez, A.D.A., Schmidt, S., Heilmann, J., Dahse, H.M., Alejandro, G.J.D. and Franzblau, S.G., 2014. Antitubercular and cytotoxic constituents from *Goniothalamus gitingensis*. *Records of Natural Products, 8*(1), p. 41.

### 2.3.1.6 *Goniothalamus laoticus* (Finet & Gagnep.) Bân

Synonym: *Mitrephora laotica* Finet & Gagnep.

Common name: Bing ya yin gou hua (China)

Habitat: Forests

Distribution: Laos and Thailand

Botanical observation: It is a small and endangered tree which grows in the rainforests of Laos, Cambodia, and Thailand. The leaves are simple, alternate, and exstipulate. The petiole is 0.5–0.9 cm long. The blade is oblong, 13–18 × 3–5 cm, leathery, glaucous below, glossy above, attenuate at base, and acuminate at apex. The flowers are cauliflorous. The calyx includes 3 sepals, which are ovate. The corolla includes 2 series of 3 petals. The outer petals are lanceolate, coriaceous, wavy, broadly elliptic, and pinkish yellow. The androecium includes several stamens, which are minute. The gynoecium comprises 10 carpels. The fruits are ripe carpels, which are ovoid and green.

Medicinal use: Fever (Thailand)

*Strong antimycobacterial amphiphilic styryl lactone:* Howiinin A (LogD = 3.1 at pH 7.4; molecular weight = 362.3 g/mol) from the flowers inhibited *Mycobacterium tuberculosis* with MIC of 6.2 μg/mL (Lekphrom et al., 2009).

*Strong antimycobacterial amphiphilic aporphine alkaloid:* Nordicentrine (LogD = 3.0 at pH 7.4; molecular weight = 325.3 g/mol) from the flowers inhibited the growth of *Mycobacterium tuberculosis* with an MIC of 12.5 μg/mL (Lekphrom et al., 2009).

*Antifungal aporphine alkaloid:* Dicentrine inhibited the growth of *Cladosporium clodosporioides* with a minimum inhibitory dose of 6 μg/spot on TLC plate (Puvanendran et al., 2008).

Commentary: Consider that antifungal principles are often antimycobacterial. This is the case for amphotericin B and azole antifungals (Ahmad et al., 2006; Mariotti et al., 2020). Howinin A presents an epoxide moiety.

**References**

Ahmad, Z., Sharma, S. and Khuller, G.K., 2006. Azole antifungals as novel chemotherapeutic agents against murine tuberculosis. *FEMS Microbiology Letters, 261*(2), pp. 181–186.

Lekphrom, R., Kanokmedhakul, S. and Kanokmedhakul, K., 2009. Bioactive styryllactones and alkaloid from flowers of *Goniothalamus laoticus. Journal of Ethnopharmacology, 125*(1), pp. 47–50.

Mariotti, S., Teloni, R., de Turris, V., Pardini, M., Peruzzu, D., Fecchi, K., Nisini, R. and Gagliardi, M.C., 2020. Amphotericin B inhibits Mycobacterium tuberculosis infection of human alveolar type II epithelial A549 cells. *Antimicrobial Agents and Chemotherapy.*

Puvanendran, S., Wickramasinghe, A., Karunaratne, D.N., Carr, G., Wijesundara, D.S.A., Andersen, R. and Karunaratne, V., 2008. Antioxidant constituents from *Xylopia championii. Pharmaceutical Biology, 46*(5), pp. 352–355.

### 2.3.1.7 *Polyalthia longifolia* (Sonn.) Thwaites

Synonyms: *Guatteria longifolia* Wall., *Uvaria longifolia* Sonn., *Unona longifolia* (Sonn.) Dunal

Common names: Indian mast tree, false asoka; debdaru (Bangladesh); nittilingam (India); chang ye an luo (China); arthaw-ka (Myanmar)

Habitat: Cultivated

Distribution: Tropical Asia and Pacific

Botanical observation: It is a fusiform tree that grows to about 10 m tall. The bark is dark brownish grey. The leaves are simple, alternate, and exstipulate. The petiole is up to 1 cm long. The blade is somewhat lanceolate, wavy, glossy, 10–15 × 2.5–5 cm, with about 20 pairs of secondary nerves, cuneate at base, and acuminate at apex. The inflorescences are axillary fascicles on about 1 cm long peduncle. The flowers are seldom seen. The calyx includes 3 sepals which are triangular and about 1 cm long. The corolla includes 6 petals which are triangular, up to 1.5 cm long, yellow and glossy. The androecium includes numerous short stamens. The gynoecium includes about 20 carpels. The fruits consist of 4–8 ripe carpels, which are purple, ovoid, glossy, fleshy, about 2 cm, long and containing a single seed.

Medicinal uses: Skin infection (Bangladesh); fever (India)

*Broad-spectrum antibacterial halo developed by clerodane diterpenes:* 16-oxocleroda-3,13*E*-15-oic acid from the bark (100 µg/disc) developed a 16 mm diameter inhibition zone against *Staphylococcus aureus*, 16.6 mm against *Shigella dysenteriae*, 11.3 mm against *Escherichia coli*, 20.3 mm against *Pseudomonas aeruginosa*, and 19.3 mm against *Salmonella typhi* A (Rashid et al., 1996). From the same plant, kolavenic acid (100 µg/disc) developed a 10.3 mm diameter inhibition zone against *Shigella dysenteriae*, and 11 mm for *Salmonella typhi* A (Rashid et al., 1996). 16β-hydroxycleroda-3,13-dien-15,16-olide (100 µg/disc) developed a 14.6 mm diameter halo against *Shigella dysenteriae*, 13.3 mm for *Escherichia coli*, 14.3 mm for *Pseudomonas aeruginosa*, and 14.3 mm for Salmonella typhi A (Rashid et al., 1996).

*Strong broad-spectrum antibacterial clerodane diterpene:* 16α-Hydroxy-cleroda-3,13 (14) Z-diene-15,16-olide and 16-oxo-cleroda-3, 13(14) E-diene-15-oic acid from the seeds inhibited the growth of *Staphylococcus aureus* with MIC of 6.2 and 12.5 µg/mL (Murthy et al., 2005). These diterpenes inhibited the growth of *Escherichia coli*, *Pseudomonas aeruginosa*, *Salmonella thyphi*, and *Klebsiella pneumonia* with MIC values of 0.7 and 1.5, 0.7 and 3.1, 0.7 and 1.5, and 1.5 and 1.5 µg/mL, respectively (Murthy et al., 2005). (-)-14, 15-bisnor-3, 11E-kolavadien-13-one and (-)-3, 12E-kolavadien-15-oic acid-16-al isolated from the leaves inhibited the growth of *Escherichia coli* and *Staphylococcus aureus* with an MIC value of 50 µg/mL (Murthy et al., 2005). (-)-16-Oxocleroda-3,13(14)Edien-15-oic acid isolated from the leaves inhibited the growth of *Escherichia coli* and *Staphylococcus aureus* with MIC values of 50 and 25 µg/mL, respectively (Koneni et al., 2009). (-)-16α-Hydroxycleroda-3,13 (14)Z-dien-15,16-olide isolated from the leaves inhibited the growth of *Staphylococcus aureus* with an MIC value of 6.2 µg/mL. (+)-(4→2)-Abeo-16(R/S)-2, 13Z-kolavadien-15, 16-olide-3-al isolated from the leaves inhibited the growth of *Staphylococcus aureus* with an MIC of 25 µg/mL (Koneni et al., 2009). (-)-3β, 16α-Dihydroxycleroda-4(18), 13(14)Z-dien-15,16-olide isolated from the leaves inhibited the growth of *Escherichia coli* and *Pseudomonas aeruginosa* with an MIC value of 50 µg/mL (Koneni et al., 2009). (-)-labd-13E-en-8-ol-15-oic acid isolated from the leaves inhibited the growth of *Escherichia coli* with an MIC value of 50 µg/mL (Koneni et al., 2009). The seeds shelter series of diterpenoids of with 16α-hydroxy-cleroda-3,13 (14)-Z-diene-15,16-olide and 16-oxo-cleroda-3, 13(14)-E-diene-15-oic acid, which demonstrated antibacterial activities (Murthy et al., 2005; Katkar et al., 2010).

16α-Hydroxy-cleroda-3,13 (14)Z-diene-15,16-olide

*Antibiotic potentiator diterpene:* 16α-Hydroxycleroda-3,13(14)-Z-dien-15,16-olide from the leaves elicited synergistic effect with norfloxacin, ciprofloxacin, and ofloxacin against several methicillin-resistant *Staphylococcus aureus* strains via inhibition of multidrug-resistant pumps (Gupta et al., 2016).

*Strong broad-spectrum antibacterial protoberberine alkaloids:* From the roots, the 8-oxoprotoberberines pendulamine A and B as well as pendulin inhibited *Staphylococcus aureus* with MIC of 0.2, 0.2, and 12.5 µg/mL, respectively (Faizi et al., 2003). Pendulamine A was active against *Klebsiella pneumonia* with MIC of 2 µg/mL and *Salmonella typhi* with MIC values of 0.02 2 and 0.2 µg/mL (Faizi et al., 2003).

Pendulamine A

*Antibacterial Gram-positive* halo developed by azafluorene alkaloid: Pendulin isolated from the roots inhibited *Bacillus subtillis, Corynebacterium hoffmanii, Staphylococcus aureus,* and *Streptococcus faecalis* with MIC values of 25, 12.5, 12.5, and 12.5 μg/disc, respectively (6 mm discs)(Faizi et al., 2008).

*Strong broad-spectrum antifungal diterpenes*: The clerodane diterpene 16-oxocleroda-3, 13*E*-dien-15-oic acid isolated from the bark exhibited antibacterial properties and inhibited *Asperigillus fumigatus, Saccharromyces caulbequence, Saccharomyces cerevaceae, Candida albicans,* and *Hensila californica* (Rashid et al., 1996). (-)-16-oxocleroda-3,13(14)Edien-15-oic acid isolated from the leaves strongly inhibited the growth of *Sporothrix schenckii* with an MIC of 50 μg/mL (Koneni et al., 2009). (-)-16α-hydroxycleroda-3,13 (14)Z-dien-15,16-olide isolated from the leaves inhibited the growth *Sporothrix schenckii* with an MIC value of 6.2 μg/mL

16-Oxocleroda-3, 13*E*-dien-15-oic

*Moderate broad-spectrum antifungal diterpenes*: 6α-Hydroxycleroda-3,13(14)Z-dien-15,16-olide isolated from the leaves inhibited *Candida albicans* (NCIM3557), *Cryptococcus neoformans* (NCIM3542 human pathogens), and *Neurospora crassa* (NCIM870, saprophyte) with $MIC_{90}$ values of 50.3, 100.6, and 201.2 μM, respectively (Bhattacharya et al., 2015). In *Candida albicans* (NCIM3557), this clerodane diterpene induced the generation of reactive oxygen species (Bhattacharya et al., 2015). The seeds shelter diterpenoids including 16α-hydroxy-cleroda-3,13 (14)-Z-diene-15,16-olide and 16-oxo-cleroda-3, 13(14)-E-diene-15-oic acid, which demonstrated antifungal activities (Murthy et al., 2015).

Commentary: Consider that clerodane diterpenes are not uncommonly able to inhibit NorA efflux pumps in *Staphylococcus aureus* (Sharma et al., 2019). Note that compounds able to inhibit the growth of both bacteria and fungi are often cytotoxic.

### References

Bhattacharya, A.K., Chand, H.R., John, J. and Deshpande, M.V., 2015. Clerodane type diterpene as a novel antifungal agent from *Polyalthia longifolia* var. pendula. *European Journal of Medicinal Chemistry, 94*, pp. 1–7.

Faizi, S., Khan, R.A., Azher, S., Khan, S.A., Tauseef, S. and Ahmad, A., 2003. New antimicrobial alkaloids from the roots of *Polyalthia longifolia* var. pendula. *Planta Medica, 69*(04), pp. 350–355.

Faizi, S., Khan, R.A., Mughal, N.R., Malik, M.S., Sajjadi, K.E.S. and Ahmad, A., 2008. Antimicrobial activity of various parts of *Polyalthia longifolia* var. pendula: isolation of active principles from the leaves and the berries. *Phytotherapy Research, 22*(7), pp. 907–912.

Gupta, V.K., Tiwari, N., Gupta, P., Verma, S., Pal, A., Srivastava, S.K. and Darokar, M.P., 2016. A clerodane diterpene from *Polyalthia longifolia* as a modifying agent of the resistance of methicillin resistant *Staphylococcus aureus*. *Phytomedicine, 23*(6), pp. 654–661.

Katkar, K.V., Suthar, A.C. and Chauhan, V.S., 2010. The chemistry, pharmacologic, and therapeutic applications of Polyalthia longifolia. *Pharmacognosy reviews, 4*(7), p. 62.

Murthy, M.M., Subramanyam, M., Bindu, M.H. and Annapurna, J., 2005. Antimicrobial activity of clerodane diterpenoids from *Polyalthia longifolia* seeds. *Fitoterapia, 76*(3–4), pp. 336–339.

Rashid, M.A., Hossain, M.A., Hasan, C.M. and Reza, M.S., 1996. Antimicrobial diterpenes from *Polyalthia longifolia* var. pendulla (Annonaceae). *Phytotherapy Research, 10*(1), pp. 79–81.

Sashidharaa, K.V., Singha. S.P. and Shuklab, P.K., 2009. Antimicrobial evaluation of clerodane diterpenes from *Polyalthia longifolia* var. pendula. *Natural Product Communications, 4*(3), pp. 327–330.

Sharma, A., Gupta, V.K. and Pathania, R., 2019. Efflux pump inhibitors for bacterial pathogens: from bench to bedside. *The Indian Journal of Medical Research, 149*(2), p. 129.

### 2.3.1.8 *Polyalthia sclerophylla* Hook. f. & Thomson

Synonym: *Monoon sclerophyllum* (Hook.f. & Thomson) B. Xue & R.M.K. Saunders

Common names: Karai, mempisang (Malaysia)

Habitat: Forests, fresh water swamps

Distribution: Thailand, Malaysia, and Indonesia

Botanical observation: It is a beautiful, yet endangered tree that grows to 15 m tall. The leaves are simple, alternate, and exstipulate. The blade is coriaceous, up to 20 cm × 5 cm wide with 9–13 pairs of secondary nerves, and elliptic. The flowers are cauliflorous and whitish in the center. The calyx comprises 3 sepals, which are valvate and 0.3–0.4 cm long. The corolla include 6 petals, which are coriaceous, oblong, glossy, and reddish green. The androecium includes numerous stamens. The fruits are ripe carpels, which are ovoid at apex, red, and 2.5–3.7 cm long.

Medicinal uses: Apparently none for antimicrobial diseases.

*Strong antiviral (enveloped monopartite linear dimeric single stranded (+) RNA) amphiphilic ent-kaurane diterpene*: ent-Kaur-16-en-19-oic acid (also known as kaurenoic acid; LogD = 3.4 at pH 7.4; molecular weight = 302.4 g/mol) from the leaves and twigs inhibited the replication of the Human immunodeficiency virus type-1 syncytium formation with an $EC_{50}$ value of 13.7 µg/mL and reverse transcriptase with an $IC_{50}$ of 34.1 µg/mL (Saepou et al., 2010).

Commentary: (i) Annonaceous kaurane-type diterpenes have often the inhibit the replication of the Human immunodeficiency virus (Singh et al., 2005) and reverse transcriptase like 16α-17-dihydroxy-ent-kauran-19-oic acid (Chang et al., 1998).

(ii) The plant produces the benzopyran derivatives (6E, 10E)-isopolycerasoidol, polycerasoidin, and polycerasoidol (González et al., 1996). (6E, 10E)-isopolycerasoidol induces mitochondrial-induced apoptosis in human breast cancer cells (Taha et al., 2015). Mitochondria in eukaryotic cells are vestigial remnants of an α-proteobacterial ancestor (Andersson & Kurland, 1999). It is probably for that reason that natural products from plants that are mitochondrial poisons are also, not uncommonly, antibacterial and antifungal. Therefore, it would be of interest to explore the antibacterial and antifungal properties of benzopyran derivatives in this plant. Note that ent-kaur-16-en-19-oic acid and from *Mitrephora celebica* Scheff. inhibited the growth of *Mycobacterium stegmatis* with the MIC of 6.2 µg/mL (Zgoda-Pols et al., 2002), and hence it would be of interest to assess members of the genus *Polyalthia* Blume for their antimycobacterial diterpenes.

(iii) 2-Methoxy-3-methyl-4,6- dihydroxy-5-(30-hydroxy)cinnamoylbenzaldehyde from a member of the genus *Desmos* Lour. inhibited the Human immunodeficiency virus replication in H9 lymphocytic cells with an $IC_{50}$ value of 12.7 µg/mL and a selectivity index of 489 (Wu et al., 2003), suggesting that simple phenols from Annonaceae should be screened for anti-Human immunodeficiency virus activities. Other anti-Human immunodeficiency virus principle in this family are lignans such as the benzodioxane neolignan (+)-4-O-demethyleusiderin C and the benzofuran lignan licarin A isolated from leaves and stems of a member of the genus *Miliusa* Lesch. ex A. DC. which inhibited the replication of the Herpes simplex virus type-1 with $IC_{50}$ of 62.5 and 66.7 µg/mL, respectively (Sawasdee et al., 2013). These lignans inhibited Herpes simplex virus type-2 (Sawasdee et al., 2013).

(iv) Diterpenes in the Annonaceae have potential against methicillin-resistant *Staphylococcus aureus*, such as ent-trachyloban-19-oic acid which inhibited methicillin-resistant *Staphylococcus aureus* with MIC of 6.2 µg/mL (Zgoda-Pols et al., 2002). Salicylsalicylic acid isolated from the roots of *Fissistigma cavaleriei* inhibited β-lactamase of *Pseudomonas aeruginosa* with an $IC_{50}$ of 71 µM (Yang et al., 2010). Phenolic potentiators for Annonaceae could be of interest to fight bacterial resistance and efforts are needed in this direction. Further, anti-HIV principles are not uncommonly able to inhibit the replication of coronaviruses, thus, anti-Severe acute respiratory syndrome-associated coronavirus molecules may await discovery in this family.

## References

Andersson, S.G. and Kurland, C.G., 1999. Origins of mitochondria and hydrogenosomes. *Current Opinion in Microbiology, 2*(5), pp. 535–541.

Chang, F.R., Yang, P.Y., Lin, J.Y., Lee, K.H. and Wu, Y.C., 1998. Bioactive kaurane diterpenoids from *Annona glabra. Journal of Natural Products, 61*(4), pp. 437–439.

González, M.C., Sentandreu, M.A., Rao, K.S., Zafra-Polo, M.C. and Cortes, D., 1996. Prenylated benzopyran derivatives from two Polyalthia species. *Phytochemistry, 43*(6), pp. 1361–1364.

Saepou, S., Pohmakotr, M., Reutrakul, V., Yoosook, C., Kasisit, J., Napaswad, C. and Tuchinda, P., 2010. Anti-HIV-1 diterpenoids from leaves and twigs of *Polyalthia sclerophylla. Planta Medica, 76*(07), pp. 721–725.

Sawasdee, K., Chaowasku, T., Lipipun, V., Dufat, T.H., Michel, S. and Likhitwitayawuid, K., 2013. New neolignans and a lignan from Miliusa fragrans, and their anti-herpetic and cytotoxic activities. *Tetrahedron Letters, 54*(32), pp. 4259–4263.

Singh, I.P., Bharate, S.B. and Bhutani, K.K., 2005. Anti-HIV natural products. *Current Science*, pp. 269–290.

Taha, H., Looi, C.Y., Arya, A., Wong, W.F., Yap, L.F., Hasanpourghadi, M., Mohd, M.A., Paterson, I.C. and Ali, H.M., 2015. (6E, 10E) isopolycerasoidol and (6E, 10E) isopolycerasoidol methyl ester, prenylated benzopyran derivatives from *Pseuduvaria monticola* induce mitochondrial-mediated apoptosis in human breast adenocarcinoma cells. *PLoS One, 10*(5), p. e0126126.

Wu, J.H., Wang, X.H., Yi, Y.H. and Lee, K.H., 2003. Anti-AIDS agents 54. A potent anti-HIV-1 chalcone and flavonoids from genus Desmos. *Bioorganic & Medicinal Chemistry Letters, 13*(10), pp. 1813–1815.

Yang, Z., Niu, Y., Le, Y., Ma, X. and Qiao, C., 2010. Beta-lactamase inhibitory component from the roots of *Fissistigma cavaleriei. Phytomedicine, 17*(2), pp. 139–141.

Zgoda-Pols, J.R., Freyer, A.J., Killmer, L.B. and Porter, J.R., 2002. Antimicrobial diterpenes from the stem bark of *Mitrephora celebica. Fitoterapia, 73*(5), pp. 434–438.

### 2.3.1.9 *Polyalthia suberosa* (Roxb.) Thwaites

Synonyms: *Guatteria suberosa* (Roxb.) Dunal; *Uvaria suberosa* Roxb.

Common names: Ashok; an luo (China); dudduga chettu (India); baling manok (the Philippines); ching klom (Thailand); quan dau vo sop (Vietnam)

Habitat: Forests

Distribution: India, Sri Lanka, Myanmar, Cambodia, Laos, Vietnam, China, Thailand, Malaysia, and the Philippines.

Botanical observation: It is a tree that grows to about 5 m tall. The bark is suberose. The leaves are simple, alternate, and exstipulate. The petiole is about 3 mm long. The blade is elliptic-oblong, 5–10 × 2–5 cm, thin, with 8–10 pairs of secondary nerves, acute at base, and obtuse to rounded at apex. The inflorescences are extra-axillary, leaf-opposed, and solitary on an about 2 cm long peduncle. The calyx includes 3 sepals, which are ovate-triangular and minute. The corolla has 6 yellowish green petals, which are oblong-lanceolate, coriaceous, and about 1 cm long. The androecium comprises numerous minute stamens. The gynoecium includes numerous minute carpels. The fruits are ripe carpels, which are up to 18, red, globose, about 5 mm in diameter, and containing 1 or 2 seeds.

Medicinal use: Puerperal fever (India)

*Broad-spectrum antibacterial halo developed by polar extract:* Ethanol extract of leaves inhibited the growth of *Streptococcus pneumoniae*, *Vibrio cholerae*, and *Pseudomonas aeruginosa* with inhibition zone diameters of 17, 19, and 18 mm, respectively (0.5 mg/disc) (Uddin et al., 2008).

*Strong antiviral (enveloped monopartite linear dimeric single-stranded (+) RNA) lipophilic lanostane triterpene*: From the stems and leaves suberosol (LogD = 7.8 at pH 7.4; molecular weight = 454.7 g/mol) inhibited the replication of the Human immunodeficiency virus in H9 lymphocyte cells with an $EC_{50}$ value of 3 µg/mL, while it inhibited uninfected H9 cell growth with an $EC_{50}$ value of 20 µg/ mL (Li et al., 1993).

Suberosol

Commentary: Lipophilic triterpenes are not uncommonly able to inhibit the replication of the HIV *in vitro*.

### References

Li, H.Y., Sun, N.J., Kashiwada, Y., Sun, L., Snider, J.V., Cosentino, L.M. and Lee, K.H., 1993. Anti-AIDS agents, 9. Suberosol, a new C31 lanostane-type triterpene and anti-HIV principle from *Polyalthia suberosa*. *Journal of Natural Products*, 56(7), pp. 1130–1133.

Uddin, S.J., Rouf, R., Shilpi, J.A., Alamgir, M., Nahar, L. and Sarker, S.D., 2008. Screening of some Bangladeshi medicinal plants for in vitro antibacterial activity. *Oriental Pharmacy and Experimental Medicine*, 8(3), pp. 316–321.

## 2.3.2 Family Magnoliaceae A.L de Jussieu (1789)

The family Magnoliaceae consists of about 15 genera and 240 species of handsome shrubs or trees. The leaves are simple, alternate, and stipulate. The flowers are often showy, fragrant, solitary terminal or axillary. The perianth includes 6–45 tepals which are not uncommonly somewhat fleshy.

The androecium consists of numerous stamens. The gynoecium is made of numerous carpels. The fruits are dehiscent or indehiscent follicles containing large seeds. Members of this family produce antimicrobial lignans, essential oils, and aporphine alkaloids.

### 2.3.2.1 *Magnolia grandiflora* L.

Synonyms: *Magnolia ferruginea* Z. Collins ex Raf., *Magnolia foetida* (L.) Sarg., *Magnolia lacunosa* Raf., *Magnolia virginiana* var. *foetida* L.

    Common names: Bull bay, Southern magnolia; he hua mu lan (China)

    Habitat: Cultivated

    Distribution: Temperate Asia

    Botanical observation: It is an ornamental tree native to North America that grows up to 30 m tall. The bark is greyish. The leaves are simple, spiral, and stipulate. The petiole is 1.5–4 cm long and channeled. The blade is oblong, 10–20 × 4–10 cm, coriaceous, glossy and dark green above, with 8–10 pairs of secondary nerves, cuneate at base, and obtuse at apex. The flowers are solitary, terminal, massive, whitish, and fragrant. The perianth includes 9–12 obovate, 6–10 × 5–7 cm, and fleshy tepals. The androecium includes numerous stamens. The gynoecium comprises numerous 1–1.5 cm long carpels. The infructescence is an oblong-ovoid, 7–10 × 4–5, aggregate of numerous numerous carpels, each carpel sheltering a about 1.5 cm long seed.

    Medicinal use: Apparently none for microbial infections.

    *Broad-spectrum antibacterial halo developed by aporphine alkaloid:* The oxoaporphine alkaloid liriodenine elicited antibacterial effect against a broad range of Gram-positive and Gram-negative bacteria with an inhibition zone diameter ≥ 12 mm (10 μg/disc) (Khan et al., 2002). Bacteria included, for instance, *Salmonella typhi*, *Escherichia coli*, and *Klebsiella pneumonia* (Khan et al., 2002).

    *Moderate antibacterial (Gram-positive) essential oil:* The essential oil from the leaves inhibited the growth of *Staphylococcus aureus* (ATCC 12598) and *Streptococcus pyogenes* with MIC values of 500 and 125 μg/mL (Guerra-Boone et al., 2013).

    *Strong broad-spectrum amphiphilic antibacterial neolignans:* Magnolol, honokiol, and 3,5′-diallyl-2′-hydroxy-4-methoxybiphenyl inhibited the growth of Gram-positive bacteria (Clark et al., 1981). Honokiol (LogD = 4.0 at pH 7.4; molecular weight = 266.3 g/mol) inhibited the growth of *Staphylococcus aureus*, *Bordetella bronchiseptica*, *Salmonella typhi*, *Bacillus subtilis*, *Micrococcus flavus*, *Bacillus cereus*, and *Serratia marcescens* with MIC values of 50, 50, 100, 25, 25, 25, and 100 μg/mL (Ho et al., 2001). Magnolol (LogD = 4.0 at pH 7.4; molecular weight = 266.3 g/mol) inhibited the growth of *Staphylococcus aureus*, *Brucella bronchiseptica*, *Bacillus subtilis*, *Micrococcus flavus*, *Bacillus cereus*, and *Serratia marcescens* with MIC values of 38, 25, 25, 25, 25, and 25 μg/mL (Ho et al., 2001).

Magnolol

Honokiol

*Strong broad-spectrum amphiphilic antifungal neolignans:* Magnolol inhibited the growth of *Trichophyton mentagrophytes* (KCTC 6077), *Microsporium gypseum* (KCTC 1252), *Epidermophyton floccosum* (KCTC 1246), *Cryptococcus neoformans* (KCTC 7224), *Aspergillus niger* (KCTC 1700), and *Candida albicans* (KCTC 1940) with MIC values of 50, 50, 25, >200, 100, and 25 µg/mL, respectively (Bang et al., 2000). Honokiol inhibited *Trichophyton mentagrophytes* (KCTC 6077), *Microsporium gypseum* (KCTC 1252), *Epidermophyton floccosum* (KCTC 1246), *Cryptococcus neoformans* (KCTC 7224), *Aspergillus niger* (KCTC 1700), and *Candida albicans* (KCTC 1940) with MIC values of 25, 25, 25, 50, 50, and 25 µg/mL, respectively (Bang et al., 2000).

*Broad-spectrum antifungal aporphine alkaloid:* Liriodenine inhibited the growth of a very broad spectrum of phytopathogen fungi (De la Cruz-Chacón et al., 2011).

*Strong antiviral (enveloped monopartite linear double-stranded DNA) polar extract:* Methanol extract of leaves at the concentration of 1.1 µg/mL inhibited the replication of the Herpes simplex virus type-1 by 76.7% (Mohamed et al., 2010).

*Strong antiviral (enveloped monopartite linear dimeric single-stranded (+) RNA) amphiphilic neolignan:* Magnolol and honokiol inhibited the replication of the Human immunodeficiency virus type-1 with the $IC_{50}$ values of 69.3 and 3.3 µg/mL, respectively (Amblard et al., 2007).

Commentaries:

(i) Antibacterial and/or antifungal compounds that are present in plants (before they get infected by phytopatogenic bacteria or fungi) are termed "phytoanticipins" (Van Etten et al., 1994). The plant constructs series of aporphine alkaloids including magnoflorine, lanuginosine, liriodenine, and anonaine (Mohamed et al., 2010) which are "phytoanticipins." Liriodenine is a planar aporphine alkaloid inhibiting topoisomerases (Woo et al., 1997). In fact natural products inhibiting the growth of both bacteria and fungi (such as liriodenine) almost always target DNA. Note that phytoanticipins are usually both antibacterial and antifungal to protect plants against bacteria and fungi at the same time. Liriodenine is common in members of the Magnoliaceae Annonaceae, and Lauraceae (González-Esquinca et al., 2014) and a well-known antibacterial (Hufford et al., 1975; 1980).

(ii) *Serratia marcescens* is an opportunistic nosocomial Gram-negative bacterium (family Enterobacteriaceae), which is responsible for pneumonia, lower respiratory tract infection, urinary tract infection, bloodstream infection, wound infection and meningitis (Khanna et al., 2013). *Bordetella bronchiseptica* (*Brucella bronchiseptica*) (family Alcaligenaceae) is a Gram-negative coccobacilli responsible for respiratory tract disease in wild and domestic mammals and, in some cases, humans (Woolfrey & Moody, 1991).

(iii) Methanol extract of seeds of *Magnolia kobus* DC at the concentration of 100 µg/mL inhibited RNA dependent DNA polymerase and ribonuclease H activities of Human immunodeficiency virus-1 reverse transcriptase by 46.7 and 33.2%, respectively (Min et al., 2001).

## References

Amblard, F., Govindarajan, B., Lefkove, B., Rapp, K.L., Detorio, M., Arbiser, J.L. and Schinazi, R.F., 2007. Synthesis, cytotoxicity, and antiviral activities of new neolignans related to honokiol and magnolol. *Bioorganic & Medicinal Chemistry Letters, 17*(16), pp. 4428–4431.

Bang, K.H., Kim, Y.K., Min, B.S., Na, M.K., Rhee, Y.H., Lee, J.P. and Bae, K.H., 2000. Antifungal activity of magnolol and honokiol. *Archives of Pharmacal Research, 23*(1), pp. 46–49.

Clark, A.M., El-Feraly, A.S. and Li, W.S., 1981. Antimicrobial activity of phenolic constituents of *Magnolia grandiflora* L. *Journal of Pharmaceutical Sciences, 70*(8), pp. 951–952.

De la Cruz-Chacón, I., González-Esquinca, A.R., Fefer, P.G. and Garcia, L.F.J., 2011. Liriodenine, early antimicrobial defence in *Annona diversifolia*. *Zeitschrift für Naturforschung C, 66*(7–8), pp. 377–384.

González-Esquinca, A.R., De-La-Cruz-Chacón, I., Castro-Moreno, M., Orozco-Castillo, J.A. and Riley-Saldaña, C.A., 2014. Alkaloids and acetogenins in Annonaceae development: biological considerations. *Revista Brasileira de Fruticultura, 36*(SPE1), pp. 01–16.

Guerra-Boone, L., Alvarez-Román, R., Salazar-Aranda, R., Torres-Cirio, A., Rivas-Galindo, V.M., Waksman de Torres, N., Gonzalez Gonzalez, G.M. and Pérez-López, L.A., 2013. Chemical compositions and antimicrobial and antioxidant activities of the essential oils from *Magnolia grandiflora*, *Chrysactinia mexicana*, and *Schinus molle* found in northeast Mexico. *Natural Product Communications, 8*(1), pp. 135–138.

Ho, K.Y., Tsai, C.C., Chen, C.P., Huang, J.S. and Lin, C.C., 2001. Antimicrobial activity of honokiol and magnolol isolated from *Magnolia officinalis*. *Phytotherapy Research, 15*(2), pp. 139–141.

Hufford, C.D., Funderburk, M.J., Morgan, J.M. and Robertson, L.W., 1975. Two antimicrobial alkaloids from heartwood of *Liriodendron tulipifera* L. *Journal of Pharmaceutical Sciences, 64*(5), pp. 789–792.

Hufford, C.D., Sharma, A.S. and Oguntimein, B.O., 1980. Antibacterial and antifungal activity of liriodenine and related oxoaporphine alkaloids. *Journal of Pharmaceutical Sciences, 69*(10), pp. 1180–1183.

Khan, M.R., Kihara, M. and Omoloso, A.D., 2002. Antimicrobial activity of *Michelia champaca*. *Fitoterapia, 73*(7), pp. 744–748.

Khanna, A., Khanna, M. and Aggarwal, A., 2013. Serratia marcescens-a rare opportunistic nosocomial pathogen and measures to limit its spread in hospitalized patients. *Journal of Clinical and Diagnostic Research: JCDR, 7*(2), p. 243.

Min, B.S., Kim, Y.H., Tomiyama, M., Nakamura, N., Miyashiro, H., Otake, T. and Hattori, M., 2001. Inhibitory effects of Korean plants on HIV -1 activities. *Phytotherapy Research, 15*(6), pp. 481–486.

Mohamed, S.M., Hassan, E.M. and Ibrahim, N.A., 2010. Cytotoxic and antiviral activities of aporphine alkaloids of *Magnolia grandiflora* L. *Natural Product Research, 24*(15), pp. 1395–1402.

VanEtten, H.D., Mansfield, J.W., Bailey, J.A. and Farmer, E.E., 1994. Two classes of plant antibiotics: phytoalexins versus" phytoanticipins". *The Plant Cell, 6*(9), p. 1191.

Woo, S.H., Reynolds, M.C., Sun, N.J., Cassady, J.M. and Snapka, R.M., 1997. Inhibition of topoisomerase II by liriodenine. *Biochemical Pharmacology, 54*(4), pp. 467–473.

Woolfrey, B.F. and Moody, J.A., 1991. Human infections associated with *Bordetella bronchiseptica*. *Clinical Microbiology Reviews, 4*(3), pp. 243–255.

### 2.3.2.2 *Magnolia officinalis* Rehder & E.H. Wilson

Synonyms: *Magnolia biloba* (Rehder & E.H. Wilson) Cheng; *Magnolia officinalis* subsp. *Biloba* (Rehder & E.H. Wilson) Cheng & Law in W. C. Cheng; *Houpoëa officinalis* (Rehder & E.H. Wilson) N.H. Xia & C.Y. Wu

Common name: Hou po (China)

Habitats: Forest, cultivated

Distribution: China

Botanical observation: It is a magnificent tree growing to 20 m tall. The bark is brownish and thick. The leaves are simple, spiral, and stipulate. The petiole is stout and 2.5–4 cm long. The blade is oblong, 22–45 cm × 10–24 cm thick, glaucous below, cuneate at base, entire at margin, and acute or bilobate at apex. The flowers are showy, solitary, terminal, and fragrant. The perianth comprises 9–17 pure white, 8–10 × 4–5 cm, fleshy, oblong-spathulate tepals. The androecium includes numerous showy stamens. The numerous carpels form an ellipsoid-ovoid 2.5–3 cm long mass. The infructescences are ellipsoid-ovoid, 9–15 cm long and include numerous dehiscent carpels, each containing a triangular and up to 1 cm long and about 1 cm long seeds.

Medicinal use: Cough and colds (China)

*Strong broad-spectrum antibacterial extract:* Ethanol extract of leaves inhibited the growth of *Staphylococcus aureus*, *Streptococcus faecalis*, *Listeria monocytogenes*, *Salmonella typhimurium*, *Bacillus anthracis*, and *Escherichia coli* with the MIC values of 0.02, 0.03, 0.5, 1.2, 0.05, and 0.1 µg/mL, respectively (Hu et al., 2011).

*Antibacterial (Gram-positive) halo developed by neolignans:* Magnolol inhibited *Staphylococcus aureus*, methicillin-resistant *Staphylococcus aureus*, *Enterococcus faecalis*, and vancomycin-resistant *Enterococcus faecium* with inhibition zone diameters of 12, 15.3, 12.7, and 14.2 mm (12.5 µg/disc) (Syu et al., 2004). Honokiol inhibited *Staphylococcus aureus*, methicillin-resistant *Staphylococcus aureus*, *Enterococcus faecalis*, and vancomycin-resistant *Enterococcus faecium* with inhibition zone diameters of 16, 14.3, 12, and 15.3 mm, respectively (12.5 µg/disc) (Syu et al., 2004).

*Strong antibacterial (Gram-positive) amphiphilic and lipophilic neolignans:* Honokiol (LogD = 4.0 at pH 7.4; molecular weight = 266.3 g/mol) inhibited the growth of *Staphylococcus aureus* (ATCC6538), *Bacillus subtilis* (ATCC6633), *Propionibacterium acnes* (ATCC6919), and *Propionibacterium granulosum* (ATCC25564) with MIC/MBC values of 13.1/26.6, 8.2/16.7, 4.1/16.7, and 8.2/16.7 µg/mL, respectively (Kim et al., 2010). Honokiol was bactericidal for methicillin-resistant *Staphylococcus aureus* with an MIC value of 12.5 µg/mL (Syu et al., 2004). Magnolol inhibited *Staphylococcus aureus*, *Staphylococcus epidermidis*, *Micrococcus luteus*, and *Bacillus subtilis* with MIC values of 25, 12.5, 12.5, and 12.5 µg/mL, respectively (He et al., 2011). Piperitylmagnolol (LogD = 7.4 at pH 7.4; molecular weight = 402.5 g/mol) was active against *Staphylococcus aureus*, methicillin-resistant *Staphylococcus aureus*, *Enterococcus faecalis*, and vancomycin-resistant *Enterococcus faecium* with MICs of 12.5, 6.2, 6.2, and 6.2 µg/mL, respectively. In time-kill assay, this alkaloid was bactericidal against vancomycin-resistant *Enterococcus faecium* (Syu et al., 2004).

Piperitylmagnolol

*Moderate broad-spectrum antifungal neolignans:* Magnolol and honokiol inhibited the growth of *Trichophyton mentagrophytes, Microsporium gypseum, Epidermophyton floccosum, Aspergillus niger, Cryptococcus neoformans,* and *Candida albicans* with MIC in the range of 25–100 µg/mL (Bang et al., 2000).

*Antiviral (enveloped monopartite linear single-stranded (+)RNA) neolignan:* Honokiol inhibited the replication of the Hepatitis virus with an $EC_{50}$ of 1.2 µM (Lan et al., 2012).

*Strong antiviral (enveloped monopartite linear single-stranded (+) RNA) neolignan:* Honokiol at a concentration of 20 µM reduction of Dengue virus production was >90% in both BHK and Huh7 cells by interfering with the endocytotic process of the virus into the host cells (Fang et al., 2015).

*Viral enzyme inhibition by polar extract:* Aqueous extract of bark at a concentration of 250 µg/mL inhibited the Human immunodeficiency virus type-1 protease by 25.7% (Xu et al., 1996).

Commentaries: (i) *Propionibacterium acnes* (family Propionibacteriaceae) is a Gram-positive, facultative, anaerobic rod that is a major colonizer and inhabitant of the human skin and is an important opportunistic pathogen (Achermann et al., 2014).

(ii) Dengue is an acute febrile disease caused by Dengue viruses (DENVs) (family Flaviviridae, genus Flavivirus), which comprise 4 serotypes (DENV 1–4). According to the World Health Organization (WHO), approximately 500,000 people suffer from dengue hemorrhagic fever requiring hospitalization, a large proportion being children from developing countries every year (Tuiskunen Bäck & Lundkvist, 2013). To date, there is no drug or vaccine for dengue.

(iii) The plant is used for the treatment of COVID-19 in China (Wang et al., 2020) and one could examine the effect of honokiol and magnolol against the Severe acute respiratory syndrome-associated coronavirus type-2.

(iv) Aqueous extract of flower buds of *Magnolia fargesii* Cheng and aqueous extract of bark of *Magnolia obovata* Thunb. at the concentration of 250 µg/mL inhibited the Human immunodeficienct virus protease by 89.2 and 25.7% (Xu et al., 1996).

### References

Achermann, Y., Goldstein, E.J., Coenye, T. and Shirtliff, M.E., 2014. Propionibacterium acnes: from commensal to opportunistic biofilm-associated implant pathogen. *Clinical Microbiology Reviews, 27*(3), pp. 419–440.

Bang, K.H., Kim, Y.K., Min, B.S., Na, M.K., Rhee, Y.H., Lee, J.P. and Bae, K.H., 2000. Antifungal activity of magnolol and honokiol. *Archives of Pharmacal Research, 23*(1), pp. 46–49.

Fang, C.Y., Chen, S.J., Wu, H.N., Ping, Y.H., Lin, C.Y., Shiuan, D., Chen, C.L., Lee, Y.R. and Huang, K.J., 2015. Honokiol, a lignan biphenol derived from the Magnolia tree, inhibits dengue virus type 2 infection. *Viruses, 7*(9), pp. 4894–4910.

He, X.F., Wang, X.N., Yin, S., Dong, L. and Yue, J.M., 2011. Ring A-seco triterpenoids with antibacterial activity from *Dysoxylum hainanense. Bioorganic & Medicinal Chemistry Letters, 21*(1), pp. 125–129.

Hu, Y., Qiao, J., Zhang, X. and Ge, C., 2011. Antimicrobial activity of *Magnolia officinalis* extracts in vitro and its effects on the preservation of chilled mutton. *Journal of Food Biochemistry, 35*(2), pp. 425–441.

Kim, Y.S., Lee, J.Y., Park, J., Hwang, W., Lee, J. and Park, D., 2010. Synthesis and microbiological evaluation of honokiol derivatives as new antimicrobial agents. *Archives of Pharmacal Research, 33*(1), pp. 61–65.

Lan, K.H., Wang, Y.W., Lee, W.P., Lan, K.L., Tseng, S.H., Hung, L.R., Yen, S.H., Lin, H.C. and Lee, S.D., 2012. Multiple effects of Honokiol on the life cycle of hepatitis C virus. *Liver International, 32*(6), pp. 989–997.

Syu Jr, W., Shen, C.C., Lu, J.J., Lee, G.H. and Sun, C.M., 2004. Antimicrobial and cytotoxic activities of neolignans from *Magnolia officinalis. Chemistry & Biodiversity, 1*(3), pp. 530–537.

Tuiskunen Bäck, A. and Lundkvist, Å., 2013. Dengue viruses–an overview. *Infection Habitat & epidemiology, 3*(1), p. 19839.

Wang, S.X., Wang, Y., Lu, Y.B., Li, J.Y., Song, Y.J., Nyamgerelt, M. and Wang, X.X., 2020. Diagnosis and treatment of novel coronavirus pneumonia based on the theory of traditional Chinese medicine. *Journal of Integrative Medicine.*

Xu, H.X., Wan, M., Loh, B.N., Kon, O.L., Chow, P.W. and Sim, K.Y., 1996. Screening of traditional medicines for their inhibitory activity against HIV-1 protease. *Phytotherapy Research, 10*(3), pp. 207–210.

### 2.3.2.3 *Manglietiastrum sinicum* Y.W. Law

Synonyms: *Magnolia sinica* (Y. W. Law) Nooteboom; *Manglietia sinica* (Y.W. Law) B.L. Chen & Noot.; *Pachylarnax sinica* (Y.W. Law) N.H. Xia & C.Y. Wu

Common name: Hua gai mu (China)

Habitat: Forests

Distribution: China

Botanical observation: It is a beautiful tree that grows to 40 m tall. The bark is pale greyish. The leaves are simple, spiral, and stipulate. The petiole is 1.5–2 cm long. The blade is obovate-spathulate, 15–26 cm × 5–8 cm, coriaceous, dark green above, glossy, marked with a raised mid-rib and 13–16 pairs of secondary nerves, cuneate at base and acuminate at apex. The flowers are terminal, showy, and solitary. The perianth comprises 9 whitish and obovate to spathulate tepals. The androecium includes numerous stamens. The gynoecium includes 13–16 carpels. The infruc-tescence is oblong, 5–8.5 × 3.5–6.5 cm, and made of numerous carpels, each containing a flat and about 1.3 cm long seed.

Medicinal use: Colds (China)

*Strong antibacterial (Gram-positive) lignans*: The eupodienone-type lignan manglisin A, the dibenzocyclooctadiene-type lignan manglisin B, and the tetrahydrodrofuran-type lignans mangli-sin C, D, and E isolated from the carpels inhibited the growth of *Staphylococcus aureus* with the MIC values of 0.1, 0.06, 0.06, 0.06, and 0.1 μM (Ding et al., 2014). Manglisin A, B, C, and D isolated from the carpels inhibited the growth of methicillin-resistant *Staphylococcus aureus* with MIC values ranging from 0.06 to 0.1 μM (Ding et al., 2014).

Manglisin A

Commentary: Consider the 2.5-cyclohexadienone moiety of manglisin A is highly reactive and form covalent bonds with the amino acids in bacterial proteins and peptides (Wipf and Jung, 1999).

**References**

Ding, J.Y., Yuan, C.M., Cao, M.M., Liu, W.W., Yu, C., Zhang, H.Y., Zhang, Y., Di, Y.T., He, H.P., Li, S.L. and Hao, X.J., 2014. Antimicrobial constituents of the mature carpels of *Manglietiastrum sinicum. Journal of Natural Products, 77*(8), pp. 1800–1805.

Wipf, P. and Jung, J.K., 1999. Nucleophilic additions to 4,4-disubstituted 2,5-cyclohexadienones: can dipole effects control facial selectivity? *Chemical Reviews, 99*(5), pp. 1469–1480.

## 2.3.2.4 *Michelia alba* DC.

Synonyms: *Michelia longifolia* Bl.; *Michelia longifolia* var. *racemosa* Bl.; *Sampacca longifolia* (Bl.) Kuntze

Common names: White champaca; bail lan (China); champaka (India); cempaka puteh (Indonesia; Malaysia); champi (Laos, Thailand); tsampakang puti (the Philippines); ngoc lan tran (Vietnam)

Habitat: Cultivated

Distribution: Tropical Asia

Botanical observation: It is a graceful tree, probably native to Indonesia, that grows up to about 15 m tall. The bark is smooth and beautifully variegated with white, grey, and light green patterns. The leaves are simple, spiral, stipulate, and fragrant. The petiole is 1.5–2 cm long. The leaf blade is narrowly elliptic, 10–27 × 4–9.5 cm, coriaceous, acute at base, acuminate at apex, and with an entire margin. The flowers are solitary, axillary, and fragrant. The perianth includes 10, white, linear-lanceolate, fleshy, and 3–4 cm × 0.3–0.5 cm tepals. The androecium presents numerous linear and flat stamens. The gynoecium comprises numerous carpels. The infructescences are some sorts of grapes of ovoids, woody, lenticelled, and about 1 cm follicles. The seeds are food for the Asian glossy starling bird.

Medicinal use: Post-partum (Malaysia)

*Broad-spectrum antibacterial halo developed by amphiphilic oxoaporphine alkaloid:* Annonaine inhibited the growth of *Bacillus cereus* (ATCC-14.579), *Escherichia coli* (ATCC-11.105), *Staphylococcus aureus* (ATCC-6538), and *Staphylococcus epidermidis* (ATCC-12.228) with inhibition zone diameters of 20, 8, 14, and 12 mm, respectively (1 mg/mL, 70 μL/well) (Paulo et al., 1992).

Annonaine

*Strong broad-spectrum antibacterial amphiphilic aporphine alkaloid:* Lysicamine (LogD = 3.6 at pH 7.4; molecular weight = 291.3 g/mL) isolated from the leaves inhibited the growth of *Staphylococcus epidermidis* with the MIC value of 50 μg/mL (Costa et al., 2010). Lysicamine inhibited the growth of *Bacillus cereus, Bacillus subtilis, Listeria monocytogenes, Micrococcus luteus, Proteus vulgaris*, methicillin-sensitive *Staphylococcus aureus, Streptococcus agalactiae, Streptococcus pneumoniae, Actinobacillus* sp., and extended-spectrum beta-lactamase-producing *Klebsiella pneumoniae* with the MIC values of 10, 10, 2.5, 10, 10, 1, 10, 4, 2.5, and 20 μg/mL, respectively (Tan et al., 2015).

Lysicamine

*Weak broad-spectrum antibacterial monoterpene:* Essential oil from the flowers and leaves contain linalool (Ueyama et al., 1992). Linalool inhibited the growth of *Streptococcus mutans* (ATCC 25175) with the MIC/MBC value of 1.6/3.2 mg/mL (Park et al., 2012). Linalool inhibited *Escherichia coli* (ATCC 25922) with an MIC of 4.5 mg/mL and an MBC of 5.4 mg/mL and *Salmonella typhimurium* (ATCC 14028) with an MIC of 5.4 mg/mL and an MBC of 5.4 mg/mL (Nguyen et al., 2016).

*Broad-spectrum antifungal monoterpene:* Linalool inhibited the growth of 10 fungi (Pattnaik et al., 1997).

*Moderate broad-spectrum antifungal aporphine alkaloids*: Lysicamine inhibited the growth of *Candida dubliniensis* (ATCC 777) with an MIC value of 100 µg/mL (Costa et al., 2010). Annonaine at 500 µg/mL inhibited the growth of *Trichophyton rubrum* (ICB-A04) and *Microsporum gypseum* (ICB-281) (Paulo et al., 1992).

*Anticandidal monoterpene*: Linalool inhibited *Candida albicans* with an MIC value of 500 ppm (Tampieri et al., 2005).

Commentary: Extended spectrum beta-lactamases (ESBLs) are enzymes produced by Gram-negative bacteria (Bradford, 2001) opening the β-lactam ring of cephalosporins, such as ceftazidime, ceftriaxone, and cefotaxime (Ghafourian et al., 2015). Gram-negative bacteria are equipped with beta-lactamase enzymes which are coded from genes in their natural chromosome or from genes acquired from other bacteria via plasmids such as the TEM-1 beta-lactamase (originally found in a single strain of *Escherichia coli* isolated from a blood culture from a patient named Temoniera in Greece, hence the designation) which has spread globally (Bradford, 2001). Over the last 40 years numerous new beta-lactam antibiotics have been developed to attempt to fight bacterial resistance and this was systematically followed by the generation of new beta-lactamase to the point that some Gram-negative bacteria are equipped with more that 150 types of extended-spectrum – lactamases which are posing huge clinical burden in hospitals (Bradford, 2001). This "plague" is responsible for increasing fatalities worldwide (Blomber et al., 2005). One could anticipate the emergence of bacteria resistant to all beta-lactam antibiotics.

### References

Blomberg, B., Jureen, R., Manji, K.P., Tamim, B.S., Mwakagile, D.S., Urassa, W.K., Fataki, M., Msangi, V., Tellevik, M.G., Maselle, S.Y. and Langeland, N., 2005. High rate of fatal cases of pediatric septicemia caused by gram-negative bacteria with extended-spectrum beta-lactamases in Dar es Salaam, Tanzania. *Journal of clinical microbiology*, 43(2), pp.745–749.

Bradford, P.A., 2001. Extended-spectrum β-lactamases in the 21st century: characterization, epidemiology, and detection of this important resistance threat. *Clinical microbiology reviews*, 14(4), pp.933–951.

Chen, C.Y., Huang, L.Y., Chen, L.J., Lo, W.L., Kuo, S.Y., Wang, Y.D., Kuo, S.H. and Hsieh, T.J., 2008. Chemical constituents from the leaves of *Michelia alba*. *Chemistry of Natural Compounds, 44*(1), pp. 137–139.

Costa, E.V., Pinheiro, M.L.B., Barison, A., Campos, F.R., Salvador, M.J., Maia, B.H.L., Cabral, E.C. and Eberlin, M.N., 2010. Alkaloids from the bark of *Guatteria hispida* and their evaluation as antioxidant and antimicrobial agents. *Journal of Natural Products, 73*(6), pp. 1180–1183.

Ghafourian, S., Sadeghifard, N., Soheili, S. and Sekawi, Z., 2015. Extended spectrum beta-lactamases: definition, classification and epidemiology. *Current Issues in Molecular Biology, 17*(1), pp. 11–22.

Nguyen, H.V., Caruso, D., Lebrun, M., Nguyen, N.T., Trinh, T.T., Meile, J.C., Chu-Ky, S. and Sarter, S., 2016. Antibacterial activity of *Litsea cubeba* (Lauraceae, May Chang) and its effects on the biological response of common carp *Cyprinus carpio* challenged with *Aeromonas hydrophila*. *Journal of Applied Microbiology, 121*(2), pp. 341–351.

Park, S.N., Lim, Y.K., Freire, M.O., Cho, E., Jin, D. and Kook, J.K., 2012. Antimicrobial effect of linalool and α-terpineol against periodontopathic and cariogenic bacteria. *Anaerobe, 18*(3), pp. 369–372.

Pattnaik, S., Subramanyam, V.R., Bapaji, M. and Kole, C.R., 1997. Antibacterial and antifungal activity of aromatic constituents of essential oils. *Microbios, 89*(358), pp. 39–46.

Paulo, M.D.Q., Barbosa-Filho, J., Lima, E.O., Maia, R.F., de Cassia, R., Barbosa, B.B.C. and Kaplan, M.A.C., 1992. Antimicrobial activity of benzylisoquinoline alkaloids from *Annona salzmanii* DC. *Journal of Ethnopharmacology, 36*(1), pp. 39–41.

Tampieri, M.P., Galuppi, R., Macchioni, F., Carelle, M.S., Falcioni, L., Cioni, P.L. and Morelli, I., 2005. The inhibition of *Candida albicans* by selected essential oils and their major components. *Mycopathologia, 159*(3), pp. 339–345.

Tan, K.K., Khoo, T.J., Rajagopal, M. and Wiart, C., 2015. Antibacterial alkaloids from *Artabotrys crassifolius* Hook. f. & Thomson. *Natural Product Research, 29*(24), pp. 2346–2349.

Ueyama, Y., Hashimoto, S., Nii, H. and Furukawa, K., 1992. The chemical composition of the flower oil and the leaf oil of *Michelia alba* DC. *Journal of Essential Oil Research, 4*(1), pp. 15–23.

Wang, H.M., Lo, W.L., Huang, L.Y., Wang, Y.D. and Chen, C.Y., 2010. Chemical constituents from the leaves of *Michelia alba*. *Natural Product Research, 24*(5), pp. 398–406.

Wink, M., 2010. Annual plant reviews, functions and biotechnology of plant secondary metabolites, 2nd. Wiley, UK.

## 2.3.3 FAMILY MYRISTICACEAE R. BROWN (1810)

The family Myristicaceae consists of about 21 genera and 50 species of handsome tropical trees yielding after incision of their trunks a characteristic somewhat blood-like sap. The inner bark has sometimes a delicate fragrance (like Annonaceae). The leaves are simple, alternate, and exstipulate. The blades in some species can be occasionally very large. The cymes or racemes are axillary. The perianth is tubular, discrete, somewhat urceolate, fleshy, glossy, and 3-lobed. The androecium includes numerous stamens. The gynoecium presents a single carpel containing a basal and anatropous ovule. The fruits are mostly dehiscent or indehicent drupes, the seed being often large, and more or less developed in a characteristic waxy aril. Members in this family produce antimicrobial essential oils, lignans, and alkylphenols.

### 2.3.3.1 *Knema angustifolia* (Roxb.) Warb.

Synonym: *Knema cinerea* var. *glauca* (Blume) Y.H. Li; *Knema cinerea* (Poir.) Warb. var. *andamanica* auct non (Warb.) J. Sinclair; *Myristica angustifolia* Roxb.; *Myristica gibbosa* Hook.f. & Thomson; Myristica laurifolia auct. non Hook.f. & Thomson

Common names: Motaa-pasuti (India); bol-lanchi (Bangladesh); ramguwaa (Nepal)

Habitat: Forests

Distribution: India, Nepal, and Bangladesh

Botanical observation: It is a tree that grows to about 6 m tall. The thrunk exudes a bright red sap upon incision. The leaves are simple, alternate, and exstipulate. The petiole is about 1 cm long. The blade is elliptic, 13.5–15.5 × 2.5–3.5 cm, papery, acute at base, acuminate at apex, glossy, and marked with 17–20 pairs of secondary nerves. The inflorescences are tuberculate, about 2 cm long, and bear minute flowers. The perianth includes 3 obovate tepals. The androecium is made of 9–10 stamens joined in a column. The ovary is pubescent, developing a short style and stigma lobes connate into a disc. The drupes are bullet-shaped, yellow, 2.5–3 × 1.8–2 cm, and open to show a seed completely imbedded on a glossy and red aril.

Medicinal uses: Mouth sores (India)

*Antibacterial (Gram-negative) halo developed by polar extract*: Ethanol extract of stems inhibited the growth of *Staphylococcus aureus* (ATTC 25923) with an inhibition zone diameter of 10 mm (5 µL of 5 mg/mL/6 mm disc) (Phadungkit et al., 2010).

Commentary: Members of the family Myristicaceae produce amphiphilic and hydrophilic alkyl phenols derived from the polyketide pathways and extractable with mid-polar or polar solvents, (Kozubek & Tyman, 1999). These alkyl phenols have strong antibacterial activity against Gram-positive bacteria. For instance, the amphiphilic alkylphenol knerachelin B (LogD = 4.4 at pH 7.4; molecular weight = 270.3 g/mol) from *Knema furfuracea* (Hook. F. & Thomson) Warb. inhibited the growth of *Staphylococcus aureus* (ATCC 25923), *Streptococcus pneumoniae* (4314-03), *Enterococcus faecalis* (ATCC 29212), and *Enterococcus faecium* (ATCC 19581) with the MIC values of 4, 8, 16, and 16 µg/mL, respectively (Zahir et al., 1993). What is the antibacterial mode of action of knerachelin B? Does it involves membrane destabilization?

Knerachelin B

### References

Kozubek, A. and Tyman, J.H., 1999. Resorcinolic lipids, the natural non-isoprenoid phenolic amphiphiles and their biological activity. *Chemical Reviews*, 99(1), pp. 1–26.

Phadungkit, M., Rattarom, R. and Rattana, S., 2010. Phytochemical screening, antioxidant, antibacterial and cytotoxic activities of *Knema angustifolia* extracts. *Journal of Medicinal Plants Research*, 4(13), pp. 1269–1272.

Zahir, A., Jossang, A., Bodo, B., Hadi, H.A., Schaller, H. and Sevenet, T., 1993. Knerachelins A and B, antibacterial phenylacylphenols from *Knema furfuracea*. *Journal of Natural Products*, 56(9), pp. 1634–1637.

### 2.3.3.2 *Myristica fragrans* Houtt.

Synonyms: *Myristica aromatica* Lam.; *Myristica aromatica* Sw.; *Myristica moschata* Thunb.; *Myristica officinalis* L. f.; *Myristica officinalis* Mart.

Common names: Common nutmeg; joitri fal (Bangladesh); rou dou kou (China); boch kak (Cambodia); ghatasha (India); bunga pala (Indonesia); chan theed (Laos); bua pala (Malaysia); zar date hpo (Myanmar); duguan (the Philippines); chan thet (Thailand); dau khau (Vietnam)

Habitat: Cultivated

Distribution: Southeast Asia

Botanical observation: It is a gradeful tree native to Indonesia that grows to about 8 m tall. The bark is light brown and exudes a dark red sap upon incision. The leaves are simple, alternate, and exstipulate. The petiole is about 1 cm long. The blade is elliptic-oblong, 4–9 cm × 2.5–4.5 cm, cuneate or rounded at base, acuminate at apex, with a pleasant green, glossy, smooth, coriaceous, and with 6–15 pairs of secondary nerves. The cymes are axillary, and 2.5–5 cm long. The perianth is urceolate, fleshy, yellow, 3-lobed, and up to about 1 cm long. The androecium includes 9–12 stamens, which are joined in a column. The ovary is ellipsoid, hairy, and develops a short style and 2 stigmas. The drupes are orange to yellow, emanating a subtile fragrance, pyriform or globose, up to 5 cm in diameter, and open to show a seed about 2 cm long and partially covered with a deeply lacerated red, glossy, and waxy aril (mace).

Medicinal uses: Cholera (India); fever, cough, ulcers (Bangladesh); cholera, pimples (Myanmar)

*Broad-spectrum antibacterial halo developed by essential oil:* Essential oil (15 µL/ 4 mm well) inhibited the growth of *Acinetobacter calcoacerica* (NICB8250), *Bacillus subtilis, Citrobcater freundii, Staphylococcus aureus, Escherichia coli, Enterococcus faecalis, Klebsiella pneumoniae,* and *Yersinia enterolytica* with the inhibition zone diameters of 12.7, 7, 12.8, 24.6, 10.4, 18.5, 16.9, and 7.3 mm, respectively (Dorman & Deans, 2000).

*Strong broad-spectrum antibacterial amphiphilic dibenzylbutane lignans*: Macelignan (LogD = 4.5 at pH 7.4; molecular weight = 328.4 g/mol) from the seeds inhibited *Streptococcus mutans* (ATCC 25175) and *Staphylococcus aureus* (TCC 12600) with the MIC values of 3.9 and 250 µg/mL and MBC of 7.8 and 250 µg/mL, respectively (Chung et al., 2006). At 20 µg/mL macelignan was bactericidal for *Streptococcus mutans* (ATCC 25175) (Chung et al., 2006). Erythro-austrobailignan-6 isolated from the seeds inhibited the growth of *Acidovorax konjaci, Agrobacterium tumefaciens,* and *Burkholderia glumae* with $IC_{50}$ values of 49, 17, and 26 µg/mL, respectively (Cho et al., 2007). Meso-dihydroguaiaretic acid inhibited the growth of *Acidovorax konjaci, Agrobacterium tumefaciens,* and *Burkholderia glumae,* and *Pseudomonas syringae* pv. *lachrymans* with $IC_{50}$ values of 17, 23, 23, and 54 µg/mL, respectively (Cho et al., 2007).

Macelignan

*Moderate antibacterial (Gram-negative) furan lignan:* Nectandrin B inhibited the growth of *Acidovorax konjaci, Agrobacterium tumefaciens,* and *Burkholderia glumae* and *Pseudomonas syringae* pv. *lachrymans* (Cho et al., 2007).

*Strong antibacterial (Gram-positive) amphiphilic alkyl phenols:* Malabaricone B (LogD = 4.5 at pH 7.4; molecular mass 328.4 g/mol) from the mace inhibited the growth of *Staphylococcus aureus, Bacillus subtilis,* and *Streptococcus durans* with MIC values of 1, 1, and 1 µg/mL, respectively (Orabi et al., 1991). Malabaricone C from the mace inhibited the growth of *Staphylococcus aureus*, *Bacillus subtilis*, and *Streptococcus durans* with MIC values of 4, 2, and 4 µg/mL, respectively (Orabi et al., 1991). Malabaricone C from the mace inhibited the growth of *Staphylococcus aureus*, *Bacillus subtilis*, and *Streptococcus durans* with MIC values of 4, 2, and 4 µg/mL, respectively (Orabi et al., 1991).

Malabaricone B

*Strong antifungal (filamentous) dibenzylbutane lignans:* Erythro-austrobailignan-6 isolated from the seeds inhibited the growth of *Alternaria alternata*, *Colletotrichum coccodes*, *Colletotrichum gloeospeoroides*, *Rhizoctoma solani*, and *Magnaporte grisea* with $IC_{50}$ values of 92, 49, 48, 19, and 89 µg/mL, respectively (Cho et al., 2007). Meso-dihydroguaiaretic acid isolated from the seeds inhibited the growth of *Alternaria alternata*, *Colletotrichum coccodes*, *Colletotrichum gloeospeoroides*, *Rhizoctoma solani*, and *Magnaporte grisea* with $IC_{50}$ values ranging from 12 to 63 µg/mL (Cho et al., 2007).

*Strong antifungal (filamentous) furan lignan:* Nectandrin B isolated from the seeds inhibited the growth of *Alternaria alternata*, *Colletotrichum coccodes*, *Colletotrichum gloeospeoroides*, *Rhizoctoma solani*, *Magnaporte grisea*, and *Pythium ultimum* (Cho et al., 2007).

*Strong antiviral (enveloped monopartite linear double-stranded DNA) dibenzylbutane lignan:* Erythro-austrobailignan-6 and meso-dihydroguaiaretic acid inhibited the replication the Herpes simplex virus type-2 with $IC_{50}$ values of 52.5 and 11.6 µg/mL and selectivity indices of 3 and 13.2, respectively (Song et al., 2013).

*Strong antiviral (non-enveloped monopartite linear double-stranded DNA) dibenzylbutane lignan:* Erythro-austrobailignan-6 and meso-dihydroguaiaretic acid inhibited the replication of Adenovirus with $IC_{50}$ values of 46.7 and 74.1 µg/mL and selectivity indices of 7.4 and 4.7, respectively (Song et al., 2013).

*Antiviral (enveloped monopartite linear single-stranded (+)RNA)* extract: Extract inhibited the replication of the mouse Hepatitis virus (신현수, 2007)

Commentaries: (i) Nordihydroguaiaretic acid isolated from a member of the family Zygophyllaceae R. Brown at the concentration of 35 µM inhibited the replication of Dengue virus by about 50% (Soto-Acosta et al., 2014) and inhibited the replication of West Nile Virus (NY99) and Zika Virus (PA259459) with $IC_{50}$ values of 7.9 and 9.1 µM and selectivity indices of 20.5 and 17.8, respectively (Merino-Ramos et al., 2017).

(ii) The lucrative trade of mace accounted to the colonization of Malaysia and Indonesia by Europeans. Further, the Islamization of Malaysia and Indonesia is a result of early Arab trade of mace in the Middle Ages.

(iii) Erythro-austrobailignan-6 is an inhibitor of topoisomerases I and II (Li et al., 2004). topoisomerase inhibitors interfere with adenovirus DNA replication because host cell topoisomerase I activity is required for the elongation of adenovirus nascent DNA (Wong & Hsu, 1990). Thus natural products from medicinal plants with topoisomerase inhibitor activity should be examined for anti-Adenovirus but also any other DNA viruses.

(iv) In the *"Tractato"* of Giovanni Manardi (1462–1536), physician and professor at the University of Ferrara small bags of herbs were carried to prevent plague containing among other things *"garyophili"* (cloves), *"nuce muscata"* (mace), and cinnamon (Vicentini et al., 2020).

(vii) The mouse Hepatitis virus is a coronavirus, thus, one could examine the anti-Severe acute respiratory syndrome properties of *Myristica frangrans* Houytt. and especially its lignans.

(viii) The mace has somewhat a graceful and positive aura and appears at first glance as some of healthy blood vessels. Although highly controversial, the Theory of Signature and other holistic approaches (Graves, 2012) are not to be laughted at by our current late capitalist era pharmaceutical and universities businesses but should receive serious consideration. After all, most of the important drugs we used today are coming from plants selected by shamans, which did not use a rational scientific mind but something far more powerful. Again, COVID-19 is an introduction for something far more sinister and the probable extinction of most humans as a consequence of environmental destruction. One could envisage that some remote tribes (which treat themselves by selecting plants in a holistic manner) in last pockets of jungle might survive the forthcoming pandemics.

## References

신현수, 2007. *Coronavirus Replication Inhibition by Herbal Extracts* (**Doctoral dissertation**).

Cho, J.Y., Choi, G.J., Son, S.W., Jang, K.S., Lim, H.K., Lee, S.O., Sung, N.D., Cho, K.Y. and Kim, J.C., 2007. Isolation and antifungal activity of lignans from *Myristica fragrans* against various plant pathogenic fungi. *Pest Management Science: Formerly Pesticide Science, 63*(9), pp. 935–940.

Chung, J.Y., Choo, J.H., Lee, M.H. and Hwang, J.K., 2006. Anticariogenic activity of macelignan isolated from *Myristica fragrans* (nutmeg) against *Streptococcus mutans*. *Phytomedicine, 13*(4), pp. 261–266.

Dorman, H.J.D. and Deans, S.G., 2000. Antimicrobial agents from plants: antibacterial activity of plant volatile oils. *Journal of Applied Microbiology, 88*(2), pp. 308–316.

Graves, J., 2012. The language of plants: a guide to the doctrine of signatures. Steiner Books.

Li, G., Lee, C.S., Woo, M.H., Lee, S.H., Chang, H.W. and Son, J.K., 2004. Lignans from the bark of *Machilus thunbergii* and their DNA topoisomerases I and II inhibition and cytotoxicity. *Biological and Pharmaceutical Bulletin, 27*(7), pp. 1147–1150.

Merino-Ramos, T., de Oya, N.J., Saiz, J.C. and Martín-Acebes, M.A., 2017. Antiviral activity of nordihydroguaiaretic acid and its derivative tetra-O-methyl nordihydroguaiaretic acid against West Nile virus and Zika virus. *Antimicrobial Agents and Chemotherapy, 61*(8), pp. e00376–17.

Orabi, K.Y., Mossa, J.S. and El-Feraly, F.S., 1991. Isolation and characterization of two antimicrobial agents from mace (*Myristica fragrans*). *Journal of Natural Products, 54*(3), pp. 856–859.

Song, Q.Y., Zhang, C.J., Li, Y., Wen, J., Zhao, X.W., Liu, Z.L. and Gao, K., 2013. Lignans from the fruit of *Schisandra sphenanthera*, and their inhibition of HSV-2 and adenovirus. *Phytochemistry Letters, 6*(2), pp. 174–178.

Soto-Acosta, R., Bautista-Carbajal, P., Syed, G.H., Siddiqui, A. and Del Angel, R.M., 2014. Nordihydroguaiaretic acid (NDGA) inhibits replication and viral morphogenesis of dengue virus. *Antiviral Research, 109*, pp. 132–140.

Vicentini, C.B., Manfredini, S., Mares, D., Bonacci, T., Scapoli, C., Chicca, M. and Pezzi, M., 2020. Empirical "integrated disease management" in Ferrara during the Italian plague (1629–1631). *Parasitology international, 75*, p.102046.

Wong, M.L. and Hsu, M.T., 1990. Involvement of topoisomerases in replication, transcription, and packaging of the linear adenovirus genome. *Journal of Virology, 64*(2), pp. 691–699.

### 2.3.3.3 *Myristica malabarica* Lam.

Common names: False nutmeg, Malabar nutmeg; malati (India)
  Habitat: Forests
  Distribution: India

Botanical observation: It is a handsome buttressed tree that grows to about 25 m tall. The bark is light brown and exudes a dark red sap upon incision. The leaves are simple, alternate, and exstipulate. The petiole is about 1 cm long. The blade is elliptic-oblong, 10–15 cm × 3.5–5.5 cm, acute at base, acuminate at apex, glossy, coriaceous, and with 8–14 pairs of secondary nerves. The cymes are axillary, umbellate, and 2.5–5 cm long. The perianth is urceolate, fleshy, yellow, 3-lobed, and up to about 1 cm long. The androecium includes 10–14 stamens, which are joined in a column. The ovary is ellipsoid, hairy, and develops a short style and a bifid stigma. The drupes are hairy, orange or yellow, pyriform or globose, up to 6.5 cm in diameter, and open to show an oblong seed about 2 cm long and partially covered with a deeply lacerated yellow and waxy aril.

Medicinal uses: Ulcers (India)

*Antibacterial (Gram-positive) halo developed by polar extract:* Ethanol extract of arils inhibited the growth of *Streptococcus pneumoniae* (50 µL of a 200 µg/mL solution/6 mm well) with a inhition zone diameter of 13 mm (Manimehalai, 2017).

*Antifungal (filamentous) alkyl phenols:* From the fruit rinds malabaricones A, B, and C inhibited the growth of plant pathogenic fungi (Choi et al., 2008).

Commentary: Malabaricone A is cytotoxic for human cells *in vitro* (Manna et al., 2015), suggesting that the antifungal mechanism is based on DNA insults because a substance which is antifungal and cytotoxic for mammalian cells often targets chromosomes. As discussed earlier, such a compound is probably antibacterial. This remains to be checked.

### References

Choi, N.H., Choi, G.J., Jang, K.S., Choi, Y.H., Lee, S.O., Choi, J.E. and Kim, J.C., 2008. Antifungal activity of the methanol extract of *Myristica malabarica* fruit rinds and the active ingredients malabaricones against phytopathogenic fungi. *The Plant Pathology Journal, 24*(3), pp. 317–321.

Manna, A., De Sarkar, S., De, S., Bauri, A.K., Chattopadhyay, S. and Chatterjee, M., 2015. The variable chemotherapeutic response of Malabaricone-A in leukemic and solid tumor cell lines depends on the degree of redox imbalance. *Phytomedicine, 22*(7–8), pp. 713–723.

Manimehalai, V., 2017. Evaluation of anthelmintic and antipneumococcal activity on seed aril of *Myristica Malabarica* Lam (Doctoral dissertation, Mohamed Sathak AJ College of Pharmacy, Chennai).

### 2.3.3.4 *Myristica simiarum* A. DC

Synonym: *Myristica discolor* Merr.; *Myristica elliptica* var. *simiarum* (A.DC.) J. Sinclair

Local names: Duguan (the Philippines)

Habitat: Forests

Distribution: Malaysia, Indonesia, and the Philippines

Botanical observation: It is a tree that grows to about 20 m tall. The leaves are simple, alternate, and exstipulate. The blade is membranous, oblong-lanceolate, wavy, pale green and hairy below, 11–17 × 3–10 cm, acute at base, and acute to acuminate at apex. Inflorescences are usually peduncled and branched. The perianth is elliptic-oblong, up to 8 mm long, and hairy. The flowers are minute. The drupes are ovoid-ellipsoid drupes, which are up to 3 cm long and 2.5 in diameter.

Medicinal uses: Skin diseases and diarrhea (the Philippines)

Commentary: No reports exit on this plant apparently for antimicrobial activity and phytochemistry. It is thus an exciting candidate to study because new chemical entities could be found. However, together with countless plants, this plant might disappear soon for palm oil-induced primary rainforest burning and with it, its potential antibacterial, antifungal, or antiviral principles.

# 3 The Clade Monocots

## 3.1 LILIIDS

### 3.1.1 ORDER ACORALES LINK (1835)

#### 3.1.1.1 Family Acoraceae Martynov (1820)

The family Acoraceae consists of the single genus *Acorus* L.

##### 3.1.1.1.1 *Acorus calamus* L.

Synonyms: *Acorus americanus* (Raf.) Raf.; *Acorus angustatus* Raf.; *Acorus angustifolius* Schott; *Acorus asiaticus* Nakai; *Acorus calamus* var. *americanus* Raf.; *Acorus calamus* var. *angustatus* Besser; *Acorus calamus* var. *angustifolius* (Schott) Engl.; *Acorus calamus* var. *spurius* (Schott) Engl.; *Acorus calamus* var. *verus* L.; *Acorus calamus* var. *vulgaris* L.; *Acorus cochinchinensis* (Lour.) Schott; *Acorus griffithii* Schott; *Acorus spurius* Schott; *Acorus triqueter* Turcz. ex Schott; *Acorus verus (L.)* Houtt.; *Calamus aromaticus* Garsault; *Orontium cochinchinense* Lour.

Common names: Sweet flag, calamus; boch (Bangladesh); gora bach, vácha, jatilá (India); jerango (Indonesia); hang khao nam (Laos); deringu (Malaysia), lin-ne, lin-lay (Myanmar); achheni (Nepal); lubigan (The Philippines); wan nam (Thailand); bo bu nep (Vietnam)

Habitat: Rivers, lakes, and ponds or cultivated

Distribution: India, China, Southeast Asia, Pacific Islands

Botanical observation: It is an aquatic herb that grows to a height of 2 m. The rhizome is up to about 20 cm long and aromatic. The leaves are dark green, glossy, grass-like, upright, reddish at the base, somewhat coriaceous, and 90 cm–2 m × 1.8–3.7 cm. The spikes are sessile, cylindrical, sessile, densely flowered, and 5–10 cm long. The perianth consists of 6 orbicular sepals. The androecium includes 6 stamens with reniform anthers. The ovary is conical, 2–3-locular, and encloses several ovules. The fruit is a berry.

Medicinal uses: Antiseptic (Indonesia), boils (China)

*Moderate broad-spectrum non-polar extract:* Petroleum ether extract of rhizomes inhibited the growth of *Bacillus subtilis* (ATCC 633), *Staphylococcus aureus* (ATCC 9144), *Escherichia coli* (ATCC 25922), and *Pseudomonas aeruginosa* (ATCC 25619) with MIC values of 250, 250, 500, and 250 µg/mL, respectively (Rani et al., 2003).

*Moderate antibacterial (Gram-positive) polar extract*: Ethanol extract of rhizome inhibited the growth of *Bacillus subtilis* (ATCC 6051) and *Staphylococcus aureus* (ATCC 12600) with MIC values of 780 and 1560 µg/mL, respectively (MacGaw et al., 2002).

*Strong antifungal (filamentous) phenylpropanoid*: β-Asarone (LogD = 2.7; molecular weight = 208.2 g/mol) abrogated the mycelial growth of *Cladosporium cucumerinum*, *Colletotrichum orbiculare*, *Magnaporthe grisea*, and *Pythium ultimum* in a range of 0.5–30 µg/mL (Lee et al., 2004). In a subsequent study, β-asarone inhibited the growth of *Microsporium gypseum*, *Trichophyton rubrum*, and *Penicillium marneffei* with $IC_{50}$ values of 200, 200, and 400 µg/mL, respectively (Phongpaichit et al., 2005). This phenylpronoid at very high concentrations (MIC/MFC values of 8 mg/mL) inhibited the growth of *Candida albicans* (ATCC 90028) via inhibition of ergosterol synthesis (Rajput Karuppayil, 2013).

β-Asarone

*Moderate antifungal (yeasts) phenylpropanoid:* β-Asarone inhibited the growth of *Candida albicans*, *Cryptococcus neoformans*, and *Saccharomyces cerevisiae* with MIC/MFC values of 120/250, 500/500, and 1000/1000 μg/mL, respectively (Phongpaichit et al., 2005).

*Strong antiviral (enveloped monopartite linear single stranded (+) RNA) polar extract:* Methanol extract of leaves inhibited the replication of the Dengue virus type-2 (strain New Guinea C) in Huh7it-1 cells by 26.6% at a concentration of 20 μg/mL (Rosmalena et al., 2019).

*Strong antiviral (enveloped monopartite linear single-stranded (+) RNA) lipophilic phenolic:* Tatanan A (LogP = 8.3; molecular wight = 624.8 g/mol) isolated from the rhizomes inhibited Dengue virus type-2 (strain New Guinea C)-induced cytopathic effects with the $IC_{50}$ of 3.9 μM via early inhibition of RNA replication (Yao et al., 2018).

Tatanan A

Commentary: (i) The plant has been used as protection during the prevalence of epidemics (Motley, 1994). (ii) Essential oil in the rhizome contains mainly the phenylpronaloid β-asarone

(Raina et al., 2003). (iii) Essential oils can be extracted with non-hydrophilic solvents like petroleum ether or hexane. β-Asarone does not have antibacterial effects (Phongpaichit et al., 2005).

### References

Lee, J.Y., Lee, J.Y., Yun, B.S. and Hwang, B.K., 2004. Antifungal activity of β-asarone from rhizomes of *Acorus gramineus*. *Journal of Agricultural and Food Chemistry*, 52(4), pp. 776–780.

McGaw, L.J., Jager, A.K. and Van Staden, J., 2002. Isolation of beta-asarone, an antibacterial and anthelmintic compound, from *Acorus calamus* in South Africa. *South African Journal of Botany*, 68(1), pp. 31–35.

Motley, T.J., 1994. The ethnobotany of sweet flag, *Acorus calamus* (Araceae). *Economic Botany*, 48(4), pp. 397–412.

Phongpaichit, S., Pujenjob, N., Rukachaisirikul, V. and Ongsakul, M., 2005. Antimicrobial activities of the crude methanol extract of *Acorus calamus* Linn. *Songklanakarin Journal of Science & Technology*, 27(2), pp. 517–523.

Raina, V.K., Srivastava, S.K. and Syamasunder, K.V., 2003. Essential oil composition of *Acorus calamus* L. from the lower region of the Himalayas. *Flavour and Fragrance Journal*, 18(1), pp. 18–20.

Rajput, S.B. and Karuppayil, S.M., 2013. β-Asarone, an active principle of *Acorus calamus* rhizome, inhibits morphogenesis, biofilm formation and ergosterol biosynthesis in *Candida albicans*. *Phytomedicine*, 20(2), pp. 139–142.

Rani, A.S., Satyakala, M., Devi, V.S. and Murty, U.S., 2003. Evaluation of antibacterial activity from rhizome extract of *Acorus calamus* Linn. 62, 623–625.

Rosmalena, R., Elya, B., Dewi, B.E., Fithriyah, F., Desti, H., Angelina, M., Hanafi, M., Lotulung, P.D., Prasasty, V.D. and Seto, D., 2019. The antiviral effect of Indonesian medicinal plant extracts against Dengue virus in vitro and in silico. *Pathogens*, 8(2), p. 85.

Yao, X., Ling, Y., Guo, S., Wu, W., He, S., Zhang, Q., Zou, M., Nandakumar, K.S., Chen, X. and Liu, S., 2018. Tatanan A from the *Acorus calamus* L. root inhibited Dengue virus proliferation and infections. *Phytomedicine*, 42, pp. 258–267.

## 3.1.2  ORDER ALISMATALES R. BROWN EX BERCHT. & J.PRESL (1820)

### 3.1.2.1  Family Alismataceae Ventenat (1799)

The family Alismataceae comprises about 13 genera and 100 species of aquatic or semiaquatic herbs with basal leaves. The petiole is elongated and sheathing. The blade presents a few longitudinal nerves. The inflorescences are racemes, panicles, or umbels. The calyx comprises 3 sepals. The corolla includes 3 ephemerous petals. The androecium is made of 3 to many stamens. The gynoecium consists of 3 to numerous free carpels, each containing 1 to several ovules. The fruits are achenes. Members of this family produce antimicrobial triterpenes and phytosterols.

#### 3.1.2.1.1  *Alisma plantago-aquatica* L.

Synonyms: *Alisma subcordatum* Raf.; *Alisma triviale* Pursh

Common names: Water plantain, frog's spoon; pani kola (Bangladesh); ze xie (China), pani (India); saji omo daka (Japan); taek sa (Korea)

Habitat: Lakes, ponds, and slow rivers

Distribution: Turkey, Iran, Afghanistan, Kazakhstan, Kyrgyzstan, Uzbekistan, Tajikistan, Pakistan, India, Nepal, Myanmar, Vietnam, Thailand, China, Mongolia, Japan, and Korea

Botanical observation: This aquatic herb grows up to a height of 1.5 m from a rhizome. The leaves are simple and in rosette. The petiole is of variable length and up to 30 cm long and fleshy. The blade is broadly lanceolate, 2–10 × 1.5–7 cm, with longitudinal secondary nerves parallel to the midrib (somewhat plantain-like). The panicles are lax and up to 50 cm long. The calyx consists of 3, broadly ovate, and minute sepals. The corolla comprises 3 white, membranous, about 1 cm long, and ephemeral petals. The androecium presents 6 stamens. and the gynoecium is made of numerous carpels. The fruits are tiny achenes.

Medicinal uses: Wounds, ulcers (Bangladesh); gonorrhea, hepatitis (China)

*Alisma plantago-aquatica* L.

*Weak antibacterial (Gram-positive) polar extract:* Aqueous extract of rhizome inhibited the growth of *Streptococcus mutans* (MT 5091) and *Streptococcus mutans* (OMZ 176) with the MIC values of 1250 and 1250 μg/mL, respectively (Chen et al., 1989).

*Very weak antifungal (filamentous) polar extract:* Methanol extract of roots inhibited the growth of *Rhizoctonia solani* by 36% at the concentration of 2000 μg/mL (Rajput et al., 2018).

*Strong antiviral (enveloped circular double stranded DNA) lipophilic protostane-type triterpenes:* Alisol A 24-acetate, 25-anhydroalisol A, 13β, 17β-epoxyalisol A, alisol B 23-acetate, alisol F (LogD = 4.5; molecular weight = 488.7 g/mol), and alisol F 24-acetate inhibited the secretion of Hepatitis B virus surface antigen by Hep G2.2.15 cell line with the $IC_{50}$ values of 2.3, 11.0, 15.4, 14.3, 0.6, and 7.7 μM, respectively, and the secretion of the Hepatitis B virus e antigen secretion with $IC_{50}$ values of 498.1, 17.6, 41.0, 19.9, 8.5 and 5.1 μM, respectively (Jiang et al., 2006). Alisol A inhibited the secretion of Hepatitis B virus surface antigen with an $IC_{50}$ value of 39 μM and a selectivity index of 1.6 (Zhang et al., 2008).

13β, 17β-epoxyalisol A

*Viral enzyme inhibition polar extracts:* Methanol extract of aerial parts inhibited the Human immunodeficiency virus type-1 protease (Park, 2003). Aqueous extract of rhizomes concentration of 250 μg/mL inhibited Human immunodeficiency virus type-1protease by 62.8% (Xu et al., 1996).

Commentaries: (i) In China, the plant is used for the treatment of COVID-19 (Lem et al., 2020), as well as *Alisma orientale* (Sam.) Juz. (Yang et al., 2020) and *Alisma plantago-aquatica* subsp. *orientale* (Sam.) Sam. (Wang et al., 2020). (ii) The plant produces a series of protostane-type triterpenes alisol E 23 acetate and 13,17-epoxialisol A, and guaiane-type sesquiterpenes (Yoshikawa et al., 1994). One could have some interest to look for anti-Severe acute respiratory syndrome-associated coronavirus 2 principles from this plant. Are protostane triterpenes potential candidates?

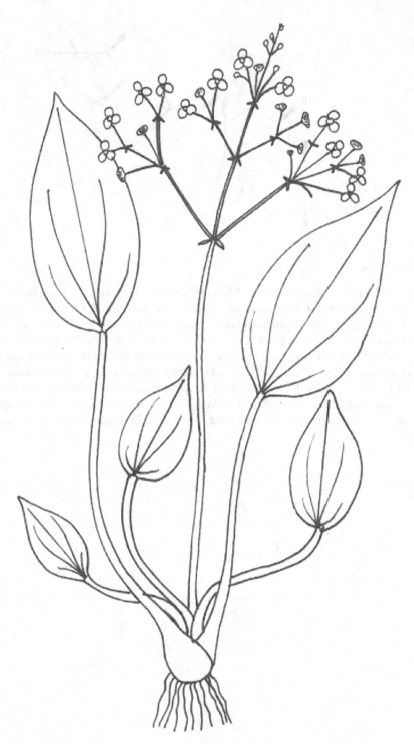

*Alisma orientale* (Sam.) Juz.

### References

Chen, C.P., Lin, C.C. and Tsuneo, N., 1989. Screening of Taiwanese crude drugs for anti-bacterial activity against *Streptococcus* mutans. *Journal of Ethnopharmacology*, 27(3), pp. 285–295.

Jiang, Z.Y., Zhang, X.M., Zhang, F.X., Liu, N., Zhao, F., Zhou, J. and Chen, J.J., 2006. A new triterpene and anti-hepatitis B virus active compounds from Alisma orientalis. *Planta medica, 72*(10), pp. 951–954.

Lem, F.F., Opook, F., Herng, D.L.J., Na, C.S., Lawson, F.P. and Tyng, C.F., 2020. Molecular mechanism of action of repurposed drugs and traditional Chinese medicine used for the treatment of patients infected with COVID-19: A systematic review. *medRxiv.*

Park, J.C., 2003. Inhibitory effects of Korean plant resources on human immunodeficiency virus type 1 protease activity. *Oriental Pharmacy and Experimental Medicine, 3*(1), pp. 1–7.

Rajput, N.A., Atiq, M., Javed, N., Ye, Y.H., Zhao, Z., Syed, R.N., Lodhi, A.M., Khan, B., Iqbal, O. and Dou, D., 2018. Antimicrobial effect of Chinese medicinal plant crude extracts against *Rhizoctonia solani* and *Phythium aphanidermatum. Fresenius Environmental Bulletin, 27*, pp. 3941–3949.

Wang, S.X., Wang, Y., Lu, Y.B., Li, J.Y., Song, Y.J., Nyamgerelt, M. and Wang, X.X., 2020. Diagnosis and treatment of novel coronavirus pneumonia based on the theory of traditional Chinese medicine. *Journal of Integrative Medicine, 18*(4), pp. 275–283.

Xu, H.X., Wan, M., Loh, B.N., Kon, O.L., Chow, P.W. and Sim, K.Y., 1996. Screening of traditional medicines for their inhibitory activity against HIV -1 Protease. *Phytotherapy Research, 10*(3), pp. 207–210.

Yang, Q.X., Zhao, T.H., Sun, C.Z., Wu, L.M., Dai, Q., Wang, S.D. and Tian, H., 2020. New thinking in the treatment of 2019 novel coronavirus pneumonia. *Complementary Therapies in Clinical Practice, 39*, p. 101131.

Yoshikawa, M., Yamaguchi, S., Matsuda, H., Kohda, Y., Ishikawa, H., Tanaka, N., Yamahara, J. and Murakami, N., 1994. Crude drugs from aquatic plants. IV. On the constituents of *Alismatis rhizoma*.(2). Stereostructures of bioactive sesquiterpenes, alismol, alismoxide, orientalols A, B, and C, from Chinese *Alismatis rhizoma. Chemical and Pharmaceutical Bulletin, 42*(9), pp. 1813–1816.

Zhang, Q., Jiang, Z.Y., Luo, J., Cheng, P., Ma, Y.B., Zhang, X.M., Zhang, F.X., Zhou, J. and Chen, J.J., 2008. Anti-Hepatitis B virus agents. Part 1: synthesis of alisol A derivatives: a new class of Hepatitis B virus inhibitors. *Bioorganic & Medicinal Chemistry Letters, 18*(16), pp. 4647–4650.

### 3.1.2.1.2 *Sagittaria trifolia* L.

Synonyms: *Sagittaria hirundinacea* Blume; *Sagittaria leucopetala* (Miq.) Bergmans; *Sagittaria sagittifolia* L.; *Sagittaria sinensis* Sims

Common names: Arrow-head; sagudana; kuka (Bangladesh); slök lumpaèng (Cambodia); ye ci gu (China); bea bea (Indonesia); omodaka (Japan); phak sob (Laos); keladi chabang (Malaysia); jazoponri (Pakistan); gauai-gauai (the Philippines); taokiat (Thailand); rau mác (Vietnam)

Habitat: Lakes, ponds, and slow rivers

Distribution: Turkey, Iran, Afghanistan, Kazakhstan, Kyrgyzstan, Uzbekistan, Tajikistan, Pakistan, India, Nepal, Myanmar, Laos, Cambodia, Vietnam, Thailand, China, Taiwan, Indonesia, the Philippines, Mongolia, Japan, and Korea

Botanical observation: This elegant aquatic herb grows from a globose and edible corm. The leaves are simple and arranged in a rosette. The petiole grows up to 60–70cm long. The blade is sagittate and up to 8–15×5–10cm. The inflorescences are racemose and present whorls of flowers. The calyx consist of 3, ovate, and minute sepals. The corolla comprises 3, white tinged, pink at base, membranous, broadly spathulate, and about 1.5cm long petals. The androecium is made of numerous tiny stamens. The achenes are minute and winged.

Medicinal uses: Wounds (Bangladesh); rabies, sores, and gonorrhea (China)

*Antibacterial essential oil*: Essential oil (mainly containing hexahydrofarnesyl acetone) exhibited antibacterial effects (Xiangwei et al., 2006).

*Moderate antibacterial (Gram-positive) ent-rosane diterpenes*: Sagittine B, C, and D, inhibited the growth of *Actinobacillus actinomycetemcomitans* (ATCC 43717) (Liu et al., 2006). Sagittines A, B, C, D, and E inhibited the growth of *Actinomyces naeslundiis* (ATCC 12104) with the MIC values of 62.5, 125, 62.5, 125, and 62.5 µg/mL, respectively (Liu et al., 2006). Sagittines A, B, C, and D inhibited the growth of *Streptococcus mutans* (ATCC 25175) with the MIC values of 62.5, 62.5, 62.5, and 125 µg/mL, respectively (Liu et al., 2006). Sagittine H inhibited the growth of *Streptococcus mutans* (ATCC 25175) and *Actinomyces naeslundiis* (ATCC 12104) (Xue-Ting et al., 2009).

Sagittine A

*Strong antibacterial (Gram-positive) kaurane diterpene*: 19-β-L-3′-acetoxyarabinofuranosyl-ent-kaur-16-ene-19-oate inhibited the growth of *Actinomyces naeslundiis* (ATCC 12104) and *Streptococcus mutans* (ATCC 25175) (Xue-Ting et al., 2009).

Commentary: The plant produces a series of kaurane diterpenes (Yoshikawa et al., 1993). Kauranes are often antiviral.

**References**

Liu, X.T., Pan, Q., Shi, Y., Williams, I.D., Sung, H.H.Y., Zhang, Q., Liang, J.Y., Ip, N.Y. and Min, Z.D., 2006. ent-Rosane and Labdane Diterpenoids from *Sagittaria sagittifolia* and their antibacterial activity against three oral pathogens. *Journal of Natural Products*, 69(2), pp. 255–260.

Xiangwei, Z., Xiaodong, W., Peng, N., Yang, Z. and JiaKuan, C., 2006. Chemical composition and antimicrobial activity of the essential oil of *Sagittaria trifolia*. *Chemistry of Natural Compounds*, 42(5), pp. 520–522.

Xue-Ting, L.I.U., Yao, S.H.I., Liang, J.Y. and Zhi-Da, M.I.N., 2009. Antibacterial ent-rosane and ent-kaurane diterpenoids from *Sagittaria trifolia* var. sinensis. *Chinese Journal of Natural Medicines*, 7(5), pp. 341–345.

Yoshikawa, M., Yamaguchi, S., Murakami, T., Matsuda, H., Yamahara, J. and Murakami, N., 1993. Absolute stereostructures of trifoliones A, B, C, and D, new biologically active diterpenes from the tuber of *Sagittaria trifolia* L. *Chemical and Pharmaceutical Bulletin*, 41(9), pp. 1677–1679.

### 3.1.2.2  Family Araceae A.L. de Jussieu (1789)

The family Araceae consists of about 110 genera and 1800 species of fleshy, not uncommonly smelly, poisonous, and often water-loving herb growing from rhizomes or corms. The leaves are solitary, alternate, and exstipulate. The petiole is often sheathing at base. The blade is simple to variously incised and occasionally very large. The inflorescence is a spadix enclosed in a spathe which can both occasionally reach gigantic proportions. The perianth consists of 4–6 lobes. The androecium is made of 2–8 stamens opposite the perianth lobes. The gynoecium presents a 1-plurilocular ovary. The berries contain 1 to several seeds. This large and interesting family has been studied very little for its antimicrobial natural products.

### 3.1.2.2.1  Aglaonema hookerianum Schott

Synonym: *Aglaonema clarkei* Hook.f.

  Common names: Sikachalal; horina shak (Bangladesh)

  Habitat: Wet, shady soils in forests and around villages

  Distribution: India, Bangladesh, and Myanmar

  Botanical observation: This fleshy climber grows to 50 cm long. The petiole is 15–25 cm long and sheathing. The blade is elliptic, 6.5–12 cm × 20–25 cm, round at base, acuminate at apex, and marked with 7–13 lateral nervations. The inflorescence peduncle is 10–20 cm long. The spathe is 3.5–6.5 cm long. The spadix is 2.5–4.5 cm long. The staminate zone is 2.0–3.5 cm long. The female portion is 0.3–0.5 cm long. The berries are red, glossy, olive shaped, 2–3 × 0.9–1.5 cm, and poisonous.

  Medicinal use: Conjunctivitis (Bangladesh)

  *Broad-spectrum antibacterial halo developed by polar extract*: Ethanol extract (500 µg/disc) inhibited the growth of *Bacillus subtilis*, *Bacillus megaterium*, *Staphylococcus aureus*, *Escherichia coli*, *Pseudomonas aeruginosa*, *Salmonella typhi*, *Salmonella paratyphi*, *Shigella dysenteriae*, and *Vibrio cholerae* with the inhibition zone diameters of 18, 16, 16, 15, 15.5, 17.4, 20.3, 18.2, and 16.1 mm, respectively (Roy et al., 2011).

  Commentary: Members of the genus *Aglaonema* Schott (also members of the genus *Commelina* L., in the family Commelinaceae) produce a series of polyhydroxylated piperidine alkaloids (also called imino sugars) including homonojirimycin (Zhang et al., 2013). Homonojirimycin (LogD = −2.3 at pH 7.4; −2. molecular weight = 193.1 g/mol) inhibited the replication of the Influenza virus A/PR/8/34 (H1N1) with the $IC_{50}$ value of 10.4 µg/mL and a selectivity index of 17.9 (Zhang et al., 2013). Given orally at a dose of 2 mg/Kg twice a day from 2 days before infection to 6 days post-infection, homonojirimycin, increased the mean survival days from about 9.7 to 11.7 post-infection days and reduced virus yield in lungs (Zhang et al., 2013a).

### References

Roy, A., Biswas, S.K., Chowdhury, A., Shill, M.C., Raihan, S.Z. and Muhit, M.A., 2011. Phytochemical screening, cytotoxicity and antibacterial activities of two Bangladeshi medicinal plants. *Pakistan Journal of Biological Sciences*, 14(19), p. 905.

Zhang, G.B., Tian, L.Q., Li, Y.M., Liao, Y.F., Li, J. and Bing, F.H., 2013a. Protective effect of homonojirimycin from Commelina communis (dayflower) on influenza virus infection in mice. *Phytomedicine*, 20(11), pp. 964–968.

Zhang, G., Zhang, B., Zhang, X. and Bing, F., 2013. Homonojirimycin, an alkaloid from dayflower inhibits the growth of influenza A virus in vitro. *Acta Virologica*, 57(1).

### 3.1.2.2.2  Alocasia cucullata (Lour.) G. Don.

Synonyms: *Alocasia rugosa* (Desf.) Schott; *Arum cucullatum* Lour.; *Caladium cucullatum* (Lour.) Pers.; *Caladium rugosum* Desf.; *Colocasia cochleata* Miq.; *Colocasia cucullata* (Lour.) Schott; *Colocasia rugosa* (Desf.) Kunth; *Panzhuyuia omeiensis* Z.Y. Zhu

  Common names: Chinese taro, Buddha's palm; bilae kochu; lipkai (Bangladesh); nai habarala (Sri Lanka); jian wei yu (China)

  Habitat: Wet, shady soils in villages and forests, cultivated

  Distribution: India, Sri Lanka, Nepal, Bangladesh, Myanmar, Laos, Thailand, and Vietnam

  Botanical observation: This fleshy herb grows to 1 m tall and is cultivated as an ornamental plant. The petiole is 25–80 cm long and sheathing. The blade broadly lanceolate-cordate, 10–40 × 7–25 cm, with 8 lateral nervations, dark green, and glossy. The inflorescence peduncle is 20–30 cm long. The spathe is 9–15 cm, green and develops a limb, which is cymbiform, 5–10×3–5 cm. The spadix is 8–15 cm long. The staminate zone is male zone 3.5 cm long. The female zone is 1.5-2.5 cm long. The berries are subglobose,about 1 cm in diameter, and dreadfully poisonous.

Medicinal use: Wounds (China)

Commentary: The plant has apparently not been studied for its potential antimicrobial effects. In China, *Alocasia odorata* (Roxb.) C. Kosch is used for the treatment of cholera. Aqueous extract of this plant at a concentration of 250 µg/mL inhibited Human immunodeficiency virus type -1 protease by 43.3% (Xu et al., 1996).

### Reference

Xu, H.X., Wan, M., Loh, B.N., Kon, O.L., Chow, P.W. and Sim, K.Y., 1996. Screening of traditional medicines for their inhibitory activity against HIV-1 protease. *Phytotherapy Research, 10*(3), pp. 207–210.

### 3.1.2.2.3   *Alocasia macrorrhizos (L.) G. Don*

Synonyms: *Alocasia indica* (Lour.) Schott; *Colocasia indica* (Lour.) Hassk.; *Alocasia indica* (Lour.) Spach; *Arum indicum* Roxb.; *Caladium indica* (Lour.) K. Koch; *Colocasia indica* (Lour.) Kunth; *Colocasia indica* (Lour.) Kunth

Common names: Giant Taro; man kachu (Bangladesh); viamila (Fiji); mankachu (India); bira (Indonesia); kaph'uk (Laos); keladi sebaring (Malaysia); pein-mohawaya (Myanmar); paragum (Papua New Guinea); aba-aba (the Philippines); kradatdam (Thailand); khoais (Vietnam)

Habitat: Moist shady soils or cultivated

Distribution: India, Bangladesh, Southeast Asia, Australia, and Pacific Islands

Botanical observation: This massive fleshy herb grows to 3 m tall. The petiole is stout, up to 1.5 m, and sheathing. The blade is sagittate, about 120×50 cm with about 9 pairs of secondary nerves. The flower peduncle is about 70 cm tall. The spathe is 15–35 cm, ovoid, the limb spoon shaped, and 10.5–30 cm long. The staminate zone of the spadic is 3–7 cm long. The female zone is 1–2 cm long. The ovary develops a 3–5-lobed stigma. The berries are olive-shaped, about 1 cm long, and red.

Medicinal uses: Herpes, wounds (Indonesia); boils (Taiwan)

*Broad-spectrum antibacterial halo developed by polar extract*: Ethanol extract of rhizome inhibited the growth of *Staphylococcus aureus*, *Streptococcus pyogenes*, *Shigella dysenteriae*, and *Salmonella typhi* with zones of inhibition of 10, 10, 12, and 10 mm, respectively (500 µg/disc) (Uddin et al., 2008).

Commentaries: (i) In Bangladesh, *Alocasia fornicata* (Roxb.) Schott is used to treat fever and diarrhea. Ethanol extract of rhizome of this plant of inhibited the growth of *Staphylococcus aureus*, *Streptococcus pyogenes*, *Shigella dysenteriae*, *Salmonella typhi*, *Escherichia coli*, *Vibrio cholerae*, *Enterobacter aerogenes*, and *Pseudomonas aeruginosa* with zones of inhibition of 12, 15, 12, 10, 11, 16, 10 and 10 mm, respectively (500 µg/disc) (Uddin et al., 2008). In China, *Alocasia longiloba* Miq. is used to treat sores. (ii) The roots contain a series of long-chain piperidine and indoles alkaloids (Huang et al., 2017; Zhu et al., 2012), which could be examined for possible antimicrobial effects.

### References

Huang, W., Yi, X., Feng, J., Wang, Y. and He, X., 2017. Piperidine alkaloids from *Alocasia macrorrhiza*. *Phytochemistry, 143*, pp. 81–86.

Uddin, S.J., Rouf, R., Shilpi, J.A., Alamgir, M., Nahar, L. and Sarker, S.D., 2008. Screening of some Bangladeshi medicinal plants for in vitro antibacterial activity. *Oriental Pharmacy and Experimental Medicine, 8*(3), pp. 316–321.

Zhu, L.H., Chen, C., Wang, H., Ye, W.C. and Zhou, G.X., 2012. Indole alkaloids from *Alocasia macrorrhiza*. *Chemical and Pharmaceutical Bulletin, 60*(5), pp. 670–673.

### 3.1.2.2.4   *Amorphophallus bulbifer (Roxb.) Bl.*

Synonyms: *Arum bulbiferum* Roxb.; *Conophallus bulbifer* (Sims) Schott; *Pythonium bulbiferum* (Roxb.) Schott

  Common names: Devil's tongue, oodoo lily; amla bela (Bangladesh); katuchena (India)

  Habitat: Roadsides, forests, and around villages

  Distribution: India, Bangladesh, Myanmar, Cambodia, Laos, Vietnam, and China

  Botanical observation: This fleshy aroid herb grows to 1.5 m tall from a globose corm. The petiole is 60–80 cm long and variegated. The blade is divided into 5–15×2–4 cm and ovate-oblong lobes. The inflorescence peduncle is about 70 cm tall. The spathe is ovate, 15–20 cm light pink. The spadix is conspicuous, lanceolate, 20–25 cm long, and pale cream-colored. The staminate zone is 4–4.5 cm long. The female zone is 2.5–3.0 cm long. The flowers are minute. The fruits are ovoid, 1.0–1.5 cm long, and red.

  Medicinal uses: Wounds (India)

  *Moderate broad-spectrum antibacterial polar extract:* Acetone extract of corms inhibited the growth of *Bacillus subtilis*, *Staphylococcus aureus*, *Pseudomonas aeruginosa*, *Escherichia coli*, *Salmonella typhi*, and *Klebsiella pneumoniae* with MIC values of 250, 200, 400, 300, 300, and 300 µg/mL, respectively (Shete et al., 2015).

#### Reference

Shete, C., Wadkar, S.U.R.Y.A.K.A.N.T., Inamdar, F.A.R.I.D.A., Gaikwad, N.I.K.H.I.L. and Patil, K.U.M.A.R., 2015. Antibacterial activity of *Amorphophallus konkanensis* and *Amorphophallus bulbifer* tuber. *Asian Journal of Pharmaceutical and Clinical Research*, 8(1), pp. 98–102.

### 3.1.2.2.5   *Amorphophallus paeoniifolius (Dennst.) Nicolson*

Synonyms: *Amorphophallus bangkokensis* Gagnep.; *Amorphophallus campanulatus* Blume ex Decne.; *Amorphophallus campanulatus* Decne.; *Amorphophallus dubius* Blume; *Amorphophallus gigantiflorus* Hayata; *Amorphophallus microappendiculatus* Engl.; *Amorphophallus paeoniifolius* var. *campanulatus (Decne.)* Sivad.; *Amorphophallus rex* Prain ex Hook. f.; *Amorphophallus sativus* Blume; *Amorphophallus virosus* N.E. Br.; *Arum campanulatum* Roxb.; *Arum campanulatum* Roxb.; *Arum decurrens* Blanco; *Arum rumphii* Gaudich.; *Candarum rumphii* (Gaudich.) Schott; *Dracontium paeoniifolium* Dennst.; *Hydrosme gigantiflora* (Hayata) S.S. Ying; *Hydrosme gigantiflorus* (Hayata) S.S. Ying; *Plesmonium nobile* Schott

  Common names: Elephant-foot yam; ol (Bangladesh); you bing mo yu (China)

  Habitat: Wet, shady soils

  Distribution: India, Sri Lanka, Bangladesh, Myanmar, Laos, Cambodia, Indonesia, Vietnam, Thailand, China, Taiwan, the Philippines, Papua New Guinea, Australia, and Pacific Islands

  Botanical observation: This strange-looking aroid grows to 2.5 tall from a massive discoid corm. The petiole is variegated and up to 1.5 m tall. The blade is dissected, about 1.5 m long, with elliptic-lanceolate leaflets, which are 10–20×2.5–12.5 cm, and acuminate at apex. The flower peduncle is stout and up to 20 cm. The spathe is showy, campanulate, 10–45×15–60 cm, spreading, purplish, wavy, and glossy. The spadix has an offensive odor, 12.5–70 cm, purplish, and manifolded. The male zone is obconic and up to 45 cm long. The female zone is up to 25 cm long. The fruiting zone is 10–50×3–10 cm with numerous reddish-yellow, olive-shaped berries, which are up to 2 cm long.

  Medicinal uses: Bronchitis, cholera (India)

  *Broad-spectrum antibacterial halo developed by a polar extract*: Ethanol extract of corms inhibited the growth of *Staphylococcus aureus*, *Shigella dysenteriae*, *Salmonella typhi*, *Escherichia coli*, *Vibrio cholerae*, and *Enterobacter aerogenes* with zones of inhibition ranging from 8 to 11 mm, respectively (500 µg/disc) (Uddin et al., 2008).

*Antifungal (filamentous) halo developed by triterpene:* Amblyone (80 μg/disc) inhibited the growth of *Aspergillus flavus, Aspergillus niger,* and *Rhizopus oryzae* with inhibition zone diameters of 7, 9, and 8 mm, respectively (Khan et al., 2008).

*Strong broad-spectrum antibacterial lipophilic triterpene:* Amblyone (LogP = 5.9; molecular weight = 430.6 g/mol) isolated from the corm inhibited the growth of *Bacillus subtilis, Bacillus megaterium, Staphylococcus aureus, Streptococcus pyogenes, Escherichia coli, Shigella dysenteriae, Shigella sonnei,* and *Shigella flexneri* with MIC values of 32, 8, 16, 16, 32, 16, 32, and 16 μg/mL, respectively (Khan et al., 2008).

Amblyone

Commentary: Amblyone presents an epoxide moiety. The presence of an epoxide moiety in compounds is often synonymous with antiviral properties.

### References

Khan, A., Rahman, M. and Islam, M.S., 2008. Antibacterial, antifungal and cytotoxic activities of amblyone isolated from *Amorphophallus campanulatus. Indian Journal of Pharmacology*, 40(1), p. 41.

Uddin, S.J., Rouf, R., Shilpi, J.A., Alamgir, M., Nahar, L. and Sarker, S.D., 2008. Screening of some Bangladeshi medicinal plants for in vitro antibacterial activity. *Oriental Pharmacy and Experimental Medicine*, 8(3), pp. 316–321.

### 3.1.2.2.6   Arisaema japonicum Bl.

Synonyms: *Amidena japonica* (Blume) Raf.; *Arisaema angustatum* Franch. & Sav. var. *peninsulae* (Nakai) Nakai.; *Arisaema peninsulae* Nakai; *Arisaema pseudojaponicum* Nakai; *Arisaema serratum* (Thunb.) Schott

   Common name: Japanese Arisaema; yamadzi-no-tennansho (Japan)

   Habitat: Riverbanks, shady and moist places

   Distribution: China, Korea, and Japan

   Botanical observation: This lugubre aroid is poisonous and grows from a tuber. The leaves are simple and alternate. The petiole is up to about 90 cm tall. The blade is lobed, dull green, somewhat fleshy, the lobes 9–17×3–9 cm, narrowly elliptic to ovate-oblong, cuneate at base, serrate at margin, and acuminate at apex. The flower peduncle is up to 80 cm tall. The spathe is greenish to purple and presents graceful white longitudinal lines, the tube about 5–8×1.5 cm, and the limb incurved, oblong, 5–6×4–5 cm, and acuminate at apex. The berries are red, glossy, up to 8 mm long, and contain 2–3 seeds.

Medicinal use: Abscesses (China)

Commentary: The antimicrobial properties of this plant have not been apparently examined. Note that a chloroform extract of *Arisaema tortuosum* (Wall.) Schott inhibited the replication of acyclovir-resistant Herpes simplex virus type-2 and Herpes simplex virus type-1 (Rittà et al., 2020). In Nepal, *Arisaema flavum* Schott (local name: *Timtry*) is used to heal wounds.

### Reference

Rittà, M., Marengo, A., Civra, A., Lembo, D., Cagliero, C., Kant, K., Lal, U.R., Rubiolo, P., Ghosh, M. and Donalisio, M., 2020. Antiviral activity of a *Arisaema tortuosum* leaf extract and some of its constituents against herpes simplex virus type 2. *Planta Medica*, 86(04), pp. 267–275.

### 3.1.2.2.7   *Homalomena aromatica* (Spreng.) Schott.

Synonyms: *Calla aromatoca* (Spreng.) Roxb.; *Zantedeschia aromatica* Spreng.

Common names: Gondhi kochu; barodaga, gandubi kachu (Bangladesh); fen fang qian nian jian (China); sugandhmantri (India)

Habitat: Forests

Distribution: India, Bangladesh, Myanmar, Thailand, Vietnam, and China

Botanical observation: This aromatic aroid herb grows to 70 cm tall from a rhizome. The petiole is 35 cm long and sheathing. The blade is sagittate, glossy, green, fleshy, smooth, 20–30×10–15 cm, with 8–12 primary lateral nervations. The flower peduncle is 10–20 cm long. The spathe is oblong, and 8–10 cm. The spadix is 7–9 cm long, or as long as the spathe. The male zone is 5.5–6 cm long. The female zone is 1.5–3 cm long. The ovary is ovoid, with a capitate stigma. The berries are orange yellow.

Medicinal use: Ulcers (Bangladesh)

*Broad-spectrum antibacterial halo developed by essential oil*: Essential oil (15 µL/6 mm paper disc) inhibited the growth of *Staphylococcus aureus*, *Klebsiella pneumoniae*, *Proteus vulgaris*, *Escherichia coli*, and *Pseudomonas aeruginosa* with inhibition zone diameters of 21, 15, 13, 11, and 8 mm, respectively (Laishram et al., 2006).

*Strong broad-spectrum antifungal essential oil*: Essential oil of rhizome (containing mainly linalool) inhibited the growth of *Microsporum fulvum* (MTCC 8478), *Microsporium gypseum* (MTCC 8469), *Trichophyton rubrum* (MTCC 8477), *Trichosporon beigelii* (NCIM 3326), and *Candida albicans* (MTCC 854) with the MIC values of 10, 12, 12, 10 and 16 µg/mL, respectively (Policegoudra et al., 2012).

Commentaries: (i) The roots contain sesquiterpenes, including (−)-α-cadinol and (−)-T-muurolol (Sung et al., 1992). Cedrelanol isolated from a member of the genus *Commiphora* Jacq. (family Bursearceae) is an stereoisomer of (−)-α-cadinol, which inhibited the growth of *Staphylococcus aureus* via cytoplasmic membrane damages and subsequent lysis (Claeson et al., 1992). Cedrelanol inhibited the growth of *Trichophyton mentagrophytes* (Claeson et al., 1992). T-muurolol inhibited the growth of *Laetiporus sulphureus*, *Lenzites betulina*, *Trametes versicolor*, and *Pcynoporus coccineus* with $IC_{50}$ values of 93.3, 74.1, 81.0, and 57.3 µg/mL, respectively (Zhang et al., 2016). (ii) *Homalomena coerulescens* Jungh. is used in Malaysia to heal sores. In Indonesia, *Homalomena javanica* Alderw. is used for the treatment of syphilis and *Homalomena rubescens* (Roxb.) Kunth. is used to heal wounds.

Cedrelanol

**References**

Claeson, P., Rådström, P., Sköld, O., Nilsson, Å. and Höglund, S., 1992. Bactericidal effect of the sesquiterpene T-cadinol on *Staphylococcus aureus*. *Phytotherapy Research*, 6(2), pp. 94–98.

Laishram, S.K.S., Nath, D.R., Bailung, B. and Baruah, I., 2006. In vitro antibacterial activity of essential oil from rhizome of homalomena aromatica against pathogenic bacteria. *Journal of Cell and Tissue Research*, 6(2), p. 849.

Policegoudra, R.S., Goswami, S., Aradhya, S.M., Chatterjee, S., Datta, S., Sivaswamy, R., Chattopadhyay, P. and Singh, L., 2012. Bioactive constituents of *Homalomena aromatica* essential oil and its antifungal activity against dermatophytes and yeasts. *Journal de Mycologie Médicale/Journal of Medical Mycology*, 22(1), pp. 83–87.

Zhang, Z., Yang, T., Mi, N., Wang, Y., Li, G., Wang, L. and Xie, Y., 2016. Antifungal activity of monoterpenes against wood white-rot fungi. *International Biodeterioration & Biodegradation*, 106, pp. 157–160.

### 3.1.2.2.8   Lasia spinosa (L.) Thw.

Synonyms: *Dracontium spinosum* L.; *Lasia aculeata* Lour.; *Lasia crassifolia* Engl.; *Lasia desciscens* Schott; *Lasia hermannii* Schott; *Lasia heterophylla* (Roxb.) Schott; *Lasia jenkinsii* Schott; *Lasia loureirii* Schott; *Lasia roxburghii* Griff.; *Lasia zollingeri* Schott; *Pothos heterophyllus* Roxb.; *Pothos heterophyllus* Roxb.; *Pothos lasia* Roxb.; *Pothos spinosus* (L.) Buch.-Ham. ex Wall.

Common names: Lasia; kata-kachu (Bangladesh); ci yu (China); gali gali (Indonesia); geli geli (Malaysia); laksmana (India); engili kohila (Sri Lanka); phak naam (Thailand); ray gai (Vietnam)

Habitat: Riverbanks, moist shady soils, and swamps

Distribution: India, Sri Lanka, Nepal, Bhutan, Bangladesh, Myanmar, Cambodia, Laos, Thailand, Vietnam, Malaysia, Indonesia, China, and Papua New Guinea

Botanical observation: This aroid herb grows to 1.5 m tall from a rhizome. The stem is creeping, stoloniferous, and spiny. The petiole is 30–100 cm long and spiny. The blade is sagittate-hastate or deeply dissected, edible (if cooked), 35–65×20–55 cm, and with 4–8 lateral nervations. The inflorescence peduncle is about 59 cm tall and spiny. The spathe is 18–35 cm, orangish yellow, and develops a slender limb about 30 cm long. The spadix is 3–5 cm long. The ovary is tiny and ovoid. The fruits are packed into an oblong mass, polygonal, and about 1 cm in diameter.

Medicinal uses: Sore throat (Bangladesh); tuberculosis (China)

Phytochemical class: Phenolics

*Broad-spectrum antibacterial halo developed by polar extract*: Ethanol extract of rhizomes inhibited the growth of *Staphylococcus aureus*, *Staphylococcus epidermidis*, *Streptococcus pyogenes*, *Shigella dysenteriae*, *Escherichia coli*, *Vibrio cholerae*, *Enterobacter aerogenes*, and *Pseudomonas aeruginosa* with inhibition zone diameters ranging from 15 to 18 mm (500 µg/disc) (Uddin et al., 2008).

*Broad-spectrum antibacterial halo developed by amphiphilic dibenzylbutyrolactone lignan:* Meridinol (LogD = 2.4 at pH 7.4; molecular weight = 370.3 g/mol) isolated from the rhizomes (100 µg/ 6 mm paper disc) inhibited the growth of *Bacillus cereus*, *Bacillus megaterium*, *Bacillus subtilis*, *Micrococcus luteus*, *Escherichia coli*, *Pseudomonas aeruginosa*, *Shigella boydii*, *Shigella dysenteriae*, and *Vibrio parahemolyticus* with inhibition zone diameters of 7, 8, 7, 8, 7, 8, 7, 8, and 8 mm, respectively (Hasan et al., 2011).

Meridinol

*Antifungal halo developed by non-polar extract:* Petroleum ether extract of rhizomes (500 µg/6 mm disc) inhibited the growth of *Candida albicans*, *Aspergillus niger*, and *Saccharomyces cerevisiae* with inhibition zone diameters of 16, 15, and 14 mm, respectively (Hasan et al., 2011).

*Broad-spectrum antifungal halo developed by dibenzylbutyrolactone lignan:* Meridinol isolated from the rhizomes (100 µg/6 mm paper disc) inhibited the growth of fungi (Hasan et al., 2011).

Commentary: As seen earlier, dibenzylbutyrolactone lignans are not uncommonly antiviral. One could have the interest to examine the antiviral properties of meridinol.

### References

Hasan, C.M., Alam, F., Haque, M., Sohrab, M.H., Monsur, M.A. and Ahmed, N., 2011. Antimicrobial and cytotoxic activity from *Lasia spinosa* and isolated lignan. *Latin American Journal of Pharmacy, 30*(3), pp. 550–553.

Uddin, S.J., Rouf, R., Shilpi, J.A., Alamgir, M., Nahar, L. and Sarker, S.D., 2008. Screening of some Bangladeshi medicinal plants for in vitro antibacterial activity. *Oriental Pharmacy and Experimental Medicine, 8*(3), pp. 316–321.

### 3.1.2.2.9   *Pinellia ternata* (Thunb.) Ten. ex Breitenb.

Synonyms: *Arisaema loureiri* Blume; *Arisaema macrourum* (Bunge) Kunth; *Arisaema ternatum* (Thunb.) Schott; *Arum atrorubens* Spreng., *Arum bulbiferum* Salisb.; *Arum bulbosum* Bl.; *Arum bulbosum* Pers. ex Kunth; *Arum fornicatum* Roth; *Arum macrourum* Bunge; *Arum subulatum* Desf.; *Arum ternatum* Thunb.; *Arum triphyllum* Houtt.; *Hemicarpurus fornicatus* (Roth) Nees; *Pinellia angustata* Schott; *Pinellia koreana* K.Tae & J.-H. Kim; *Pinellia ternata* (Thunb.) Druce; *Pinellia tuberifera* Ten.; *Pinellia tuberifera* var. *subpandurata* Engl.; *Pinellia zinguiensis* H. Li; *Typhonium tuberculigerum* Schott

Common names: East-African Arum; ban xia (China); karasubishaku (Japan)

Habitat: Wastelands and open forests

Distribution: China, Korea, and Japan

Botanical observation: This aroid grows up to about 1 m tall from a globose tuber. The leaves are compound and exstipulate. The petiole is up to 20 cm long and sheathing at base. The blade presents 3 folioles, which are lanceolate, 3–10×1–3 cm, cuneate at base, somewhat fleshy, and acuminate at apex, with 7–9 pairs of secondary nerves merging by the margin. The inflorescence peduncle is up to 35 cm tall. The spathe is greenish to purplish, purplish within, 6–7 cm long, and with an oblong limb about 4 cm. The spadix is about 10 cm long, slender, the female zone 2 cm long, and the male zone is about 6 mm long. The drupes are glossy and greenish yellow.

Medicinal use: Abscesses, cough (China)

*Strong antibacterial (Gram-positive) cerebroside*: Pinelloside isolated from the tubers inhibited the growth of *Bacillus subtilis* and *Staphylococcus aureus* with the MIC values of 20 and 50 μg/mL, respectively (Chen et al., 2003).

Pinelloside

*Strong broad-spectrum antifungal cerebroside*: Pinelloside isolated from the tubers inhibited the growth of *Aspergillus niger* and *Candida albicans* with the MIC values of 30 and 10 μg/mL, respectively (Chen et al., 2003).

*In vivo antiviral (enveloped non-segmented linear single-stranded (–) RNA virus) polar extract*: Aqueous extract of tubers given orally to mice at the dose of 20 mg/mice from 9 days before to 16 days after intranasal vaccination with Influenza virus A/PR/8/34 HA increased the bronchoalveolar anti-Influenza virus IgG Ab titers (Nagai et al., 2002).

*In vivo antiviral (enveloped non-segmented linear single-stranded (–) RNA virus) fatty acid*: Pinellic acid (LogD = 0.1 at pH 7.4; molecular weight = 330.4 g/mol) at a dose of 1 μg/mouse at 3-week intervals increased the bronchoalveolar anti-Influenza virus IgG Ab titers (Nagai et al., 2002).

Pinellic acid

*Viral enzyme inhibition by polar extract*: Methanol extract at a concentration of 100 μg/mL inhibited RNA-dependent DNA polymerase activity of Human immunodeficiency virus type-1 reverse transcriptase and Human immunodeficiency virus type-1 protease by 2 and 15.2%, respectively (Min et al., 2001).

Commentary: The plant is used in China for the treatment of COVID-19 (Luo, 2020; Wang et al., 2020). Isolation of anti-Severe acute respiratory syndrome-associated coronavirus 2 constituents on this plant is warranted.

### References

Chen, J.H., Cui, G.Y., Liu, J.Y. and Tan, R.X., 2003. Pinelloside, an antimicrobial cerebroside from *Pinellia ternata*. *Phytochemistry, 64*(4), pp. 903–906.

Luo, A., 2020. Positive SARS-Cov-2 test in a woman with COVID-19 at 22 days after hospital discharge: a case report. *Journal of Traditional Chinese Medical Sciences, 7*(4), pp. 413–417.

Min, B.S., Kim, Y.H., Tomiyama, M., Nakamura, N., Miyashiro, H., Otake, T. and Hattori, M., 2001. Inhibitory effects of Korean plants on HIV-1 activities. *Phytotherapy Research*, *15*(6), pp. 481–486.

Nagai, T., Kiyohara, H., Munakata, K., Shirahata, T., Sunazuka, T., Harigaya, Y. and Yamada, H., 2002. Pinellic acid from the tuber of *Pinellia ternata* Breitenbach as an effective oral adjuvant for nasal influenza vaccine. *International Immunopharmacology*, *2*(8), pp. 1183–1193.

Wang, S.X., Wang, Y., Lu, Y.B., Li, J.Y., Song, Y.J., Nyamgerelt, M. and Wang, X.X., 2020. Diagnosis and treatment of novel coronavirus pneumonia based on the theory of traditional Chinese medicine. *Journal of Integrative Medicine, 18*(4), pp. 275–283.

### 3.1.2.2.10 *Pistia stratiotes* L.

Synonyms: *Pistia crispata* Bl.; *Pistia minor* Blume; *Pistia obcordata* Schleid.; *Zala asiatica* Lour.

Common names: Tropical duckweed, water lettuce; takapana (Bangladesh); da piao (China); jalkumbhi (India); kayu apu (Indonesia); kiambang (Malaysia); kiapo (the Philippines); phak kok (Thailand); beo cai (Vietnam)

Habitat: Lakes, ponds, slow streams

Distribution: Tropical Asia and Pacific

Botanical observation: This fast multiplying floating herbs develops fibrous roots. The leaves are simple, spiral, in rosettes, and sessile. The blade is obovate, fleshy, somewhat glaucous, hydrophobic, 1.3–10×1.5–6 cm, round or somewhat notched at apex, and with about 11 longitudinal nervation. The spathe is white, lanceolate, hairy, and about 1 cm long. The fruits are membranous and shelter a few tiny seeds.

Medicinal uses: Syphilis, urinary tract infection (China)

*Broad-spectrum antibacterial halo developed by polar extract*: Aqueous extract (impregnated with a solution at 50 mg/mL/6.2 mm paper disc) inhibited the growth of *Staphylococcus aureus* and *Salmonella typhi* with inhibition zone diameters of 16 and 6.5 mm, respectively (Mukhtar & Tukur, 2019).

*Moderate antifungal (filamentous) polar extract*: Methanol extract of leaves was fungicidal for *Trichophyton rubrum*, *Trichopyton mentagrophytes*, *Microsporium gypseum*, *Microsporium nanum*, and *Epidermophyton floccosum* with MIC/MFC values of 250/250, 250/250, 125/125, 125/125, and 250/250 μg/mL, respectively (Premkumar & Shyamsundar, 2005).

Commentary: The amphiphilic stigmastane steroid 1 α-hydroxy-24S-ethyl-5α-cholest-22-en-3,6-dione (Monaco & Previtera, 1991) has cyclohexanone moieties and therefore probably antibacterial and/or antifungal. Amphiphilic triterpenes are not uncommonly antifungal for filamentous fungi. In most fungal species the inner cell wall consists of a core of covalently attached branched β-(1,3) glucan and chitin (Gow et al., 2017). Yeast cells have wall bud scars that tend to have fewer outer cell wall layers and therefore have exposed inner wall chitin and β-(1,3) glucan (Gow et al., 2017). Yeasts have an outer cell wall comprising highly mannosylated glycoproteins that cover the inner wall. In filamentous fungi, the wall is thicker and much more complex hence more resistance to antifungal natural products (Gow et al., 2017).

### References

Gow, N.A., Latge, J.P. and Munro, C.A., 2017. The fungal cell wall: structure, biosynthesis, and function. *The fungal kingdom*, pp. 267–292.

Monaco, P. and Previtera, L., 1991. A steroid from *Pistia stratiotes*. *Phytochemistry*, *30*(7), pp. 2420–2422.

Mukhtar, M.D. and Tukur, A., 2019. Antibacterial activities of aqueous and ethanolic extracts of *Pistia stratiotes* L. *NISEB Journal*, *1*(1).51–59.

Premkumar, V.G. and Shyamsundar, D., 2005. Antidermatophytic activity of *Pistia stratiotes*. *Indian Journal of Pharmacology*, 37(2), p. 127.

### 3.1.2.2.11    Pothos scandens L.

Synonyms: *Batis hermaphrodita* Blanco; *Podospadix angustifolia* Raf.; *Pothos angustifolius* (Raf.) C. Presl; *Pothos angustifolius* C. Presl; *Pothos chapelieri* Schott; *Pothos cognatus* Schott; *Pothos decipiens* Schott; *Pothos exiguiflorus* Schott; *Pothos fallax* Schott; *Pothos hermaphroditus* (Blanco) Merr.; *Pothos horsfeldii* Miq.; *Pothos leptospadix* de Vriese; *Pothos longifolius* C. Presl; *Pothos longipedunculatus* Engl.; *Pothos microphyllus* C. Presl; *Pothos roxburghii* de Vriese; *Pothos vrieseanus* Schott; *Pothos zollingeri* Schott; *Pothos zollingerianus* Schott; *Tapanava indica* Raf.; *Tapanava rheedii* Hassk.

   Common names: Climbing aroid, pothos; bendarli (India); tang lang die da (China)
   Habitat: Forests
   Distribution: India, Sri Lanka, Nepal, Comora, Myanmar, Bangladesh, Cambodia, Laos, Vietnam, China, Thailand, Malaysia, Indonesia, and the Philippines
   Botanical observation: This poisonous climber grows on trunks up to a length of 6 m. The stem is angular, somewhat fleshy, and up to about 1 cm across. The leaves are simple, alternate, and exstipulate. The blade is winged, up to about 15 cm long, lanceolate, 2–10×3–14 cm, attenuate at apex, with about 2 pairs of secondary nerves and scalariform tertiary nerves. The inflorescence pedicel is up to 15 cm long. The spathe is greenish, ovate, up to 8 mm long. The spadix is slender, curved or bent, somewhat purplish, and presents a fertile zone, which is globose, white, and about 1 cm across. The berries are red to orange, glossy, and up to about 1.5 cm long.
   Medicinal uses: Abscesses (India); smallpox (India; Malaysia)
   *Weak antiviral (enveloped monopartite linear (+) RNA) kaurane diterpene*: Methyl pothoscandensate inhibited the replication of the Porcine reproductive and Respiratory syndrome virus (YN-1 strain) with an $IC_{50}$ value of 40.3 µM and a selectivity index of 15.7 (Liu et al., 2012).
   Commentaries: (i) Very few plants are used for the treatment of smallpox in Asia Pacific. Examination of the anti-Variola virus properties of methyl pothoscandensate is warranted. Being active against an *enveloped monopartite linear (+) RNA) virus,* methyl pothoscandensate could be able to inhibit the growth of coronaviruses. (ii) The Porcine reproductive and Respiratory syndrome virus (PRRSV) is an enveloped RNA virus from the genus *Betaarterivirus* in the family Arteriviridae, which affect the swine industry (Pringproa et al., 2014).

**Reference**

Liu, H.X., Bi, J.L., Wang, Y.H., Gu, W., Su, Y., Liu, F., Yang, S.X., Hu, G.W., Luo, J.F., Yin, G.F. and Long, C.L., 2012. Methyl Pothoscandensate, a New ent-18 (4→ 3)-Abeokaurane from *Pothos scandens. Helvetica Chimica Acta,* 95(7), pp. 1231–1237.
Pringproa, K., Khonghiran, O., Kunanoppadol, S., Potha, T. and Chuammitri, P., 2014. In vitro virucidal and virustatic properties of the crude extract of *Cynodon dactylon* against porcine reproductive and respiratory syndrome virus. *Veterinary Medicine International.*

### 3.1.2.2.12    Remusatia vivipara Schott

Synonyms: *Arum viviparum* Roxb.; *Caladium viviparum* (Roxb.) Lodd.; *Caladium viviparum* (Roxb.) Nees; *Colocasia vivipara* (Roxb.) Thwaites; *Remusatia bulbifera* Vilm.; *Remusatia formosana* Hayata

   Common names: Hitchhiker elephant ear; yan yu (China); marathali (India)
   Habitat: Trees and rocks in forests
   Description: India, Nepal, Sri Lanka, Buthan, Bangladesh, Cambodia, Laos, Vietnam, Thailand, Australia, and Pacific Islands
   Botanical observation: This aroid climber grows to 1.5 m tall from a poisonous globose tuber. The petiole is up to 30 cm long and sheathing at base. The blade is graceful, dark green, glossy, peltate, 12–40×7.5–30 cm, acuminate, cordate, and with 3–4 nerves. The inflorescence peduncle is 6–20 cm long. The spathe is about 15 cm long, first erect, later reflexed, yellow, and apiculate. The staminate and female portions are separated by a 1.5–2 cm long neuter zone. The male zone is 1–1.5 cm long. The female zone is 2 cm long. The ovary is ovoid, unilocular, with discoid stigma.

Medicinal uses: Gonorrhea (India)

Commentary: The antimicrobial properties of this plant have apparently not been examined. Its medicinal use is a suggestion that it many contain antibacterial constituents.

### 3.1.2.2.13 *Scindapsus officinalis* (Roxb.) Schott

Synonym: *Pothos officinalis* Roxb.; *Monstera officinalis* (Roxb.) Schott

Common names: Gajakrishna, gajapipal, bor gumunia (India); gaj pipul (Bangladesh)

Habitat: Trees and rocks in forests

Distribution: Himalaya, India, Nepal, Bangladesh, Myanmar, Andaman Islands, and China.

Botanical observation: This large climber that grows up to a length of 10 m. The leaves are simple and spiral. The petiole is up to 15 cm long and winged. The blade is broadly lanceolate, 12.5–25×6.5–15 cm, somewhat fleshy, with about 9 pairs of secondary nerves, dark green, glossy, cordate at base, and acuminate at apex. The spathe is funnel shaped, about 6×1.5–2.5 cm, pale green with a limb that is 7×3 cm, glossy dark green, ending in an acuminate–filiform tip to 1 cm long. The spadix is 3.5 cm×2–3 mm. The fertile portion is up to 4 cm long. The androecium includes 4 stamens. The gynoecium is angular and minute. The infructescence are cylindrical, up to 6.5×3 cm, and made of numerous berries, which are lobed and containing up to 2 seeds, which are reniform and up to 5 mm long.

Medicinal uses: Fever (India); tuberculosis (Himalaya); dysentery (Nepal); diarrhea (India, Nepal).

Commentary: The stems contain chromone glycosides (Yu et al., 2017) and alkaloids, including quinolines (Yu et al., 2018) and no reports for the antimicrobial properties of this plant apparently exist.

**Reference**

Yu, J., Song, X., Wang, D., Wang, X. and Wang, X., 2017. Five new chromone glycosides from *Scindapsus officinalis* (Roxb.) Schott. *Fitoterapia*, *122*, pp. 101–106.

### 3.1.2.2.14 *Typhonium trilobatum* (L.) Schott.

Synonyms: *Arum orixense* Roxb.; *Arum trilobatum* L.; *Typhonium orixense* (Roxb. ex Andrews) Schott; *Typhonium siamense* Engl.; *Typhonium triste* Griff.

Common names: Bankachu; get kuchu (Bangladesh); ma ti li tou jian (China)

Habitat: Forests, roadsides, wastelands, and villages

Distribution: India, Nepal, Sri Lanka, Bhutan, Bangladesh, Myanmar, Cambodia, Laos, Vietnam, China, Thailand, and Malaysia

Botanical observation: This aroid grows from a globose tuber. The leaves are simple. The petiole is up to 40 cm tall. The blade is 3-lobed, the middle lobe ovate, 10–15×6–11 cm, acuminate at apex, and the lateral lobes asymmetrical and up to about 13 cm long. The inflorescence peduncle is up to 10 cm tall. The spathe is up to 30 cm long, green outside, dark purple to burgundy within, acuminate at apex. The spadix female zone is 1 cm long and the male zone up to 2 cm long. The berries are ellipsoid green with purple spots.

Medicinal uses: Boils (India, Thailand)

Commentary: The plant has apparently not been studied for its antibacterial and antiviral effects. The use of this plant by the traditional system of medicine of two countries for boils is a strong indication that it produces antibacterial principles.

## 3.1.3 ORDER DIOSCOREALES MART. (1835)

### 3.1.3.1 Family Dioscoreaceae R. Brown (1810)

The family Dioscoreaceae consists of about 6 genera and about 650 species of climbers growing from massive rhizomes or tubers. The stems are terete, flexuous, lignose and not uncommonly prickly. The leaves are simple or compound, spiral, and exstipulate. The blade is often coriaceous

and cordate to hastate with secondary nerves characteristically originating from the base and somewhat parallel to the midrib. Leaflets can be strongly asymmetrical. The inflorescences are racemes
or panicles. The perianth consists of 6 tepals. The androecium comprises 6 stamens. The gynoecium presents 3 carpels united to form a compound, inferior, 3-locular ovary, each locules containing 2 to many ovules on axile placentas and a trifid stigma. The fruits are dehiscent capsules, which
are 3-lobed. The seeds are winged. Plants in this family produce antimicrobial steroidal saponins,
stilbenes, and phenanthrenes.

### 3.1.3.1.1 *Dioscorea batatas* Decne.

Synonyms: *Dioscorea oppositifolia* L.; *Dioscorea polystachya* Turcz.

Common names: Chinese yam; chupri alu (Bangladesh); han yao, shu yu (China); uvi (Fiji);
verilaivali (India); uwi (Indonesia); yama imo (Japan); houo (Laos); myauk u ni (Myanmar); kukur
tarul (Nepal); ubi (the Philippines); huai sua (Thailand); man han (Vietnam)

Habitat: Cultivated

Ditribution: Tropical Asia and Pacific Islands

Botanical observation: This climber grows from an elongated and edible (if cooked) tuber, which
can reach up to about 1 m long. The stems are glabrous, twining, purplish, and smooth. The leaves
are simple, alternate or opposite, and exstipulate. The petiole is up to about 3 cm long. The blade
is broadly and deeply cordate to somewhat trilobed, triangular, glossy, dark green, 3–7×2–7 cm,
glabrous, and marked with 4 longitudinal pairs of secondary nerves. The inflorescence is axillary,
slender spikes, which are up to 8 cm long. The perianth includes 6 yellowish, dotted, purple lobes
about 2 mm long. The androecium includes 6 stamens. The fruits are about 2 cm long, oblong,
smooth, and winged.

Medicinal use: Abscesses (China)

*Broad-spectrum antibacterial amphiphilic stilbene*: The phytoalexin dihydropinosylvin
(LogD = 3.2 at pH 7,4; molecular weight = 214.2 g/mol) isolated from the tubers (infected with
*Pseudomonas cichorii*) at 5% of culture media (liquid broth) inhibited the growth of *Bacillus subti-
lis* and *Xanthomonas campestris* by 100 and 50%, respectively (Takasugi et al., 1987).

Dihydropinosylvin

*Antifungal (filamentous) stilbene*: Dihydropinosylvin at 5% of culture media inhibited the
growth of *Alternaria japonica, Botrytis alli, Calonectria graminicola, Pyricularia oryzae*, and
*Fusarium solani* (Takasugi et al., 1987).

*Strong broad-spectrum anticandidal spirostane-type amphiphilic steroid saponin:* Dioscin (LogP
= 1.3; molecular weight = 869.0 g/mol) inhibited the growth of *Candida albicans, Candida glabrata*,
and *Candida tropicalis* with the MIC values of 12.5, 12.5 and 25 µg/mL, respectively (Sautour et al.,
2004). In a subsequent study, dioscin inhibited the growth of *Candida albicans* and *Candida glabrata*
with MIC/MFC values of 2.5/2.5 and 5/>20 µg/mL, respectively (Yang et al., 2006). Dioscin
inhibited the growth of *Candida albicans* (ATCC 90028), *Candida parapsilosis* (ATCC 22019),

*Trichophytum beugellii* (KCTC 7707), and *Malassezia furfur* (KCTC 7744) with the MIC values of 22.5, 11.3, 11.3, and 22.5 µg/mL, respectively (Cho et al., 2013), and evoked cytoplasmic insults in *Candida albicans* (Cho et al., 2013).

Dioscin

*Strong antiviral (enveloped linear single stranded (+) RNA) polar extract*: Aqueous extract of tubers inhibited the replication of the Severe acute respiratory syndrome-associated coronavirus in Vero cells with the $IC_{50}$ value of 8 µg/mL and a selectivity index above 62 (Wen et al., 2011).

*Strong antiviral (enveloped monopartite linear single-stranded (+) RNA) lipophilic spirostane steroid*: Diosgenin (LogD = 5.7 at pH 7.4; molecular weight = 414.6 g/mol) inhibited the replication of the Hepatitis C virus with an $EC_{50}$ value of 3.8 µM (Wang et al., 2011).

Diosgenin

Commentary: (i) The plant produces antibacterial and antifungal stilbenes and phenantheres (Takasugi et al., 1987), as well as the steroid diosgenin (Edwards et al., 2002). Is diosgenin active against COVID-19? (ii) The plant is used for the treatment of COVID-19 in China (Lem et al., 2020; Yang et al., 2020).

**References**

Cho, J., Choi, H., Lee, J., Kim, M.S., Sohn, H.Y. and Lee, D.G., 2013. The antifungal activity and membrane-disruptive action of dioscin extracted from *Dioscorea nipponica*. *Biochimica et Biophysica Acta (BBA)-Biomembranes, 1828*(3), pp. 1153–1158.

Edwards, A.L., Jenkins, R.L., Davenport, L.J. and Duke, J.A., 2002. Presence of diosgenin in *Dioscorea batatas* (Dioscoreaceae). *Economic Botany, 56*(2), pp. 204–206.

Lem, F.F., Opook, F., Herng, D.L.J., Na, C.S., Lawson, F.P. and Tyng, C.F., 2020. Molecular mechanism of action of repurposed drugs and traditional Chinese medicine used for the treatment of patients infected with COVID-19: a systematic review. *medRxiv*.1–29.

Sautour, M., Mitaine-Offer, A.C., Miyamoto, T., Dongmo, A. and Lacaille-Dubois, M.A., 2004. Antifungal steroid saponins from *Dioscorea cayenensis*. *Planta Medica, 70*(01), pp. 90–92.

Takasugi, M., Kawashima, S., Monde, K., Katsui, N., Masamune, T. and Shirata, A., 1987. Antifungal compounds from *Dioscorea batatas* inoculated with *Pseudomonas cichorii*. *Phytochemistry, 26*(2), pp. 371–375.

Wang, Y.J., Pan, K.L., Hsieh, T.C., Chang, T.Y., Lin, W.H. and Hsu, J.T.A., 2011. Diosgenin, a plant-derived sapogenin, exhibits antiviral activity in vitro against Hepatitis C virus. *Journal of Natural Products, 74*(4), pp. 580–584.

Wen, C.C., Shyur, L.F., Jan, J.T., Liang, P.H., Kuo, C.J., Arulselvan, P., Wu, J.B., Kuo, S.C. and Yang, N.S., 2011. Traditional Chinese medicine herbal extracts of *Cibotium barometz, Gentiana scabra, Dioscorea batatas, Cassia tora*, and *Taxillus chinensis* inhibit SARS-CoV replication. *Journal of Traditional and Complementary Medicine, 1*(1), pp. 41–50.

Yang, L., Liu, X., Zhong, L., Sui, Y., Quan, G., Huang, Y., Wang, F. and Ma, T., 2018. Dioscin inhibits virulence factors of *Candida albicans*. *BioMed Research International, 2018*.

Yang, Q.X., Zhao, T.H., Sun, C.Z., Wu, L.M., Dai, Q., Wang, S.D. and Tian, H., 2020. New thinking in the treatment of 2019 novel coronavirus pneumonia. *Complementary Therapies in Clinical Practice, 39*, p. 101131.

*3.1.3.1.2   Dioscorea bulbifera L.*

Synonyms: *Dioscorea anthropophagorum* A. Chev. ex Jum.; *Dioscorea hoffa* Cordem.; *Dioscorea hofika* Jum. & H. Perrier; *Dioscorea latifolia* Benth.; *Dioscorea longipetiolata* Baudon; *Dioscorea perrieri* R. Knuth; *Dioscorea sativa* L.; *Dioscorea sativa* Thunb.; *Dioscorea violacea* Baudon; *Helmia bulbifera* (L.) Kunth

Common names: Air potato; aerial yam; kukuralu (Bangladesh); damlo ng sdam pre (Cambodia); huwi buah (Indonesia); man pauz (Laos); sarau (Fiji); ubi atas (Malaysia); lapma (Papua New Guinea); ubi-ubingan (the Philippines); man soen (Thailand); khoai dai (Vietnam)

Habitat: Cultivated

Distribution: Tropical Asia and Pacific

Botanical observation: This climber grows up to a length of 6 m from a large tuber. The stems are glabrous, twining, and smooth. The leaves are simple, alternate, and exstipulate. The petiole is up to 5 cm long. The blade is broadly and deeply cordate, glossy, dark green, 8–15×2–14 cm, glabrous, marked with about 6–10 parallel secondary nerves and scalariform tertiary nerves. The spikes are axillary, slender, pendulous, and 6.5–25 cm long. Pendulous bulbils are present at leaf axil, somewhat potato-shaped or globose, up to 5 cm across, smooth or rough, greyish, and edible (after cooking). The androecium includes 6 stamens. The berries are about 2 cm long, oblong, smooth, dark brown, and contain winged seeds.

Medicinal uses: Boils (China; India); ulcers, syphilitic sores (India); sore throat (China); AIDS (Thailand)

*Strong antimycobacterial clerodane amphiphilic diterpenes:* Bafoudiosbulbins B, C, F (LogP = 1.3; molecular weight = 404.4 g/mol), and G isolated from the bulbils inhibited the growth of *Mycobacterium smegmatis* (ATCC 700084) with the MIC/MBC values of 32/128, 8/16, 32/128, and 64/128 µg/mL, respectively (Kuete et al., 2012). Bafoudiosbulbins B, C, F, and G inhibited the growth of *Mycobacterium tuberculosis* (ATCC 27294) with the MIC values of 32/64, 8/8, 64/64 and 64/256 µg/mL, respectively (Kuete et al., 2012).

Bafoudiosbulbin F

*Moderate broad-spectrum antibacterial clerodane diterpenes*: Bafoudiosbulbin A isolated from the tubers inhibited the growth of *Salmonella typhi, Salmonella paratyphi* A, *Salmonella paratyphi* B, and *Pseudomonas aeruginosa* with MIC/MBC values of 50/200, 50/100, 25/100, and 50/200 µg/mL (Teponno et al., 2006). Bafoudiosbulbin B isolated from the tubers inhibited the growth of *Salmonella typhi, Salmonella paratyphi* A, and *Salmonella paratyphi* B, and *Pseudomonas aeruginosa* with the MIC/MBC values of 50/200, 25/100, 25/100, and 25/200 µg/mL (Teponno et al., 2006). Bafoudiosbulbins A, B, C, F, and G inhibited the growth of *Escherichia coli* (ATCC 8739) with the MIC/MBC values of 256/>256, 64/128, 64/128, 64/128, and 64/128 µg/mL, respectively (Kuete et al., 2012). Bafoudiosbulbins A, B, C, F, and G inhibited the growth of *Enterobacter aerogenes* (ATCC 13048) with the MIC values of 256/>256, 128/256, 64/128, 128/256, and 128/256 µg/mL, respectively (Kuete et al., 2012). Bafoudiosbulbins B, C, F, and G isolated from the bulbils inhibited the growth of *Pseudomonas aeruginosa* (PA01) with MIC values of 128/512, 126/256, 128/256, and 256/512 µg/mL, respectively (Kuete et al., 2012).

*Moderate antibacterial (Gram-negative) diterpene*: 8-Epidiosbulbin E acetate inhibited the growth of multidrug-resistant *Escherichia coli* and *Pseudomonas aeruginosa* with the MIC values of 200 and 400 µg/mL, respectively (Shriram et al., 2008).

*Moderate broad-spectrum antibacterial phenanthrene*: 2,7-Dihydroxy-4-methoxyphenanthrene isolated from the bulbils inhibited the growth of *Escherichia coli* (ATCC 8739) and *Klebsiella pneumoniae* (ATCC 11296) (Kuete et al., 2012).

*Moderate broad-spectrum antibacterial dihydrostilbene*: Demethylbatatasin IV isolated from the bulbis (experimentally infected with *Botrytis cinerea*) inhibited the growth of *Bacillus cereus,*

*Staphylococcus aureus*, *Pseudomonas aeruginosa*, and *Escherichia coli* with the MIC values of 100, 50, 10, and 50 µg/mL, respectively (Fagboun et al., 1987; Adesanya et al., 1989).

*Moderate antifugal (filamentous) stilbene*: Demethylbatatasin IV isolated from the bulbis (experimentally infected with *Botrytis cinerea*) inhibited germ tube elongation of *Aspergillus niger*, *Botryodiplodia theobromae*, *Penicillium sclerotigenum*, and *Cladosporium cladosporioides* with $EC_{50}$ values of 46, 59, 54, and 53 µg/mL, respectively (Fagboun et al., 1987; Adesanya et al., 1989).

*Antifungal (filamentous) piperidine alkaloid*: Dihydrodioscorine (0.1% of agar) inhibited the mycelial growth and spore production growth of *Sclerotium rolfsii*, *Curvularia lunata*, *Fusarium moniliforme*, *Botryodiplodia theobromae*, and *Macrophomina phaseolina* (Adeleye & Ikotun, 1989).

*Viral enzyme inhibition by hydrophilic flavonol and flavonol glycosides*: Myricetin, quercetin-3-O-β-D-glucopyranoside, and quercetin-3-O-β-D-galactopyranoside isolated from the bulbils inhibited the replication of Human immunodeficiency virus type-1 integrase with $IC_{50}$ values of 3.1, 19.3, and 21.8 µM, respectively (Chaniad et al., 2016).

*Viral inhibition by hydrophilic phenanthrene*: 2,4,6,7-Tetrahydroxy-9,10-dihydrophenanthrene inhibited Human immunodeficiency virus type-1 integrase activity with an $IC_{50}$ value of 14.2 µM (Chaniad et al., 2016).

*Viral enzyme inhibition by flavonol*: Myricetin at a concentration of 300 µM inhibited aminopeptidase N by 48% and angiotensin-converting enzyme by 26% (Bauvois & Dauzonne, 2006).

Commentaries: (i) MIC and MBC values recorded by Kuete et al. (2012) (including extracts) were lower in the presence of the efflux-pump inhibitor phenylalanine arginine β-naphthylamide, indicating that efflux pumps not only expel antibiotics but also natural products from plants.(ii) Methanol extract of fruits of *Dioscorea septemloba* Thunb. at a concentration of 100 µg/mL inhibited RNA-dependent DNA polymerase activity of Human immunodeficiency virus type-1 reverse transcriptase and Human immunodeficiency virus type-1 protease by 48.5 and 19%, respectively (Min et al., 2001). (iii) Myricetin being a substrate of the angiotensine-converting enzyme might be of value to block the entry of the Severe acute respiratory syndrome-associated coronavirus 2 in host cells.

### References

Adeleye, A. and Ikotun, T., 1989. Antifungal activity of dihydrodioscorine extracted from a wild variety of *Dioscorea bulbifera* L. *Journal of Basic Microbiology*, 29(5), pp. 265–267.

Adesanya, S.A., Ogundana, S.K. and Roberts, M.F., 1989. Dihydrostilbene phytoalexins from *Dioscorea bulbifera* and D. dumentorum. *Phytochemistry*, 28(3), pp. 773–774.

Bauvois, B. and Dauzonne, D., 2006. Aminopeptidase-N/CD13 (EC 3.4. 11.2) inhibitors: Chemistry, biological evaluations, and therapeutic prospects. *Medicinal Research Reviews*, 26(1), pp. 88–130.

Chaniad, P., Wattanapiromsakul, C., Pianwanit, S. and Tewtrakul, S., 2016. Anti-HIV integrase compounds from *Dioscorea bulbifera* and molecular docking study. *Pharmaceutical Biology*, 54(6), pp. 1077–1085.

Fagboun, D.E., Ogundana, S.K., Adesanya, S.A. and Roberts, M.F., 1987. Dihydrostilbene phytoalexins from Dioscorea rotundata. *Phytochemistry*, 26(12), pp. 3187–3189.

Kuete, V., BetrandTeponno, R., Mbaveng, A.T., Tapondjou, L.A., Meyer, J.J.M., Barboni, L. and Lall, N., 2012. Antibacterial activities of the extracts, fractions and compounds from *Dioscorea bulbifera*. *BMC Complementary and Alternative Medicine*, 12(1), p. 228.

Min, B.S., Kim, Y.H., Tomiyama, M., Nakamura, N., Miyashiro, H., Otake, T. and Hattori, M., 2001. Inhibitory effects of Korean plants on HIV-1 activities. *Phytotherapy Research*, 15(6), pp. 481–486.

Shriram, V., Jahagirdar, S., Latha, C., Kumar, V., Puranik, V., Rojatkar, S., Dhakephalkar, P.K. and Shitole, M.G., 2008. A potential plasmid-curing agent, 8-epidiosbulbin E acetate, from *Dioscorea bulbifera* L. against multidrug-resistant bacteria. *International Journal of Antimicrobial Agents*, *32*, pp. 405–410. 10.1016/j.ijantimicag.2008.05.013.

Teponno, R.B., Tapondjou, A.L., Gatsing, D., Djoukeng, J.D., Abou-Mansour, E., Tabacchi, R., Tane, P., Stoekli-Evans, H. and Lontsi, D., 2006. Bafoudiosbulbins A, and B, two anti-salmonellal clerodane diterpenoids from *Dioscorea bulbifera* L. var sativa. *Phytochemistry*, *67*(17), pp. 1957–1963.

### 3.1.3.1.3  *Dioscorea hispida* Dennst.

Synonyms: *Dioscorea daemona* Roxb; *Helmia daemona* (Roxb.) Kunth

Common names: Wild yam, intoxicating yam, Indian three-leaved yam; jangli-lota-alu, bon-alu, kulu (Bangladesh); damlong kduoch (Cambodia); bai shu liang (China); hastyaaluka, peiperendai (India); gadung (Indonesia); hwa koy (Laos); ubi arak (Malaysia); kywe (Myanmar); gayos (the Philippines); kloi (Thailand); cune (Vietnam)

Distribution: India, Bangladesh, Sikkim, Bhutan, Laos, Vietnam, Cambodia, Vietnam, China, Thailand, Indonesia, and the Philippines

Habitat: Openforests

Botanical observation: This climber grows up to about 3 m long from somewhat potato-shaped, hispid, monstruous, and poisonous tubers. The stems are prickly. The leaves are palmately 3-foliolate, alternate, and exstipulate. The petiole is hairy and up to about 30 cm long. The folioles are ovate to elliptic, 6–17.5×4–12 cm, hispid below, with 3–4 longitudinal nerves, asymmetrical, glossy, and acuminate at apex. The inflorescences are either panicles of spikes, axillary and hairy and up to 50 cm long, or about 40 cm long. The perianth comprises 6 minute hairy lobes. The androecium is made of 6 stamens. The capsules are oblong, 3-winged, glossy, yellowish green, coriaceous, and about 3.5–7 cm long.

Medicinal uses: Small pox, dermatitis, syphilis (Bangladesh)

*Broad-spectrum antibacterial halo developed by polar extract:* Methanol extract inhibited the growth of *Bacillus cereus*, *Bacillus megaterium*, *Bacillus subtilis*, *Sarcina lutea*, *Staphylococcus aureus*, *Salmonella paratyphi*, *Salmonella typhi*, *Shigella boydii*, *Shigella dysenteriae*, *Pseudomonas aeruginosa*, *Vibrio mimicus*, and *Vibrio parahaemolyticus* (Miah et al., 2018).

*Broad-spectrum antifungal halo developed by polar extract:* Methanol extract inhibited the growth of *Saccharomyces cerevisiae*, *Candida albicans*, and *Aspergillus niger* with inhibition zone of 10.5, 9.9, and 10.0 mm, respectively (Miah et al., 2018).

*Weak antiviral (non-enveloped monopartite linear single-stranded (+) RNA) polar extract:* Ethanol extract of tubers at a concentration of 700 µg/mL inhibited the replication the Foot and Mouth Disease virus (FMDV) (type O, local strain KPS/005/2545) (Chungsamarnyart et al., 2007).

*Viral enzyme inhibition by polar extract:* Aqueous extract of tubers at a concentration of 250 µg/mL inhibited Human immunodeficiency virus type-1 protease by 53.6% (Xu et al., 1996).

Commentary: The tubers contain the poisonous piperidine alkaloid dioscorin (Leete & Michelson, 1988; Kumoro & Hartati, 2015). As mentioned earlier, piperidine alkaloids have a tendency to have interesting antiviral properties. Is dioscorine active against the Severe acute respiratory syndrome-associated coronavirus?

### References

Chungsamarnyart, N., Sirinarumitr, T., Chumsing, W. and Wajjawalku, W., 2007. In vitro study of antiviral activity of plant crude-extracts against the foot and mouth disease virus. *Kasetsart Journal*, *41*, pp. 97–103.

Kumoro, A.C. and Hartati, I., 2015. Microwave assisted extraction of dioscorin from Gadung (*Dioscorea hispida* Dennst) tuber flour. *Procedia Chemistry*, *14*, pp. 47–55.

Leete, E. and Michelson, R.H., 1988. Biosynthesis of dioscorine from trigonelline in *Dioscorea hispida*. *Phytochemistry*, *27*(12), pp. 3793–3798.

Miah, M.M., Das, P., Ibrahim, Y., Shajib, M.S. and Rashid, M.A., 2018. In vitro antioxidant, antimicrobial, membrane stabilization and thrombolytic activities of *Dioscorea hispida* Dennst. *European Journal of Integrative Medicine*, *19*, pp. 121–127.

Xu, H.X., Wan, M., Loh, B.N., Kon, O.L., Chow, P.W. and Sim, K.Y., 1996. Screening of traditional medicines for their inhibitory activity against HIV-1 protease. *Phytotherapy Research*, *10*(3), pp. 207–210.

### 3.1.4   ORDER PANDANALES R. BROWN EX BERCHT. & J. PRESL (1820)

#### 3.1.4.1   Family Pandanaceae R. Brown (1810)

The family Pandanaceae consists of 3 genera and about 700 species of trees, often with stilt roots. The leaves are simple, narrow, coriaceous, and spiny. The inflorescence is a spadix (often massive) more or less surrounded by a leaf-like spathe, somewhat like the Araceae. The perianth is either abscent or comprising up to 4 tepals. The androecium is made of numerous stamens. The gynoecium includes numerous, free or fused carpels. The syncarps are massive and contain minute seeds.

*3.1.4.1.1   Benstonea foetida (Roxb.) Callm. & Buerki*

Synonyms: *Fisquetia macrocarpa* Gaudich.; *Pandanus foetidus* Roxb.; *Pandanus pseudofoeti*dus Martelli; *Pandanus wallichianus* Martelli

   Common name: Keya kanta (Bangladesh)

   Habitat: Mangroves

   Distribution: India, Bangladesh, Myanmar, Thailand, and Malaysia

   Botanical observation: This shrub grows to about 3 m tall and develops a few aerial roots. The leaves are simple, spiral, sessile, exstipulate, and gathered at apex of stems. The blade is linear-ensiform, up to 3.5 m × 6 cm, spiny, and coriaceous. The bracteated spikes are terminal, up to 10 cm long, and nauseating The androecium includes numerous stamens. The syncarp is massive, woody, and made of numerous phalanges

   Medicinal use: Skin diseases (Bangladesh)

   Pharmacology: Antibacterial

   *Broad-spectrum antibacterial halo developed by polar extract:* Ethanol extract of leaves inhibited the growth of *Staphylococcus aureus*, *Streptococcus pneumoniae*, *Shigella dysenteriae*, *Shigella sonnei*, *Salmonella thyphi*, *Escherichia coli*, *Vibrio cholerae*, *Enterobacter aerogenes*, and *Pseudomonas aeruginosa* (500 µg/disc) (Uddin et al., 2008).

   Commentary: The potential antimicrobial properties of this plant have apparently not been examined.

#### Reference

Uddin, S.J., Rouf, R., Shilpi, J.A., Alamgir, M., Nahar, L. and Sarker, S.D., 2008. Screening of some Bangladeshi medicinal plants for in vitro antibacterial activity. *Oriental Pharmacy and Experimental Medicine*, *8*(3), pp. 316–321.

*3.1.4.1.2   Pandanus odorus Ridl.*

Synonyms: *Pandanus amaryllifolius Roxb.*; *Pandanus latifolius Perr.*

   Common names: Fragrant Pandan, fragrant screw pine; shanlha (Bangladesh); tey (Cambodia); xiang lu dou (China); keora (India); pandan bebau (Indonesia); tey hom (Laos); pandan (Malaysia; the Philippines); karukai (Papua New Guinea); pandai (Sri Lanka); toei Hom (Thailand); la dua (Vietnam)

   Habitat: Cultivated

   Distribution: India, Sri Lanka, Bangladesh, Vietnam, Thailand, Malaysia, Indonesia, and the Philippines

Botanical observation: This herb grows to about 2 m tall. The leaves are simple, spiral, sessile, and exstipulate. The blade is linear-ensiform, up to 25–75 cm×2–5 cm, dark green, aromatic, and somewhat fleshy. The inflorescences are pendulous bracteated spikes of minute flowers. The androecium includes up to 6 stamens.

Medicinal uses: Boils (Bangladesh); flu (India); putrefied wounds (Malaysia); gonorrhea, syphilis (Bangladesh; Malaysia)

*Broad-spectrum antibacterial halo developed by mid-polar extract:* Ethyl acetate extract of leaves (500 µg/disc) inhibited the growth of *Escherichia coli, Sarcinia lutea, Staphylococcus aureus,* and *Shigella dysenteriae* with the inhibition zone diameters of 9, 11.5, 11.5, and 11 mm, respectively (Hamid et al., 2011).

*Broad-spectrum antifungal halo developed by mid-polar extract:* Ethyl acetate extract of leaves (500 µg/disc) inhibited the growth *Aspergillus niger, Candida albicans,* and *Saccharomyces cerevisiae* with inhibition zone diameters of 11.5, 11, and 10 mm, respectively (Hamid et al., 2011).

*Very weak broad-spectrum antibacterial simple phenol:* 4-Hydroxybenzoic acid (also known as para-hydroxybenzoic acid) (Peungvicha et al., 2001) inhibited the growth of *Staphylococcus aureus* (ATCC 5838), *Bacillus subtilis* (ATCC 6633), *Escherichia coli* (ATCC10536), *Pseudomonas aeruginosa* (KCTC 1628), and *Salmonella typhimurium* (ATCC 19430) with $IC_{50}$ values of 926, 956, 458, 619, and 688 (Cho et al., 1998).

*Strong antibacterial (Gram-negative) hydrophilic pyrrolidine alkaloid lactones:* Pandamarilactonine A (LogD = 1.7 at pH 0.7; molecular weight = 317.3 g/mol) isolated from the leaves inhibited the growth of *Escherichia coli, Pseudomonas aeruginosa,* and *Staphylococcus aureus* with the MIC/MBC values of 62.5/125, 15.6/31.2, and 250/500 µg/mL (Laluces et al., 2015).

Pandamarilactonine A

*Weak broad-spectrum antibacterial piperidine alkaloid lactones:* Pandamarilactone-1 isolated from the leaves inhibited the growth of *Escherichia coli, Pseudomonas aeruginosa,* and *Staphylococcus aureus* with the MIC/MBC values of 500/>500, 500/>500, and 250/500 µg/mL (Laluces et al., 2015). Pandamarilactone-32 isolated from the leaves inhibited the growth of *Escherichia coli, Pseudomonas aeruginosa,* and *Staphylococcus aureus* with the MIC/MBC values of 125/250, 500/>500, and 250/500 µg/mL (Laluces et al., 2015). Pandamarilactonine B isolated from the leaves inhibited the growth of *Escherichia coli, Pseudomonas aeruginosa,* and *Staphylococcus aureus* with MIC/MBC values of 500/>500, 500/>500, and 250/500 µg/mL (Laluces et al., 2015).

Commentary: (i) The plant produces pyrrolidine alkaloids with furanone moieties (Takayama et al., 2001). Furanones in general highly react with amino acids in peptides and proteins (Luo et al., 2019). The traditional system of medicine in Bangladesh uses several members of the genus *Pandanus* Parkinson to treat microbial infections: *Pandanus fascicularis* (local name: *Keowa phol*) is used to treat tongue infection, *Pandanus furcatus* Roxb. (local name: *Pathor shila*) is used to treat urinary tract infection, and *Pandanus pygmaeus* Thouars (local name: *Keya*) is used there to treat cough. *Pandanus furcatus* Roxb. is used for cough, diarrhea, and dysentery in Indonesia

### References

Cho, J.Y., Moon, J.H., Seong, K.Y. and Park, K.H., 1998. Antimicrobial activity of 4-hydroxybenzoic acid and trans 4-hydroxycinnamic acid isolated and identified from rice hull. *Bioscience, Biotechnology, and Biochemistry*, 62(11), pp. 2273–2276.

Hamid, K., Urmi, K.F., Saha, M.R., Zulfiker, A.H.M. and Rahman, M.M., 2011. Screening of different parts of the plant *Pandanus odorus* for its cytotoxic and antimicrobial activity. *Journal of Pharmaceutical Sciences and Research*, 3(1), p. 1025.

Laluces, H.M.C., Nakayama, A., Nonato, M.G., dela Cruz, T.E. and Tan, M.A., 2015. Antimicrobial alkaloids from the leaves of *Pandanus amaryllifolius*. *Journal of Applied Pharmaceutical Science*, 5(10), pp. 151–153.

Luo, S.H., Yang, K., Lin, J.Y., Gao, J.J., Wu, X.Y. and Wang, Z.Y., 2019. Synthesis of amino acid derivatives of 5-alkoxy-3, 4-dihalo-2 (5 H)-furanones and their preliminary bioactivity investigation as linkers. *Organic & Biomolecular Chemistry*, 17(20), pp. 5138–5147.

Peungvicha, P., Thirawarapan, S.S. and Watanabe, H., 2001. Possible mechanism of hypoglycemic effect of 4-hydroxybenzoic acid, a constituent of *Pandanus odorus* root. *The Japanese Journal of Pharmacology*, 78(3), pp. 395–398.

Takayama, H., Ichikawa, T., Kitajima, M., Nonato, M.G. and Aimi, N., 2001. Isolation and characterization of two new alkaloids, Norpandamarilactonine-A and-B, from *Pandanus amaryllifolius* by spectroscopic and synthetic methods. *Journal of Natural Products*, 64(9), pp. 1224–1225.

#### 3.1.4.1.3 *Pandanus tectorius* Parkinson

Synonyms: *Pandanus fascicularis* Lam.; *Pandanus odoratissimus* L. f.; *Pandanus tectorius* Sol. ex Balf. f.

Common names: Screw pine; keya (Bangladesh); lu dou chu (China); kafu (Guam); pu hala (Haiwai); ketaki (India); pandan pudak (Indonesia); pandan pasir (Malaysia); panhaka (the Philippines); karaket (Thailand); dua go (Vietnam)

Habitat: Seashores

Distribution: Tropical Asia and Pacific

Botanical observation: This tree grows to about 4 m tall anddevelops numerous aerial roots. The leaves are simple, spiral, sessile, exstipulate, and at gathered at the apex of stems. The blade is linear-ensiform, up to 180×10 cm, serrate, and coriaceous. The spikes are up to 60 cm long and bracteated. The androecium includes 10 stamens. The syncarp is massive, woody, pendulous, globose, or cylindric, and consists ofaggregates of phalanges.

Medicinal uses: Leprosy, flu, chicken pox (Bangladesh); wounds (Indonesia); boils, blennorrhea (the Philippines)

*Moderate antimycobacterial lipophilic tirucallane-type triterpene:* 24,24-dimethyl-5β-tirucall-9(11), 25-dien-3-one isolated from the leaves inhibited the growth of *Mycobacterium tuberculosis* ($H_{37}$Rv) with the MIC value of 64 µg/mL (Tan et al., 2008).

24,24-dimethyl-5β-tirucall-9(11), 25-dien-3-one

Commentary: Zhang et al. (2012) isolated from the fruits series of flavones and hydroxycinnamic acids. What is the antimycobacterial mechanism of action of 24,24-dimethyl-5β-tirucall-9(11), 25-dien-3-one?

**References**

Tan, M.A., Takayama, H., Aimi, N., Kitajima, M., Franzblau, S.G. and Nonato, M.G., 2008. Antitubercular triterpenes and phytosterols from *Pandanus tectorius* Soland. var. laevis. *Journal of Natural Medicines*, 62(2), pp. 232–235.

Zhang, X., Guo, P., Sun, G., Chen, S., Yang, M., Fu, N., Wu, H. and Xu, X., 2012. Phenolic compounds and flavonoids from the fruits of *Pandanus tectorius* Soland. *Journal of Medicinal Plants Research*, 6(13), pp. 2622–2626.

### 3.1.5   Order Liliales Perleb (1828)

#### 3.1.5.1   Family Liliaceae A.L. de Jussieu (1789)

The family Liliaceae consists of about 250 genera and 3500 species of herbs growing from rhizomes, bulbs, or corms. The leaves are simple, basal, spiral, alternate, opposite, or whorled. The blade is often somewhat narrowly lanceolate and with parallel nerves. The inflorescence is a raceme, panicle, spike, umbel, panicle, or solitary. The perianth consists of 6 tepals in 2 whorls. The androecium comprises 6 stamens. The gynoecium presents 3 carpels fused in a 3-locular ovary, each locule sheltering numerous ovules growing on axile placentas. The fruit is a capsule or berry.

##### 3.1.5.1.1   *Erythronium japonicum* Decne

Synonyms: *Erythronium dens-canis* var. *japonicum* Baker

Common names: Japanese dogtooth violet; zhu ya hua (China); katakuriko (Japan); eolleji (Korea)

Habitat: Moist, shady soils in forests

Distribution: China, Korea, and Japan

Botanical observation: It is a beautiful and edible herb that grows up to 20 cm tall from a bulb that is up to about 5 cm across. The leaves are simple, cauline, and in pairs. The petiole is 3–4 cm long. The blade is elliptic to broadly lanceolate, 10–11×2.5–6.5 cm, cuneate at base, and acute or obtuse at apex. The flower is terminal, nodding, and solitary. The perianth includes 6 tepals, which are purple, somewhat penciled in dark purple at the middle, recurved, lanceolate, and up to 5 cm long. The androecium includes 6 stamens with linear filaments and dark purple linear anthers (resembling small incense sticks), which are up to about 7 mm long. The style is protuding. The capsules are 3-lobed.

Medicinal uses: Abscesses, tuberculosis (Korea, Malaysia); diarrhea (China)

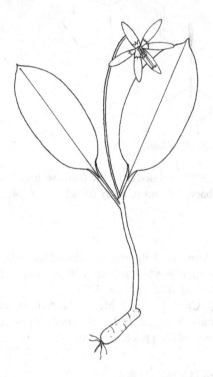

*Erythronium japonicum* Decne

Commentary: The antimicrobial properties of this plant have apparently not been examined. Note that members of the genus *Erythronium* L. synthetize series of α-methylene γ-lactones (α-methylene butyrolactone) (Cavallito & Haskell, 1946). Consider that α-methylene γ-lactones have the tendency to alkylate DNA, RNA, peptides, and proteins (Diamond et al., 1985) of bacteria of fungi via the enone moiety by Michael-type reaction with nucleophilic groups (Huang & Lee, 2011)

**References**

Cavallito, C.J. and Haskell, T.H., 1946. a-Methylene hutyrolactone from *Erythronium americanum*. *Journal of the American Chemical Society, 68*, pp. 2332–2334.

Diamond, K.B., Warren, G.R. and Cardellina II, J.H., 1985. Native American food and medicinal plants. 3. α-methylene butyrolactone from *Erythronium grandiflorum* Pursh. *Journal of Ethnopharmacology, 14*(1), pp. 99–101.

Huang, P.J. and Lee, K.H., 2011. Synthesis and antibacterial evaluation of 6-azapyrimidines with α-methylene-γ-(4-substituted phenyl)-γ-butyrolactone pharmacophores. *Medicinal Chemistry Research, 20*(7), pp. 1081–1090.

### 3.1.5.1.2   *Fritillaria thunbergii* Miq

Synonyms: *Fritillaria verticillata* var. *thunbergii* (Miq.) Baker

Common names: Zhe bei mu (China); bai mo (Japan)

Habitat: Moist, shady soils in forests

Distribuition: China and Japan

Botanical observation: This somewhat eery herb grows up to 80 cm tall from a white bulb that is up to about 3 cm across. The leaves are simple, cauline, sessile, and opposite or verticillate. The blade is linear-lanceolate, 7–11 cm × 1–2.5 cm, cuneate at base, and characteristcally curled at apex. The inflorescence is a fascicle of nodding flowers. The perianth includes 6 tepals, which are light yellow with purplish lines, oblong, and up to 3.5 cm long. The androecium is made of 6 stamens, which are up to about 1.5 cm long with whitish and linear anthers. The style is 3-lobed. The capsules are winged.

Medicinal uses: Cough (China); tuberculosis (Korea, Taiwan)

*Moderate antibacterial (Gram-positive) phytosterol saponin:* β-Sitosterol α-glucopyranoside inhibited the growth of *Bacillus subtilis*, *Staphylococcus aureus*, and *Micrococcus luteus* with the MIC values of 50, 200, and 400 µg/mL, respectively, and inhibited the enzymatic activity of sortase with an $IC_{50}$ value of 18.3 µg/mL (Kim et al., 2003).

*Moderate antiviral (enveloped non-segmented linear single-stranded (+) RNA virus) polar extract:* Aqueous extract protected Madin-Darby canine kidney cells against Human Influenza virus type A/PR/8/1934 (H1N1) with an $IC_{50}$ value of 148.2 µg/mL and a selectivity index of 50.6 (Kim et al., 2020).

*In vivo antiviral polar extract:* Aqueous extract given orally at a dose of 6 mg/mice for 10 days evoked a 40% survival rate in BALB/c mice infected with the H1N1 virus (Kim et al., 2020).

*Viral enzyme inhibition by polar extract:* Aqueous extract of rhizomes at a concentration of 250 µg/mL inhibited Human immunodeficiency virus type-1 protease by 20.8% (Xu et al., 1996).

Commentary:

(i) In Gram-positive bacteria, sortases (A–F) are enzymes involved in adhesion, iron acquisition, and spore formation (Schmohl & Schwarzer, 2014). For instance, the reaction that attaches surface proteins (like protein A which permits the escape of bacteria from our immune system) to the peptidoglycan wall is catalyzed by sortase A (Schmohl & Schwarzer, 2014).

(ii) The plant synthetizes series of steroidal quinolizidine alkaloids: dongbeinine (Zhang et al., 1993) and peimine (verticine), as well as iso-pimarane and *ent*-kaurane diterpenes (Kitajima et al., 1982). Note that verticine is an inhibitor of angiotensin-converting enzyme (Oh et al., 2003) and as such a probable inhibitor of the Severe acute respiratory syndrome-associated coronavirus binding to host cells (Zhang et al., 2020). Hence, *Fritillaria thunbergii* Miq., could be examined for its anti-COVID-19 properties.

## References

Kim, M., Nguyen, D.V., Heo, Y., Park, K.H., Paik, H.D. and Kim, Y.B., 2020. Antiviral activity of *Fritillaria thunbergii* extract against human influenza virus H1N1 (PR8) in vitro, in ovo and in vivo. *Journal of Microbiology and Biotechnology*, 30(2), pp. 172–177.

Kim, S.H., Shin, D.S., Oh, M.N., Chung, S.C., Lee, J.S., Chang, I.M. and Oh, K.B., 2003. Inhibition of sortase, a bacterial surface protein anchoring transpeptidase, by β-sitosterol-3-O-glucopyranoside from *Fritillaria verticillata*. *Bioscience, Biotechnology, and Biochemistry*, 67(11), pp. 2477–2479.

Kitajima, J., Komori, T. and Kawasaki, T., 1982. Studies on the constituents of the crude drug" Fritillariae bulbus." III. On the diterpenoid constituents of fresh bulbs of Fritillaria thunbergii Miq. *Chemical and Pharmaceutical Bulletin*, 30(11), pp. 3912–3921.

Oh, H., Kang, D.G., Lee, S., Lee, Y. and Lee, H.S., 2003. Angiotensin converting enzyme (ACE) inhibitory alkaloids from *Fritillaria ussuriensis*. *Planta Medica*, 69(06), pp. 564–565.

Schmohl, L. and Schwarzer, D., 2014. Sortase-mediated ligations for the site-specific modification of proteins. *Current Opinion in Chemical Biology*, 22, pp. 122–128.

Xu, H.X., Wan, M., Loh, B.N., Kon, O.L., Chow, P.W. and Sim, K.Y., 1996. Screening of traditional medicines for their inhibitory activity against HIV -1 protease. *Phytotherapy Research*, *10*(3), pp. 207–210.

Zhang, J., Lao, A. and Xu, R., 1993. Steroidal alkaloids from *Fritillaria thunbergii* var. chekiangensis. *Phytochemistry*, *33*(4), pp. 946–947.

Zhang, H., Penninger, J.M., Li, Y., Zhong, N. and Slutsky, A.S., 2020. Angiotensin-converting enzyme 2 (ACE2) as a SARS-CoV-2 receptor: molecular mechanisms and potential therapeutic target. *Intensive Care Medicine*, *46*(4), pp. 586–590.

### 3.1.5.1.3   *Lilium brownii* F.E. Brown ex Miellez

Common names: Brown's lily, Hong Kong lily; aye bai he (China)

Habitat: waste lands, grassy lands, forest edges, villages

Distribution: China

Botanical observation: This graceful herb grows up to 2 m tall from a scaly, edible (if cooked), globose, and up to about 4.5 cm across bulb. The leaves are simple, cauline, and alternate. The petiole is 3–4 cm long. The blade is linear-lanceolate or obovate, 7–15 cm × 1–2 cm, and acute at apex. The flowers are showy, fragrant, nodding, terminal, and solitary or arranged in fascicles. The perianth is made of 6 pure white and up to 18×5 cm tepals. The 6 stamens can reach 10 cm long. The style is protuding and up to about 10 cm long. The capsules are 3-lobed and up to about 5 cm long.

Medicinal uses: Ulcers, cuts, and bronchitis (China)

*Very weak antifungal (filamentous) polar extract:* Methanol extract of bulbs inhibited the growth of *Rhizoctonia solani* by 51.3% at the concentration of 2000 µg/mL (Rajput et al., 2018).

*Strong antiviral (enveloped monopartite linear single stranded (+) RNA) hydrophilic flavonol glycosides:* Isoquercitrin (quercetin-3-*O*-glucoside; LogD = 1.1 at pH 7.4; molecular weight = 464.3 g/mol) inhibited the replication of the Zika virus in SH-SY5Y cells with the $IC_{50}$ value of 9.7 µM and a selectivity index of 60 (Gaudry et al., 2018). Rutin (also known as quercetin-3-rutinoside; LogD = −1.7 at pH 7.4; molecular weight = 610.5 g/mol) reduced Avian Influenza strain H5N1 plaque formation Madin-Darby canine kidney cells by 73.2%, at 1 ng/mL (Ibrahim et al., 2013).

Isoquercitrin

Rutin

*Strong antiviral (enveloped monopartite linear single-stranded (−) RNA virus) flavonol:* Quercetin (molecular weight = 320.2 g/mol) reduced Avian Influenza strain H5N1 plaque formation Madin-Darby canine kidney cells by 68% at 1 ng/mL (Ibrahim et al., 2013).

*Antiviral (non-enveloped monopartite linear single-stranded (+) RNA) hydroxycinnamic acid:* Ferulic acid at 500 µ/mL reduced Feline calicivirus (strain F9) (FCV-F9) titers to undetectable levels after 3 hours (Joshi et al., 2015).

*Viral enzyme inhibition by flavonol:* Quercetin inhibited the Severe acute respiratory syndrome-associated coronavirus chymotrypsin-like protease with an $IC_{50}$ value of 23.8 µM (Ryu et al., 2010).

Commentaries:

(i) The plant brings to being a panel of spirostane-type steroidal saponins and steroidal glycoalkaloids, including solasodine glycosides (Mimaki & Sashida, 1990; Munafo & Gianfagna, 2015). The tepals contain isoquercitrin, quercetin, rutin, ferulic acid, and caffeic acid (Chen et al., 2015).

(ii) Strong anti-Respiratory Syncytial Virus phenolics are produced by the bulbs of *Lilium lancifolium* Thunb. (Zhou et al., 2014), which is used in China for the treatment of COVID-19 in combination with oseltamivir phosphate (Luo, 2020).

(iii) Isoquercitrin is found in other plants used to treat COVID-19. Is it of value against COVID-19?

### References

Chen, G.L., Chen, S.G., Xie, Y.Q., Chen, F., Zhao, Y.Y., Luo, C.X. and Gao, Y.Q., 2015. Total phenolic, flavonoid and antioxidant activity of 23 edible flowers subjected to in vitro digestion. *Journal of Functional Foods*, 17, pp. 243–259.

Gaudry, A., Bos, S., Viranaicken, W., Roche, M., Krejbich-Trotot, P., Gadea, G., Desprès, P. and El-Kalamouni, C., 2018. The flavonoid isoquercitrin precludes initiation of Zika virus infection in human cells. *International Journal of Molecular Sciences*, 19(4), p. 1093.

Ibrahim, A.K., Youssef, A.I., Arafa, A.S. and Ahmed, S.A., 2013. Anti-H5N1 virus flavonoids from *Capparis sinaica* Veill. *Natural Product Research*, 27(22), pp. 2149–2153.

Joshi, S.S., Dice, L. and D'Souza, D.H., 2015. Aqueous extracts of *Hibiscus sabdariffa* calyces decrease hepatitis A virus and human norovirus surrogate titers. *Food and Environmental Virology*, 7(4), pp. 366–373.

Luo, A., 2020. Positive SARS-Cov-2 test in a woman with COVID-19 at 22 days after hospital discharge: a case report. *Journal of Traditional Chinese Medical Sciences, 7*(4), pp. 413–417.

Mimaki, Y. and Sashida, Y., 1990. Sterodial saponins and alkaloids from the bulbs of *Lilium brownii* var. colchesteri. *Chemical and Pharmaceutical Bulletin, 38*(11), pp. 3055–3059.

Munafo Jr, J.P. and Gianfagna, T.J., 2015. Chemistry and biological activity of steroidal glycosides from the *Lilium genus*. *Natural Product Reports, 32*(3), pp. 454–477.

Rajput, N.A., Atiq, M., Javed, N., Ye, Y.H., Zhao, Z., Syed, R.N., Lodhi, A.M., Khan, B., Iqbal, O. and Dou, D., 2018. Antimicrobial effect of Chinese medicinal plant crude extracts against *Rhizoctonia solani* and *Phythium aphanidermatum*. *Fresenius Environmental Bulletin, 27*, pp. 3941–3949.

Ryu, Y.B., Jeong, H.J., Kim, J.H., Kim, Y.M., Park, J.Y., Kim, D., Naguyen, T.T.H., Park, S.J., Chang, J.S., Park, K.H. and Rho, M.C., 2010. Biflavonoids from *Torreya nucifera* displaying SARS-CoV 3CLpro inhibition. *Bioorganic & Medicinal Chemistry, 18*(22), pp. 7940–7947.

Zhou, Z.L., Lin, S.Q., Yang, H.Y., Zhang, H.L. and Xia, J.M., 2014. Antiviral constituents from the bulbs of *Lilium lancifolium*. *Asian Journal of Chemistry, 26*(22).

### 3.1.5.1.4    *Liriope graminifolia* (L.) Baker

Synonyms: *Asparagus graminifolius* L.; *Dracaena graminifolia* (L.) L.; *Liriope angustissima* Ohwi; *Liriope crassiuscula* Ohwi; *Mondo graminifolium* (L.) Koidz.

Common names: Creeping Liriope, lilyturf; he ye shan mai dong (China)

Habitat: Wet and shady spots in grassy lands and thickets

Distribution: China and Taiwan

Botanical observation: This herb grows up to 50 cm tall from tuberous roots. The leaves are simple, basal, and spiral. The blade is linear, grass-like, up to about 60 cm×4 mm, serrulate, and acute at apex. The panicles are terminal on a peduncle, which is up to about 40 cm tall. The 6 tepals are purplish and up to 4 mm long. The androecium includes 6 stamens. The berries are blue, glossy, somewhat globose, and up to about 5 mm long.

Medicinal use: Tuberculosis (China)

Commentaries:

(i) The plant produces series of homoisoflavonoids, methylophiopogonanone B, and 5,7-dihydroxy-3-(4-methoxybenzyl)-6-methyl-chroman-4-one (Wang et al., 2011). Homoisoflavonoids are not uncommonly antiviral (Jeong et al., 2012; Tait et al., 2006). For instance, *Liriope platyphylla* F.T. Wang & Tang (used in China for the treatment of tuberculosis) produces the homoisoflavanone (3*R*)-3-(4′-hydroxybenzyl)-5,7-dihydroxy-6-methyl-chroman-4-one, which inhibited the secretions of Hepatitis B virus surface antigen and e-antigen by HepG2.2.15 cells with the $IC_{50}$ values of 3.8 and 2.5 µg/mL, respectively (Huang et al., 2014).

(ii) The antimycobacterial potentials of *Liriope graminifolia* (L.) Baker have apparently not been examined. Note that the catabolism of cholesterol in *Mycobacterium tuberculosis* needs a number of enzymes, including 3β-hydroxysteroid dehydrogenase, 3-ketosteroid-1Δ-dehydrogenase, and steroid 9α-hydroxylase (Capyk et al., 2009). The plant assembles series of spirostane-type steroids (Wang et al., 2011), and one could wonder if these spirostane-type steroids or steroidal alkaloids could inhibit (because of their steroid framework) the enzymes responsible for the catabolism of the steroid cholesterol.

### References

Capyk, J.K., D'Angelo, I., Strynadka, N.C. and Eltis, L.D., 2009. Characterization of 3-ketosteroid 9α-hydroxylase, a Rieske oxygenase in the cholesterol degradation pathway of *Mycobacterium tuberculosis*. *Journal of Biological Chemistry, 284*(15), pp. 9937–9946.

Huang, T.J., Tsai, Y.C., Chiang, S.Y., Wang, G.J., Kuo, Y.C., Chang, Y.C., Wu, Y.Y. and Wu, Y.C., 2014. Anti-viral effect of a compound isolated from *Liriope platyphylla* against Hepatitis B virus in vitro. *Virus Research*, *192*, pp. 16–24.

Jeong, H.J., Kim, Y.M., Kim, J.H., Kim, J.Y., Park, J.Y., Park, S.J., Ryu, Y.B. and Lee, W.S., 2012. Homoisoflavonoids from *Caesalpinia sappan* displaying viral neuraminidases inhibition. *Biological and Pharmaceutical Bulletin*, *35*(5), pp. 786–790.

Tait, S., Salvati, A.L., Desideri, N. and Fiore, L., 2006. Antiviral activity of substituted homoisoflavonoids on enteroviruses. *Antiviral Research*, *72*(3), pp. 252–255.

Wang, K.W., Zhang, H., Shen, L.Q. and Wang, W., 2011. Novel steroidal saponins from *Liriope graminifolia* (Linn.) Baker with anti-tumor activities. *Carbohydrate Research*, *346*(2), pp. 253–258.

### 3.1.5.1.5   *Rohdea japonica* (Thunb.) Roth

Synonyms: *Orontium japonicum* Thunb.; *Rohdea esquirolii* H. Lév.; *Rohdea sinensis* H. Lév.

Common names: Japanese sacred lily; wan nian qing (China); omoto (Japan)

Habitat: Wet and shady spots in grassy lands and forests, cultivated

Distribution: China, Korea, and Japan

Botanical observation: This ornamental herb grows up to about 50 cm tall from a rhizome. The leaves are simple, basal, and spiral. The blade is 15–50×2.5–7 cm, with about 15 longitudinal nervations, wavy, fleshy, lanceolate-oblong, and acute at apex. The spikes are terminal on a peduncle that is up to about 10 cm tall. The perianth is tubular, yellowish, up to 4 mm long, 6-lobed, the lobes round and inconspicuous. The androecium includes 6 stamens adnate to the corolla. The stigma is 3-lobed. The berries are red, glossy, and somewhat globose, up to about 8 mm across and dreadfully poisonous. The infructescence has somewhat an Araceous look.

Medicinal uses: Sores, abscesses, boils (China, Japan); fever (China)

Commentary: The plant produces the cardenolide rhodexin A (Masuda et al., 2003). Apparently no reports on the antimicrobial properties of this plant exist.

### Reference

Masuda, T., Oyama, Y., Yamamoto, N., Umebayashi, C., Nakao, H., Toi, Y., Takeda, Y., Nakamoto, K., Kuninaga, H., Nishizato, Y. and Nonaka, A., 2003. Cytotoxic screening of medicinal and edible plants in Okinawa, Japan, and identification of the main toxic constituent of *Rhodea japonica* (Omoto). *Bioscience, Biotechnology, and Biochemistry*, *67*(6), pp. 1401–1404.

### 3.1.5.1.6   *Tulipa edulis* (Miq.) Baker

Synonyms: *Amana edulis* (Miq.) Honda; *Amana graminifolia* (Baker ex S. Moore) A.D. Hall; *Gagea argyi* H. Lév.; *Gagea coreana* H. Lév.; *Gagea hypoxioides* H. Lév.; *Orithyia edulis* Miq.; *Tulipa graminifolia* Baker ex S. Moore; *Tulipa minifolia* Baker ex S. Moore

Common name: Lao ya ban (China)

Habitat: Grassy lands

Distribution: China, Korea, and Japan

Botanical observation: This graceful herb grows up to about 30 cm tall from an ovoid bulb covered by a tunic, edible, and up to about 4 cm across. The pseudostem is about 25 cm tall. The leaves are simple, cauline, and in pairs. The blade is 15–25 cm×5–9 mm, somewhat glaucous, and linear. The flowers are terminal and solitary. The perianth comprises 6 tepals, which are white penciled purple, lanceolate, and up to about 3 cm long. The androecium includes 6 stamens. The stigma is 3-lobed. The capsules are up to about 7 mm across, subglobose, and beaked at apex.

Medicinal uses: Abscesses (China)

*Strong antiviral (enveloped monopartite linear dimeric single stranded (+) RNA) polar extract:* Butanol extract at the concentration of 0.3 μg/mL abrogated the replication of the Human immuno-deficiency virus type-1 (Eum et al., 2019).

*Viral enzyme inhibition by polar extract:* Butanol extract inhibited Human immunodeficiency virus type-1 reverse transcriptase by 19.5% at a concentration of 200 μg/mL (Eum et al., 2019).

Commentaries: (i) Members of the genus *Tulipa* L. synthetize an ester of glucose with α-methylene-γ-hydroxybutyric acid termed tuliposide A which, upon enzymatic catalysis during tissue injuries, yields an aglycone, which lactonizes into α-methylene butyrolactones (Slob et al., 1975), which are antibacterial and antifungal as described earlier. (ii) Butanol is a polar solvent that extracts glycosides and saponins.

### References
Eum, J.S., Park, Y.D. and Hong, S.K., 2009. Activities of natural plant extracts against HIV-1. *Journal of Information and Communication Convergence Engineering, 7*(4), pp. 576–579.
Slob, A., Jekel, B., de Jong, B. and Schlatmann, E., 1975. On the occurrence of tuliposides in the Liliiflorae. *Phytochemistry, 14*(9), pp. 1997–2005.

### 3.1.5.2 Family Melanthiaceae Batsch ex Borkh. (1797)
The family Melanthiaceae consists of about 20 genera and 100 species of herbs growing from bulbs, rhizomes, or corms. The leaves are simple, alternate, or verticillate, and often fleshy with narrow parallel secondary nerves, basal or cauline. The inflorescence is a raceme, umbel, panicle, or a spike. The perianth comprises 6 tepals. The androecium includes 6 stamens. The gynoecium consists of 3 carpels, which are free, each carpel concealing 2 to numerous ovules on axile placentas or all carpels are fused into a unilocular ovary. The capsules are dehiscent and more or less 3-lobed. Plants in this family bring into being extremely toxic steroidal alkaloids.

#### 3.1.5.2.1 *Paris chinensis* Franch.
Synonyms: *Daiswa chinensis* (Franch.) Takht.; *Paris brachysepala* Pamp.; *Paris brevipetala* Y.K. Yang; *Paris formosana* Hayata; *Paris polyphylla* var. *chinensis* (Franch.) H. Hara
Common name: Hua chong lou (China)
Habitat: Bamboo forests
Distribution: Nepal, Myanmar, Laos, Cambodia, Vietnam, Thailand, China, and Taiwan
Botanical observation: This odd-looking poisonous herb grows to a height of 1.3 m from a stout rhizome. The leaves are whorled, by up to 10, and terminal. The petiole is 1–6 cm long. The blade is lanceolate, 6–15 cm×0.5–5 cm, and acute at apex. The perianth comprises 6 yellowish green tepals, the inner shorter, leaf-like, narrowly ovate, and up to 7 cm long. The 6 stamens are up to 2 cm long with bright yellow and linear anthers. The ovary is globose, ribbed, and develop a 5-lobed stigma. The capsules are globose and shelter numerous seeds embedded in red arils.
Medicinal use: Cough, fever (Nepal); ringworm (China); wounds (Taiwan)
*Broad-spectrum antifungal spirostane-type steroidal saponin:* (25R)-spirost-5-ene-3 β, 17 α-diol 3-O-{O-α-L-rhamnopyranosyl-(1→2)-O-[O-β-xylopyranosyl-(1→5)-α-L-arabinofuranosyl-(1→4)]-β-D-glucopyranoside isolated from the rhizome inhibited the growth of *Cladosporium cladosporioides* and *Candida sp.* (Deng et al., 2008).
*Strong antiviral (non-enveloped linear monopartite (+) RNA virus) polar extract:* Ethanol extract inhibited the replication of the Coxsackie virus B3 as potently as the antiviral drug ribavirin *in vitro* (Wang et al., 2011).
*Strong (enveloped monopartite linear single-stranded (+) RNA) polar extract:* Ethanol extract of leaves inhibited the replication of the Chikungunya virus with the $IC_{50}$ value of 8.7 μg/mL and a selectivity index of 1.7 (Joshi et al., 2020).

*Strong antiviral (enveloped monopartite linear single-stranded (–) RNA virus) spirostane-type steroidal saponin:* Polyphylla saponin I at a concentration of 5 μg/mL inhibited the replication of Influenza A virus (A/PR/8/1934) by 77.3% and was virucidal by 62.4% (Pu et al., 2015).

*In vivo antiviral (enveloped non-segmented linear single stranded (–) RNA virus) spirostane-type steroidal saponin:* Polyphylla saponin I given orally at the dose of 10 mg/Kg twice a day for 5 days to mice experimentally infected with Influenza A virus (A/PR/8/1934) increased the mean survival rate from 8.5 to 13.2 days (Oseltamivir at 3 mg/Kg twice a day for 5 days: 12.1 days) and inhibited thickening of alveolar wall and infiltrative inflammatory cells in the lungs (Pu et al., 2015).

Commentaries: (i) The rhizome produces dioscin (Wang et al., 2010), which is a strong broad-spectrum antifungal steroidal saponin as seen earlier. Another spirostane-type steroidal saponin in the rhizome is polyphyllin D (Wang et al., 2010), which has been reported to have antibacterial effects (Man et al., 2013). (ii) Methanol extract of *Paris verticillata* (Local name in China: *Bei chong lou*) at a concentration of 100 μg/mL inhibited Human immunodeficiency virus type-1 protease by 18%, respectively (Min et al., 2001). (iii) *Paris quadrifolia* L. is applied to itchy sores in China (local name: *Si ye chong lou*). (iv) The plant is used to treat COVID-19 in China (Gu et al., 2020).

### References

Deng, D., Lauren, D.R., Cooney, J.M., Jensen, D.J., Wurms, K.V., Upritchard, J.E., Cannon, R.D., Wang, M.Z. and Li, M.Z., 2008. Antifungal saponins from *Paris polyphylla* Smith. *Planta Medica*, *74*(11), pp. 1397–1402.

Gu, M., Liu, J., Shi, N.N., Li, X.D., Huang, Z.D., Wu, J.K., Wang, Y.G., Wang, Y.P., Zhai, H.Q. and Wang, Y.Y., 2020. Analysis of property and efficacy of traditional Chinese medicine in staging revention and treatment of coronavirus disease 2019. *Zhongguo Zhong yao za zhi= Zhongguo zhongyao zazhi= China Journal of Chinese Materia Medica*, *45*(6), pp. 1253–1258.

Joshi, B., Panda, S.K., Jouneghani, R.S., Liu, M., Parajuli, N., Leyssen, P., Neyts, J. and Luyten, W., 2020. Antibacterial, antifungal, antiviral, and anthelmintic activities of medicinal plants of Nepal selected based on ethnobotanical evidence. *Evidence-Based Complementary and Alternative Medicine*.

Man, S.L., Wang, Y.L., Li, Y.Y., Gao, W.Y. and Huang, X.X., 2013. Phytochemistry, pharmacology, toxicology, and structure-cytotoxicity relationship of Paridis Rhizome Saponin. *Chinese Herbal Medicines*, *5*(1), pp. 33–46.

Min, B.S., Kim, Y.H., Tomiyama, M., Nakamura, N., Miyashiro, H., Otake, T. and Hattori, M., 2001. Inhibitory effects of Korean plants on HIV -1 activities. *Phytotherapy Research*, *15*(6), pp. 481–486.

Pu, X., Ren, J., Ma, X., Liu, L., Yu, S., Li, X. and Li, H., 2015. Polyphylla saponin I has antiviral activity against influenza A virus. *International Journal of Clinical and Experimental Medicine*, *8*(10), p. 18963.

Wang, G.X., Han, J., Zhao, L.W., Jiang, D.X., Liu, Y.T. and Liu, X.L., 2010. Anthelmintic activity of steroidal saponins from *Paris polyphylla*. *Phytomedicine*, *17*(14), pp. 1102–1105.

Wang, Y.C., Yi, T.Y. and Lin, K.H., 2011. In vitro activity of *Paris polyphylla* Smith against enterovirus 71 and coxsackievirus B3 and its immune modulation. *The American Journal of Chinese Medicine*, *39*(06), pp. 1219–1234.

### 3.1.5.2.2 *Veratrum nigrum* L.

Synonyms: *Veratrum bracteatum* Batalin, *Veratrum nigrum* subsp. *ussuriense* (Loes.) Vorosch., *Veratrum nigrum* var. *microcarpum* Loes., *Veratrum nigrum* var. *ussuriense* Loes., *Veratrum ussuriense* (Loes.) Nakai

Common names: Black hellebore; li lu (China); cham yeo ro (Korea)
Habitat: Grassy hills, forests
Distribution: Kazakhstan, China, Korea, and Mongolia
Botanical observation: This massive, sinister, frightfully poisonous herb grows upright to about 1 m tall. The stems are terete and stout. The leaves are cauline, spiral, and sessile. The blade is broadly elliptic, up to about 25 cm × 10 cm, glabrous, with longitudinal nerves, and acute at apex (somewhat Orchidaceous). The inflorescence is a massive raceme of spikes. The 6 tepals are black-purple, oblong, glossy, up to about 8 mm long, fleshy, and recurved. The 6 stamens are about 4 mm long. The gynoecium consists of 3 partially free carpels. The capsules are about 1.5 cm × 1 cm.
Medicinal uses: Boils, bronchitis, dysentery, and sore throat (China)
Commentaries:

(i) Members of the genus *Veratrum* L. have antiviral properties. Methanol extract of roots of *Veratrum patulum* Loes. (local name in China: *Jian bei li lu*) at a concentration of 100 μg/mL inhibited RNA-dependent DNA polymerase and ribonuclease H activity of Human immunodeficiency virus type-1 reverse transcriptase and Human immunodeficiency virus type-1 protease by 15, 1.4, and 9%, respectively (Min et al., 2001). Ethanol extract of rhizome of *Veratrum album* L. inhibited the replication of Herpes simplex virus type-1 and Sindbis virus with minimum concentrations of 250 and 500 μg/mL, respectively (Hudson et al., 2000).

(ii) *Veratrum nigrum* L. generates steroidal alkaloids such as veratramine evoking, if orally ingested, bradycardia, hypotension, cardiac conduction abnormalities, and death (Schep et al., 2006). *Veratrum japonicum* (Baker) Loes is used in Japan to treat skin infection, and *Veratrum formosum* O. Loes. is a remedy for ringworms in Taiwan. Members of the genus *Veratrum* L., like any poisonous plants, are of no therapeutic value and obviously never to be taken internally and even externally. However, several species in the genus *Veratrum* L. are used in China for boils, bronchitis, dysentery, and sore throat. In this context, one needs to remember that plants are medicinal not only because of their constituents but also because of an appropriate collection, preparation, and dosage from the traditional healers (knowledge often transmitted from one generation to another since the beginning of the human era).

(iii) Traditional healers will probably all disappear within this 50 years if the current global dysfunctional late capitalist acculturation (Nederveen Pieterse, 2007; Schmitt, M., 2017) continues, hence it is of critical importance for the well-being of humanity to learn from the remaining healers and to have toward them an open minded, careful, and humble approach. The pharmaceutical industry are managing (thanks to accreditation boards under their control) to transform the schools of pharmacy into centers to form docile dispensers of pills (Brezis, 2008). These same graduate, so-called pharmacists will then be replaced some days, under the same lobbies of the pharmaceutical industry, by some sort of dispensing machines or robots (Rodriguez-Gonzalez et al., 2019).

(iv) The current COVID-19 tragedy imposes us to completely rethink the current strategies of drug discovery as well as the current management of so-called dysfunctional neoliberal universities (Zabrodska et al., 2011; Smyth, 2020).

Without pharmacy graduates being fully knowledgeable on medicinal plants, botany, phytochemistry, and pharmacotoxicology, or pharmacognosy, how will antiviral drugs be discovered from plants? As for universities, the main goal of faculties in many parts of the world has become to make money (Bennis & O'Toole, J., 2005), and academes are now spending most of their time to fill forms and attend meetings, promoted based on administration skills and have no time to read books or papers, to do research, and in brief to get deep knowledge in a time when humans need, more than ever, brains to discover antiviral drugs (Osbaldiston et al., 2019). Could have Louis Pasteur, Ernest Duschesne, Albert Einstein, or Flemings survived in neoliberal academia? Meanwhile, COVID-19 can been seen as among the first global zoonotic pandemic viral wave shaking the world before the arrival of other waves of other new microorganisms that eradicate humans as negative feedback loops sent by nature to put an end to the current ecogenocide or what is called the "Anthropocene"

(Crutzen, 2006). Another thought is the fact that current drug development is under private sector and it is perhaps time to let public universities and hospitals do the job with more flexible and economical guidelines. Big pharmas, like most universities, need financial benefits, and developing drugs is an investment (Morgan et al., 2011). All of this is madness.

## References

Bennis, W.G. and O'Toole, J., 2005. How business schools have lost their way. *Harvard business review*, *83*(5), pp. 96–104.

Brezis, M., 2008. Big pharma and health care: unsolvable conflict of interests between private enterprise and public health. *Israel Journal of Psychiatry and Related Sciences*, *45*(2), p. 83.

Crutzen, P.J., 2006. The "anthropocene". In *Earth system science in the anthropocene* (pp. 13–18). Springer, Berlin, Heidelberg.

Hudson, J.B., Lee, M.K., Sener, B. and Erdemoglu, N., 2000. Antiviral activities in extracts of Turkish medicinal plants. *Pharmaceutical Biology*, *38*(3), pp. 171–175.

Min, B.S., Kim, Y.H., Tomiyama, M., Nakamura, N., Miyashiro, H., Otake, T. and Hattori, M., 2001. Inhibitory effects of Korean plants on HIV -1 activities. *Phytotherapy Research*, *15*(6), pp. 481–486.

Morgan, S., Grootendorst, P., Lexchin, J., Cunningham, C. and Greyson, D., 2011. The cost of drug development: a systematic review. *Health policy*, *100*(1), pp. 4–17.

Nederveen Pieterse, J., 2007. Global multiculture, flexible acculturation. *Globalizations*, *4*(1), pp. 65–79.

Osbaldiston, N., Cannizzo, F. and Mauri, C., 2019. 'I love my work but I hate my job'— Early career academic perspective on academic times in Australia. *Time & Society*, *28*(2), pp. 743–762.

Schep, L.J., Schmierer, D.M. and Fountain, J.S., 2006. Veratrum poisoning. *Toxicological Reviews*, *25*(2), pp. 73–78.

Rodriguez-Gonzalez, C.G., Herranz-Alonso, A., Escudero-Vilaplana, V., Ais-Larisgoitia, M.A., Iglesias-Peinado, I. and Sanjurjo-Saez, M., 2019. Robotic dispensing improves patient safety, inventory management, and staff satisfaction in an outpatient hospital pharmacy. *Journal of evaluation in clinical practice*, *25*(1), pp. 28–35.

Schmitt, M., 2017. Dysfunctional capitalism: Mental illness, schizoanalysis and the epistemology of the negative in contemporary cultural studies. *Psychoanalysis, Culture & Society*, *22*(3), pp. 298–316.

Smyth, J., The Making of Bullshit Leadership and Toxic Management in the Neoliberal University. Social Epistemology Review and Reply Collective 9 (5), pp. 9–18.

Zabrodska, K., Linnell, S., Laws, C. and Davies, B., 2011. Bullying as intra-active process in neoliberal universities. *Qualitative Inquiry*, *17*(8), pp. 709–719.

### 3.1.5.3 Family Smilacaceae Ventenat (1799)

The family Smilacaceae consists of 4 genera and about 375 species of mostly climbers growing from rhizomes. The leaves are simple, alternate or opposite, and exstipulate. The blade is often coriaceous, glossy, and often marked with 3 longitudinal nerves. The androecium comprises 6 stamens. The gynoecium consists of 3 carpels fused in a bilocular ovary, each locule sheltering up to 2 ovules. The fruits are berries. Plants in this family produce antimicrobial stilbenes and flavonoids.

#### 3.1.5.3.1 *Smilax calophylla* Wall.

Common name: Dedawai betina, kethart (Malaysia)
  Habitat: Forests
  Distribution: Malaysia and Thailand

Botanical observation: This climber grows from a massive tuber. The leaves are simple, alternate, and exstipulate. The petiole is about 8 mm long and curved. The blade is lanceolate to elliptic, coriaceous, glossy, 2.5–12 cm×5–18 cm, and with 3 longitudinal nervations. The inflorescences are sessile and axillary umbels. The perianth consists of 6 obovate tepals, which are yellowish and hooded. The androecium consists of 6 stamens. A disc is present. The ovary is minute and the stigma is sessile. The berries are globose, red, glossy, and about 1 cm in diameter.

Medicinal use: Gonorrhea (Malaysia, Thailand)

Commentary: The potential antimicrobial activities of this endangered plant have not been apparently examined.

### 3.1.5.3.2 *Smilax china* L.

Synonyms: *Coprosmanthus japonicus* Kunth; Smilax japonica (Kunth) A. Gray; Smilax pteropus Miq.; Smilax taiheiensis Hayata

Common names: Chinese sarsaparilla; a qia, jingangteng (China); gadong cina (Indonesia; Malaysia); biri (Pakistan); sarsaparillang china (the Philippines); kim chang trung quoe (Vietnam)

Habitat: Forests, thickets

Distribution: Myanmar, the Philippines, Thailand, Vietnam, and China

Botanical observation: This climber grows to a length of 5 m from a rhizome. The stems are woody, wiry, and thorny. Tendrils are present. The leaves are simple, alternate, and exstipulate. The petiole is about 1.5 cm long and winged. The blade is orbicular (at very first glance resembles *Ziziphus jujuba* Mill.), coriaceous, glossy, wavy, 3–10 cm×1.5–6 cm, and with 3 longitudinal nervations. The inflorescences are axillary umbels of numerous minute flowers. The perianth consists of 6 tepals, which are yellowish green and up to about 5 mm long. The androecium consists of 6 stamens. The berries are globose, dull red and up to about 1.5 cm in diameter.

Medicinal use: Dysentery, nephritis, syphilis, tuberculosis (China); venereal diseases (Malaysia)

*Antifungal (yeast) polar extract:* Acetone extract inhibited the growth of *Saccharomyces cerevisiae* (BY4741) and *Candida albicans* (SC5314) (Liu et al., 2012).

*Weak antibacterial (Gram-positive) flavanone glycosides:* Astilbin, neoastilbin, and isoastilbin inhibited the growth of *Streptococcus sobrinus* (Kuspradini et al., 2009).

*Strong antiviral (enveloped monopartite linear dimeric single-stranded (+) RNA) polar extract:* Butanol extract of rhizome at a concentration of 20 µg/mL inhibited the replication of the Human immunodeficiency virus type-1 by 64.3% (Wang et al., 2014).

*Strong antiviral (enveloped monopartite linear dimeric single stranded (+) RNA) amphiphilic flavanonol:* Dihydrokaempferol (also known as aromadendrin; LogD = 1.5 at pH 7.4; molecular weight = 288.2 g/mol) isolated from the rhizome at a concentration of 20 µg/mL inhibited the replication of the Human immunodeficiency virus type-1 by 30.5% (Wang et al., 2014).

(+)-Dihydrokaempferol

*Strong antiviral (enveloped monopartite linear dimeric single-stranded (+) RNA) hydrophilic flavonol glycoside:* Kaempferol-7-*O*-glucoside (LogD = −1.5 at pH 7.4; molecular mass = 448.3 g/mol) isolated from the rhizome at a concentration of 20 µg/mL inhibited the replication of the Human immunodeficiency virus type-1 by 38% (Wang et al., 2014).

Kaempferol-7-*O*-glucoside

*Strong antiviral (enveloped monopartite linear dimeric single-stranded (+) RNA) flavanol glycoside:* Dihydrokaempferol-3-*O*-rhamnoside isolated from the rhizome at a concentration of 20 µg/mL inhibited the replication of the Human immunodeficiency virus type-1 by 27.7% (Wang et al., 2014).

*Strong antiviral (enveloped monopartite linear dimeric single-stranded (+) RNA) amphiphilic stilbenes:* Resveratrol (LogD = 2.8 at pH 7.4; molecular weight = 228.2 g/mol) and oxyresveratrol (LogD = 2.4 at pH 7.4; molecular weight = 244.2 g/mol) isolated from the rhizome at a concentration of 20 µg/mL inhibited the replication of the Human immunodeficiency virus type-1 by 46.8 and 46.9%, respectively (Wang et al., 2014). Oxyresveratrol inhibited Human immunodeficiency virus type-1 (-1/LAI) with an $IC_{50}$ value of 28.2 µM (Likhitwitayawuid et al., 2005).

Resveratrol

*Strong antiviral (enveloped monopartite linear dimeric single-stranded (+) RNA) stilbenes:* Oxyresveratrol inhibited the replication of the Herpes simplex virus type-1 and type-2 with $EC_{50}$ values of 63.5 and 55.3 µM, respectively (acyclovir 0.2 µM). Resveratrol inhibited the replication of the Herpes simplex virus with an $IC_{50}$ value of 12.8 µg/mL (Sasivimolphan et al., 2009).

Oxyresveratrol

*Viral enzyme inhibitor simple phenol:* Protocatechuic acid at a concentration of 10 μM inhibited Human immunodeficiency virus type-1 integrase by 31.9% (Panthong et al., 2015).

*Viral enzyme inhibitor flavanone glycoside:* Engeletin inhibited H1N1 neuraminidase with an $IC_{50}$ value of 5.4 μM (Grienke et al., 2012).

Commentaries:

(i) The plant synthetizes series of steroidal saponins (Shao et al., 2007) as well as flavonoids and flavonoid glycosides (Wu et al., 2010), proanthocyanidins (cinchonain IIa and cinchonain IIb), flavanols (cinchonain Ia and cinchonain Ib), chlorogenic acid, protocatechuic acid, flavanone glycosides (engeletin, astilbin, neoastilbin, isoastilbin isoneoastilbin) (-)-epicatechin and the stilbene resveratrol (Zhong et al., 2017), which most probably account for the antitubercular use. This could be examined.

(ii) Note that steroidal saponins and genins in this plant are anti-inflammatory, such as the spirostane-type steroid sieboldogenin (Khan et al., 2009), and one could suggest that the use of the plant could be at least partially explained by symptomatic relief of syphilis. *Smilax lanceifolia* Roxb is also used in Vietnam for the treatment of syphilis.

Syphilis is a sexually transmissible bacterial infection (zoonosis from llamas) due to the Gram-negative bacteria *Treponema pallidum* brought from Central America by the navigators in Columbus fleet in 1493 and disseminated by both French and Italian mercenaries to entire Europe and later, with colonization, to the world. Until the end of the nineteenth century the syphilis pandemic was treated with mercury or the anti-inflammatory wood of *Guaiacum officinale* L. (Duwiejua et al., 1994; Sarbu et al., 2014). Paul Ehrlich (1854–1915) received the Nobel Prize in Physiology and Medicine in 1908 for his discovery of the synthetic organoarsenic compound arsphenamine (Salvarsan) used to treat syphilis, later eradicated by penicillin from 1940 onwards. *Treponema pallidum* is becoming increasingly resistant to antibiotics (Stamm, 2010), and new anti-syphilitic agents are needed.

### References

Duwiejua, M., Zeitlin, I.J., Waterman, P.G. and Gray, A.I., 1994. Anti-inflammatory activity of *Polygonum bistorta, Guaiacum officinale* and *Hamamelis virginiana* in rats. *Journal of Pharmacy and Pharmacology, 46*(4), pp. 286–290.

Grienke, U., Schmidtke, M., von Grafenstein, S., Kirchmair, J., Liedl, K.R. and Rollinger, J.M., 2012. Influenza neuraminidase: a druggable target for natural products. *Natural Product Reports, 29*(1), pp. 11–36.

Khan, I., Nisar, M., Ebad, F., Nadeem, S., Saeed, M., Khan, H., Khuda, F., Karim, N. and Ahmad, Z., 2009. Anti-inflammatory activities of Sieboldogenin from *Smilax china* Linn.: experimental and computational studies. *Journal of Ethnopharmacology, 121*(1), pp. 175–177.

Kuspradini, H., Mitsunaga, T. and Ohashi, H., 2009. Antimicrobial activity against *Streptococcus sobrinus* and glucosyltransferase inhibitory activity of taxifolin and some flavanonol rhamnosides from kempas (*Koompassia malaccensis*) extracts. *Journal of Wood Science*, 55(4), pp. 308–313.

Likhitwitayawuid, K., Sritularak, B., Benchanak, K., Lipipun, V., Mathew, J. and Schinazi, R.F., 2005. Phenolics with antiviral activity from *Millettia erythrocalyx* and *Artocarpus lakoocha*. *Natural Product Research*, 19(2), pp. 177–182.

Liu, Q., Luyten, W., Pellens, K., Wang, Y., Wang, W., Thevissen, K., Liang, Q., Cammue, B.P., Schoofs, L. and Luo, G., 2012. Antifungal activity in plants from Chinese traditional and folk medicine. *Journal of Ethnopharmacology*, 143(3), pp. 772–778.

Panthong, P., Bunluepuech, K., Boonnak, N., Chaniad, P., Pianwanit, S., Wattanapiromsakul, C. and Tewtrakul, S., 2015. Anti-HIV-1 integrase activity and molecular docking of compounds from *Albizia procera* bark. *Pharmaceutical Biology*, 53(12), pp. 1861–1866.

Sarbu, I., Matei, C., Benea, V. and Georgescu, S.R., 2014. Brief history of syphilis. *Journal of Medicine and Life*, 7(1), p. 4.

Sasivimolphan, P., Lipipun, V., Likhitwitayawuid, K., Takemoto, M., Pramyothin, P., Hattori, M. and Shiraki, K., 2009. Inhibitory activity of oxyresveratrol on wild-type and drug-resistant varicella-zoster virus replication in vitro. *Antiviral Research*, 84(1), pp. 95–97.

Shao, B., Guo, H., Cui, Y., Ye, M., Han, J. and Guo, D., 2007. Steroidal saponins from *Smilax china* and their anti-inflammatory activities. *Phytochemistry*, 68(5), pp. 623–630.

Stamm, L.V., 2010. Global challenge of antibiotic-resistant *Treponema pallidum*. *Antimicrobial Agents and Chemotherapy*, 54(2), pp. 583–589.

Wang, W.X., Qian, J.Y., Wang, X.J., Jiang, A.P. and Jia, A.Q., 2014. Anti-HIV -1 activities of extracts and phenolics from *Smilax china* L. *Pakistan Journal of Pharmaceutical Sciences*, 27, pp. 147–151.

Wu, L.S., Wang, X.J., Wang, H., Yang, H.W., Jia, A.Q. and Ding, Q., 2010. Cytotoxic polyphenols against breast tumor cell in *Smilax china* L. *Journal of Ethnopharmacology*, 130(3), pp. 460–464.

Zhong, C., Hu, D., Hou, L.B., Song, L.Y., Zhang, Y.J., Xie, Y. and Tian, L.W., 2017. Phenolic compounds from the rhizomes of *Smilax china* L. and their anti-inflammatory activity. *Molecules*, 22(4), p. 515.

### 3.1.5.3.3  *Smilax glabra* Roxb.

Synonyms: *Smilax blinii* H. Lév.; *Smilax calophylla* var. *concolor* C.H. Wright; *Smilax dunniana* H. Lév.; *Smilax hookeri* Kunth; *Smilax mengmaensis* R.H. Miao; Smilax trigona Warb.

Common name: Tu fu ling (China)

Habitats: Forests and thickets

Distribution: India, Myanmar, Thailand, Vietnam, and China

Botanical observation: This climber grows to a length of 4 m from a rhizome. The stems are woody, wiry, and smooth. The leaves are simple, alternate, and exstipulate. The petiole is about 1.5 cm long and winged. The blade is elliptic, coriaceous, glossy, wavy, 6–15 cm × 1–6.5 cm, and with 3 longitudinal nervations. The inflorescences are axillary umbels of numerous minute flowers. The perianth consists of 6 tepals, which are yellowish green, hooded, and up to about 3 mm long. The androecium presents 6 stamens. The berries are globose, dark blue, glossy, and up to about 1 cm in diameter.

Medicinal uses: Abscesses (China); bronchitis (Laos)

Pharmacology: Antibacterial

*Very weak antibacterial (Gram-positive) amphiphilic flavanol:* Smiglabrone A (LogP = 2.5; molecular weight = 496.5 g/mol) isolated from the rhizome inhibited the growth of methicillin-resistant *Staphylococcus aureus*, *Staphylococcus aureus* (ATCC 6538), and *Enterococcus faecalis* (Xu et al., 2013).

*Very weak antibacterial (Gram-positive) stilbenes:* Smiglastilbene isolated from the rhizome inhibited the growth of methicillin-resistant *Staphylococcus aureus* and *Staphylococcus aureus* (ATCC 6538) with MIC values of 409 and 205 μM, respectively (Xu et al., 2013).

*Weak antibacterial (Gram-positive) stilbene:* Resveratrol from the rhizome inhibited the growth of methicillin-resistant *Staphylococcus aureus*, *Staphylococcus aureus* (ATCC 6538), and *Enterococcus faecalis* with MIC values of 159, 79, and 159 μM, respectively (Xu et al., 2013).

*Moderate antibacterial (Gram-positive) flavanol:* Cinchonain Ib isolated from the rhizome inhibited the growth of *Pseudomonas aeruginosa*, methicillin-resistant *Staphylococcus aureus*, *Staphylococcus aureus* (ATCC 6538), and *Enterococcus faecalis* with MIC values of 663, 80.1, 80.1, and 160 μM, respectively (Xu et al., 2013).

*Very weak antifungal (yeast) flavanols:* Smiglabrone A isolated from the rhizome inhibited the growth of *Candida albicans* (SC5314) with an MIC value of 146 μM et al., 2013).

Cinchonain Ib isolated from the rhizome inhibited the growth of *Candida albicans* (SC5314) with an MIC value of 160 μM (Xu et al., 2013).

*Very weak (enveloped linear single-stranded (+) RNA) antiviral hydrophilic stilbene:* Resveratrol at a concentration of 250 μM reduced the cytopathic effects of Middle East respiratory syndrome (MERS) coronavirus (HCoV-EMC/2012) against Vero cells to about 25% via inhibition of RNA replication (Lin et al., 2017). At a concentration of 250 μM, resveratrol decreased Chikungunya virus titer by about 2 $\log_{10}$ in Vero cells (Lin et al., 2017).

At the concentration of 50 μM, resveratrol decreased virus titter of Zika virus from about $4 \times 10^6$ to $4 \times 10^6$ FFU/mL by virucidal effect and inhibition surface attachment to host cells (Mohd et al., 2019). Is resveratrol of usefulness against COVID-19?

### References

Lin, S.C., Ho, C.T., Chuo, W.H., Li, S., Wang, T.T. and Lin, C.C., 2017. Effective inhibition of MERS-CoV infection by resveratrol. *BMC Infectious Diseases*, 17(1), p. 144.

Mohd, A., Zainal, N., Tan, K.K. and AbuBakar, S., 2019. Resveratrol affects Zika virus replication in vitro. *Scientific Reports*, 9(1), pp. 1–11.

Xu, S., Shang, M.Y., Liu, G.X., Xu, F., Wang, X., Shou, C.C. and Cai, S.Q., 2013. Chemical constituents from the rhizomes of *Smilax glabra* and their antimicrobial activity. *Molecules*, 18(5), pp. 5265–5287.

### 3.1.5.3.4    *Smilax zeylanica* L.

Synonyms: *Smilax elliptica* Desv. ex Ham.

Common names: Antikinari (Bangladesh); kayu cina hutan (Indonesia, Malaysia)

Habitat: Forests

Distribution: India, Sri Lanka, Bangladesh, Myanmar, Thailand, Malaysia, and Indonesia

Botanical observation: This climber grows to about 3 m long. The stems are lignose and coriaceous. The leaves are simple, alternate, and exstipulate. The petiole is about 1.5 cm long and curved. The blade is broadly elliptic, coriaceous, glossy, 6.5–15 cm × 4–12.5 cm, base and apex rounded, and with 3–5 longitudinal nervations. The inflorescences are axillary and solitary umbels on a 3 cm long peduncle. The perianth consists of 6 whitish and about 5 mm long tepals. The androecium presents 6 stamens. The ovary is minute and the stigma is sessile. The berries are globose, green, glossy, and about 1 cm in diameter.

Medicinal uses: Boils, leucorrhea, abscesses, diarrhea, tooth infection, small pox, cholera, syphilis, gonorrhea (Bangladesh)

*Broad-spectrum antibacterial halo developed by polar extract:* Ethanol extract of leaves (250 μg/disc) inhibited the growth of *Escherichia coli*, *Shigella dysenteriae*, *Salmonella typhi*, *Salmonella paratyphi*, and *Staphylococcus aureus* with inhibition zone diameters of 5.3, 8.5, 8.7, 9.2, and 7.4 mm, respectively (Hossain et al., 2013).

**Reference**
Hossain, A.M., Saha, S., Asadujjaman, M. and Kahan, A.S., 2013. Analgesic, antioxidant and antibacterial activity of *Smilax zeylanica* Linn. (family-Smilacaceae). *Pharmacologyonline*, *1*, pp. 244–250.

### 3.1.6 ORDER ASPARAGALES LINK (1829)

#### 3.1.6.1 Family Amaryllidaceae J.St.-Hil. (1805)

The family Amaryllidaceae consists of about 100 species and 1200 species of herbs growing from bulbs, corms, rhizomes, or tubers. The leaves are simple, basal or cauline, and often elongated. The inflorescences are terminal spikes, umbels, racemes, panicles, or solitary. The perianth comprises 6 tepals in 2 whorls. The androecium presents 6 stamens. The gynoecium consists of 3 carpels fused into a 3-locular ovary each locule sheltering few to numerous ovules growing from axile placentas. The fruit is a dehiscent capsule or a berry. Members in this family produce antimicrobial alky sulfur compounds and Amaryllidaceae alkaloids of tremendous antiviral interest.

*3.1.6.1.1  Allium cepa L.*

Common names: Onion; piyanj (Bangladesh); khtüm barang (Cambodia); yang cong (China); Vengayam (India); Bawang merah (Indonesia; Malaysia); pyaz (Iran); tamanegi (Japan); hum bwax (Laos); piaz (Pakistan); anian (Papua New Guinea); bauang pula (the Philippines); hom-huayai (Thailand); sogna (Turkey); hanh tay (Vietnam)

   Habitat: Cultivated

   Distribution: Asia and Pacific

   Botanical observation: This herb grows to about 80 cm tall from a pungent bulb, which is ovoid, solitary, papery, and brownish, glossy, tunic. The leaves are simple, entire, linear, terete, and fistulose. The scapes are about 1 m tall, terete, and fistulose. The inflorescences are globose umbels. The perianth includes 6 tepals, which are oblong ovate, white, and about 4 mm long. The androecium includes 6 stamens. The ovary is subglobose and minute.

   Medicinal uses: Antiseptic (Bangladesh); burns (China)

   *Broad-spectrum antibacterial halo developed by essential oil:* Essential oil of bulbs (50 µL/mL/5 mm paper disc \) inhibited the growth of *Staphylococcus aureus* (ATCC 11522) and *Salmonella enteritidis* (ATCC 13076) with inhibition zone diameters of 6.1 and 7.1 mm, respectively (Benkeblia, 2004).

   *Weak antibacterial (Gram-positive) essential oil:* Essential oil inhibited the growth of *Staphylococcus aureus* (B31) with an MIC value of 100 ppm (Kim et al., 2004).

   *Antifungal (filamentous) essential oil:* Essential oil of bulbs (200 µL/mL) inhibited the mycelial growth of *Aspergillus niger* (ATCC 10575), *Penicillum cyclopium* (ATTC 26165), and *Fusarium oxysporum* (ATCC 11850) (Benkeblia, 2004).

   *Strong anticandidal essential oil:* Essential oil inhibited the growth of *Candida utilis* (ATCC 42416) with an MIC value of 25 ppm (Kim et al., 2004).

   *Strong broad-spectrum antifungal polar extract:* Aqueous extract of bulb inhibited the growth of *Malassezia furfur, Candida albicans, Candida glabrata, Candida tropicalis, Candida parapsilosis, Trichophyton mentagrophytes, Trichophyton rubrum, Microsporium canis, Microsporium gypseum,* and *Epidermophyton floccosum* with $MIC_{50}$ values of 2, 0.1, 0.2, 0.1, 0.2, 0.5, 0.5, 0.5, 0.5, and 0.5 µg/mL, respectively (Shams-Ghahfarokhi et al., 2006).

   *Very weak antiviral (enveloped monopartite linear double-stranded DNA) essential oil:* Essential oil inhibited the replication of Herpes simplex virus type-1 with an $IC_{50}$ value of 1060 µg/mL (Romeilah et al., 2010).

   *Broad-spectrum antibacterial halo developed by flavonol:* Quercetin from bulb peel (6 mm diameter paper disc, 10 µL/disc of a solution at a concertation of 10 mg/mL) inhibited the growth of methicillin-resistant *Staphylococcus aureus* and *Helicobacter pylori* (ATCC 43504) with inhibition zone diameters of 10 and 13 mm, respectively (Ramos et al., 2006).

*Moderate anticandidal alkyl sulfur:* Dipropyldisulfide inhibited the growth of *Candida utilis* (ATCC 42416) with an MIC value of 130 ppm (Kim et al., 2004).

Commentaries: (i) Upon physical injury, the nonprotein amino acid (+)-S-(1-propenyl)-L-cysteine sulfoxide (isoalliin) in the bulb is converted by contact with the enzyme alliinase into alkenyl polysulfides (Lanzotti, 2006), including propyl mercaptan, dimethylthiophene, dipropyldisulfide, and (*E*)-1-propenyl propyl disulfide (Storsberg et al., 2004) which are antibacterial (Cavallito et al., 1945). (ii) This plant has immunostimulating properties (Goodarzi et al., 2013). The plant has anti-Influenza properties (Lee et al., 2012).

### References

Benkeblia, N., 2004. Antimicrobial activity of essential oil extracts of various onions (*Allium cepa*) and garlic (*Allium sativum*). *LWT-Food Science and Technology, 37*(2), pp. 263–268.

Cavallito, C.J., Bailey, J.H., Haskell, T.H., McCormick, J.R. and Warner, W.F., 1945. The inactivation of antibacterial agents and their mechanism of action. *Journal of Bacteriology, 50*(1), p. 61.

Goodarzi, M., Landy, N. and Nanekarani, S., 2013. Effect of onion (*Allium cepa* L.) as an antibiotic growth promoter substitution on performance, immune responses and serum biochemical parameters in broiler chicks. *Health, 5*(8), p. 1210.

Kim, J.W., Huh, J.E., Kyung, S.H. and Kyung, K.H., 2004. Antimicrobial activity of alk (en) yl sulfides found in essential oils of garlic and onion. *Food Science and Biotechnology, 13*(2), pp. 235–239.

Lanzotti, V., 2006. The analysis of onion and garlic. *Journal of Chromatography A, 1112*, pp. 3–22.

Lee, J.B., Miyake, S., Umetsu, R., Hayashi, K., Chijimatsu, T. and Hayashi, T., 2012. Anti-influenza A virus effects of fructan from Welsh onion (*Allium fistulosum* L.). *Food Chemistry, 134*(4), pp. 2164–2168.

Ramos, F.A., Takaishi, Y., Shirotori, M., Kawaguchi, Y., Tsuchiya, K., Shibata, H., Higuti, T., Tadokoro, T. and Takeuchi, M., 2006. Antibacterial and antioxidant activities of quercetin oxidation products from yellow onion (*Allium cepa*) skin. *Journal of Agricultural and Food Chemistry, 54*(10), pp. 3551–3557.

Romeilah, R.M., Fayed, S.A. and Mahmoud, G.I., 2010. Chemical compositions, antiviral and antioxidant activities of seven essential oils. *Journal of Applied Sciences Research, 6*(1), pp. 50–62.

Shams-Ghahfarokhi, M., Shokoohamiri, M.R., Amirrajab, N., Moghadasi, B., Ghajari, A., Zeini, F., Sadeghi, G. and Razzaghi-Abyaneh, M., 2006. In vitro antifungal activities of *Allium cepa*, *Allium sativum* and ketoconazole against some pathogenic yeasts and dermatophytes. *Fitoterapia, 77*(4), pp. 321–323.

Storsberg, J., Schulz, H., Keusgen, M., Tannous, F., Dehmer, K.J. and Keller, E.J., 2004. Chemical characterization of interspecific hybrids between *Allium cepa* L. and *Allium kermesinum* Rchb. *Journal of Agricultural and Food Chemistry, 52*(17), pp. 5499–5505.

*3.1.6.1.2   Allium sativum L.*

Synonyms: *Allium pekinense* Prokhanov; *Porrum sativum* (L.) Rchb.

Common names: Garlic; rashun; rasun; krachaaipru (Bangladesh); khtum sa (Cambodia); suan (China); arishtha (India); bawang puteh (Indonesia; Malaysia); seer (Iran); kathiem (Laos); lassan (Pakistan); ahos (the Philippines); sarimsak (Turkey); krathiam (Thailand); toi (Vietnam)

Habitat: Cultivated

Distribution: Asia and Pacific

Botanical observation: This herb grows to about 50 cm tall from a bulb with 6–10 scaly pungent bulblets covered with a white papery tunic. The leaves are simple, entire, linear, flattened, and sheathing. The scapes 25–50 cm, longer than the leaves, and terete. The inflorescences are globose umbels. The perianth includes 6 tepals, which are lanceolate, acuminate, whitish, and about 4 mm long. The androecium includes 6 stamens. The ovary is globose and minute.

Medicinal uses: Dysentery, tuberculosis, ulcers, leucorrhea, leprosy, cough (Bangladesh); cough, infection (Indonesia); flu (China)

*Weak antibacterial (Gram-positive) essential oil:* Essential oil inhibited the growth of *Staphylococcus aureus* (B31) with an MIC value of 100 ppm (Kim et al., 2004).

*Antibacterial halo developed by diallyl thiosulfinate:* Allicin exhibited antibacterial properties (Cavallito & Bailey, 1944).

*Strong antibacterial (Gram-positive) amphiphilic alkyl sulfurs:* Diethyl trisulfide, diallyl trisulfide, dimethyl tetratsulfide, diethyl tetrasulfide, and diallyl tetrasulfide (LogD = 4.5 at pH 7.4; molecular mass = 210.4 g/mol) inhibited the growth *Staphylococcus aureus* (B31) with MIC values of 40, 40, 30, 50, and 30 ppm, respectively (Kim et al., 2004).

Diallyl tetrasulfide

*Strong antibacterial (Gram-positive) amphiphilic diallyl thiosulfinate:* Allicin inhibited the growth of *Staphylococcus aureus* (NCTC 6571) with an MIC of 32 µg/mL and an MBC of 256 µg/mL (Cutler & Wilson, 2004).

*Very weak antifungal (yeast) hydrophilic extract:* Ethanol extract of bulbs inhibited the growth of *Candida albicans* (clinical strain) with an MIC/MFC value of 1.1/1.1 mg/mL (Vaijayanthimala et al., 2000).

*Strong broad-spectrum antifungal polar extract:* Aqueous extract of bulb inhibited the growth of *Malassezia furfur*, *Candida albicans*, *Candida glabrata*, *Candida tropicalis*, *Candida parapsilosis*, *Trichophyton mentagrophytes*, *Trichophyton rubrum*, *Microsporium canis*, *Microsporium gypseum*, and *Epidermophyton floccosum* with $MIC_{50}$ values of 0.03, 0.03, 0.03, 0.03, 0.03, 0.007, 0.007, 0.007, 0.003, and 0.03 µg/mL, respectively (Shams-Ghahfarokhi et al., 2006).

*Strong anticandidal essential oil:* Essential oil inhibited the growth of *Candida utilis* (ATCC 42416) with an MIC value of 25 ppm (Kim et al., 2004).

*Strong antifungal (yeast) lipophylic diallyl thiosulfinate:* Allicin (LogD = 6.4 at pH 7.4; molecular weight = 162.2 g/mol) inhibited the growth of *Candida albicans*, *Cryptococcus neoformans*, *Candida parapsilosis*, *Candida tropicalis*, *Candida krusei*, and *Torulopsis glabrata* with MIC values of 0.3, 0.3, 0.1, 0.3, 0.3, and 0.3 µg/mL, respectively (Ankri & Mirelman, 1999).

Allicin

*Strong anticandidal alkyl sulfurs:* Diethyl trisulfide, diallyl trisulfide, dimethyl tetratsulfide, diethyl tetrasulfide (LogD = 3.8 at pH 7.4; molecular weight = 186.3 g/mol), and diallyl tetrasulfide inhibited the growth of *Candida utilis* (ATCC 42416) (Kim et al., 2004).

*Weak antiviral (enveloped monopartite linear double-stranded DNA) essential oil:* Essential oil inhibited the replication of the Herpes simplex virus type-1 with an $IC_{50}$ value of 320 μg/mL (Romeilah et al., 2010).

*Moderate antiviral (enveloped monopartite linear double-stranded DNA) extract:* Extract of bulbs at a concentration of 150 μg/mL inhibited the replication of Herpes simplex virus type-1 by about 50 and 25%, respectively (Tsai et al., 1985).

*Strong antiviral (enveloped monopartite linear dimeric single stranded (+) RNA) amphiphilic sulfoxide:* Ajoene (LogD = 2.8 at pH 7.4; molecular weight = 234.4 g/mol) inhibited the fusion of H9 cells with Human immunodeficiency virus-infected H9:RF cells with an $IC_{50}$ of 45 μM and inhibited Human immunodeficiency virus type-1 replication with an $IC_{50}$ value 5 μM (Tatarintsev et al., 1992).

(E)-Ajoene

*In vivo antiviral amphiphilic diallyl thiosulfinate:* Capsules containing allicin (180 mg) given once a day to 73 volunteers for 90 days evoked a decreased in cold cases from 65 to 24 and decreased the average numbers of days to recover from cold from of 5.6 to 4.6 days (Josling, 2001).

*Antibiotic potentiator diallyl thiosulfinate:* Allicin increased the sensitivity of clinical isolates of *Pseudomonas aeruginosa* to cefoperazone, *Staphylococcus aureus* and *Staphylococcus epidermidis* to Cefalozin and Oxacillin.

Commentaries:

(i) Upon physical injury of bulbs, the nonprotein amino acid alliin is converted by contact with the enzyme alliinase into diallyl thiosulfinate (allicine) from which derives *E*-ajoene, *Z*-ajoene, diallyl disulfide, and diallyl trisulfide (Suleria et al., 2015; Yamaguchi et al., 2020). Allicin inhibits in bacteria and yeasts the enzyme acetyl-CoA synthetase (Focke et al., 1990).

(ii) In Bangladesh, *Allium sativum* L. is being used for protection against COVID-19.

(iii) In Europe, *Allium sativum* L. has been used during plagues (*la theriaque du pauvre*), notably during the infamous Great Plague of London of 1665 (Aaron, 1996).

(iv) This plant stimulates the immune system (Kyo et al., 1999).

### References

Aaron, C., 1996. Garlic & life. *The North American Review, 281*(2), pp. 14–23.

Ankri, S. and Mirelman, D., 1999. Antimicrobial properties of allicin from garlic. *Microbes and Infection, 1*(2), pp. 125–129.

Cavallito, C.J. and Bailey, J.H., 1944. Allicin, the antibacterial principle of *Allium sativum*. I. Isolation, physical properties and antibacterial action. *Journal of the American Chemical Society, 66*(11), pp. 1950–1951.

Cutler, R.R. and Wilson, P., 2004. Antibacterial activity of a new, stable, aqueous extract of allicin against methicillin-resistant *Staphylococcus aureus*. *British Journal of Biomedical Science, 61*(2), pp. 71–74.

Focke, M., Feld, A. and Lichtenthaler, H.K., 1990. Allicin, a naturally occurring antibiotic from garlic, specifically inhibits acetyl-CoA synthetase. *FEBS Letters, 261*(1), pp. 106–108.

Josling, P., 2001. Preventing the common cold with a garlic supplement: a double-blind, placebo-controlled survey. *Advances in Therapy, 18*(4), pp. 189–193.

Kim, J.W., Huh, J.E., Kyung, S.H. and Kyung, K.H., 2004. Antimicrobial activity of alk (en) yl sulfides found in essential oils of garlic and onion. *Food Science and Biotechnology*, *13*(2), pp. 235–239.

Kyo, E., Uda, N., Kasuga, S., Itakura, Y. and Sumiyoshi, H., 1999. Garlic as an immunostimulant. In *Immunomodulatory Agents from Plants* (pp. 273–288). Basel: Birkhäuser.

Romeilah, R.M., Fayed, S.A. and Mahmoud, G.I., 2010. Chemical compositions, antiviral and antioxidant activities of seven essential oils. *Journal of Applied Sciences Research*, 6(1), pp. 50–62.

Shams-Ghahfarokhi, M., Shokoohamiri, M.R., Amirrajab, N., Moghadasi, B., Ghajari, A., Zeini, F., Sadeghi, G. and Razzaghi-Abyaneh, M., 2006. In vitro antifungal activities of *Allium cepa*, *Allium sativum* and ketoconazole against some pathogenic yeasts and dermatophytes. *Fitoterapia*, *77*(4), pp. 321–323.

Suleria, H.A., Butt, M.S., Anjum, F.M., Saeed, F. and Khalid, N., 2015. Onion: nature protection against physiological threats. *Critical Reviews in Food Science and Nutrition*, *55*, pp. 50–66.

Tatarintsev, A.V., Vrzhets, P.V., Ershov, D.E., Shchegolev, A.A., Turgiev, A.S., Karamov, E.V., Kornilaeva, G.V., Makarova, T.V., Fedorov, N.A. and Varfolomeev, S.D., 1992. The ajoene blockade of integrin-dependent processes in an HIV -infected cell system. *Vestnik Rossiiskoi akademii meditsinskikh nauk*, (11–12), pp. 6–10.

Tsai, Y., Cole, L.L., Davis, L.E., Lockwood, S.J., Simmons, V. and Wild, G.C., 1985. Antiviral properties of garlic: in vitro effects on influenza B, herpes simplex and coxsackie viruses. *Planta Medica*, *51*(05), pp. 460–461.

Vaijayanthimala, J., Anandi, C., Udhaya, V. and Pugalendi, K.V., 2000. Anticandidal activity of certain South Indian medicinal plants. *Phytotherapy Research*, *14*(3), pp. 207–209.

Yamaguchi, Y. and Kumagai, H., 2020. Characteristics, biosynthesis, decomposition, metabolism and functions of the garlic odour precursor, S-allyl-l-cysteine sulfoxide. *Experimental and Therapeutic Medicine*. *19*(2), pp. 1528–1535.

### 3.1.6.1.3 *Crinum asiaticum* L.

Synonyms: *Bulbine asiatica* (L.) Gaertn.; *Crinum brevifolium* Roxb.; *Crinum firmifolium* var. *hygrophilum* H. Perrier

Common names: Crinum lily; poison bulb, spider lily; nadagamani (India); baking, kajang kajang (Indonesia); bakong (Malaysia); didil (Papua New Guinea); phlaphueng (Thailand); la nang (Vietnam)

Habitat: Cultivated

Distribution: Tropical Asia and Pacific

Botanical observation: This dreadfully poisonous (yet graceful) and bulbiferous herb is grown as ornamental. The leaves are simple, basal, and spiral. The blade is linear-oblong, somewhat fleshy, wavy, dull light green, smooth, pendulous, up to about 1 m × 12 cm and acute at apex. The inflorescences are many-flowered umbels at the apex of a pedicel, which is up to about 50 cm tall. The perianth is infundibuliform, up to about 15 cm long, membranous, ephemeral, pure white, and develops 6 linear and pendulous lobes. The 6 stamens are slender, greenish white, slightly recurved, up to 12 cm long with brownish, linear, and versatile saffron-colored anthers (move when the wind blows). The gynoecium consists of an ovary, which is 3-locular and a slender style. The capsule is somewhat globose, glossy, 3-lobed, fleshy and contain numerous seeds.

Medicinal uses: Cough (Bangladesh; India); bronchitis, gonorrhea (India); fever (Bangladesh; India; Malaysia); gonorrhea (Papua new Guinea); wounds, abscesses (China; India)

*Broad-spectrum antibacterial halo developed by polar extract:* Methanol extract of bulbs (500 μg/4 mm paper disc) inhibited the growth of *Bacillus cereus*, *Bacillus subtilis*, *Bacillus megaterium*, *Staphylococcus aureus*, *Escherichia coli*, *Salmonella typhi*, *Salmonella paratyphi*, and *Shigella sonnei* with the inhibition zone diameters of 11, 12, 11, 12, 10, 13, 11, and 10 mm, respectively (Rahman et al., 2011).

*Antibacterial crinine-type Amaryllidaceae alkaloid:* Crinamine (Adesanya et al., 1992).

*Strong broad-spectrum antifungal amphiphilic lycorine-type Amaryllidaceae alkaloid:* Lycorine (LogD = 3.1 at pH 7.4; molecular weight = 287.3 g/mol) at the concentration of 100 μg/ mL inhibited the growth of *Alternaria oleracea, Colletotrichum gloeosporioides, Fusarium graminearum, Colletotrichum ophiopogonis,* and *Pleospora lycopersici* by 61.9, 57.2, 63.7, 63.2, and 52.6%, respectively (Shen et al., 2014). Lycorine inhibited the growth of *Candida albicans, Candida dubium, Candida glabrata, Lodderomyces elongisporus,* and *Saccharomyces cerevisiae* with MIC values of 39, 32, 512, 64, and 97.3 μg/mL, respectively (Nair & van Staden, 2018).

(-)-Lycorine

*Strong broad-spectrum (enveloped monopartite linear single-stranded (+) RNA) antiviral amphiphilic lycorine-type Amaryllidaceae alkaloid:* Lycorine (LogD = 3.1 at pH 7.4; molecular weight = 287.3 g/mol) inhibited the replication of the Japanese encephalitis virus, Yellow fever virus, Dengue-4 virus, Punta toro virus, and Rift valley fever with $IC_{50}$ values of 0.3, 0.2, 0.2, 0.5, and 0.9 μg/mL, respectively (Gabrielsen et al., 1992).

*Strong antiviral (enveloped monopartite linear single-stranded (+) RNA) amphiphilic lycorine-type Amaryllidaceae alkaloid:* Lycorine (LogD = 3.1 at pH 7.4; molecular weight = 287.3 g/mol) inhibited the replication of the Severe acute respiratory syndrome-associated coronavirus (BJ-001) in Vero cells with the $EC_{50}$ value of 15.7 nM and a selectivity index of 954 (Li et al., 2005).

*Strong antiviral (non-enveloped monopartite linear single-stranded (+) RNA) amphiphilic lycorine-type Amaryllidaceae alkaloid:* Lycorine inhibited the replication of Enterovirus 71 in human muscular cells with an $IC_{50}$ of 0.4 μg/mL ($CC_{50}$: 48.5 μg/mL) via inhibition viral protein synthesis in host cells att the concentration of 10 μM. Lycorine inhibited the replication of the Poliovirus (Sabin Type 1) with an IC50 value of 0.06 μg/mL and a selectivity index of 24 (Oluyemisi et al., 2015)

*Antiviral (enveloped monopartite linear dimeric single-stranded (+) RNA) lycorine-type Amaryllidaceae alkaloid:* Lycorine inhibited the Human immunodeficiency virus type-1 replication in CD4 T cells (Peng et al., 2014).

*Strong antiviral (enveloped monopartite linear dimeric single-stranded (+) RNA) amphiphilic lycorine-type Amaryllidaceae alkaloid:* Lycorine inhibited the replication of the West Nile Virus (epidemic strain) with an $IC_{50}$ value of 0.2 μM via inhibition of viral RNA replication and at a concentration of 1.5 μM inhibited Dengue virus type-1 replication (Zou et al., 2009).

*In vivo antiviral lycorine-type Amaryllidaceae alkaloid:* Lycorine given intraperitoneally at a dose of 1 mg/Kg twice for 7 days to ICR mice experimentally infected with Enterovirus 71 prolonged the survival time of mice by 45% (Liu et al., 2011).

*Strong antiviral (enveloped segmented linear single stranded (−) RNA) amphiphilic lycorine-type Amaryllidaceae alkaloid:* Lycorine inhibited the replication of Influenza virus H5N1 with an IC90 value of 0.5 μM and selectivity index > 45 via inhibition of nuclear-to-cytoplasmic export of the viral ribonucleoprotein (He et al., 2013).

*Viral enzyme inhibitor polar extract:* Methanol extract at a concentration of 100 µg/mL inhibited RNA-dependent DNA polymerase and ribonuclease H activities of Human immunodeficiency virus type-1 reverse transcriptase and Human immunodeficiency virus type-1 protease by 70.8, 6.7, and 7.6%, respectively (Min et al., 2001).

Commentaries:

(i) *Crinum macrantherum* Engl. is used in Papua New Guinea to heal wounds and burns.

(ii) Lycorine could be examined for its possible anti-COVID-19 properties *in vitro* and *in vivo*. What is the anti-Severe acute repiratory syndrome-associated coronavirus mechanism of lycorine, has it anything to do with RNA polymerase? DNA polymerase? The somewhat planar structure of lycorine and inhibition of RNA viruses points to that possibility.

(iii) The genus flavivirus in the family Flaviviridae includes the arthropod-borne RNA and envelopes four serotypes of Dengue viruses, West nile virus, Japanese encephalitis virus, yellow fever virus, and tick-borne encephalitis virus (Gubler et al., 2007). More than 50 million, 200,000, and 50,000 human cases are reported annually for the Dengue, Yellow fever, and Japanese encephalitis viruses infections, respectively (Gubler et al., 2007). As for the West nile virus, an outbreak occurred in New York in 1999, causing thousands of infections spreading throughout North America. To date, no effective drug exists for the treatment of these infections (Zou et al., 2009).

## References

Adesanya, S.A., Olugbade, T.A., Odebiyl, O.O. and Aladesanmi, J.A., 1992. Antibacterial alkaloids in *Crinum jagus*. *International Journal of Pharmacognosy*, 30(4), pp. 303–307.

Gabrielsen, B., Monath, T.P., Huggins, J.W., Kefauver, D.F., Pettit, G.R., Groszek, G., Hollingshead, M., Kirsi, J.J., Shannon, W.M., Schubert, E.M. and DaRe, J., 1992. Antiviral (RNA) activity of selected Amaryllidaceae isoquinoline constituents and synthesis of related substances. *Journal of Natural Products*, 55(11), pp. 1569–1581.

Gubler, D., Kuno, G., Flaviviruses, M.L., Knipe, D.M. and Howley, P.M. (Eds.). 2007, *Fields Virology*, vol. 1. Philadelphia, PA: Lippincott William and Wilkins, pp. 1153–1253.

He, J., Qi, W.B., Wang, L., Tian, J., Jiao, P.R., Liu, G.Q., Ye, W.C. and Liao, M., 2013. Amaryllidaceae alkaloids inhibit nuclear-to-cytoplasmic export of ribonucleoprotein (RNP) complex of highly pathogenic avian influenza virus H5N1. *Influenza and Other Respiratory Viruses*, 7(6), pp. 922–931.

Li, S.Y., Chen, C., Zhang, H.Q., Guo, H.Y., Wang, H., Wang, L., Zhang, X., Hua, S.N., Yu, J., Xiao, P.G. and Li, R.S., 2005. Identification of natural compounds with antiviral activities against SARS-associated coronavirus. *Antiviral Research*, 67(1), pp. 18–23.

Liu, J., Yang, Y., Xu, Y., Ma, C., Qin, C. and Zhang, L., 2011. Lycorine reduces mortality of human enterovirus 71-infected mice by inhibiting virus replication. *Virology Journal*, 8(1), p. 483.

Nair, J.J. and van Staden, J., 2018. Antifungal constituents of the plant family Amaryllidaceae. *Phytotherapy Research*, 32(6), pp. 976–984.

Oluyemisi, O.O., Oriabure, A.E., Adekunle, A.J., Ramsay, K.S.T., Shyyaula, S. and Choudhary, M.I., 2015. Bioassay-guided isolation of poliovirus-inhibiting constituents from *Zephyranthes candida*. *Pharmaceutical Biology*, 53(6), pp. 882–887.

Peng, X., Sova, P., Green, R.R., Thomas, M.J., Korth, M.J., Proll, S., Xu, J., Cheng, Y., Yi, K., Chen, L. and Peng, Z., 2014. Deep sequencing of HIV-infected cells: insights into nascent transcription and host-directed therapy. *Journal of Virology*, 88(16), pp. 8768–8782.

Rahman, M.A., Sharmin, R., Uddin, M.N., Rana, M.S. and Ahmed, N.U., 2011. Antibacterial, antioxidant and cytotoxic properties of *Crinum asiaticum* bulb extract. *Bangladesh Journal of Microbiology*, 28(1), pp. 1–5.

Shen, J.W., Ruan, Y., Ren, W., Ma, B.J., Wang, X.L. and Zheng, C.F., 2014. Lycorine: a potential broad-spectrum agent against crop pathogenic fungi. *Journal of Microbiology, Biotechnology*, 24(3), pp. 354–358.

Zou, G., Puig-Basagoiti, F., Zhang, B., Qing, M., Chen, L., Pankiewicz, K.W., Felczak, K., Yuan, Z. and Shi, P.Y., 2009. A single-amino acid substitution in West Nile virus 2K peptide between NS4A and NS4B confers resistance to lycorine, a flavivirus inhibitor. *Virology*, 384(1), pp. 242–252.

### 3.1.6.1.4   *Lycoris radiata* (L'Hér.) Herb.

Synonym: *Amaryllis radiata* L'Hér.

Common names: Red spider lily; shi suan (China)

Habitat: Shady and moist soils, cultivated

Description: Nepal, China, Korea, and Japan

Botanical observation: This poisonous and bulbiferous herb is grown as ornamental. The leaves are simple, basal, and spiral. The blade is linear-oblong, somewhat fleshy, wavy, pendulous, and up to about 15 cm×5 mm and round at apex. The inflorescences are many-flowered umbels at the apex of pedicel a, which are up to about 30 cm tall. The perianth is infundibuliform, dull red, and develops 6 linear, wavy, and recurved lobes, which are up to about 3 cm×5 mm. The 6 stamens are slender, slightly recurved, red, up to 12 cm long and with brownish and linear to saffron-colored and linear anthers (At first glance these stamen have a Caesalpinious look). The gynoecium consists of an ovary, which is 3-locular and a slender style. The capsule is somewhat globose, glossy, 3-lobed, fleshy and contain numerous seeds.

Medicinal use: Ulcer, cough, burns, abscesses (China)

*Broad-spectrum antibacterial halo developed by Amaryllidaceae alkaloids*: O-Methyllycorenine, lycorenine, lycoricidinol, lycorine, and lycoricidine (1 mg/mL solution/8 mm paper discs) isolated from the bulbs inhibited the growth of *Escherichia coli* with inhibition zone diameters of 14, 15, 13, 13, and 12 mm, respectively (Lee et al., 2014). O-Methyllycorenine, lycorenine, lycoricidinol, lycorine, and lycoricidine (1 mg/mL solution/8 mm paper discs) isolated from the bulbs inhibited the growth of *Staphylococcus aureus* with inhibition zone diameters of 14, 14, 11, 11, and 11 mm, respectively (Lee et al., 2014).

*Antibacterial narcilasine-type Amaryllidaceae alkaloids:* Narciclasine inhibited the growth of *Corynebacterium fascians* (Pettit et al., 2002).

*Antifungal narcilasine-type Amaryllidaceae alkaloids:* Narciclasine inhibited the growth of *Cryptococcus neoformans* (Pettit et al., 2002).

*Antifungal lycorenine-type Amaryllidaceae alkaloid:* Hippeastrine inhibited the growth of *Candida albicans* with the MIC value of 125 µg/mL (Nair & van Staden, 2018).

*Strong antiviral (enveloped monopartite linear single-stranded (+) RNA) hydrophilic polar extract:* Ethanol extract of stems inhibited the replication of the Severe acute respiratory syndrome-associated coronavirus BJ-001 and BJ-006 with the $IC_{50}$ values of 2.4 and 2.1 µg/mL and selectivity indices of 370 and 422, respectively (Li et al., 2005).

*Strong (enveloped monopartite linear single-stranded (+) RNA) alkaloid fraction:* Alkaloid fraction inhibited the replication of the Severe acute respiratory syndrome-associated coronavirus (BJ-001) with the $IC_{50}$ value of 1 µg/mL and a selectivity index of 94 (Li et al., 2005).

*Strong antiviral (enveloped monopartite linear single-stranded (+) RNA) hydrophilic lycorine-type Amaryllidaceae alkaloid:* Lycorine isolated from the stems inhibited the replication of the Severe acute respiratory syndrome-associated coronavirus (BJ-001) with an $IC_{50}$ value of 15.7 nM and a selectivity index of 954 (Li et al., 2005).Hippeastrine (LogD = 0.8; molecular weight = 315.3 g/mol) inhibited the replication of the Zika virus with an $IC_{50}$ value of 1.9 µM (Zhou et al., 2017).

*Antiviral (enveloped monopartite linear double-stranded DNA) lycorine-type Amaryllidaceae alkaloid:* Hippeastrine (LogD = 0.8; molecular weight = 315.3 g/mol) inhibited the replication of Herpes simplex virus type-1 (Renard-Nozaki et al., 1989).

Hippeastrine

*Strong-antiviral (enveloped RNA) hydrophilic narcilasine-type Amaryllidaceae alkaloids:* Narciclasine (LogD = 0.09 at pH 7.4; molecular weight = 307.2 g/mol) inhibited the replication of Japanese encephalitis virus, Yellow fever virus, Dengue type 4 virus, and Punta toro virus with $IC_{50}$ values of 0.008, 0.006, 0.001, and 0.007 µg/mL, respectively (Gabrielsen et al., 1992). Lycoricidine (7-deoxynarciclasine; LogD = −0.1 at pH 7.4; molecular weight = 291.2 g/mol) inhibited the replication of Japanese Encephalitis virus, Yellow fever virus, Dengue type 4 virus, Punta Toro, Rift Valley Fever, and Sandfly fever-Sicilian viruses with $IC_{50}$ values of 0.005, 0.005, 0.005, 0.004, 0.1, and 0.005 µg/mL, respectively (Gabrielsen et al., 1992).

Narciclasine

Lycoricidine

*Moderate antiviral (enveloped segmented linear single-stranded (–) RNA virus) lycorine-type Amaryllidaceae alkaloid:* Lycorine inhibited the replication of Influenza virus H5N1 with an $IC_{90}$ value of 82 μM and a selectivity index >6.7 (He et al., 2013).

*In vivo antiviral lycorenine-type Amaryllidaceae alkaloid:* Hippeastrine given subcutaneously at a dose of 100 mg/Kg/day to mice experimentally infected with the Zika virus (MR766) for 7 days decreased brain levels of Zika virus RNA (Zhou et al., 2017).

Commentary: Preclinical trials for hippeastrine as anti-Zika agent are warranted. Is hippeastrine anti-COVID-19? In China, *Lycoris aurea* is applied to burns. In Korea, *Lycoris squamigera* Maxim, is used to treat abscesses and tuberculosis.

## References

Gabrielsen, B., Monath, T.P., Huggins, J.W., Kefauver, D.F., Pettit, G.R., Groszek, G., Hollingshead, M., Kirsi, J.J., Shannon, W.M., Schubert, E.M. and DaRe, J., 1992. Antiviral (RNA) activity of selected Amaryllidaceae isoquinoline constituents and synthesis of related substances. *Journal of Natural Products*, 55(11), pp. 1569–1581.

He, J., Qi, W.B., Wang, L., Tian, J., Jiao, P.R., Liu, G.Q., Ye, W.C. and Liao, M., 2013. Amaryllidaceae alkaloids inhibit nuclear-to-cytoplasmic export of ribonucleoprotein (RNP) complex of highly pathogenic avian influenza virus H5N1. *Influenza and Other Respiratory Viruses*, 7(6), pp. 922–931.

Lee, D.G., Lee, A.Y., Kim, S.J., Lee, S., Cho, E.J. and Lee, S., 2014. Antibacterial phytosterols and alkaloids from *Lycoris radiata*. *Natural Product Sciences*, 20, pp. 107–112.

Li, S.Y., Chen, C., Zhang, H.Q., Guo, H.Y., Wang, H., Wang, L., Zhang, X., Hua, S.N., Yu, J., Xiao, P.G. and Li, R.S., 2005. Identification of natural compounds with antiviral activities against SARS-associated coronavirus. *Antiviral Research*, 67(1), pp. 18–23.

Nair, J.J. and van Staden, J., 2018. Antifungal constituents of the plant family Amaryllidaceae. *Phytotherapy Research*, 32(6), pp. 976–984.

Pettit, G., Melody, N., Herald, D.L., Schmidt, J.M., Pettit, R. and Chapuis, J., 2002. Synthesis of 10b (R)-hydroxypancratistatin, 10b (S)-hydroxy-1-epipancratistatin, 10b (S)-hydroxy-1, 2-diepipancratistatin and related isocarbostyrils. *Heterocycles*, 56(1–2), pp. 139–155.

Renard-Nozaki, J., Kim, T., Imakura, Y., Kihara, M. and Kobayashi, S., 1989. Effect of alkaloids isolated from Amaryllidaceae on herpes simplex virus. *Research in Virology*, 140, pp. 115–128.

Zhou, T., Tan, L., Cederquist, G.Y., Fan, Y., Hartley, B.J., Mukherjee, S., Tomishima, M., Brennand, K.J., Zhang, Q., Schwartz, R.E. and Evans, T., 2017. High-content screening in hPSC-neural progenitors identifies drug candidates that inhibit Zika virus infection in fetal-like organoids and adult brain. *Cell Stem Cell*, 21(2), pp. 274–283.

### 3.1.6.1.5  *Narcissus tazetta* L.

Synonyms: *Hermione tazetta* (L.) Haw.; *Jonquilla tazetta* (L.) Raf.; *Narcissus tazetta var. chinensis* M. Roem.

Common names: Shui xian (China); Nargis (Pakistan)

Habitat: Wastelands, cultivated

Distribution: Turkey, Iran, Kashmir, India, China, and Japan

Botanical observation: This very poisonous, graceful, and bulbiferous herb, is grown as ornamental. The leaves are simple, basal, and spiral. The blade is linear-oblong, somewhat fleshy, 20–40 cm × 1–1.5 cm, and obtuse at apex. The inflorescences are many-flowered and fragrant umbels at the apex of pedicels, which are up to about 40 cm tall. The perianth is tubular, the tube green and up to about 2 cm long and develops 6 broadly lanceolate, pure white or whitish, and up to about 2 cm long lobes as well as a corona that is yellowish to bright orangish and up to about 1 cm long.

The androecium comprises 6 up to 4 mm long stamens attached to the perianth tube. The gynoecium consists of a 3-locular ovary and a slender style. The capsules are somewhat globose, glossy, 3-lobed, and containing numerous black seeds.

Medicinal uses: Abscesses, boils, mastitis (China); fever (Indonesia)

*Moderate anticandidal polar extract:* Butanol extract of aerial part inhibited the growth of *Candida albicans* with an MIC value of 125 µg/mL (Talib & Mahasneh, 2010).

*Very weak antifungal tazettine-type Amaryllidaceae alkaloid:* Tazettine inhibited the growth of *Candida dubliniensis* and *Lodderomyces elongisporus* (Nair & van Staden, 2018).

*Strong antiviral (enveloped, RNA) hydrophilic tazettine-type Amaryllidaceae alkaloid:* Pretazettine (LogD = 0.2 at pH 7.4; molecular weight = 331.3 g/mol) inhibited the replication of Japanese encephalitis, Yellow fever virus, Punta toro virus, Rift valley fever, and Sandfly fever-Sicilian virus, with $IC_{50}$ values of 0.6, 0.5, 0.6, 2.9, and 0.8 µg/mL, respectively (Gabrielsen et al., 1992).

Pretazettine

*Strong antiviral (enveloped RNA) hydrophilic lycorine-type Amaryllidaceae alkaloid:* Pseudolycorine (LogD = 0.3 at pH 7.4; molecular weight = 289.3 g/mol) inhibited the replication of Japanese encephalitis, Yellow fever virus, Dengue type 4 virus, Punta toro virus, and Rift valley fever virus with $IC_{50}$ values of 0.2, 0.3, 0.3, 0.6, and 0.6 µg/mL, respectively (Gabrielsen et al., 1992).

Pseudolycorine

*Viral enzyme inhibition by alkaloidal extract:* Alkaloid extract inhibited DNA polymerase from Avian myeloblastosis virus (Papas et al., 1973).

Commentary: The plant generates the Amaryllidaceae alkaloids tazettine, lycorine, pseudolycorine, and pretazettine (Furusawa et al., 1976). Pseudolycorine and pretazettine inhibit protein synthesis and are active against murine Rauscher leukemia virus and neurotropic RNA viruses. Note that Amaryllidaceae alkaloids have the tendency to inhibit topoisomerase (Chen et al., 2016) and as such are an interesting class of antibacterial agents. These alkaloids most probably inhibit virus RNA and DNA replication in host cells.

### References
Chen, G.L., Tian, Y.Q., Wu, J.L., Li, N. and Guo, M.Q., 2016. Antiproliferative activities of Amaryllidaceae alkaloids from *Lycoris radiata* targeting DNA topoisomerase I. *Scientific Reports*, 6, p. 38284.

Furusawa, E., Furusawa, S., Tani, S., Irie, H., Kitamura, K. and Wildman, W.C., 1976. Isolation of pretazettine from *Narcissus tazetta* L. *Chemical and Pharmaceutical Bulletin*, 24(2), pp. 336–338.

Gabrielsen, B., Monath, T.P., Huggins, J.W., Kefauver, D.F., Pettit, G.R., Groszek, G., Hollingshead, M., Kirsi, J.J., Shannon, W.M., Schubert, E.M. and DaRe, J., 1992. Antiviral (RNA) activity of selected Amaryllidaceae isoquinoline constituents and synthesis of related substances. *Journal of Natural Products*, 55(11), pp. 1569–1581.

Nair, J.J. and van Staden, J., 2018. Antifungal constituents of the plant family Amaryllidaceae. *Phytotherapy Research*, 32(6), pp. 976–984.

Papas, T.S., Sandhaus, L., Chirigos, M.A. and Furusawa, E., 1973. Inhibition of DNA polymerase of avian myeloblastosis virus by an alkaloid extract from *Narcissus tazetta* L. *Biochemical and Biophysical Research Communications*, 52(1), pp. 88–92.

Talib, W.H. and Mahasneh, A.M., 2010. Antimicrobial, cytotoxicity and phytochemical screening of Jordanian plants used in traditional medicine. *Molecules*, 15(3), pp. 1811–1824.

### 3.1.6.1.6    *Pancratium triflorum* Roxb.

Synonyms: *Crinum pauciflorum* Miq.; *Pancratium malabathricum* Herb.

Common name: Kaattulli pola (India)

Habitat: Moist, rocky, and shady soils

Distribution: India, Bangladesh, and Sri Lanka

Botanical observation: This magnificent yet dreadfully poisonous herb grows from a somewhat globose bulb which is about 5 cm across. The leaves are simple, basal, and spiral. The blade is linear-lanceolate, 12.5–30 cm × 1.5–3 cm and obtuse at apex. The inflorescences are many-umbels of up to about 8 flowers at the apex of a peduncle. The plant is considered sacred in India and used in Temples as offering. The perianth is pure white, tubular, membranous, ephemeral, the tube is up to about 2.5 cm long, and develops 6 lobes, which are elliptic-lanceolate, up to about 2.5 cm long, as well as a dentate corona. The 6 stamens grow up to up to 6 mm long with versatile anthers. The ovary is 3-locular with a slender style reaching a length of about 1.5 cm long. The capsule contain numerous seeds.

Medicinal use: Boils, cough (India)

Commentary: The plant has apparently not been studied for its potential antimicrobial properties. Note that the narcilasine-type Amaryllidaceae alkaloid pancratistatin isolated from the roots of *Pancratium littorale* Jacq. inhibited the replication of the Japanese encephalitis virus, Yellow fever virus, Dengue type 4 virus, and Rift valley fever with the $IC_{50}$ values of 0.02, 0.01, 0.06, and 0.1 μg/mL, respectively (Gabrielsen et al., 1992). This alkaloid given subcutaneously at a dose of 4 mg/Kg/day once daily for 7 days evoked 100% protection against the Japanese encephalitis virus (Gabrielsen et al., 1992).

## Reference

Gabrielsen, B., Monath, T.P., Huggins, J.W., Kefauver, D.F., Pettit, G.R., Groszek, G., Hollingshead, M., Kirsi, J.J., Shannon, W.M., Schubert, E.M. and DaRe, J., 1992. Antiviral (RNA) activity of selected Amaryllidaceae isoquinoline constituents and synthesis of related substances. *Journal of Natural Products*, 55(11), pp. 1569–1581.

### 3.1.6.1.7  *Zephyranthes carinata* Herb.

Synonym: *Amaryllis carinata* (Herb.) Spreng.

Common name: Rain lily; jiu lian (China); safuranmodoki (Japan)

Habitat: Cultivated

Distribution: Tropical Asia and Pacific

Botanical observation: This herb grows up to about 30 cm from a tunicate bulb which is about 3 cm across. The leaves are simple, basal, and spiral. The blade is linear-lanceolate, 15–30 cm × 6–8 mm, and obtuse at apex. The inflorescences are solitary and terminal on a slender pedicel. The 6 lobes of the periath are heavenly pink, broadly lanceolate to elliptic and up to about 6 cm long. The 6 stamens have versatile anthers. The gynoecium consists of an ovary which is 3-locular and a slender style. The capsule is 3-lobed, subglobose, and contain numerous black seeds.

Medicinal uses: Abscesses, fever (China)

*Anticandidal lycorine-type Amaryllidaceae alkaloid:* Galanthine inhibited the growth of *Candida dubliniensis*, *Candida glabrata*, and *Lodderomyces elongisporus* (Nair & van Staden, 2018).

Commentary: Antimicrobial activities have apparently not been fully examined from this plant. Note the presence of Amaryllidaceae alkaloids in the bulbs: pretazettine, carinatine, galanthine, haemanthamine, and lycorine (Kobayashi et al., 1977).

### References

Kobayashi, S., Ishikawa, H., Kihara, M., Shingu, T. and Hashimoto, T., 1977. Isolation of carinatine and pretazettine from the bulbs of *Zephyranthes carinate* Herb. (Amaryllidaceae). *Chemical and Pharmaceutical Bulletin*, 25(9), pp. 2244–2248.

Nair, J.J. and van Staden, J., 2018. Antifungal constituents of the plant family Amaryllidaceae. *Phytotherapy Research*, 32(6), pp. 976–984.

### 3.1.6.2  Family Asparagaceae A.L. de Jussieu (1789)

The family Asparagaceae consists of more than 15 genera and 300 species of herbs. The leaves are absent or simple, linear, and spiral. The inflorescences are solitary or racemes or panicles. The perianth comprises 6 tepals. The androecium includes 6 stamens. The gynoecium consists of 3 carpels fused into a 3-locular ovary, each locule containing numerous ovules on axile placentas. The fruit is a berry or a capsule. Plants in this family produce antimicrobial steroidal saponins, homoisoflavonoids, and lignans.

### 3.1.6.2.1  *Agave americana* L.

Synonyms: *Agave complicata* Trel. ex Ochot.; *Agave felina* Trel.; *Agave gracilispina* Engelm. ex Trel.; *Agave melliflua* Trel.; *Agave rasconensis* Trel.; *Agave subzonata* Trel.; *Agave zonata* Trel. ex Bailey

Common names: American Aloe; century plant; murga muji (Bangladesh); long she lan (China); kantala (India); magey (the Philippines)

Habitat: Cultivated

Distribution: Tropical and subtemperate Asia and Pacific

Botanical observation: This large fleshy herb present a rosette of 30 or more lanceolate and massive leaves. The blades are spiny at margin and apex, somewhat glaucous, 1–2 m × 10–20 cm light green, sappy, fibrous, stout, and coriaceous. The inflorescence is a panicle that is about 6 m tall and at first glance looks like a monstrous asparagus. The flowers are numerous, 7–10.5 cm long, with a bright yellow perianth consisting of 6 lobes. The androecium consist of conspicuous and filiform stamens protruding out of the perianth and with versatile anthers. The gynoecium comprises a 3-locular ovary, a slender style protruding out of the perianth, and a 3-lobed stigma. The capsules are 3-lobed and contain numerous black and flat seeds.

Medicinal uses: Gonorrhea (Bangladesh); fever, sores (Cambodia, Laos, Vietnam); syphilis (India)

*Moderate anticandidal polar extract:* Methanol extract of stem inhibited the growth of *Candida albicans* (ATCC 90029) with MIC/MFC values of 250/500 µg/mL (Chea et al., 2007).

Commentary: The plant has apparently not been fully examined for its antibacterial properties. It synthetizes series of spirostane steroidal saponins (Yokosuka et al., 2000). Note that steroidal saponins have antifungal effects (Yang et al., 2006) via complexation with ergosterol in the cell membrane, leading to pore formation and consequent loss of membrane integrity (Yang et al., 2006). An example is a strong broad-spectrum antifungal spirostane saponin is degalactotigonin (from a member of the genus *Solanum* L. in the family Solanaceae), which inhibited the growth of *Candida albicans*, *Candida glabrata*, *Cryptococcus neoformans*, and *Aspergillus fumigatus* with MIC/MFC values of 5/5, 5/5, 0.6/0.6, and 2.5/10 µg/mL, respectively (Yang et al., 2006). Tetratriacontanol derivatives isolated from this plant exhibited antibacterial effects (Parmar et al., 1992).

### References

Chea, A., Jonville, M.C., Bun, S.S., Laget, M., Elias, R., Duménil, G. and Balansard, G., 2007. In vitro antimicrobial activity of plants used in Cambodian traditional medicine. *The American Journal of Chinese Medicine*, 35(05), pp. 867–873.

Parmar, V.S., Jha, H.N., Gupta, A.K., Prasad, A.K., Gupta, S., Boll, P.M. and Tyagi, O.D., 1992. New antibacterial tetratriacontanol derivatives from *Agave americana* L. *Tetrahedron*, 48(7), pp. 1281–1284.

Yang, C.R., Zhang, Y., Jacob, M.R., Khan, S.I., Zhang, Y.J. and Li, X.C., 2006. Antifungal activity of C-27 steroidal saponins. *Antimicrobial Agents and Chemotherapy*, 50(5), pp. 1710–1714.

Yokosuka, A., Mimaki, Y., Kuroda, M. and Sashida, Y., 2000. A new steroidal saponin from the leaves of *Agave americana*. *Planta medica*, 66(04), pp. 393–396.

### 3.1.6.2.2 *Anemarrhena asphodeloides* Bunge

Common name: Zhi mu (China)

Habitat: Steppes and sandy lands

Distribution: Korea, China, and Taiwan

Botanical observation: This herb grows up to about 1 m tall from an elongated rhizome. The leaves are simple and spiral. The blade is linear, up to 60 cm × 1 cm, dull green, and grass-like. The inflorescence is a raceme, which is 10–50 cm long. The 6 lobes of the perianth are linear and up to 1 cm long. The gynoecium comprises an ovoid ovary. The capsule up to 1.5 cm long, 6-angled, beaked, and contains numerous black seeds.

Medicinal uses: Influenza, pneumonia (China); tuberculosis, dysentery (Vietnam); fever (China, Vietnam)

*Strong antifungal (filamentous) amphiphilic norlignan:* Nyasol (cis-hinokiresinol) (LogD = 3.4 at pH 7.4; molecular weight = 252.3 g/mol) isolated from the rhizome inhibited the growth of *Cladosporium cucumerinum*, *Colletotrichum orbiculare Fusarium oxysporum* f.sp. *lycopersici*, *Phytophthora capsici*, *Pythium ultimum*, *Rhizoctonia solani*, with MIC values of 1, 50, 50, 50, 5, and 10 µg/mL, respectively (Park et al., 2003).

Nyasol

*Antifungal agent potentiator norlignan:* Nyasol increased the sensitivity of *Candida albicans* to miconazole, ketoconazole, and clotrimazole (Lida et al., 2000).

*Strong antiviral (enveloped linear monopartite single-stranded (-)RNA virus) amphiphilic norlignans:* Nyasol and 4′-O-methylnyasol isolated from the rhizome inhibited the replication of the Respiratory syncytial virus with $IC_{50}$ values of 0.8 and 0.3 µM, respectively (Bae et al., 2007).

*Strong antiviral (enveloped linear monopartite single-stranded (-)RNA virus) amphiphilic diarylpropane:* Broussonin A (LogD = 3.3 at pH 7.3; molecular weight = 258.3 g/mol) isolated from the rhizome inhibited the replication of the Respiratory syncytial virus with an $IC_{50}$ value of 0.6 µM (Bae et al., 2007).

Broussonin A

*Strong antiviral (enveloped linear monopartite single-stranded (-)RNA virus) amphiphilic spirostane saponin:* Timosaponin A-III (LogD = 3.1 at pH 7.4; molecular weight = 740.9 g/mol) inhibited the replication of Respiratory syncytial virus (A2 strain) with an $IC_{50}$ value of 1 µM (Youn et al., 2011).

Timosaponin A-III

*Weak antiviral (enveloped linear monopartite single-stranded (-)RNA virus) hydrophilic xanthone C-glycoside:* Mangiferin inhibited the Respiratory syncytial virus (Long strain in Hep-2 cells) plaque formation with an $IC_{50}$ value of 40 μM and a selectivity index >12.5 (Zhang et al., 2015).

*Weak antiviral (enveloped monopartite linear double-stranded DNA) hydrophilic xanthone C-glycoside:* Mangiferin inhibited the replication of Herpes simplex virus type-2 with an $IC_{50}$ value of 111.7 μg/mL (Zhu et al., 1993).

*Viral enzymatic inhibition by polar extract:* Aqueous extract of rhizomes at a concentration of 250 μg/mL inhibited Human immunodeficiency virus type-1 protease by 76.2% (Xu et al., 1996).

Commentary: Nyasol has a broad antifungal spectrum, could it be a phytoalexin? Broussonin A is a phytoalexin (Takasugi et al., 1980). One could examine the anti-coronavirus properties of nyasol and mangiferin. There is a need to constitute an armamentarium of first-line natural products from medicinal plants to be ready to test against the coming viral, bacterial, or fungal pandemic waves.

### References

Bae, G., Yu, J.R., Lee, J., Chang, J. and Seo, E.K., 2007. Identification of nyasol and structurally related compounds as the active principles from *Anemarrhena asphodeloides* against respiratory syncytial virus (RSV). *Chemistry & Biodiversity*, 4(9), pp. 2231–2235.

Iida, Y., Oh, K.B., Saito, M., Matsuoka, H. and Kurata, H., 2000. In vitro synergism between nyasol, an active compound isolated from *Anemarrhena asphodeloides*, and azole agents against *Candida albicans*. *Planta Medica*, 66(05), pp. 435–438.

Park, H.J., Lee, J.Y., Moon, S.S. and Hwang, B.K., 2003. Isolation and anti-oomycete activity of nyasol from *Anemarrhena asphodeloides* rhizomes. *Phytochemistry*, *64*(5), pp. 997–1001.

Takasugi, M., Anetai, M., Masamune, T., Shirata, A. and Takahashi, K., 1980. Broussonins A and B, new phytoalexins from diseased paper mulberry. *Chemistry Letters*, *9*(3), pp. 339–340.

Xu, H.X., Wan, M., Loh, B.N., Kon, O.L., Chow, P.W. and Sim, K.Y., 1996. Screening of traditional medicines for their inhibitory activity against HIV -1 protease. *Phytotherapy Research*, *10*(3), pp. 207–210.

Youn, U.J., Jang, J.E., Nam, J.W., Lee, Y.J., Son, Y.M., Shin, H.J., Han, A.R., Chang, J. and Seo, E.K., 2011. Anti-respiratory syncytial virus (RSV) activity of timosaponin A-III from the rhizomes of *Anemarrhena asphodeloides*. *Journal of Medicinal Plant Research*, *5*(7), pp. 1062–1065.

Zhang, Y.B., Wu, P., Zhang, X.L., Xia, C., Li, G.Q., Ye, W.C., Wang, G.C. and Li, Y.L., 2015. Phenolic compounds from the flowers of *Bombax malabaricum* and their antioxidant and antiviral activities. *Molecules*, *20*(11), pp. 19947–19957.

Zhu, X.M., Song, J.X., Huang, Z.Z., Wu, Y.M. and Yu, M.J., 1993. Antiviral activity of mangiferin against herpes simplex virus type 2 in vitro. *Zhongguo yao li xue bao = Acta pharmacologica Sinica*, *14*(5), pp. 452–454.

### 3.1.6.2.3   *Asparagus cochinchinensis* (Lour.) Merr.

Synonyms: *Asparagopsis sinica* Miq.; *Asparagus gaudichaudianus* Kunth; *Asparagus insularis* Hance; *Asparagus lucidus* Lindl.; *Asparagus sinicus* (Miq.) C.H. Wright; *Melanthium cochinchinense* Lour.

   Common names: Tian men dong (China); kheua ya nang xang (Laos)

   Habitat: Roadside, waste lands, and forests

   Distribution: Laos, Vietnam, China, Korea, and Japan

   Botanical observation: This diffuse herb grows to about 3 m tall from tubers. The stem is angled, develops spiny, woody, and sharp leaf spurs, which are straight, and minute. The cladodes are arranged in fascicles of 3, linear, up to 8 cm long, and flat. The inflorescences are clusters. The perianth is tubular, greenish to pure white, 3 mm long, and develops 6 lobes. The 6 stamens have bright orangish-yellow anthers. The ovary minute. The berries are green, glossy, and about 7 mm in diameter.

   Medicinal uses: Cough, fever, and tuberculosis (China)

   *Very weak antibacterial (Gram-positive) polar extract:* Aqueous extract of rhizome inhibited the growth of *Streptococcus mutans* (MT 5091) and *Streptococcus mutans* (OMZ 176) with MIC values of 2500 and 2500 µg/mL, respectively (Chen et al., 1989).

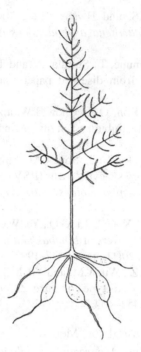

*Asparagus cochinchinensis* (Lour.) Merr.

*Strong antiviral (enveloped monopartite linear dimeric single-stranded (+)RNA) amphiphilic norlignans:* Nyasol (LogD = 3.4 at pH 7.4; molecular weight = 252.3 g/mol) and 1,3-bis-di-p-hydroxyphenyl-4-penten-1-one inhibited Human immunodeficiency virus type-1 replication with $IC_{50}$ values of 11.7 and 20 μg/mL, respectively (Zhang et al., 2004).

Commentary: The plant produces spirostane and furostane-type steroidal saponins (Zhu et al., 2014).

### References

Chen, C.P., Lin, C.C. and Tsuneo, N., 1989. Screening of Taiwanese crude drugs for antibacterial activity against *Streptococcus* mutans. *Journal of Ethnopharmacology*, 27(3), pp. 285–295.

Zhang, H.J., Sydara, K., Tan, G.T., Ma, C., Southavong, B., Soejarto, D.D., Pezzuto, J.M. and Fong, H.H., 2004. Bioactive constituents from *Asparagus cochinchinensis*. *Journal of Natural Products*, 67(2), pp. 194–200.

Zhu, G.L., Hao, Q., Li, R.T. and Li, H.Z., 2014. Steroidal saponins from the roots of *Asparagus cochinchinensis*. *Chinese Journal of Natural Medicines*, 12(3), pp. 213–217.

### 3.1.6.2.4    *Asparagus racemosus* Willd.

Synonyms: *Asparagopsis javanica* Kunth; *Asparagus dubius* Decaisne; A. *schoberioides* Kunth.; *Protasparagus racemosus* Oberm.

Common names: Indian asparagus, satáver white, satáver yellow, wild asparagus; sattis chara gach (Bangladesh); mem sam sob (Cambodia); chang ci tian men dong (China); shatavari (India); sangga langit (Indonesia); satavari (Nepal); satavor, tilora (Pakistan); samsip (Thailand)

Habitat: Forests, riverbanks

Distribution: Pakistan, India, Nepal, Bangladesh, Myanmar, Cambodia, Thailand, Malaysia, and Indonesia

Botanical observation: This herb grows to about 2 m tall from tubers. The stem is erect and develops spiny, woody, and sharp leaf spurs, which are straight, 1.5–2 cm on main stems, and 5–10 mm on branches. The cladodes are arranged in fascicles of 3–6, linear, 1–2.5 cm long, and flat. The inflorescences are up to 4 cm long clusters. The perianth is tubular, white, 3 mm long and develops 6 lobes. The androecium includes 6 minute and white stamens with bright yellow anthers. The ovary is obovoid, minute, and somewhat yellowish and develops a trilobed stigma. The berries are red, glossy, about 8 mm in diameter, and trilobed.

Medicinal uses: Leucorrhea, vaginitis (Bangladesh); dysentery, bronchitis (India); diarrhea, dysentery (Nepal)

*Broad-spectrum antibacterial halo developed by polar extract:* Methanol extract of tubers (10 mm diameter disc impregnated with 50 µg/mL solution) inhibited the growth of *Escherichia coli, Shigella dysenteriae, Shigella sonnei, Shigella flexneri, Vibrio cholerae, Salmonella typhi, Staphylococcus aureus*, and *Bacillus subtilis* (Mandal et al., 2000).

*Moderate anticandidal polar extract:* Methanol extract of tubers inhibited the growth of *Candida albicans, Candida tropicalis, Candida krusei, Candida guillermondii, Candida parapsilosis*, and *Candida stellatoidea* with MIC/MFC values of 312/625, 625/1250, 625/1250, 625/1250, 625/1250, and 625/1250 µg/mL, respectively (Uma et al., 2009).

*Strong antiviral (enveloped monopartite linear dimeric single-stranded (+)RNA) polar extract:* Butanol fraction of tubers inhibited the replication of the Human immunodeficiency virus type-1 (NL4.3 strain) in Human CD4+ cells by 87.2% at the maximum noncytotoxic concentration of 40 µg/mL (Sabde et al., 2011).

Commentaries: (i) Butanol is an organic solvent that dissolves and extracts very hydrophilic compounds of which glycosides and saponins. It has a high boiling point and its evaporation under pressure takes time. (ii) The plant produces spirostane and furostane-type steroidal saponins (Mandal et al., 2000), the pyrrolozidine alkaloid asparagamine A, the phenanthrene racemosol A, and the benzofuran lignan racemofuran (Wiboonpun et al., 2004), as well as immunostimulating spirostane steroidal saponins (Sharma et al., 2011). A strong immune system allows, depending on the type of virus attacking the body, increased chances of survival (Perelson, 2002). In the case of COVID-19 pandemic, for which no effective treatment exists as of yet, it has been demonstrated that patients with strong immune systems had better survival rates (Tay et al., 2020). In this light, it would appear that the intake of plant extracts or infusions (as traditionally done) would permit the synergistic activity of antiviral compounds and immunostimulating principles. Do we need to rethink the current strategy of drug discovery from medicinal plants? Also, it can be said that industrial food and lifestyle in the so-called developed countries are both responsible for weakening our immune system (Colon & Bird, 2015).

## References

Conlon, M.A. and Bird, A.R., 2015. The impact of diet and lifestyle on gut microbiota and human health. *Nutrients, 7*(1), pp. 17–44.

Mandal, S.C., Nandy, A., Pal, M. and Saha, B.P., 2000. Evaluation of antibacterial activity of *Asparagus racemosus* Willd. root. *Phytotherapy Research: An International Journal Devoted to Pharmacological and Toxicological Evaluation of Natural Product Derivatives, 14*(2), pp. 118–119.

Perelson, A.S., 2002. Modelling viral and immune system dynamics. *Nature Reviews Immunology, 2*(1), pp. 28–36.

Sabde, S., Bodiwala, H.S., Karmase, A., Deshpande, P.J., Kaur, A., Ahmed, N., Chauthe, S.K., Brahmbhatt, K.G., Phadke, R.U., Mitra, D. and Bhutani, K.K., 2011. Anti-HIV-1 activity of Indian medicinal plants. *Journal of Natural Medicines, 65*(3–4), pp. 662–669.

Sharma, P., Chauhan, P.S., Dutt, P., Amina, M., Suri, K.A., Gupta, B.D., Suri, O.P., Dhar, K.L., Sharma, D., Gupta, V. and Satti, N.K., 2011. A unique immuno-stimulant steroidal sapogenin acid from the roots of *Asparagus racemosus. Steroids, 76*(4), pp. 358–364.

Tay, M.Z., Poh, C.M., Rénia, L., MacAry, P.A. and Ng, L.F., 2020. The trinity of COVID-19: immunity, inflammation and intervention. *Nature Reviews Immunology*, pp. 1–12.

Uma, B., Prabhakar, K. and Rajendran, S., 2009. Anticandidal activity of *Asparagus racemo-sus*. *Indian Journal of Pharmaceutical Sciences, 71*(3), p. 342.

Wiboonpun, N., Phuwapraisirisan, P. and Tip-pyang, S., 2004. Identification of antioxi-dant compound from *Asparagus racemosus*. *Phytotherapy Research: An International Journal Devoted to Pharmacological and Toxicological Evaluation of Natural Product Derivatives, 18*(9), pp. 771–773.

### 3.1.6.2.5   *Cordyline fruticosa* (L.) A. Chev.

Synonyms: *Aletris chinensis* Lam.; *Asparagus terminalis* L.; *Convallaria fruticosa* L.; *Cordyline terminalis (L.)* Kunth; *Dracaena ferrea* L.; *Dracaena fruticosa* K. Koch; *Dracaena stricta* Sims; *Dracaena terminalis* L.; *Taetsia ferrea* Medik.; *Taetsia fruticosa* (L.) Merr.; *Taetsia fruticosa var. ferrea* Standl.; *Taetsia stricta* Standl.; *Taetsia terminalis* (L.) W. Wight ex Saff.

Common names: Ti plant; zhu jiao (China); ti (Fiji); andong (Indonesia; Malaysia); aegop (Papua New Guinea); kila (the Philippines); lau ti (Samoa); mak mia (Thailand); si si tongotongo (Tonga Island)

Habitat: Cultivated

Distribution: Tropical Asia and Pacific

Botanical observation: This showy and common ornamental shrub grows up to 3 tall. The stems are woody, solitary, and marked with annular leaf-scars. The leaves are simple, spiral, and at apex of stem. The petiole is up to 30 cm long, channeled, clasping the stem and sheathing at base, pur-ple, pink, or green. The blade is bright purple-pink and variegated, wavy, oblong, glossy, 25–50 cm × 5–10 cm, and aristate, acuminate, or acute at apex. The panicles are up to 60 cm long. The peri-anth is light pink or white, tubular, the tube 6 mm long, and develops 6 lobes, which are glossy and recurved. The 6 stamens present versatile and yellow anthers. The ovary develops a straight, light pink or white style that protrudes out of the corolla. The berries are globose, about 6 mm across, red, glossy, and shelter a few black and glossy seeds.

Medicinal uses: Dysentery (China, Malaysia); wounds (Indonesia); fever (Papua New Guinea)

*Antifungal (yeast) halo developed by non-polar extract:* Hexane extract of leaves (400 μg/well, 7 mm diameter well) inhibited the growth of *Candida albicans* with an inhibition zone diameter of 15 mm (Kusuma et al., 2016).

*Weak antibacterial (Gram-positive) spirostane steroidal saponin:* Fructicoside I isolated from the leaves inhibited the growth of *Enterococcus faecalis* with the MIC value of 128 μg/mL (Fouedjou et al., 2014).

### References

Fouedjou, R.T., Teponno, R.B., Quassinti, L., Bramucci, M., Petrelli, D., Vitali, L.A., Fiorini, D., Tapondjou, L.A. and Barboni, L., 2014. Steroidal saponins from the leaves of *Cordyline fruticosa* (L.) A. Chev. and their cytotoxic and antimicrobial activity. *Phytochemistry Letters, 7*, pp. 62–68.

Kusuma, I.W., Sari, N.M., Murdiyanto and Kuspradini, H., 2016. Anticandidal Activity of Several Plants Used by Bentian Tribe in East Kalimantan, Indonesia. *AIP Conference Proceedings* (Vol. 1755, No. 1, p. 040002). AIP Publishing LLC.

### 3.1.6.2.6   *Dracaena spicata* Roxb.

Synonyms: *Draco spicata* (Roxb.) Kuntze; *Pleomele spicata* (Roxb.) N.E. Br.

Common names: Kadorateng (Bangladesh)

Habitat: Forests

Distribution: India, Bangladesh, Andamans, Myanmar, Laos, Thailand, Vietnam, and Malaysia

Botanical observation: Thus shrub grows to about 1.2 m tall. The stem is terete. The leaves gath-ered at apex of stems. The blades are lanceolate, dull dark green, without conspicuous longitudinal nervations, and about 15–30 cm long. The flowers are arranged in fascicles. The 6 perianth lobes are linear and greenish yellow. The 6 stamens are inserted at the throat of the perianth. The ovary produces a 3-lobed style. The berries are reddish orange and contain 3 seeds.

Medicinal use: Dysentery (Bangladesh)

Commentary: This plant has apparently not been examined for its antimicrobial activities. *Dracaena angustifolia* Roxb. is traditionally used to treat blennorrhea in Cambodia, Laos, and Vietnam.

### 3.1.6.2.7  *Ophiopogon japonicus* (L. f.) Ker Gawl.

Synonyms: *Anemarrhena cavaleriei* H. Lév.; *Convallaria japonica* L. f.; *Flueggea japonica* (L. f.) Rich.; *Mondo japonicum* (L. f.) Farw.; *Mondo stolonifer* (H. Lév. & Vaniot) Farw.; *Ophiopogon argyi* H. Lév.; *Ophiopogon chekiangensis* Koiti Kimura & Migo; *Ophiopogon stolonifer* H. Lév. & Vaniot; *Slateria japonica* (L. f.) Desv.

Common names: Dwarf liliturf, mondo grass; mai dong (China); janohige (Japan)

Habitat: Forests, rocky and moist soils, cultivated

Distribution: China, Korea, Japan, Pacific

Botanical observation: It is an ornamental, grass-like plant that grows from tubers. The stems are stoloniferous. The leaves are simple, basal, and spiral. The blade is linear 10–50 cm×2–4 mm, serrulate, and a dull dark green. The panicle is at apex of a 15 cm tall peduncle. The 6 tepals are pure white, lanceolate, and about 5 mm long. The androecium comprises 6 stamens. The gynoecium includes a conical style which is white and about 4 mm long. The berries are blue, globose, glossy, and up to about 8 mm across. Except for the inflorescence the plant has some resemblance with *Liriope graminifolia* (L.) Baker described earlier.

Medicinal use: Tuberculosis (China)

Commentaries:

(i) The tubers contain series of homoisoflavonoids (Anh et al., 2003), borneol glycosides, the spirostane steroid ophiogenin, and spirostane steroidal saponins (Adinolfi et al., 1990; Chen et al., 2016). The plant produces immunostimulating fructans (Wu et al., 2006).

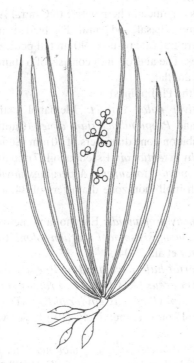

*Ophiopogon japonicus* (L. f.) Ker Gawl.

(ii) The plant is used in China for the treatment of COVID-19 (Luo, 2020; Wang et al., 2020).

**References**

Adinolfi, M., Parrilli, M. and Zhu, Y., 1990. Terpenoid glycosides from *Ophiopogon japonicus* roots. *Phytochemistry, 29*(5), pp. 1696–1699.

Anh, N.T.H., Van Sung, T., Porzel, A., Franke, K. and Wessjohann, L.A., 2003. Homoisoflavonoids from *Ophiopogon japonicus* Ker-Gawler. *Phytochemistry, 62*(7), pp. 1153–1158.

Chen, M.H., Chen, X.J., Wang, M., Lin, L.G. and Wang, Y.T., 2016. Ophiopogon japonicus—a phytochemical, ethnomedicinal and pharmacological review. *Journal of Ethnopharmacology, 181*, pp. 193–213.

Luo, A., 2020. Positive SARS-Cov-2 test in a woman with COVID-19 at 22 days after hospital discharge: a case report. *Journal of Traditional Chinese Medical Sciences, 7*(4), pp. 413–417.

Wang, S.X., Wang, Y., Lu, Y.B., Li, J.Y., Song, Y.J., Nyamgerelt, M. and Wang, X.X., 2020. Diagnosis and treatment of novel coronavirus pneumonia based on the theory of traditional Chinese medicine. *Journal of Integrative Medicine, 18*(4), pp. 275–283.

Wu, X., Dai, H., Huang, L., Gao, X., Tsim, K.W. and Tu, P., 2006. A fructan, from Radix ophiopogonis, stimulates the proliferation of cultured lymphocytes: structural and functional analyses. *Journal of Natural Products, 69*(9), pp. 1257–1260.

### 3.1.6.2.8   *Polianthes tuberosa* L.

Synonym: *Crinum angustifolium* Houtt.

Common names: Tuberose; ye lai xiang (China); rajnigandha (India); sundal malam (Indonesia, Malaysia) nardo (the Philippines)

Habitat: Cultivated

Distribution: Asia and Pacific

Botanical observation: This ornamental herb native to Central America grows up about 1 m tall from a bulb. The leaves are simple, basal, and spiral. The blade is up to 30–50 cm long, linear, and reddish at base. The racemes are fragrant on up to 90 cm tall pedicel. The perianth is tubular, whitish cream, and develops 6 lobes. The androecium consists of 6 stamens. The ovary is 3-locular and develops 3 stigmas. The seeds are flat.

Medicinal use: Gonorrhea (the Philippines)

*Antibacterial (Gram-positive) polar extract:* Methanol extract inhibited the growth of *Staphylococcus epidermidis* and *Propionibacterium acnes* (6 mm paper disc impregnated with 250 mg/mL solution) with inhibition zone diameters of 10 mm (Setiani et al., 2020).

*Antifungal halo developed by essential oil:* Essential oil (7 mm diameter wells containing 40 μL) inhibited the growth of *Trichophyton tonsurans, Aspergillus flavus, Epidermophyton floccosum,* and *Microsporum nanum* with inhibition zone diameters of 34, 15, 20, and 22 mm, respectively (Singh et al., 2011).

*Antifungal (filamentous) phenylpropanoids:* Eugenol and methyl eugenol inhibited the growth of *Drechslera sorokiniana, Phomompsis sojae, Fusarium solani, Colletotrichum graminicola,* and *Macrophomina phaseolina* (Dev et al., 2004).

*Broad-spectrum antibacterial halo developed by linear monoterpene:* Geraniol inhibited the growth of *Staphylococcus aureus* (ATCC 9144), *Proteus vulgaris* (ATCC 13315), *Bacillus cereus* (NCIMB 6349), and *Staphylococcus epidermidis* (ATCC 12228) with inhibition zone diameters of 16, 15, 13, and 17 mm, respectively (10 μL per well; 4 mm well) (Lis-Balchin et al., 1998).

*Moderate broad-spectrum antibacterial linear monoterpene:* Geraniol inhibited the growth of *Bacillus subtilis* (ATCC 6633), *Bacillus cereus* (ATCC 11778), *Staphylococcus aureus* (ATCC 6538), *Staphylococcus aureus* (ATCC 29213), and *Escherichia coli* (ATCC 35218) with MIC values of 80, 700, 700, 0.7, and 1400 μg/mL, respectively (Rosato et al., 2007).

*Strong broad-spectrum antibacterial amphiphilic phenylpropanoid:* Eugenol (LogD = 2.4 at pH 7.4; molecular weight = 164.2 g/mol) inhibited the growth of *Staphylococcus aureus, Bacillus cereus, Listeria monocytogenes, Proteus mirabilis, Escherichia coli, Klebsiella pneumoniae,* and *Pseudomonas aeruginosa* with MIC values of 31.2, 15.6, 15.6, 31.2, 31.2, 15.6, and >500 µg/mL, respectively (Mohammed et al., 2009).

Commentary: Essential oil in flowers comprises benzyl acetate, geraniol, methyl anthranilate, benzyl benzoate, methyl eugenol, eugenol, and methyl isoeugenol (Chandravadana et al., 1994), which may all work synergistically. The plant produces spirostane and furostane steroidal saponins (Jin et al., 2004). As described earlier, spirsotane-type steroidal saponins are not uncommonly antimicrobial.

### References
Chandravadana, M.V., Srinivas, M. and Murthy, N., 1994. Indole in tuberose (*Polianthes tuberosa*) varieties. *Journal of Essential Oil Research, 6*(6), pp. 653–655.

Dev, U., Devakumar, C., Mohan, J. and Agarwal, P.C., 2004. Antifungal activity of aroma chemicals against seed-borne fungi. *Journal of Essential Oil Research, 16*(5), pp. 496–499.

Jin, J.M., Zhang, Y.J. and Yang. C.R., 2004. Spirostanol and furostanol glycosides from the fresh tubers of *Polianthes tuberosa. Journal of Natural Products, 67,* pp. 5–9.

Lis-Balchin, M., Buchbauer, G., Ribisch, K. and Wenger, M.T., 1998. Comparative antibacterial effects of novel Pelargonium essential oils and solvent extracts. *Letters in Applied Microbiology, 27*(3), pp. 135–141.

Mohammed, M.J. and Al-Bayati, F.A., 2009. Isolation and identification of antibacterial compounds from *Thymus kotschyanus* aerial parts and *Dianthus caryophyllus* flower buds. *Phytomedicine, 16*(6), pp. 632–637.

Rosato, A., Vitali, C., De Laurentis, N., Armenise, D. and Milillo, M.A., 2007. Antibacterial effect of some essential oils administered alone or in combination with Norfloxacin. *Phytomedicine, 14*(11), pp. 727–732.

Setiani, N.A., Aulifa, D.L. and Septiningsih, E., 2020. Phytochemical Screening and Antibacterial Activity of Flower, Stem, and Tuber of *Polianthes tuberosa* L. against Acne-Inducing Bacteria. *2nd Bakti Tunas Husada-Health Science International Conference (BTH-HSIC 2019)* (pp. 92–95). Atlantis Press.

Singh, P.R.A.B.A.L., Bundiwale, R.U.C.H.I. and Dwivedi, L.K., 2011. In-vitro study of antifungal activity of various commercially available itra (Volatile plant oil) against the keratinophilic fungi isolated from soil. *International Journal of Pharma and Bio Sciences, 2*(3), pp. 178–184.

### 3.1.6.2.9   *Polygonatum officinale* All.
Synonyms: *Polygonatum odoratum* (Mill.) Druce; *Polygonatum japonicum* C. Morren & Decne.

Common names: Solomon's seal; yu zhu (China); gak si dung gul re (Korea)

Habitat: Forests

Ditribution: China, Korea, and Japan

Botanical observation: This herb grows up to about 50cm tall from a somewhat ginger-like rhizome. The stem is terete, smooth, somewhat bending and reddish, and inconspicuously zigzag shaped. The leaves are simple, sessile, sheathing at base, and alternate. The blade is elliptic, glaucous below, elliptic 5–15×3–7cm, obtuse to acuminate at apex, with about up to 10 longitudinal nervations, and wavy. The inflorescence is axillary, pendulous from a slender pedicel, and solitary. The perianth is tubular (somewhat Ericaceous), greenish white, elongated, up to about 2cm long, and developing 6 lobes, which are about 3mm long. The 6 stamens have 4mm long anthers and inserted at perianth. The ovary is globose, 4mm long, and develops a slender style, which is up to about 1.5cm long. The berry is globose, dark blue (blue berry-like), and about 1cm across.

Medicinal uses: Abscesses (Armenia); tuberculosis (China, Japan); bronchitis (Russia); Influenza, infection of the breast, shingles (China)

*Broad-spectrum Antibacterial halo developed by mid-polar extract:* Chloroform extract of rhizome (8 mm wells, 50 µg/well) inhibited the growth of *Pseudomonas aeruginosa*, *Bacillus subtilis*, *Salmonella typhimurium*, and *Staphylococcus aureus* with inhibition zone diameters of 23, 12, 10, and 13 mm, respectively (Ginovyan & Trchounian, 2017).

*Antibacterial halo developed by polar extract:* Ethanol extract of rhizome (400 µg/ 6 mm disc) inhibited the growth of *Staphylococcus aureus*, *Bacillus subtilis*, *Mycobacterium smegmatis*, *Shigella sonnei*, and *Shigella flexneri* with inhibition zone diameters below 10 mm (Moskalenko, 1986).

*Antibacterial homoisoflavones:* 3-(4'-methoxy-benzyl)-5,7-dihydroxy-6-methyl-8-methoxy-chroman-4-one and 3-(4'-hydroxy-benzyl)-5,7-dihydroxy-6,8-dimethyl-chroman-4-one exhibited antibacterial effects (Wang et al., 2009).

*Antibacterial triterpene:* 9,19-cyclolart-25-en-3$\beta$, 24 (*R*)-diol exhibited antibacterial effects (Wang et al., 2009).

*Antifungal homoisoflavones:* 3-(4'-methoxy-benzyl)-5,7-dihydroxy-6-methyl-8-methoxy-chroman-4-one and 3-(4'-hydroxy-benzyl)-5,7-dihydroxy-6,8-dimethyl-chroman-4-one exhibited antibacterial effects (Wang et al., 2009).

*Anticandidal halo (yeast) developed by mid-polar extract:* Chloroform extract of rhizome (8 mm wells, 50 µg/well) inhibited the growth of *Candida guilliermondii* with an inhibition zone diameter of 9 mm (Ginovyan & Trchounian, 2017).

*Strong broad-spectrum antifungal (yeast) spirostane-type saponin:* 3-*O*-$\beta$-D-glucopyranosyl-(1 → 2)-[$\beta$-D-xylopyranosyl-(1 → 3)]-$\beta$-D-glucopyranosyl (1 → 4)-$\beta$-D-galacopyranosyl-yamogenin inhibited the growth of *Candida albicans* (JCM 1542) and *Aspergillus fumigatus* (JCM 1738) with MIC the values of 3.1 and 6.3 µg/mL, respectively (Bai et al., 2014).

*Antifungal triterpene:* 9,19-cyclolart-25-en-3$\beta$, 24 (*R*)-diol exhibited antifungal effects (Wang et al., 2009).

Commentaries:

(i) *Polygonatum cirrhifolium* (Wall.) Royle, *Polygonatum erythrocarpum* Hua, *Polygonatum filipes* Merr. ex C. Jeffrey & McEwan, *Polygonatum kingianum* Collett & Hemsl., *Polygonatum multiflorum* (L.) All., and *Polygonatum chinense* Kunth are used in China for the treatment of tuberculosis. Apparently, no antimycobacterial principle has been isolated yet from these plants, including *Polygonatum officinale*. All, but indolizinone alkaloids and/or homoisoflavones could be involved.

(ii) Members of the genus *Polygonatum* Mill. often generate indolizinone alkaloids (Wang et al., 2003) and such compounds are known for their antimycobacterial effects (Dannhardt et al., 1987) via interaction with RNA and DNA (Ou et al., 2007). The homoisoflavonone 3-(4'-methoxybenzyl)-7,8-methylenedioxy-chroman-4-one isolated from *Chlorophytum inornatum* Ker Gawl. (family Asparagaceae) inhibited the growth of *Mycobacteria* with MIC ranging from 16–256 µg/mL (Lin et al., 2014).

(ii) The plant is used in China for the treatment of COVID-19 (Luo, 2020). Note that plants with the ability to inhibit RNA or DNA polymerase are often active against Severe acquired respiratory syndrome-associated coronavirus.

### References

Bai, H., Li, W., Zhao, H., Anzai, Y., Li, H., Guo, H., Kato, F. and Koike, K., 2014. Isolation and structural elucidation of novel cholestane glycosides and spirostane saponins from *Polygonatum odoratum*. *Steroids, 80*, pp. 7–14.

Dannhardt, G., Meindl, W., Gussmann, S., Ajili, S. and Kappe, T., 1987. Anti-mycobacterial 7-hydroxy-2, 3-dihydro-1H-indolizin-5-ones. *European Journal of Medicinal Chemistry*, 22(6), pp. 505–510.

Ginovyan, M.M. and Trchounian, A.H., 2017. Screening of some plant materials used in arme-
nian traditional medicine for their antimicrobial activity. *Chemical Biology*, *51*, pp. 44–53.

Lin, L.G., Liu, Q.Y. and Ye, Y., 2014. Naturally occurring homoisoflavonoids and their pharmacological activities. *Planta Medica*, *80*(13), pp. 1053–1066.

Luo, A., 2020. Positive SARS-Cov-2 test in a woman with COVID-19 at 22 days after hospital
discharge: a case report. *Journal of Traditional Chinese Medical Sciences*, *7*(4), pp. 413–417.

Min, B.S., Kim, Y.H., Tomiyama, M., Nakamura, N., Miyashiro, H., Otake, T. and Hattori,
M., 2001. Inhibitory effects of Korean plants on HIV-1 activities. *Phytotherapy Research*,
*15*(6), pp. 481–486.

Moskalenko, S.A., 1986. Preliminary screening of far-eastern ethnomedicinal plants for antibacterial activity. *Journal of Ethnopharmacology*, *15*(3), pp. 231–259.

Ou, T.M., Lu, Y.J., Zhang, C., Huang, Z.S., Wang, X.D., Tan, J.H., Chen, Y., Ma, D.L., Wong,
K.Y., Tang, J.C.O. and Chan, A.S.C., 2007. Stabilization of G-quadruplex DNA and down-
regulation of oncogene c-myc by quindoline derivatives. *Journal of Medicinal Chemistry*,
*50*(7), pp. 1465–1474.

Wang, D., Li, D., Zhu, W. and Peng, P., 2009. A new C-methylated homoisoflavanone and tri-
terpenoid from the rhizomes of *Polygonatum odoratum*. *Natural Product Research*, *23*(6),
pp. 580–589.

Wang, Y.F., Lu, C.H., Lai, G.F., Cao, J.X. and Luo, S.D., 2003. A new indolizinone from
*Polygonatum kingianum*. *Planta Medica*, *69*(11), pp. 1066–1068.

### 3.1.6.2.10   *Smilacina japonica* A. Gray

Synonyms: *Maianthemum japonicum* (A. Gray) La Frankie; *Tovaria japonica* (A. Gray) Baker;
*Vagnera japonica* (A. Gray) Makino

Common names: Japanese false Solomon's seal; lu yao, pian tou qi (China)

Habitat: Shady and moist places in forests, cultivated

Distributions: China, Korea, and Japan

Botanical observation: It is an ornamental herb, which grows up to about 60 cm tall from a rhizome. The stem is terete, somewhat light purplish, slightly zigzag, and hairy. The leaves are simple,
sessile, and alternate. The blade is ovate-elliptic, edible, glossy, 6–15×3–7 cm, wavy, with about 5
longitudinal nerves sunken above, and acute to acuminate at apex. The panicles are terminal and
many flowered. The 5 tepals are connate at base, recurved, oblong, and up to about 5 mm long. The
androecium consists of 6 stamens. The ovary is minute and develops a columnar style. The berries
are red and up to 6 mm across and globose.

Medicinal uses: Boils, mastitis (China)

*Moderate antifungal (yeast) polar extract:* Ethanol extract of rhizome inhibited the growth of
fluconazole-resistant *Candida albicans* (103), *Candida albicans* (SC5314), *Candida parapsilosis*
(ATCC 22019), *Cryptococcus neoformans* (32609), *Candida krusei* (ATCC 2340), *Candida glabrata* (ATCC 1182), and *Candida tropicalis* (2718) with the MIC values of 416, 416, 416, 208, 1665,
416, and 208 μg/mL, respectively (Liu et al., 2019).

*Antifungal drug synergistic polar extract:* Ethanol extract of rhizome increased the susceptibility
of *Candida krusei* (ATCC 2340) and *Cryptococcus neoformans* (32609) to fluconazole (Liu et al.,
2019).

*Antiviral (non-enveloped linear monopartite (+) RNA virus) flavanonol:* Dihydroquercetin at a
concentration of 100 μg/mL inhibited the replication of Coxsackievirus B4 virus by acting mainly
at an early stage of the virus life cycle (Bakay et al., 1968).

*In vivo antiviral (non-enveloped monopartite (+) RNA virus) amphiphilic flavonol:*
Dihydroquercetin (also known as taxifolin; LogD = 1.1 at pH 7.4; molecular weight = 304.5 g/mol)
administered intraperitoneally at a dose of 150 mg/Kg/day for 5 days reduced Coxsackievirus B4
virus pancreatic titer (Galochkina et al., 2016).

Dihydroquercetin

*Antiviral (enveloped segmented linear single-stranded (–) RNA virus)*: Dihydroquercetin given orally protected mice against Influenza virus A/Aichi/2/68 (H3N2) (Zarubaev et al., 2010).

*Antiviral (enveloped monopartite linear double-stranded DNA) flavonol:* Dihydroquercetin was virucidal against Herpes simplex virus type-1 (Beladi et al., 1965).

Commentary: The plant produces series of spirostane-type steroidal saponins (Liu et al., 2012) (which are cytotoxic and therefore most probably toxic for yeasts), as well as the flavonols dihydroquercetin and 6,8-di-C-methylquercetin-3-methyl ether and the simple phenolic fareanol (Cui et al., 2019).

### References

Bakay, M., Mucsi, I., Beladi, I. and Gabor, M., 1968. Effect of flavonoids and related substances. II. Antiviral effect of quercetin, dihydroquercetin and dihydrofisetin. *Acta Microbiologica Academiae Scientiarum Hungaricae, 15*(3), p. 223.

Beladi, I., Pusztai, R. and Bakai, M., 1965. Inhibitory activity of tannic acid and flavonols on the infectivity of herpesvirus hominis and herpesvirus suis. *Die Naturwissenschaften, 52*, p. 402.

Cui, Y., Li, Y., Fan, H., Huang, W., Zhang, H., Zhang, D., Song, X. and Qin, B., 2019. Chemical constituents isolated from the roots and rhizomes of *Smilacina japonica*. *Biochemical Systematics and Habitat, 86*, p. 103920.

Galochkina, A.V., Anikin, V.B., Babkin, V.A., Ostrouhova, L.A. and Zarubaev, V.V., 2016. Virus-inhibiting activity of dihydroquercetin, a flavonoid from *Larix sibirica*, against coxsackievirus B4 in a model of viral pancreatitis. *Archives of Virology, 161*(4), pp. 929–938.

Liu, W., Sun, B., Yang, M., Zhang, Z., Zhang, X., Pang, T. and Wang, S., 2019. Antifungal activity of crude extract from the rhizome and root of *Smilacina japonica* A. Gray. *Evidence-Based Complementary and Alternative Medicine*.

Liu, X., Zhang, H., Niu, X.F., Xin, W. and Qi, L., 2012. Steroidal saponins from *Smilacina japonica*. *Fitoterapia, 83*(4), pp. 812–816.

Zarubaev, V., Garshinina, A., Kalinina, N., Anikin, V., Babkin, V., Ostroukhova, L. and Kiselev, O., 2010. Anti-influenza activity of dihydroquercetin against lethal influenza virus infection. *Antiviral Research, 1*(86), p. A50.

### 3.1.6.3　Family Hypoxidaceae R. Brown (1814)

The family Hypoxidaceae consists of 9 genera and about 120 species of herbs growing from rhizomes or corms. The leaves are simple, spiral, and basal. The inflorescences are solitary or racemes, spikes, or umbels. The perianth consists of 6 tepals. The androecium includes 6 stamens. The gynoecium includes 3 carpels fused into a 3-locular ovary, each locule containing numerous ovules growing from axil placentas. The fruits are fleshy capsules containing numerous seeds. Plants in this family are known so far to produce phenolic glycosides, triterpene, saponins, and long chain aliphatic ketones.

### 3.1.6.3.1   *Curculigo disticha* Gagnep.

Habitat: Forests
   Distribution: Cambodia, Laos, and Vietnam
   Botanical observation: It is a herb that grows up to 80 cm tall from fibrous roots. The leaves are simple and basal. The petiole is slender and 15–20 cm long. The blade is lanceolate, with numerous and conspicuous longitudinal nervations, tapering at base, acuminate to caudate at apex, and about 30–40×4.5–6 cm. The flowers are basal and arranged in 4 cm long fascicles at apex of up to about 3 cm tall and hairy peduncles. The perianth comprises 6 bright yellow oblong lobes, which are ovate, 3.5 mm long, hairy below, and obtuse at apex. The androecium includes 6 stamens, which are subsessile. The gynoecium comprises a 6 mm long ovary, which develops a beak, a slender style, and a punctiform stigma. The berry is beaked and contain numerous tiny seeds.
   Medicinal uses: Sore throat (Cambodia, Laos, Vietnam)
   Commentary: Apparently the plant has not been examined for antimicrobial effects. Consider that the use of this plant by the traditional systems of medicine of 3 different countries point to the presence of strong antibacterial or antiviral principles.

### 3.1.6.3.2   *Curculigo latifolia* Dryand. ex W.T. Aiton

Synonym: *Molineria latifolia* (Dryand. ex W.T. Aiton) Kurz
   Common name: Lembah (Indonesia, Malaysia)
   Habitat: Forests
   Distribution: Bangladesh, Myanmar, Cambodia, Laos, Vietnam, Thailand, Malaysia, Indonesia, and the Philippines
   Botanical observation: It is a herb that grows up to 1 m tall from fibrous roots. The leaves are simple and basal. The petiole is slender. The blade is elliptic, with numerous and conspicuous longitudinal nervations, tapering at base, acuminate to caudate at apex, and about 30–100×4.5–10 cm. The flowers are arranged in fascicles, which are up to about 6 cm long at apex of up to 10 cm tall peduncle. The perianth comprises 6 bright yellow oblong lobes. The berry is beaked, up to about 2.5 cm long, and contain numerous tiny seeds.
   Medicinal use: Fever (Malaysia)
   *Broad-spectrum antibacterial halo developed by polar extract:* Methanol extract of root inhibited the growth of *Bacillus cereus, Bacillus subtilis, Enterobacter aerogenes, Erwinia* sp., *Klebsiella* sp., *Pseudomonas* sp., and *Staphylococcus aureus* (Hong & Ibrahim, 2012).
   *Anticandidal halo developed by polar extract:* Methanol extract of root inhibited the growth of *Candida albicans* and *Cryptococcus neoformans* (Hong & Ibrahim, 2012).
   Commentaries: (i) Extracts are not uncommonly specifically toxic for yeasts and inactive against filamentous fungi or the opposite. This specificity could be attributed to the presence of specific targets or permeability differences of yeast and filamentous fungi to antifungal agents. (ii) In Indonesia, some unidentified members of the genus *Curculigo* Gaertn. are used to treat fever.

### Reference

Hong, L.S. and Ibrahim, D., 2012. Studies on Antibiotic Compounds of Methanol Extract of *Curculigo latifolia* Dryand. *Proceedings of the Annual International Conference, Syiah Kuala University-Life Sciences & Engineering Chapter* (Vol. 2, No. 1).

*3.1.6.3.3    Curculigo orchioides* Gaertn.

Synonyms: *Curculigo brevifolia* W.T. Aiton; *Curculigo malabarica* Wight; *Hypoxis orchioides* (Gaertn.) Kurz

Common names: Black musli; xian mao (China); kalimusli (Pakistan); kali musali (India); lemba (Malaysia); kalo musali, dhusilo (Nepal); kenemote (Papua New Guinea); tataluangi (The Philippines); waan phraao (Thailand); ngai cau (Vietnam)

Habitat: Forests

Distribution: Pakistan, India, Nepal, Bangladesh, Myanmar, Cambodia, Laos, Thailand, Vietnam, Malaysia, Indonesia, the Philippines, Japan, and Papua New Guinea

Botanical observation: It is a herb which grows up to 50 cm tall from a rhizome. The leaves are simple and basal. The blade is narrowly lanceolate, with numerous and conspicuous longitudinal nervations, tapering at base, acuminate at apex, and about 10–45×0.5–3 cm. The flowers are arranged in fascicles at apex of up to about 7 cm tall on hairy peduncles. The plant is sacred in India. The perianth comprises 6 bright yellow oblong lobes, which are ovate, 1.2 long, hairy below, and obtuse at apex. The androecium includes 6 stamens, which are about 7 mm long and as yellow as the perianth. The gynoecium comprises an 8 mm long ovary, which is oblong, and a stigma with slender lobes. The berry is elongated, up to about 1.5 cm long, beaked, and containing numerous tiny seeds.

Medicinal use: Blennorrhea, fever (China); diarrhea, dysentery (Nepal); gonorrhea (India); jaundice (India, Nepal); skin diseases (the Philippines)

*Broad-spectrum antibacterial essential oil:* Essential oil of rhizomes (2 mg/4 mm paper disc) inhibited the growth of *Staphylococcus aureus* (ATCC 6571), *Staphylococcus epidermidis* (clinical isolate), *Escherichia coli* (ATCC 10418), *Salmonella typhimurium*, and *Pseudomonas aeruginosa* (ATCC 10662) with inhibition zone diameters of 13.6, 18.3, 9.6, 13.6, and 10.3 mm, respectively (Nagesh & Shanthamma, 2009).

*Broad-spectrum antibacterial non-polar extract:* Petroleum ether extract of rhizomes (4 mm paper disc impregnated with solution at a concentration of 0.2 mg/0.1 mL) inhibited the growth of *Staphylococcus aureus*, *Staphylococcus albus*, *Bacillus subtilis*, *Bacillus anthracis*, *Klebsiella* spp., *Salmonella newport*, *Proteus vulgaris*, *Escherichia coli*, and *Pseudomonas aeruginosa* with inhibition zone diameters of 9, 8, 9, 11, 7, 9, 8, 8, and 7 mm, respectively (Jaiswal et al., 1984).

*Strong antibacterial (Gram-positive) polar extract:* Ethanol extract of rhizomes inhibited the growth of *Streptococcus pyogenes* with MIC value of 49 μg/mL (Marasini et al., 2015).

*Antifungal (filamentous) non-polar extract:* Petroleum ether extract of rhizomes (4 mm paper disc impregnated with solution at a concentration of 0.2 mg/0.1 mL) inhibited the growth of *Aspergillus niger*, *Aspergillus flavus*, *Fusarium solani*, *Fusarium moniliforme*, *Cladosporium* sp. and *Pericenia* spp. with inhibition zone diameters of 8, 14, 14, 16, 14, and 8 mm, respectively (Jaiswal et al., 1984).

*Antiviral (enveloped double stranded circular DNA) phenolic glycoside:* Curculigoside F isolated from the rhizome inhibited Hepatitis B virus e antigen secretion by HepG2.2.15 cell line with an $IC_{50}$ value of 2 mM (Zuo et al., 2010).

Commentary: (i) The plant has immunostimulant properties (Lakshmi et al., 2003). It produces cycloartane-type triterpenes and saponins (Xu & Xu, 1992), glucosides of syringic acid, 2,6-dimethoxybenzoic acid, and orcinol (Dall'Acqua et al., 2009), as well as 27-hydroxytriacontan-6-one and 23-hydroxytriacontan-2-one (Misra et al., 1984). Note that long-chain aliphatic ketones are not uncommonly antibacterial. One further example is 16-hentriacontanone isolated from *Annona squamosa* L., which inhibited the growth of *Staphylococcus aureus*, *Escherichia coli*, and *Klebsiella* (Sharma, 1993).

The antiviral activity of phenolic glycosides in this plant could be examined.

**References**

Dall'Acqua, S., Shrestha, B.B., Comai, S., Innocenti, G., Gewali, M.B. and Jha, P.K., 2009. Two phenolic glycosides from *Curculigo orchioides* Gaertn. *Fitoterapia*, *80*(5), pp. 279–282.

Jaiswal, K., Batra, K.A. and Mehta, B.K., 1984. The antimicrobial efficiency of root oil against human pathogenic bacteria and phytopathogenic fungi. *Journal of Phytopathology*, *109*(1), pp. 90–93.

Lakshmi, V., Pandey, K., Puri, A., Saxena, R.P. and Saxena, K.C., 2003. Immunostimulant principles from *Curculigo orchioides*. *Journal of Ethnopharmacology*, *89*(2–3), pp. 181–184.

Marasini, B.P., Baral, P., Aryal, P., Ghimire, K.R., Neupane, S., Dahal, N., Singh, A., Ghimire, L. and Shrestha, K., 2015. Evaluation of antibacterial activity of some traditionally used medicinal plants against human pathogenic bacteria. *BioMed Research International*.

Misra, T.N., Singh, R.S., Upadhyay, J. and Tripathi, D.N.M., 1984. Aliphatic hydroxy-ketones from *Curculigo orchioides* rhizomes. *Phytochemistry*, *23*(8), pp. 1643–1645.

Nagesh, K.S. and Shanthamma, C., 2009. Antibacterial activity of *Curculigo orchioides* rhizome extract on pathogenic bacteria. *African Journal of Microbiology Research*, *3*(1), pp. 5–9.

Sharma, R.K., 1993. Phytosterols: wide-spectrum antibacterial agents. *Bioorganic Chemistry*, *21*(1), pp. 49–60.

Xu, J.P. and Xu, R.S., 1992. Cycloartane-type sapogenins and their glycosides from *Curculigo orchioides*. *Phytochemistry*, *31*(7), pp. 2455–2458.

Zuo, A.X., Shen, Y., Jiang, Z.Y., Zhang, X.M., Zhou, J., Lü, J. and Chen, J.J., 2010. Three new phenolic glycosides from *Curculigo orchioides* G. *Fitoterapia*, *81*(7), pp. 910–913.

### 3.1.6.4 Family Iridaceae A.L. de Jussieu (1789)

The family Iridaceae consists of about 90 genera and 1800 species of herbs growing from rhizomes, bulbs, or corms. The leaves are simple and alternate. The flowers are terminal racemes, panicles, or cymes. The perianth comprises 6 tepals, which are of equal length or not. The androecium includes 3 stamens. The gynoecium comprises a 3-locular ovary, each locule containing numerous ovules growing from axile placentas, and a slender style. The fruit is a trilobed capsule which is dehiscent and sheltering numerous seeds. The name of the family originates from the Greek *irida* = rainbow, on probable account of the beauty of the flowers of most members in this taxon.

#### 3.1.6.4.1 *Belamcanda chinensis* (L.) Redouté

Common names: Leopard flower; she gan (China)

Synonyms: *Belamcanda pampaninii* H. Lév.; *Belamcanda punctata* Moench; *Gemmingia chinensis* (L.) Kuntze; *Ixia chinensis* L.; *Pardanthus chinensis* (L.) Ker Gawl.

Habitat: Grassy lands, cultivated

Distribution: India, Nepal, Bhutan, Myanmar, Vietnam, China, Korea, and Japan

Botanical observation: This graceful herb grows up to a height of 1 m from a rhizome. The leaves are simple and alternate. The blade is ensiform, dull green, somewhat fleshy, 20–60 cm×2–4 cm. The inflorescence is a terminal cyme. The perianth is tubular and develops 6 lobes, which are spathulate to elliptic, dull orangish, and marked with dark spots at base; the 3 outer lobes are up to 2.5 cm long. The androecium comprises 6 stamens, which are up to about 2 cm long. The style is slender, about 2 cm long, and develops a 3-lobed stigma. The fruit is an elongated, up to 3 cm long capsule, containing numerous back and glossy seeds. The capsule dries up, opens, and the seeds are exposed in longitudinal grapes.

Medicinal uses: Boils, fever, tonsilitis (China); post-partum (Malaysia)

*Moderate broad-spectrum antibacterial isoflavone:* Tectorigenin (5,7,4′-trihydroxy-6-methoxy-isoflavone) isolated from the rhizome inhibited the growth of *Pseudomonas aeruginosa* (ATCC 9027), *Proteus vulgaris* (ATCC 3851), *Micrococcus luteus* (ATCC 9341), and *Staphylococcus aureus* (ATCC 6538) with MIC values of 100, 50, 100, and 50 µg/mL, respectively (Oh et al., 2001). In a subsequent experiment, tectorigenin inhibited the growth of *Staphylococcus aureus* (ATCC 25923) and *Staphylococcus aureus* (ATCC 33591) with an MIC value of 125 µg/mL (Joung et al., 2014).

*Broad-spectrum antifungal polar extract:* Aqueous extract inhibited the growth of *Trichophyton rubrum* (11788), *Trichophyton mentagrophytes* (11115), *Microsporium canis* (11883), *Epidermophyton floccosum* (10342), *Trichophyton schoenleinii* (0300), *Microsporium gypseum* (10079), *Trichophyton tonsurans* (11542), and *Trichophyton violaceum* (8001) with MIC values of 150, 400, 150, 250, 150, 300, 100, and 50 µg/mL, respectively (Yang et al., 2015).

*Strong broad-spectrum antifungal amphiphilic isoflavone:* Tectorigenin (LogD = 1.5 at pH 7.4; molecular weight = 300.2 g/mol) isolated from the rhizome inhibited the growth of *Candida albicans* (ATCC 10231), *Candida tropicalis* (ATCC 750), *Mucor* sp. (HIC 101), *Neurospora crassa* (HIC 5674), *Saccharomyces cerevisiae* (IFO 565), *Trichophyton mentagrophytes* (IFO 40996), and *Trichophyton rubrum* (IFO 9185) with MIC values of 25, 50, 50, 100, 25, 3.1, and 3.1 µg/mL, respectively (Oh et al., 2001).

Tectorigenin

*Antiviral (enveloped monopartite linear single-stranded (+) RNA) polar extract:* Ethanol extract of rhizome inhibited the replication of the Sindbis virus with an MIC value of 125 µg/mL (Yip et al., 1991).

*Viral enzyme inhibition by polar extract:* Aqueous extract of rhizomes at the concentration of 250 µg/mL inhibited Human immunodeficiency virus protease by 48.8% (Xu et al., 1996).

Commentary:

(i) The plant produces series of isoflavones: irisflorentin, tectorigenin, irilin D, tectoridin, iristectorin A, and iristectorin B, as well as resveratrol and the benzophenone iriflophenone (Monthakantirat et al., 2005).

(ii) The Sindbis virus is an RNA-enveloped virus belonging to the genus *Alphavirus* in the family Togaviridae (together with the Semliki forest virus and the Chikungunya virus). Note that compounds or extract active against Alphaviruses are not uncommonly active against Coronavirus and especially COVID-19. This is the case with *Belamcanda chinensis* (L.) Redouté, which is used in China for the treatment of COVID-19 (Yang et al., 2020).

(iii) Log P is equal to the ratio of concentrations of a compound between octanol and water. Hydrophilic compounds have low or negative values (about -3) (compounds are mainly found in the water phase). Amphiphilic compounds have a logP near to 0 (the compound is equally partitioned between the octanol and water layers). Non-hydrophilic, (hydrophobic, liposoluble) compounds have a high logP (up to about 7) (note that lipophilic substances tend to remain in the cytoplasmic membranes of bacteria and fungi and destabilize them). In Gram-negative bacteria, hydrophilic compounds of low molecular weight can cross the outer membrane via the aquaporins (porins). Non-polar compounds will tend to remain in the membranes, the core of which is strongly lipophilic. (iv) However, logP is only of value for non-ionizable principles and for ionized substance a logD is preferable but the pH must be fixed. In some database, one can find for the ionic alkaloid berberine a LogP of 3, which would suggest a lipophilic substance, which is not sensible (JF Weber, personal communication). Since compounds destined to pharmaceutical developments are mainly exposed to physiological pH, and are often weak bases or weak acids, we define at pH 7.4 hydrophilic compounds with a negative LogD value up to about 1, mid-polar (amphiphilic) compounds for Log D above 1 and below about 5, and lipophilic compounds for a LogD above about 5. Note that LogD values given here are predicted values.

(v) Here, low molecular weight molecules are defined with a molecular mass below 200 g/mol, medium molecular mass from 200 to 400 g/mol, and high mass weight above 400 g/mol.

### References
Joung, D.K., Mun, S.H., Lee, K.S., Kang, O.H., Choi, J.G., Kim, S.B., Gong, R., Chong, M.S., Kim, Y.C., Lee, D.S. and Shin, D.W., 2014. The antibacterial assay of tectorigenin with detergents or ATPase inhibitors against methicillin-resistant *Staphylococcus aureus*. *Evidence-Based Complementary and Alternative Medicine*.

Monthakantirat, O., De-Eknamkul, W., Umehara, K., Yoshinaga, Y., Miyase, T., Warashina, T. and Noguchi, H., 2005. Phenolic constituents of the rhizomes of the Thai medicinal plant *Belamcanda chinensis* with proliferative activity for two breast cancer cell lines. *Journal of Natural Products*, 68(3), pp. 361–364.

Oh, K.B., Kang, H. and Matsuoka, H., 2001. Detection of antifungal activity in *Belamcanda chinensis* by a single-cell bioassay method and isolation of its active compound, tectorigenin. *Bioscience, Biotechnology, and Biochemistry*, 65(4), pp. 939–942.

Xu, H.X., Wan, M., Loh, B.N., Kon, O.L., Chow, P.W. and Sim, K.Y., 1996. Screening of traditional medicines for their inhibitory activity against HIV -1 protease. *Phytotherapy Research*, 10(3), pp. 207–210.

Yang, F., Ding, S., Liu, W., Liu, J., Zhang, W., Zhao, Q. and Ma, X., 2015. Antifungal activity of 40 TCMs used individually and in combination for treatment of superficial fungal infections. *Journal of Ethnopharmacology*, 163, pp. 88–93.

Yang, Q.X., Zhao, T.H., Sun, C.Z., Wu, L.M., Dai, Q., Wang, S.D. and Tian, H., 2020. New thinking in the treatment of 2019 novel coronavirus pneumonia. *Complementary Therapies in Clinical Practice*, 39, p. 101131.

Yip, L., Pei, S., Hudson, J.B. and Towers, G.H.N., 1991. Screening of medicinal plants from Yunnan Province in southwest China for antiviral activity. *Journal of Ethnopharmacology*, 34(1), pp. 1–6.

### 3.1.6.4.2  *Crocus sativus* L

Common names: Saffron; jafran (Bangladesh); romiet (Cambodia); fan hong hua (China); kashmirajanman (India); safuran (Japan); kuma kuma (Indonesia, Malaysia); karkam, zafaran (Iran); yafaran (Thailand); zafera (Turkey); nghe tai (Vietnam)

Habitat: Cultivated

Distribution: China

Botanical observation: This magnificent herb grows up to 30 cm tall from a tunicate corm. The leaves are simple, grass-like, basal, and spiral. The blade is 15–20 cm×2–3 mm. The flowers are basal and solitary. The perianth is somewhat purple, tubular, and comprises 6 oblanceolate, up to 5 cm long lobes. The androecium comprises 3 stamens, which are up to 2.5 cm long with elongated and bright yellow anthers. The gynoecium presents a trifid and edible stigma, which is bright red, slender, and about 2.5 cm long.

Medicinal uses: Cough, measles (Iran)

*Moderate broad-spectrum antibacterial polar extracts:* Aqueous of petals inhibited the growth of *Staphylococcus aureus* and *Escherichia coli* with MIC/MBC values of 250/500 and 62.5/250 µg/mL, respectively (Ahmadi Shadmehri et al., 2019). Methanol extract of stigmas inhibited the growth of *Helicobacter pylori* with an MIC value of 677 µg/mL (Moghaddam, 2010).

*Strong antibacterial (Gram-negative) amphiphilic cyclic monoterpene aldehyde:* Safranal (LogD = 2.6 at pH 7.4; molecular weight = 150.2 g/mol) isolated from the stigmas inhibited the growth of *Helicobacter pylori* with an MIC value of 16.6 µg/mL (Moghaddam, 2010).

Safranal

*Moderate antibacterial amphiphilic linear diterpene glycoside:* Crocin isolated from the stigmas inhibited the growth of *Helicobacter pylori* with MIC/MBC values of 263/300 µg/mL (Moghaddam, 2010).

*Strong broad-spectrum antifungal (filamentous) mid-polar extract:* Ethyl ether fraction of stigma inhibited the growth of *Pyricularia oryzae*, *Candida albicans* (ATCC 76615), *Cryptococcus neoformans* (32609), and *Trichophyton rubrum* with MIC values of about 15.6, 64, 32, and 32 µg/mL, respectively (Zheng et al., 2011).

Commentary: Genoese and Venetian sailors imported saffron from the South Mediterranean and Rhodes as a remedy for the Black Plague in the fourteenth century (Yildirim et al., 2020). In the *Reggimento contra peste* of Pietro Castagno published in 1572, saffron is mentioned as one of the remedies against plague (Vicentini et al., 2020). The plant is recommended for COVID-19 in Unani medicine (Nikhat & Fazil, 2020).

### References

Ahmadi Shadmehri, A., Miri, H.R., Namvar, F., Nakhaei Moghaddam, M. and Yaghmaei, P., 2019. Cytotoxicity, antioxidant and antibacterial activities of crocus sativus petal extract. *International Journal of Research in Applied and Basic Medical Sciences*, 5(1), pp. 69–76.

Moghaddam, M.N., 2010. In vitro antibacterial activity of saffron (*Crocus sativus* L.) extract and its two major constituents against *Helicobacter pylori*. *Planta Medica*, 76(12), p. P496.

Nikhat, S. and Fazil, M., 2020. Overview of Covid-19; its prevention and management in the light of Unani medicine. *Science of the Total Environment*, p. 138859.

Vicentini, C.B., Manfredini, S., Mares, D., Bonacci, T., Scapoli, C., Chicca, M. and Pezzi, M., 2020. Empirical "integrated disease management" in Ferrara during the Italian plague (1629–1631). *Parasitology International*, 75, p. 102046.

Yildirim, M.U., Sarihan, E.O. and Khawar, K.M., 2020. Ethnomedicinal and traditional usage of saffron (*Crocus sativus* L.) in Turkey. In *Saffron* (pp. 21–31). Academic Press.

Zheng, C.J., Li, L., Ma, W.H., Han, T. and Qin, L.P., 2011. Chemical constituents and bio-activities of the liposoluble fraction from different medicinal parts of *Crocus sativus*. *Pharmaceutical Biology*, 49(7), pp. 756–763.

*3.1.6.4.3   Eleutherine americana* (Aubl.) Merr. ex K. Heyne

Synomyms: *Eleutherine bulbosa* (Mill.) Urb.; *Eleutherine palmifolia (L.)* Merr.; *Eleutherine plicata* Herb. ex Klatt; *Galatea americana* Kuntze; *Galatea bulbosa* (Mill.) Britton; *Ixia americana* Aubl.; *Marica plicata* Ker Gawl.; *Moraea plicata* Sw.; *Sisyrinchium americanum* (Aubl.) *Lemée*; *Sisyrinchium bulbosum* Mill.; *Sisyrinchium palmifolium* L.

Common names: American eleutherine, dayak onion; bawang kapal, bawang tiwai (Indonesia); bawang dayak (Malaysia); ponnulli, vishanarayani (India)

Habitat: Cultivated

Distribution: India, China, and Indonesia

Botanical observation: This beautiful herb is native to tropical America and grows up to a height of 60 cm from red elongated and edible (?) onion-like bulbs. The leaves are simple, spiral, and basal. The blade is narrowly elliptic, white at base, plicate, and 25–60×1–2.5 cm. The inflorescences are terminal cymes of ephemeral flowers. The perianth is pure white and develops 6 obovate lobes (somewhat vaguely like a white *Gardenia tubiflora* flower). The androecium comprises 3 stamens, which are sessile and yellowish orange with elongated anthers up to about 1 cm long. The gynoecium includes an ellipsoid ovary, which is minute. The capsule is globose, about 6 mm across, and contain numerous tiny and dark brown seeds.

Medicinal uses: Diarrhea (India); dysentery (Indonesia); flu (Thailand); boils, wounds (the Philippines)

*Moderate broad-spectrum antibacterial polar extract:* Ethanol extract of bulbs inhibited the growth of *Staphylococcus aureus* (ATCC 23235) and *Staphylococcus aureus* (ATCC 27664) via cytoplasmic membrane damages (Ifesan et al., 2009). Ethanol extract of bulbs inhibited the growth of *Bacillus cereus, Listeria monocytogenes, Staphylococcus aureus, Streptococcus mutans,* and *Streptococcus pyogenes* with MIC/MBC of 250/1000, 500/1000, 250/250, 125/500, and 250/500 μg/mL, respectively (Limsuwan et al., 2009). Ethanol extract of bulbs inhibited the growth of *Bacillus cereus* (10876), *Bacillus licheniformis* (12759), *Bacillus subtilis, Erwinia* sp., *Staphylococcus aureus* (12600), and *Streptococcus* sp. with MIC values of 125, 125, 250, 125, 250, and 250 μg/mL, respectively (Ifesan et al., 2010).

*Moderate antifungal (filamentous) polar extract:* Ethanol extract of bulbs of inhibited the growth of *Aspergillus niger, Penicillium* sp., and *Rhizopus* sp. with MIC values of 1000, 250, and 250 μg/mL, respectively (Ifesan et al., 2010).

*Antibacterial (Gram-positive) pyranonaphthoquinone:* Eleutherin inhibited the growth of *Staphylococcus aureus, Staphylococcus haemolyticus* A, and *Bacillus subtilis* (Schmid & Ebnöther, 1951).

*Strong antifungal (filamentous) amphiphilic pyranonaphthoquinones:* Eleutherin (LogD = 3.3 at pH 7.4; molecular weight = 272.2 g/mol) and isoeleutherin (LogD = 3.3; molecular weight = 272.2 g/mol) inhibited the growth of *Pyricularia oryzae* with the minimum morphological deformation concentration of 30 μg/mL (Insanu et al., 2014).

(+)-Eleutherin

*Moderate antifungal (filamentous) naphthalenes:* Hongconin and dihydroeleutherinol inhibited the growth of *Pyricularia oryzae* with the MIC values of 130 and 60 µg/mL (Insanu et al., 2014).

*Moderate antifungal (filamentous) anthraquinone:* 1,3,6-Trihydroxy-8-methyl-anthraquinone inhibited the growth of *Pyricularia oryzae* with the MIC value of 150 µg/mL (Insanu et al., 2014).

*Strong antiviral (enveloped monopartite linear dimeric single-stranded (+) RNA) pyranonaphthoquinone:* Isoeleutherin (also known as (-)-eleutherin) inhibited the replication of Human immunodeficiency virus with an $IC_{50}$ value of 8.5 µg/mL (Insanu et al., 2014).

*Moderate antiviral (enveloped monopartite linear dimeric single-stranded (+) RNA) naphthalene:* Isoeleutherol inhibited the replication of Human immunodeficiency virus with an $IC_{50}$ value of 100 µg/mL (Insanu et al., 2014).

Commentary: Eleutherin is a DNA topoisomerase II inhibitor (Bachu et al., 2008). A closely related pyronaphthoquinone to eleutherin and isoeleutherin is thysanone, isolated from the fungi *Thysanophora penicilloides*, which inhibited Human rhinovirus 3C-protease (Bachu et al., 2008) with an $IC_{50}$ value 47 µM (Jeong et al., 2014). The Human rhinovirus belongs to the genus *Enterovirus* in the family Picornaviridae. It is a single-stranded-RNA and non-enveloped virus responsible for sore throat, cough, headache, subjective fevers, and malaise that can evolve into bronchitis and pneumonia (Jacobs et al., 2013). One could examine the effect of eleutherin and congeners in this plant against Severe acute respiratory syndrome-associated coronavirus 3C-like protease. Compounds intercalating into DNA (toposisomerase inhibitors and others) are often heterocyclic and planar and antiviral.

### References

Bachu, P., Sperry, J. and Brimble, M.A., 2008. Chemoenzymatic synthesis of deoxy analogues of the DNA topoisomerase II inhibitor eleutherin and the 3C-protease inhibitor thysanone. *Tetrahedron, 64*(21), pp. 4827–4834.

Ifesan, B.O., Ibrahim, D. and Voravuthikunchai, S.P., 2010. Antimicrobial activity of crude ethanolic extract from *Eleutherine americana. Journal of Food, Agriculture & Environment, 8*(3&4), p. 1233.

Ifesan, B.O.T., Joycharat, N. and Voravuthikunchai, S.P., 2009. The mode of antistaphylococcal action of *Eleutherine americana. FEMS Immunology & Medical Microbiology, 57*(2), pp. 193–201.

Insanu, M., Kusmardiyani, S. and Hartati, R., 2014. Recent studies on phytochemicals and pharmacological effects of *Eleutherine americana* Merr. *Procedia Chemistry, 13*, pp. 221–228.

Jacobs, S.E., Lamson, D.M., George, K.S. and Walsh, T.J., 2013. Human rhinoviruses. *Clinical Microbiology Reviews, 26*(1), pp. 135–162.

Jeong, J.Y., Sperry, J., Taylor, J.A. and Brimble, M.A., 2014. Synthesis and evaluation of 9-deoxy analogues of (–)-thysanone, an inhibitor of HRV 3C protease. *European Journal of Medicinal Chemistry, 87*, pp. 220–227.

Limsuwan, S., Subhadhirasakul, S. and Voravuthikunchai, S.P., 2009. Medicinal plants with significant activity against important pathogenic bacteria. *Pharmaceutical Biology, 47*(8), pp. 683–689.

Schmid, H. and Ebnöther, A., 1951. Über die Konfiguration der Eleutherine-Chinone (Inhaltsstoffe aus *Eleutherine bulbosa* (Mill.) Urb. V). *Helvetica Chimica Acta, 34*(4), pp. 1041–1049.

### 3.1.6.4.4   *Iris ensata* Thunb.

Synonyms: *Iris kaempferi* Siebold ex Lem.; *Limniris ensata* (Thunb.) Rodion.

Common names: Japanese iris; yu chan hua (China); Thema (Himalaya); Roghan e Sosan (India)

Habitat: Shady and moist places by lakes, riverbanks

Distribution: China, Korea, and Japan

Botanical observation: This magnificent herb grows from a rhizome. The leaves are simple, basal, and alternate. The blade is linear, ensiform, 30–80 cm×0.5–1.2 cm, and acuminate at apex. The inflorescence is terminal, solitary, and on a peduncle, which is up to 1 m tall. The perianth tube is up to 2 cm long, 3 outer lobes obovate, up to 8 cm long, and the 3 inner lobes erect, narrowly lanceolate, and up to 5 cm long. The androecium consists of 3 stamens, which are up to 3.5 cm long, with flat filaments, opposed to style branches and appressed to them. The gynoecium comprises a cylindrical ovary, which is 2 cm long and develops purple branches. The capsule is elongated, about 5 cm long, and contain numerous angled seeds.

Medicinal uses: Abscesses (Korea); fever, wounds, ulcers (China); laryngitis (Japan)

*Strong antiviral (enveloped monopartite linear dimeric single-stranded (+)RNA) hydrophilic xanthone C-glycoside:* Mangiferin (LogD = −1.7 at pH 7.4; molecular weight = 422.3 g/mol) has the ability to inhibit the replication of Human immunodeficiency virus type-1 (strain IIIB) and Human immunodeficiency virus type-1 (strain RF) with $IC_{50}$ values of 7.1 and 15.4 μg/mL, respectively via, at least partially, inhibition of protease (Wang et al., 2011).

Mangiferin

*Antiviral (enveloped monopartite linear double-stranded DNA) xanthone C-glycosides:* Mangiferin and isomangiferin inhibited the replication of the Herpes simplex virus type-1 by inhibiting virus replication (Zheng & Lu, 1990).

*Moderate antiviral (enveloped circular double stranded DNA) flavone C-glycoside:* Isovitexin (apigenin-8-C-glucoside) at a concentration of 50 μg/mL evoked about 20% inhibition of replication of Hepatitis B virus (Japanese Fulminant Hepatitis-1 (JFH1) strain) in hepatoma cells (Rashed et al., 2014).

Commentaries:

(i) The plant produces series of flavonoids (Iwashina, 2000), anthocyanins (Imayama & Yabuya, 2003), and xanthone glycosides (Blinova & Kalyupanova, 1974).

(ii) Japanese iris is used in Unani medicine for COVID-19 (Nikhat & Fazil, 2020) as well as *Iris domestica* (L.) Goldblatt & Mabb., which is used for the treatment of COVID-19 in China (Lem et al., 2020). Is there an anti-COVID-19 principle in this plant? *Iris confusa* Sealy (local name *bian zhu lan*) is traditionally used to treat bronchitis and produces iridal-type triterpenoids (unique constituents in the member *Iris* L.), including spirioiridoconfal C and 28-deacetyl-belamcandal, which inhibited Hepatitis B virus DNA replication with $IC_{50}$ values of 84.6 and 58.6 μM, respectively (Chen et al., 2018).

**References**

Blinova, K.F. and Kalyupanova, N.I., 1974. Xanthone glycosides of *Iris ensata*. *Chemistry of Natural Compounds*, *10*(4), pp. 551–551.

Chen, X., Zhang, X., Ma, Y., Deng, Z., Geng, C. and Chen, J., 2018. Iridal-type triterpenoids with anti-hepatitis B virus activity from *Iris confusa*. *Fitoterapia*, *129*, pp. 126–132.

Imayama, T. and Yabuya, T., 2003. Characterization of anthocyanins in flowers of Japanese garden iris, *Iris ensata* Thunb. *Cytologia*, *68*(2), pp. 205–210.

Iwashina, T., 2000. The structure and distribution of the flavonoids in plants. *Journal of Plant Research*, *113*(3), p. 287.

Lem, F.F., Opook, F., Herng, D.L.J., Na, C.S., Lawson, F.P. and Tyng, C.F., 2020. Molecular mechanism of action of repurposed drugs and traditional Chinese medicine used for the treatment of patients infected with COVID-19: a systematic review. *medRxiv*.

Nikhat, S. and Fazil, M., 2020. Overview of Covid-19; its prevention and management in the light of Unani medicine. *Science of the Total Environment*, p. 138859.

Rashed, K., Sahuc, M.E., Deloison, G., Calland, N., Brodin, P., Rouillé, Y. and Séron, K., 2014. Potent antiviral activity of *Solanum rantonnetii* and the isolated compounds against hepatitis C virus in vitro. *Journal of Functional Foods*, 11, pp. 185–191.

Wang, R.R., Gao, Y.D., Ma, C.H., Zhang, X.J., Huang, C.G., Huang, J.F. and Zheng, Y.T., 2011. Mangiferin, an anti-HIV -1 agent targeting protease and effective against resistant strains. *Molecules*, 16(5), pp. 4264–4277.

Zheng, M.S. and Lu, Z.Y., 1990. Antiviral effect of mangiferin and isomangiferin on herpes simplex virus. *Chinese Medical Journal*, *103*(2), p. 160.

### 3.1.6.5   Family Orchidaceae A.L. de Jussieu (1789)

The family Orchidaceae consists of about 800 genera and 30,000 species of herbs, often magnificent, delicate, growing on trees, rocks, or terrestrial from pseudobulbs. Orchidaceae in Southeast Asia are disappearing in the wake of palm oil cultivation and other deforestation in the name of business. Consider that disturbing the fine biodiversity system of primary rain forest triggers zoonotic pandemics. The leaves are simple, not uncommonly fleshy, basal or alternate, when alternate the stem is not uncommonly fleshy. The flowers exists in myriads of breathtaking forms, colors, patterns, sometime mimicking insects or other life forms, including birds, and are especially beautiful in the early morning. The inflorescences are solitary or in panicles. The flowers are very characteristic with a perianth comprising 2 whorls of 6 tepals: the outer 3 tepals are of equal length, the inner 3 tepals comprise 2 lateral tepals and one median tepal (labellum). The androecium consists of up to 2 anthers and the gynoecium consists of up to 3 stigmas, which are fused into a column (gynostemium). The ovary consists of 3 carpels merged in a single or 3-locular ovary, each locule sheltering numerous ovules growing from parietal placentas. The fruit is a dehiscent capsule containing numerous tiny seeds. Plants in this family are known so far to produce interesting series of phenolic phytoalexins. Medicinal Orchidaceae in Asia Pacific have apparently not been much studied yet for their antimicrobial properties.

#### 3.1.6.5.1   *Acriopsis javanica* Reinw. ex Bl.

Common names: Sakat bawang (Malaysia); bosur bosur hau (Indonesia); sakko (Papua New Guinea); ruuhinee (Thailand); to yen. (Vietnam)

Habitat: Forests, grows on trees

Distribution: Myanmar, Cambodia, Laos, Vietnam, Thailand, Malaysia, Indonesia, Papua New Guinea, Solomon Islands, and Australia

Botanical observation: This epiphytic herb grows up to about 1 tall, from ovoid pseudobulbs. The leaves are simple and basal. The blade is narrowly elliptic to oblanceolate,

5–30 cm×0.5–2 cm, and fleshy. The inflorescences are panicles on up to 60 cm tall slender peduncles. The perianth is up to about 1.5 cm across, somewhat cross shaped, graceful greenish white to yellowish, penciled with purple or light pink. The column is straight. The fruit is an elongated and dehiscent capsule, which is up to 2.5 cm long and contains tiny seeds.

Medicinal use: Fever (Indonesia, Malaysia)

Commentary: Nothing is apparently known about the antimicrobial properties of this orchid.

### 3.1.6.5.2   *Aerides odorata* Lour.

Synonyms: *Epidendrum odoratum* (Lour.) Poir.

Common names: Pargasa (Bangladesh); xiang hua zhi jia lan (China)

Habitat: Forests, grows on trees, ornamental

Distribution: India, Nepal, Bhutan, Bangladesh, Myanmar, Cambodia, Laos, Vietnam, China, Thailand, Malaysia, Indonesia, Papua New Guinea, Solomon Islands, and Australia

Botanical observation: The leaves are simple and alternate. The blade is narrowly elliptic, 15–20×2.5–4.6 cm, fleshy, obtuse, and unequally bilobed. The inflorescence is racemose, showy, pendulous up to 30 cm long, and presents up to 30 fragrant flowers, which are whitish tinged pink and 1.5–3 cm in diameter. The perianth includes a whorl of 3 sepals, which are elliptic, about 1 cm long, and a second whorl of 2 tepals, which are broadly ovate, about 1.2 cm long, and a hooked labellum.

Medicinal uses: Wounds, tuberculosis, boils (Bangladesh)

*Broad-spectrum antibacterial halo developed by polar extract:* Aqueous extract inhibited the growth of *Escherichia coli*, ampicillin-resistant *Escherichia coli*, and kanamycin-resistant *Escherichia coli* (Paul et al., 2013).

*Broad-spectrum antibacterial halo developed by non-polar extract:* Petroleum ether extract inhibited the growth of *Bacillus subtilis*, *Staphylococcus aureus*, *Shigella dysenteriae*, *Salmonella typhi*, and *Vibrio* sp. (Paul et al., 2013).

*Antifungal (filamentous) polar extract:* Aqueous extract inhibited the growth of *Alternaria alternaria*, *Curvularia lunata*, *Colletotrichum corchori*, *Fusarium equiseti*, *Macrophotimina phaeolina*, and *Botryodiplodia theobromae* by 57.1, 71.4, 57.1, 64.2, 64.2, and 57.1%, respectively (Hoque et al., 2016).

Commentary: Nothing is apparently known about the antimicrobial constituents of this orchid.

### References

Hoque, M.M., Khaleda, L. and Al-Forkan, M., 2016. Evaluation of pharmaceutical properties on microbial activities of some important medicinal orchids of Bangladesh. *Journal of Pharmacognosy and Phytochemistry*, 5(2), p. 265.

Paul, P.A.U.L.O.M.I., Chowdhury, A.B.H.I.S.H.E.K., Nath, D.E.E.P.A. and Bhattacharjee, M.K., 2013. Antimicrobial efficacy of orchid extracts as potential inhibitors of antibiotic resistant strains of *Escherichia coli*. *Asian Journal of Pharmaceutical and Clinical Research*, 6(3), pp. 108–111.

### 3.1.6.5.3   *Apostasia nuda* R. Brown ex Wall.

Synonym: *Adactylus brunonis* (Griff.) Cretz.; *Adactylus lobbii* (Rchb.f.) Rolfe; *Adactylus nudus* (R. Brown ex Wall.) Rolfe; *Apostasia lobbii* Rchb.f.

Common names: Si sarsar bulung (Indonesia); pokok pelampas budak (Malaysia); co lan tran (Vietnam)

Habitat: Forests

Distribution: India, Bangladesh, Myanmar, Cambodia, Thailand, Vietnam, Malaysia, and Indonesia

Botanical observation: This erect orchid grows up to 60 cm tall from a scaly rhizome. The leaves are simple, sessile, and spiral. The blade is narrowly lanceolate, sheathing at base, somewhat wavy, dark green, glossy, with numerous parallel nerves, 15–30 cm×0.5–1.5 cm, and attenuate at apex. The inflorescence is a terminal raceme. The perianth comprises 6 tepals, which are up to about 4 mm long, recurved, and pure white to cream. The androecium consists of 2 stamens merged in a column with the style. The capsule elongated and somewhat 3-lobed, papery, and sheltering numerous tiny seeds.

Medicinal uses: Sore eyes (Malaysia)

Commentary: The antimicrobial activities of this endangered orchid have apparently not been examined.

### 3.1.6.5.4   Bletilla striata (Thunb.) Rchb. f.

Synonyms: *Bletia gebina* Lindl.; *Bletia hyacinthina* (Sm.) R. Brown; *Bletia striata* (Thunb.) Druce; *Bletilla elegantula* (Kraenzl.) Garay & Romero; *Bletilla gebina* (Lindl.) Rchb. f.; *Bletilla hyacinthina* (Sm.) Rchb. f.; *Bletilla striata* var. *gebina* (Lindl.) Rchb. f.; *Calanthe gebina* Lodd. ex Lindl.; *Coelogyne elegantula* Kraenzl.; *Cymbidium hyacinthinum* Sm.; *Cymbidium striatum* (Thunb.) Sw.; *Epidendrum striatum* (Thunb.) Thunb.; *Jimensia nervosa* Raf.; *Jimensia striata* (Thunb.) Garay & R.E. Schult.; *Limodorum hyacinthinum* (Sm.) Donn; *Limodorum striatum* Thunb.

Common name: Bai ji (China)

Habitat: Forests, cultivated

Distribution: Myanmar, China, Korea, and Japan

Botanical observation: This orchid grows up to about 60 cm tall from a characteristically white, smooth, penciled reddish pseudobulb developing some fibrous roots. The leaves are simple, spiral, and sessile. The blade is elliptic to lanceolate, light green, dull, 8–29×1.5–4 cm, sheathing at base, with numerous longitudinal nerves, and acuminate at apex. The inflorescence is a terminal raceme on a peduncle, which is up to about 30 cm tall. The perianth includes 6 tepals, which are pink or white, up to 3 cm long, acute at apex, the labellum penciled with purplish lines, obovate-elliptic, 3-lobed, and wavy. The column is up to 2 cm long and slender. The capsule is up to about 3.5 cm long.

Medicinal uses: Cough, fever, thrush, tuberculosis, wounds (China)

Phytochemical class: Phenolics

*Antibacterial (Gram-positive) biphenanthrenes:* 4,7,7′-Trimethoxy-9′, 10′-dihydro(1,3′-biphenanthrene)-2,2′, 5′- triol isolated from the rhizome inhibited the growth of *Staphylococcus aureus* (ATCC 25293), *Staphylococcus epidermidis* (CMCC 26069), *Enterococcus faecalis* (ATCC 29212), and *Bacillus subtilis* with MIC values of 8, 8, 64, and 16 µg/mL, respectively (Qian et al., 2015). 4,7,4′-Trimethoxy-9′, 10′-dihydro(1,1′-biphenanthrene)-2,2′,7′-triol isolated from the rhizome inhibited the growth of *Staphylococcus aureus* (ATCC 25293), *Staphylococcus epidermidis* (CMCC 26069), *Enterococcus faecalis* (ATCC 29212), and *Bacillus subtilis* (Qian et al., 2015). 4,7,3′,5′-Tetramethoxy-9′, 10′-dihydro(1,1′-biphenanthrene)-2,2′, 7′-triol isolated from the rhizome inhibited the growth of *Staphylococcus aureus* (ATCC 25293), *Staphylococcus epidermidis* (CMCC 26069), and *Bacillus subtilis* with MIC values of 64, 8, and 84 µg/mL, respectively (Qian et al., 2015).

4,7,4′-Trimethoxy-9′, 10′-dihydro(1,1′-biphenanthrene)-2,2′, 7′- triol

*Strong antibacterial (Gram-positive) stilbenes:* Bulbocol, shanciguol, and shancigusin B isolated from the rhizome inhibited the growth of *Staphylococcus aureus* (ATCC 6538) with MIC values of 9, 7, and 3 μg/mL, respectively (Jiang et al., 2019). Bulbocol and shancigusin B isolated from the rhizome inhibited the growth of methicillin-resistant *Staphylococcus aureus* (ATCC 43300) with MIC values of 35 and 13 μg/mL, respectively (Jiang et al., 2019). Shanciguol and shancigusin B isolated from the rhizome inhibited the growth of *Bacillus subtilis* (ATCC 6538) with MIC values of 55 and 26 μg/mL, respectively (Jiang et al., 2019).

Shancigusin B

*Strong antiviral (enveloped segmented linear single stranded (–) RNA) phenanthrene:* 4,4′, 7,7′-tetrahydroxy-2,2′, 8,8′-tetramethoxy-1,1′-di-phenanthrene inhibited the replication of Influenza virus A/Sydney/5/97 (H3N2) with an IC$_{50}$ value of 14.6 μM, via a mechanism involving, at least partially, RNA synthesis (Shi et al., 2017).

4′, 7,7′-Tetrahydroxy-2,2′, 8,8′-tetramethoxy-1,1′-di-phenanthrene

*Weak antiviral (enveloped segmented linear single-stranded (–) RNA virus) phenanthrenes:* 2,2,7′-trihydroxy-4,4′, 7-trimethoxy-9′, 10′-dihydro-1,1′-diphenanthrene, 2,2′, 7′-trihydroxy-3′, 4,5′, 7-tetramethoxy-9′, 10′-dihydro-1,1′-di-phenanthrene, and 4,4′, 7-trihydroxy-2,2′, 7′-trimethoxy-1,1′-di-phenanthrene inhibited the replication of Influenza virus A/Sydney/5/97 (H3N2) with IC$_{50}$ values of 28.6, 20.4, and 28.5 μM, respectively, via a mechanism involving, at least partially, RNA synthesis (Shi et al., 2017).

*Viral enzyme inhibition by phenanthrene:* 2,2′, 7′-trihydroxy-3′, 4,5′, 7-tetramethoxy-9′, 10′-dihydro-1,1′-di-phenanthrene inhibited Influenza virus neuraminidase with an IC$_{50}$ value of 16.8 μM (Shi et al., 2017).

Commentary: The plant produces stilbenes and dihydrophenanthrenes (Bai et al., 1993), and bisbenzyl derivatives (Feng et al., 2008) and phytoalexins.

### References

Bai, L., Kato, T., Inoue, K., Yamaki, M. and Takagi, S., 1993. Stilbenoids from *Bletilla striata*. *Phytochemistry*, *33*(6), pp. 1481–1483.

Feng, J.Q., Zhang, R.J. and Zhao, W.M., 2008. Novel bibenzyl derivatives from the tubers of *Bletilla striata*. *Helvetica Chimica Acta*, *91*(3), pp. 520–525.

Jiang, S., Wan, K., Lou, H.Y., Yi, P., Zhang, N., Zhou, M., Song, Z.Q., Wang, W., Wu, M.K. and Pan, W.D., 2019. Antibacterial bibenzyl derivatives from the tubers of *Bletilla striata*. *Phytochemistry*, *162*, pp. 216–223.

Qian, C.D., Jiang, F.S., Yu, H.S., Shen, Y., Fu, Y.H., Cheng, D.Q., Gan, L.S. and Ding, Z.S., 2015. Antibacterial Biphenanthrenes from the fibrous roots of *Bletilla striata*. *Journal of Natural Products*, *78*(4), pp. 939–943.

Shi, Y., Zhang, B., Lu, Y., Qian, C., Feng, Y., Fang, L., Ding, Z. and Cheng, D., 2017. Antiviral activity of phenanthrenes from the medicinal plant *Bletilla striata* against influenza A virus. *BMC Complementary and Alternative Medicine*, *17*(1), p. 273.

### 3.1.6.5.5    *Calanthe triplicata* (Willemet) Ames

Synonyms: *Alismorkis furcata* (Bateman ex Lindl.) *Kuntze*; *Alismorkis veratrifolia* (Willd.) *Kuntze*; *Amblyglottis veratrifolia* Bl.; *Calanthe furcata* Bateman ex Lindl.; *Calanthe rubicallosa* Masam.; *Calanthe veratrifolia* Ker Gawl.; *Limodorum veratrifolium* Willd.; *Orchis triplicata* Willemet

Common names: Christmas orchid; san zhe xia ji lan (China)

Habitat: Swamp forests, cultivated

Distribution: India, Sri Lanka, Nepal, Bhutan, Bangladesh, Myanmar, Cambodia, Laos, Thailand, China, Taiwan, Japan, Malaysia, Indonesia, the Philippines, Papua New Guinea, Australia, and Pacific Islands

Botanical observation: This erect orchid grows up to about 1 m tall from pseudobulbs, which are somewhat ovoid and up to 3 cm long. The leaves are basal, simple, and spiral. The blade is elliptic, 20–60×5–15 cm, dark green, glossy, presents numerous conspicuous longitudinal nerves, and acute at apex. The inflorescences are showy and many flowered, somewhat conical panicles at the apex of up to about 45 cm tall peduncles. The perianth is up to about 2.5 cm across, pure white with a central yellow to orangish spot, and with a labellum, which is somewhat anthropomorphic. The column is 5 mm long. The capsule is obovoid, glossy, ribbed, elongated and contain numerous tiny seeds.

Medicinal use: Caries and diarrhea (Indonesia)

Commentary: No reports on the antimicrobial properties of the constituents of this orchid are apparently available. *Calanthe discolor* Lindl. is used in China (local name: *ia ji lan*) to treat infections, and has been reported by Yoshikawa et al. (1998) to produce phenanthrenes as well as the indole glycosides calanthoside and glucoindican precursors of indican, the indoloquinazoline alkaloid tryptanthrin, and the indole alkaloids isatin and indirubin. Tryptanthrin, indirubin, and indican are antimicrobial (see *Isatis tinctoria* L.)

### Reference

Yoshikawa, M., Murakami, T., Kishi, A., Sakurama, T., Matsuda, H., Nomura, M., Matsuda, H. and Kubo, M., 1998. Novel indole S, O-bisdesmoside, calanthoside, the precursor glycoside of tryptanthrin, indirubin, and isatin, with increasing skin blood flow promoting effects, from two *Calanthe species* (Orchidaceae). *Chemical and Pharmaceutical Bulletin*, 46(5), pp. 886–888.

*3.1.6.5.6 Cymbidium aloifolium* (L.) Sw.

Synonyms: *Cymbidium pendulum* (Roxb.) Sw.; *Cymbidium simulans* Rolfe; *Epidendrum aloifolium* L.; *Epidendrum pendulum* Roxb.

Common names: Aloe leaf Cymbidium, flat orchid; Surimas (Bangladesh); wen ban lan (China); harjor (Nepal); kare karon (Thailand); doan kiem (Vietnam)

Habitat: Forests, roadsides, grows on large trees and cliffs

Distribution: India, Sri Lanka, Nepal, Bangladesh, Myanmar, Andaman Islands, Cambodia, Laos, Vietnam, China, Taiwan, Thailand, Malaysia, Indonesia, and the Philippines

Botanical observation: This graceful orchid grows from pseudobulbs, which are ovoid and about 10 cm long. The leaves are simple and exstipulate. The blade is linear-oblong, 40–90×1.5–3.5 cm, somewhat fleshy, pendulous, dull green, smooth, with a conspicuous midrib, which is sunken above, obtuse, and unequally 2-lobed at apex. The inflorescences are pendulous and laxly flowered racemes, which are up to 60 cm long, slender, and yellowish. The perianth is about 4 cm across, with a discrete fragrance, somewhat yellow with brownish and longitudinal bands, comprises of an outer series of 3 tepals, which are narrowly oblong up to 2 cm long, and a whorl of 3 inner shorter tepals including a labellum, which is oblong and curved. The column is slightly curved and about 1 cm long. The capsules are oblong-ellipsoid, up to 6 cm long, and containing minute seeds.

Medicinal uses: Boils, fever (Bangladesh); burns, sores, colds (Vietnam); cuts, fever, (India); otitis (Thailand); wounds (India, Cambodia, Laos, and Vietnam)

*Moderate antifungal mid-polar extract:* Chloroform extract of pseudobulbs inhibited the growth of *Trichophyton mentagrophytes* with an MIC value of 250 µg/mL (Juneja et al., 1987). From this extract were isolated the stilbenes aloifols I and II, gigantol, coelonin, batatasin III, and the phenanthrene 6-*O*-methylcoelonin (Juneja et al., 1987).

*Weak antibacterial (Gram-positive) stilbene:* Batatasin III inhibited the growth of *Staphylococcus aureus* (ATCC 6538), *Bacillus subtilis* (ATCC 6051), and methicillin-resistant *Staphylococcus aureus* (ATCC 43300) with MIC values of 250, 500, and 500 µg/mL, respectively (Jiang et al., 2019).

*Weak antiviral (enveloped monopartite linear double-stranded DNA) stilbenes:* Gigantol inhibited the replication of Herpes simplex virus type-1 and Herpes simplex virus type-2 with the $IC_{50}$ values of 304.1 and 319.3 µM, respectively (Sukphan et al., 2014). Batatasin III inhibited the replication of Herpes simplex virus type-1 and Herpes simplex virus type-2 with $IC_{50}$ values of 341.5 and 384.2 µM, respectively (Sukphan et al., 2014).

Commentary: *Cymbidium ensifolium* (L.) Sw. is used to treat venereal diseases in China (local name: jian lan)

### References
Jiang, S., Wan, K., Lou, H.Y., Yi, P., Zhang, N., Zhou, M., Song, Z.Q., Wang, W., Wu, M.K. and Pan, W.D., 2019. Antibacterial bibenzyl derivatives from the tubers of *Bletilla striata*. *Phytochemistry, 162*, pp. 216–223.

Juneja, R.K., Sharma, S.C. and Tandon, J.S., 1987. Two substituted bibenzyls and a dihydrophenanthrene from *Cymbidium aloifolium*. *Phytochemistry, 26*(4), pp. 1123–1125.

Sukphan, P., Sritularak, B., Mekboonsonglarp, W., Lipipun, V. and Likhitwitayawuid, K., 2014. Chemical constituents of Dendrobium venustum and their antimalarial and antiherpetic properties. *Natural Product Communications, 9*(6), p. 1934578X1400900625.

### 3.1.6.5.7 *Dendrobium nobile* Lindl.
Common names: Noble Dendromium; shi hu (China); ueang khao kiu (Thailand); hoàng thảo (Vietnam)

Synonyms: *Callista nobilis* (Lindl.) Kuntze; *Dendrobium coerulescens* Wall. ex Lindl.; *Dendrobium formosanum* (Rchb. f.) Masam.; *Dendrobium lindleyanum* Griff.

Habitat: Forests, grows on trees or rocks, cultivated

Distribution: India, Bhutan, Laos, Myanmar, Thailand, Vietnam, China, and Taiwan

Botanical observation: This magnificent yet dreadfully poisonous orchid grows up to about 50 cm tall. The stems are somewhat erect, terete, fleshy, glossy, stout, and regularly swollen at internodes. The leaves are simple, sessile, and alternate. The blade is oblong to elliptic, fleshy, 5.5–10.5 cm × 1–3 cm, obtuse and somewhat bilobed at apex. The inflorescences are axillary and few-flowered racemes. The perianth is pure white or reddish to pinkish. Out of 6 tepals, the 3 outer are oblong, and about 3 cm long, and the 3 inner tepals are broadly ovate, up to about 3.5 cm long, including a broadly ovate labellum with a darkish throat. The column is about 5 mm long.

Medicinal use: Fever (China, Laos, Vietnam)

*Strong broad-spectrum antibacterial coumarins:* Dendrocoumarin inhibited the growth of *Staphylococcus aureus* (ATCC 8799), *Escherichia coli* (ATCC 25922), *Micrococcus tetragenus* (ATCC 13623), *Kocuria rhizophila* (ATCC 9341), and *Bacillus cereus* (ATCC 14579), with MIC values of 2.5, 0.6, 5, 5, and 2.5 µg/mL, respectively (Zhou et al., 2018). Itolide A inhibited the growth of *Staphylococcus aureus* (ATCC 8799), *Escherichia coli* (ATCC 25922), *Micrococcus tetragenus* (ATCC 13623), *Kocuria rhizophila* (ATCC 9341), and *Bacillus cereus* (ATCC 14579), with MIC values of 2.5, 1.2, 5, 10, and 1.2 µg/mL, respectively (Zhou et al., 2018).

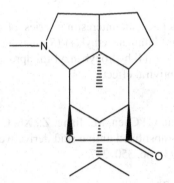

Dendrocoumarin

*Very weak antifungal (filamentous) coumarin:* Itolide A inhibited the growth of *Sclerotium rolfssi* and *Rhizoctonia solani* with MIC values of 240 and 132.2 μM, respectively (Zhou et al., 2018).

*Strong antiviral (enveloped segmented linear single-stranded (−) RNA virus) hydrophilic sesquiterpene alkaloid:* Dendrobine (LogD = −0.3 at pH 7.4; molecular weight = 263.3 g/mol) inhibited the replication of Influenza A virus (A/Aichi/2/68) (H3N2), Influenza A virus (A/FM1/47 (H1N1), and Influenza A virus (A/Puerto Rica/8/34 H274Y) (H1N1) with $IC_{50}$ values of 5.3, 3.3, and 2.1 μg/mL, respectively, by inhibiting early steps in the viral replication cycles and binding to viral nucleoprotein (Li et al., 2017).

Commentaries:

Dendrobine

(i) Members of the genus *Dendrobium* L. are not uncommonly used to treat bacterial infection in Asia. *Dendrobium crumenatum* Sw. is used to treat boils and pimples in Malaysia. *Dendrobium draconis* Reichb. F is used to treat fever in Thailand (local name: *Ueang ngoen*), as well as *Dendrobium trigonopus* Reichb. F (local name: *Ueang kham liam*). Dendrocoumarin and congeners are planar and amphiphilic and probably target microbial DNA. The amphiphilicity and low molecular weight permit porin penetration in Gram-negative bacteria.

(ii) Dendrobin is a neurotoxin (Chen & Chen, 1935) that can't be used as antiviral agents but offers an original framework from which viral nucleoprotein inhibitors can be developed. This sesquiterpene inhibits synaptic transmission in the frog spinal cord (Kudo et al., 1983). It has a lactone moiety which might react with bionucleophiles or sulfhydryl amino acids. Further, neuroactive substances often inhibit bacterial or fungal efflux pumps.

**References**

Chen, K.K. and Chen, A.L., 1935. The pharmacological action of dendrobine. The alkaloid of Chin-shih-hu. *Journal of Pharmacology and Experimental Therapeutics, 55*(3), pp. 319–325.

Kudo, Y., Tanaka, A. and Yamada, K., 1983. Dendrobine, an antagonist of β-alanine, taurine and of presynaptic inhibition in the frog spinal cord. *British Journal of Pharmacology, 78*(4), pp. 709–715.

Li, R., Liu, T., Liu, M., Chen, F., Liu, S. and Yang, J., 2017. Anti-influenza A virus activity of dendrobine and its mechanism of action. *Journal of Agricultural and Food Chemistry, 65*(18), pp. 3665–3674.

Zhou, X.M., Zhang, B., Chen, G.Y., Han, C.R., Jiang, K.C., Luo, M.Y., Meng, B.Z., Li, W.X. and Lin, S.D., 2018. Dendrocoumarin: a new benzocoumarin derivative from the stem of *Dendrobium nobile. Natural Product Research, 32*(20), pp. 2464–2467.

### 3.1.6.5.8   Gastrodia elata Bl.

Common name: Tian ma (China)

Synonyms: *Gastrodia mairei* Schltr.; *Gastrodia viridis* Makino

Habitat: Forest margins

Distribution: India, Bhutan, Nepal, China, Korea, and Japan

Botanical observation: This orchid has an orobanchaceous look and grows up to about 1 m tall from a somewhat oblong, fleshy, and light brown rhizome. The scape is yellowish brown, up to about 80 cm tall, and ends in a many-flowered spike. The flowers are orangish to pale yellow, tubular, up to 1 cm long, curved, and develop a labellum, which is minutely 3-lobed. The column is up to about 7 mm. The capsule is ellipsoid and up to about 2 cm long.

Medicinal use: Cold (China)

Commentary: The plant generates an interesting series of phenolic compounds, including (+)-(*S*)-[*N*-(4′-hydroxybenzyl)]pyroglutamate, ethyl (+)-(*S*)- and (−)-(*R*)-[*N*-(4′-hydroxybenzyl)]pyroglutamates, and (+)-(*S*)-2-hydroxy-3-[(4′-hydroxybenzyl)thio]propanoate (Guo et al., 2015), which could be examined for possible antiviral effects.

**Reference**

Guo, Q., Wang, Y., Lin, S., Zhu, C., Chen, M., Jiang, Z., Xu, C., Zhang, D., Wei, H. and Shi, J., 2015. 4-Hydroxybenzyl-substituted amino acid derivatives from *Gastrodia elata. Acta Pharmaceutica Sinica B, 5*(4), pp. 350–357.

### 3.1.6.5.9   Geodorum nutans (Presl.) Ames

Common names: Shepherd's crook orchid; di bao lan (China); daun kora kora (Indonesia); bandabok, kula (the Philippines); wan nang (Thailand)

Synonyms: *Arethusa glutinosa* Blco; *Callista nutans* (C. Presl) Kuntze; *Dendrobium nutans* C. Presl; *Limodorum densiflorum* Lam.; *Cistella cernua* (Willd.) Bl.; *Cymbidium pictum* R. Brown; *Dendrobium nutans* C. Presl; *Geodorum densiflorum* (Lam.) Schltr.; *Geodorum formosanum* Rolfe; *Geodorum fucatum* Lindl.; *Geodorum pacificum* Rolfe; *Geodorum pictum* (R. Brown) Lindl.; *Geodorum purpureum* R. Brown; *Geodorum semicristatum* Lindl.; *Malaxis cernua* Willd.; *Ortmannia cernua* (Willd.) Opiz; *Otandra cernua* (Willd.) Salisb.; *Tropidia grandis* Hance.

Habitat: Forests, grasslands

Distribution: India, Sri Lanka, Myanmar, Cambodia, Laos, Thailand, Vietnam, China, Korea, Taiwan, Japan, Malaysia, Indonesia, the Philippines, Papua New Guinea, and Australia

Botanical observation: This beautiful terrestrial orchid grows up to a height of 50 cm from ovoid pseudobulbs. The pseudostem is terete, smooth, and somewhat fleshy. The leaves are simple, and 2 or 3. The blade is elliptic, oblong-lanceolate, light green, 15–35 × 3–9 cm, somewhat plicate, and acuminate or acute at apex. The inflorescences are terminal clusters. The perianth comprises 5 tepals which are oblong, about 1 cm long, heavenly light pink, acuminate, and a broadly ovate labellum penciled with purple lines and minutely incised. The column is 5 mm long. The capsules are oblong, ribbed, and up to about 5 cm long.

Medicinal uses: Sores (Indonesia); abscesses, boils (the Philippines)

Pharmacology: Antibacterial and antifungal

*Broad-spectrum antibacterial halo developed by polar extract:* Methanol extract of pseudo-bulbs (400 µg/disc, 5 mm diameter paper disc) inhibited the growth of *Bacillus cereus, Bacillus megaterium, Bacillus subtilis, Staphylococcus aureus, Sarcinia lutea, Salmonella paratyphi, Salmonella typhi, Vibrio parahaemolyticus, Vibrio minicus, Escherichia coli, Shigella dysenteriae, Pseudomonas aeruginosa,* and *Shigella boydii* with inhibition zone diameters of 13, 12, 14, 15, 15, 13, 14, 11, 12, 15, 14, 13, and 14 mm, respectively (Akter et al., 2010).

*Broad-spectrum antifungal halo developed by polar extract:* Methanol extract of pseudobulbs (400 µg/disc, 5 mm diameter paper disc) inhibited the growth of *Saccharomyces cerevisiae, Candida albicans,* and *Aspergillus niger* with inhibition zone diameters of 16, 16, and 15 mm, respectively (Akter et al., 2010).

Commentary: One could have the curiosity to look for antibacterial and antifungal principles from this orchid.

**Reference**

Akter, S., Imam, M.Z. and Akter, T., 2010. Antimicrobial activity of different extracts of *Geodorum densiflorum* (Lam) Schltr. Pseudobulb. *Stamford Journal of Pharmaceutical Sciences, 3*(2), pp. 49–50.

*3.1.6.5.10   Liparis plicata Franch. & Sav.*

Common names: lian chi yang er suan (China)

Synonyms: *Cestichis plicata* (Franch. & Sav.) F. Maek.; *Leptorchis plicata* (Franch. & Sav.) Kuntze; *Leptorkis plicata* (Franch. & Sav.) Kuntze; *Liparis bootanensis* Griff.

Habitat: Forests, cliffs

Distribution: India, Bhutan, Sri Lanka, Myanmar, Cambodia, Laos, Thailand, Vietnam, China, Taiwan, Malaysia, and Indonesia

Botanical description: It is an orchid that grows on trees from ovoid pseudobulbs. The leaves are simple and 1 or 2. The petiole is up to 7 cm long. The blade is elliptic-oblong, 8–20 cm×1–3 cm, somewhat fleshy, and acuminate at apex. The inflorescences are racemes on top of a scape, which is about 20 cm tall. The perianth comprises 5 linear tepals and a labellum, which is broadly obovate, about 6 mm long, emarginated, brownish green, recurved, and glossy. The column is arcuate and 3 mm long. The capsules are elliptic and up to about 1 cm long.

Medicinal uses: Cough, fever, furuncles, gonorrhea, ulcers (China)

Commentary: Members of the genus *Liparis* Rich are not uncommonly used to treat microbial infections in China: *Liparis viridiflora* (Bl.) Lindl. (local name: *chang jing yang er suan*), *Liparis olivacea* Lindl., *Liparis odorata* (Willd.) Lindl. (local name: *xiang hua yang er suan*), and *Liparis japonica* Maxim. are traditionally used to treat leucorrhea. Other examples are *Liparis nervosa* (Thunb.) Lindl. (local name: *jian xue qing*), *Liparis petiolata* (D. Don) P.F. Hunt & Summerh. (local name: *bing ye yang er suan*), *Liparis stricklandiana* Rchb. f., (local name: *shan chun yang er suan*), *Liparis chloroxantha* Hance, *Liparis distans* C.B. Clarke (local name: *da hua yang er suan*), *Liparis fargesii* Finet (local name: *xiao yang er suan*), and *Liparis inaperta* Finet (local name: *chang bao yang er suan*). Isolation of antibacterial and/or antiviral principles from the genus *Liparis* Rich. is warranted.

*3.1.6.5.11   Nervilia aragoana Gaudich.*

Common names: Guang bu yu lan (China); Olirathamara (Nepal); I tiam ong (Taiwan)

Synonyms: *Aplostellis flabelliformis* (Lindl.) Ridl.; *Epipactis carinata* Roxb.; *Nervilia carinata* (Roxb.) Schltr.; *Nervilia flabelliformis* (Lindl.) Tang & F.T. Wang; *Nervilia scottii* (Rchb. f.) Schltr.;

*Nervilia tibetensis* Rolfe; *Nervilia yaeyamensis* Hayata; *Pogonia carinata* Lindl.; *Pogonia flabelliformis Lindl.*; *Pogonia gracilis Blume*; *Pogonia nervilia Blume*; *Pogonia scottii Rchb. f.*

Habitat: Forests, shady and moist soils

Distribution: India, Nepal, Bhutan, Bangladesh, Sri Lanka, Myanmar, Cambodia, Laos, Thailand, Vietnam, China, Taiwan, Japan, Malaysia, Indonesia, the Philippines, Papua New Guinea, Australia, and Pacific Islands

Botanical observation: This terrestrial orchid grows from a tuber. The leaves are basal and solitary. The petiole is erect and up to about 10 cm tall. The blade is characteristically broadly cordate-, 9–15 × 10–18 cm, somewhat plicate, glossy, and acute or rounded at apex. The inflorescences are racemes on top of 15–45 cm tall scapes. The perianth includes 5 tepals, which are light green, linear, about 2 cm long, acuminate or acute at apex, and a labellum, which is penciled with purplish, wavy lines. The column is about 1 cm long and white. The capsules are elongated and nodding.

Medicinal use: Post-partum (India, Malaysia)

*Very weak broad-spectrum antifungal mid-polar extract:* Ethyl acetate extract of the whole plant inhibited the growth of *Aspergillus niger*, *Aspergillus fumigatus*, *Cryptococcus neoformans*, and *Saccharomyces cerevisiae* with the MIC values of 1200, 950, 750, and 1400 µg/mL, respectively (Reddy et al., 2010).

Commentary: So far the plant is known to generate series of phytosterols, including 24-*epi*-brassicasterol (Kikuchi et al., 1985).

### References

Kikuchi, T., kadota, S., Suehara, H. and Namba, T., 1985. Studies on the constituents of orchidaceous plants. III. Isolation of non-conventional side chain sterols from *Nervilia purpurea* SCHLECHTER and structure determination of nervisterol. *Chemical and Pharmaceutical Bulletin*, 33(6), pp. 2235–2242.

Reddy, K.H., Sharma, P.V.G.K. and Reddy, O.V.S., 2010. A comparative in vitro study on antifungal and antioxidant activities of *Nervilia aragoana* and *Atlantia monophylla*. *Pharmaceutical Biology*, 48(5), pp. 595–602.

*3.1.6.5.12   Peristylus constrictus* (Lindl.) Lindl.

Synonyms: *Habenaria constricta* (Lindl.) Hook. f.; *Herminium constrictum* Lindl.

Comme names: Constricted Peristylus; bhuinora (Bangladesh); da hua kuo rui lan (China)

Habitat: Grassy lands

Distribution: India, Bhutan, Nepal, Myanmar, Cambodia, Thailand, and Vietnam

Botanical observation: This magnificent terrestrial, erect, orchid grows from ovoid tubers, to about 80 cm tall. The stem is terete and upright. The leaves are simple and spiral. The blade is broadly elliptic, sheathing at base, 5–13 × 3.5–6.5 cm, dull light green with numerous parallel nervations, coriaceous, tapering at base and acute at apex. The inflorescence is erect, up to about 50 cm tall, and many flowered. The perianth is about 1 cm long and includes a series of 3 tepals, which are greenish, and a second whorl of 3 tepals, which are immaculately white, including a labellum, which is deeply 3 lobed. The fruits are ellipsoid capsules.

Medicinal uses: Jaundice (Bangladesh), boils (India, Bangladesh)

Commentary: The plant has apparently not been investigated for possible antimicrobial effects.

*3.1.6.5.13   Pholidota chinensis* Lindl.

Common names: shi xian tao (China)

Synonyms: *Coelogyne chinensis* (Lindl.) Rchb. f.; *Coelogyne pholas* Rchb. f.

Habitat: Forests on trees or rocks, cultivated, ornamental

Distribution: Myanmar, Vietnam, and China

Botanical observation: This orchid grows from pseudobulbs. The leaves are simple and spiral. The petiole is up to 2 cm long. The petiole is up to about 5 cm long. The blade is elliptic, 5–22×2–6 cm, with 3 conspicuous nervations, acuminate at apex, and somewhat wavy. The inflorescence is a pendulous and showy raceme that grows up to about 50 cm long. The perianth comprises 6 yellowish-brown tepals, which are up to about 1 cm long and membranous, including a broadly ovate and somewhat bilobed labellum. The column is about 5 mm long. The capsule is ellipsoid, about 1.5 cm long, and ridged.

Medicinal use: Bronchitis, tuberculosis (China)

*Strong broad-spectrum antifungal amphiphilic dihydrophenanthrene:* Hircinol (4-methoxy-9,10-dihydrophenanthrene-2,5-diol; LogD= 3.4; molecular weight = 242.2 g/mol) at 100 ppm completely inhibited the growth of *Candida lipolytica* (Fisch et al., 1973; Wang et al., 2007) and inhibited the germ tube growth of *Botryodiplodia theobromae, Fusarium moniliforme, Penicillium sclerotigenum, Aspergillus niger, Botrytis cinerea,* and *Cladosporium cladoporioides* with the $EC_{50}$ values of 16, 65, 41, 42, 44, and 26 μg/mL, respectively (Coxon et al., 1982).

Hircinol

Commentaries: (i) The plant brings to being an interesting array of antibacterial and antifungal stilbenes and dihydrophenanthrenes, which could be examined for their antiviral properties. Consider that dihydrophenanthrenes and phenanthrenes are antifungal phytoalexins in Orchidaceae. What is the antifungal mode of action of dihydrophenanthrenes? (ii) The dihydrophenanthrene juncusol isolated from a member of the genus *Juncus* L. (family Juncaceae) inhibited the growth of *Mycobacterium phlei* (ATCC 11758) with an inhibition zone diameter of 129 mm (900 ppm/disc, 127 mm diameter paper disc) (Chapatwala et al., 1981). Dihydrophenanthrenes in *Pholidota chinensis* Lindl., may, at least partially, account for the antitubercular traditional use.

### References

Chapatwala, K.D., Armando, A. and Miles, D.H., 1981. Antimicrobial activity of juncusol, a novel 9–10-dihydrophenanthrene from the marsh plant *Juncus roemerianus. Life Sciences,* 29(19), pp. 1997–2001.

Coxon, D.T., Ogundana, S.K. and Dennis, C., 1982. Antifungal phenanthrenes in yam tubers. *Phytochemistry,* 21(6), pp. 1389–1392.

Fisch, M.H., Flick, B.H. and Arditti, J., 1973. Structure and antifungal activity of hircinol, loroglossol and orchinol. *Phytochemistry,* 12(2), pp. 437–441.

*3.1.6.5.14　Vanda tessellata* (Roxb.) Hook. ex G. Don

Common names: Nakuli, rasna (India)

　　Synonyms: *Aerides tessellata* (Roxb.) Wight ex Wall.; *Epidendrum tessellatum* Roxb.

　　Habitat: Forests on trees, cultivated

　　Distribution: India, Sri Lanka, Bangladesh, Myanmar, and China

　　Botanical observation: This orchid grows up to a height of 1 m. The leaves are simple, alternate, and sessile. The blade is oblong-linear, up to about 1.5 cm×30 cm, and round at apex. The inflorescence is an axillary raceme, which grows up to about 25 cm long. The perianth comprises 5 spathulate and wavy tepals stained brownish to greenish and up to about 3 cm long, and a purple to bright blue labellum. The column grows up to 5 mm long. The capsules are obovate and up to about 8 cm long.

　　Medicinal use: Syphilis (India)

　　*Broad-spectrum antibacterial halo developed by polar extract:* Ethanol extract inhibited the growth of *Bacillus subtilis, Bacillus cereus, Micrococcus luteus, Staphylococcus aureus, Escherichia coli, Vibrio cholerae, Shigella dysenteriae, Shigella flexneri, Serratia marcescens, Salmonella typhi,* and *Salmonella enterica* (Biswas & Sinha, 2020).

　　*Anticandidal, mid-polar extract:* Ethyl acetate extract of leaves inhibited the growth of *Candida albicans* (Biswas & Sinha, 2020).

　　*Weak antibacterial broad-spectrum mid-polar extract:* Ethylacetate extract of leaves inhibited the growth of *Escherichia coli, Proteus mirabilis, Bacillus subtilis, Staphylococcus aureus,* and *Klebsiella pneumoniae* with MIC values of 78, 312, 1250, 156, and 1250 µg/mL, respectively (Gupta & Katewa, 2012.).

### References

Biswas, K. and Sinha, S.N., 2020. Evaluation of phytoconstituents and antibacterial activity of *Vanda tessellata* using in vitro model. In *Orchid Biology: Recent Trends & Challenges* (pp. 473–480). Singapore: Springer.

Gupta, C.H.H.A.V.I. and Katewa, S.S., 2012. Antimicrobial testing of *Vanda tessellate* leaf extracts. *Journal of Medicinal and Aromatic Plant Science, 34,* pp. 158–162.

## 3.2　COMMELINIDS

### 3.2.1　ORDER ARECALES BROMHEAD (1840)

The family Arecaceae consists of about 180 genera and 2500 species of palm trees, shrubs, or climbers. The bole is often smooth, greyish, and regularly marked with scars of fallen leaves. The leaves are palmate and coriaceous or fleshy. The inflorescence is cauliflorous fascicles of minute flowers. The perianth comprises 3 sepals and 3 petals. The androecium includes 6 stamens. The gynoecium comprises 3 carpels united into a 3-locular ovary with a few ovules. The fruit is a nut encapsulated in a fibrous husk or a berry.

#### 3.2.1.1　Family Arecaceae Schultz-Schultzenstein (1832)

*3.2.1.1.1　Areca catechu* L.

Common names: Betel palm, Areca palm; betel-nut palm; gooa, supari (Bangladesh); sla (Cambodia); kramuka (India); buah pinang (Indonesia, Malaysia); kok hmak (Laos); bunga (the Philippines); mak (Thailand); binh lang (Vietnam)

　　Synonyms: *Areca catechu* Willd.; *Areca faufel* Gaertn.; *Areca himalayana* Griff. ex H. Wendl.; *Areca hortensis* Lour.; *Areca nigra* Giseke ex H. Wendl.; *Sublimia areca* Comm. ex Mart.

　　Habitat: Cultivated

　　Distribution: Tropical Asia and Pacific

Botanical observation: This resinous palm tree grows to 20 m tall. The bole is straight, slender, 10–20 cm in diameter, greyish, and with conspicuous marks. The leaves' sheath forms a crown-shaft, which is up to 1 m. The petiole is up to 5 cm long. The rachis grows up to 2 m, and supports 20–30 pinnae per side, which are 30–60×3–7 cm. The inflorescence is up to 25 cm. The male flowers comprise 3 sepals and 3 petals, with stamens 6. The female flowers are larger than male flowers. The drupes are yellow, orange, or red, ovoid, hard, and about 7 cm×4 cm with a persistent calyx.

Medicinal uses: Syphilis, cholera, and diarrhea (Bangladesh)

*Broad-spectrum antibacterial halo developed by polar extract:* Methanol extract of seeds (6 mg; agar-well diffusion method) inhibited the growth of *Bacillus cereus*, *Listeria monocytogenes*, *Staphylococcus aureus*, *Escherichia coli*, and *Salmonella anatum* with inhibition zone diameters of 18.3, 17.9, 21.1, 8.6 and 12 mm, respectively (Shan et al., 2007).

*Broad-spectrum antibacterial halo developed by polar extract:* Aqueous extract of drupe (1 mg/mL in 6 mm agar wells) inhibited the growth of *Mucor* sp., *Aspergillus niger*, *Cladosporium* sp., and *Candida albicans* with inhibition zone diameters of 12, 14, 13, and 18 mm, respectively (Anthikat et al., 2014).

*Moderate antibacterial (Gram-positive) polar extract:* Aqueous extract of seeds inhibited the growth of *Streptococcus mutans* (MT 5091) and *Streptococcus mutans* (OMZ 176) with MIC values of 625 and 1250 μg/mL, respectively (Chen et al., 1989).

*Moderate antibacterial (Gram-negative) polar extract:* Methanol extract of seeds (containing catechin, epicatechin, and proanthocyanidins) inhibited the growth of *Streptococcus mutans* (strains MT 5091) (serotype *c*) with an MIC value of 100 μg/mL (Hada et al., 1989).

*Antifungal (filamentous) triterpene:* The triterpene fernenol isolated from the fruits inhibited the mycelial growth of *Colletotrichum gloeosporioides* by 50% when at a concentration of 36.7 mg/mL (Yenjit et al., 2010).

*Strong antiviral (enveloped monopartite single-stranded (+) RNA virus) hydrophilic flavanol:* (+)-Catechin (LogD = 0.5; molecular weight = 290.2 g/mol) at a concentration of 20 μM protected Swine testicle cells against the transmissible Gastroenteritis virus by about 85% and inhibited virus replication by about 50% (Liang et al., 2015).

(+)-catechin

*Strong antiviral (enveloped monopartite linear single-stranded (+) RNA) hydrophilic piperidine alkaloid:* Arecoline (LogD = 0.2 at pH 7.4; molecular weight = 155.1 g/mol) inhibited the replication of Japanese encephalitis virus in TE-671 cells with an $IC_{50}$ value of 32.5 μg/mL and selectivity index above 92.3 (Lin et al., 2008).

Arecoline

Commentary: The Transmissible gastroenteritis virus (TGEV) belongs to the family Coronaviridae, genus *Alphacoronavirus*. *Areca catechu* L. is used for the treatment of COVID-19 in China (Wang et al., 2020), and one could examine the anti-COVID-19 and other human pathogenic coronavirus properties of catechin and arecoline. Being neuroactive, arecoline is probably able to inhibit bacterial or fungal efflux pumps.

### References
Anthikat, R.R.N., Michael, A., Kinsalin, V.A. and Ignacimuthu, S., 2014. Antifungal activity of *Areca catechu* L. *International Journal of Clinical Pharmacy*, 4(1), pp. 1–3.
Chen, C.P., Lin, C.C. and Tsuneo, N., 1989. Screening of Taiwanese crude drugs for antibacterial activity against *Streptococcus* mutans. *Journal of Ethnopharmacology*, 27(3), pp. 285–295.
Hada, L.S., Kakiuchi, N., Hattori, M. and Namba, T., 1989. Identification of antibacterial principles against *Streptococcus* mutans and inhibitory principles against glucosyltransferase from the seed of *Areca catechu* L. *Phytotherapy Research*, 3(4), pp. 140–144.
Liang, W., He, L., Ning, P., Lin, J., Li, H., Lin, Z., Kang, K. and Zhang, Y., 2015. (+)-Catechin inhibition of transmissible gastroenteritis coronavirus in swine testicular cells is involved its antioxidation. *Research in Veterinary Science*, 103, pp. 28–33.
Lin, C.W., Wu, C.F., Hsiao, N.W., Chang, C.Y., Li, S.W., Wan, L., Lin, Y.J. and Lin, W.Y., 2008. Aloe-emodin is an interferon-inducing agent with antiviral activity against Japanese encephalitis virus and enterovirus 71. *International Journal of Antimicrobial Agents*, 32(4), pp. 355–359.
Shan, B., Cai, Y.Z., Brooks, J.D. and Corke, H., 2007. The in vitro antibacterial activity of dietary spice and medicinal herb extracts. *International Journal of Food Microbiology*, 117(1), pp. 112–119.
Wang, S.X., Wang, Y., Lu, Y.B., Li, J.Y., Song, Y.J., Nyamgerelt, M. and Wang, X.X., 2020. Diagnosis and treatment of novel coronavirus pneumonia based on the theory of traditional Chinese medicine. *Journal of Integrative Medicine*, 18(4), pp. 275–283.
Yenjit, P., Issarakraisila, M., Intana, W. and Chantrapromma, K., 2010. Fungicidal activity of compounds extracted from the pericarp of *Areca catechu* against *Colletotrichum gloeosporioides* in vitro and in mango fruit. *Postharvest Biology and Technology*, 55(2), pp. 129–132.

*3.2.1.1.2   Borassus flabellifer* L.

Synonyms: *Borassus flabelliformis* Murray; *Borassus flabelliformis* Roxb.

Common names: Palmyra palm, toddy palm; tar (Bangladesh, India); tnaot (Cambodia); talah (India); tal (Indonesia); tan (Laos, Thailand); lontar (Malaysia); tan bin (Myanmar); thot lot (Vietnam)

Habitat: Cultivated

Distribution: India, Sri Lanka, Bangladesh, and Southeast Asia

Botanical observation: This palm tree grows to 20 m tall. The bole is rough and upright. The petiole is 60–120 cm long. The blade is 60–120 cm long, with 60–80 linear-lanceolate, and induplicate segments. The male inflorescence is 30–150 cm long. The calyx comprises 3 sepals. The corolla includes 3 petals. The androecium includes 6 stamens. The ovary is globose and minute. The nuts are globose, edible, purple, 15–20 cm in diameter, and contain 3 seeds.

Medicinal uses: Abscesses, cuts, dysentery, and fever (India)

*Strong broad-spectrum antibacterial polar extract:* Methanol extract of leaves inhibited the growth of *Bacillus subtilis* (B28) *Staphylococcus aureus* (MTC96), *Escherichia coli* (MTCC 170), and *Pseudomonas aeruginosa* (CC 488) with MIC/MBC values of 60/70, 50/60, 70/80, and 60/70 μg/mL, respectively (Jamkhande et al., 2016).

*Strong broad-spectrum antifungal polar extract:* Methanol extract of leaves inhibited the growth of *Aspergillus flavus* (MTCC 873), *Microsporium canis* (MTCC 2520), *Saccharomyces cerevisiae* (MCIM 170), and *Vestilago myditis* (MCIM 983) with MIC/MFC values of 50/50, 50/70, 60/60, and 60/0 6μg/mL, respectively (Jamkhande et al., 2016).

*In vivo broad-spectrum antibacterial steroidal saponin:* An ointment consisting of white soft paraffin containing 2% of a steroidal saponin isolated from the fruits (impregnated on 4 cm×4 cm piece of gauze) evoked wound healing in patients. This steroidal saponin abrogated the growth of *Staphylococcus* spp., *Pseudomonas* spp., and Gram-positive bacilli from the infected wounds at a concentration of 4 mg/mL (Keerthi et al., 2007).

Commentary: The plant generates an interesting series of spirostane saponins (Yoshikawa et al., 2007). Spirostane saponins have the tendency to be antiviral.

**References**

Jamkhande, P.G., Suryawanshi, V.A., Kaylankar, T.M. and Patwekar, S.L., 2016. Biological activities of leaves of ethnomedicinal plant, *Borassus flabellifer* Linn. (Palmyra palm): an antibacterial, antifungal and antioxidant evaluation. *Bulletin of Faculty of Pharmacy, Cairo University*, 54(1), pp. 59–66.

Keerthi, A.A.P., Mendis, W.S.J., Jansz, E.R., Ekanayake, S. and Perera, M.S.A., 2007. A preliminary study on the effects of an antibacterial steroidal saponin from *Borassus flabellifer* L. fruit, on wound healing. *Journal of the National Science Foundation of Sri Lanka*, 35.

Yoshikawa, M., Xu, F., Morikawa, T., Pongpiriyadacha, Y., Nakamura, S., Asao, Y., Kumahara, A. and Matsuda, H., 2007. Medicinal flowers. XII. 1) New spirostane-type steroid saponins with antidiabetogenic activity from *Borassus flabellifer*. *Chemical and Pharmaceutical Bulletin*, 55(2), pp. 308–316.

### 3.2.1.2   Family Xanthorrhoeaceae Dumortier (1829)

The family Xanthorrhoeaceae consists of about 60 genera and 1450 species of trees or rhizomatous herbs not uncommonly laticiferous. The stems or leaves secrete a yellow, red, or brown resin. The leaves are simple, spiral, and coriaceous or fleshy. The inflorescence is a spike. The corolla consists of 3 sepals which are petaloid. The perianth comprises 6 tepals. The androecium includes 6 stamens. The gynoecium comprises 3 carpels united into a 3-locular ovary with few ovules. The fruit is a capsule.

### 3.2.1.2.1   *Aloe vera (L.) Burm.f.*

Common names: Indian Aloe, Barbados Aloe, Curaçao aloe; ghrita koomari (Bangladesh); Kathalai (India); Lidah buaya (Indonesia, Malaysia); dilang-halo (The Philippines); waan faimai (Thailand); nha dầm (Vietnam)

Synonyms: *Aloe barbadensis* Mill.; *Aloe barbadensis* var. *chinensis* Haw.; *Aloe chinensis* (Haw.) Baker; *Aloe perfoliata* var. vera L.; *Aloe vera* var. *chinensis* (Haw.) A. Berger; *Aloe vulgaris* Lam.

Habitat: Cultivated

Distribution: Tropical and subtropical Asia and Pacific

Botanical examination: This fleshy herb forming a rosette, which grows to 80 cm tall. The leaves are linear lanceolate, spiny at margin, fleshy, 15–35×4–5 cm, produce a yellow gum when incised, and accumulate an abundant translucent and crystal-clear aqueous gel. The inflorescence is an erect raceme, which is 60–90 cm tall. The perianth is 2.5–3 cm long, tubular, light-yellow to orangish, with 6 petals. The ovary is ovoid and minute, the style elongated, and the stigma 3-lobed. The capsules are 1.5 cm long.

Medicinal uses: Cough (Indonesia); dysentery (Bangladesh, the Philippines); wounds (Malaysia)

*Moderate broad-spectrum polar extract:* Ethanol extract of leaves inhibited the growth of clinical isolates of *Enterococcus bovis, Staphylococcus aureus, Proteus vulgaris, Proteus mirabilis, Pseudomonas aeruginosa*, and *Morganella morganii* with MIC values of 500, 500, 500, 500, 100, and 300 µg/mL, respectively (Panday & Mishra, 2010).

*Moderate broad-spectrum antibacterial anthraquinone:* Aloe-emodin (also known as rhabarberone) inhibited the growth of *Bacillus subtilis, Micrococcus kristinae, Bacillus cereus, Staphylococcus aureus, Staphylococcus epidermidis, Escherichia coli, Proteus vulgaris, Enterobacter aerogenes*, and *Shigella sonnei* with MIC values below 250 µg/mL (Coopoosamy & Magwa, 2006).

*Moderate broad-spectrum antibacterial C-glycosyl anthrone:* Aloin A (also known as aloin or barbaloin) inhibited the growth of *Bacillus subtilis, Micrococcus kristinae, Bacillus cereus, Staphylococcus aureus, Staphylococcus epidermidis, Escherichia coli, Proteus vulgaris, Enterobacter aerogenes*, and *Shigella sonnei* with MIC values of 62.5, 125, 62.5, 62.5, 125, 125, 125, 250, and 250 µg/mL, respectively (Coopoosamy & Magwa, 2006).

*Moderate antimycobacterial anthraquinones:* Aloe-emodin inhibited the growth of *Mycobacterium tuberculosis* (H37Ra) and *Mycobacterium bovis* with MIC/MBC values of 64/128 and 64/128 µg/mL, respectively (Smolarz et al., 2013). Chrysophanol inhibited the growth of *Mycobacterium tuberculosis* (H37Ra) and *Mycobacterium bovis* with MIC/MBC values of 64/128 and 64/128 µg/mL, respectively (Smolarz et al., 2013).

*Strong antimycobacterial hydrophilic C-glycosyl anthrone:* Aloin A (LogD = 1.0; molecular weight = 418.3 g/mol) inhibited the growth of *Mycobacterium tuberculosis* (H37Ra) and *Mycobacterium bovis* with MIC/MBC values of 32/64 and 128/256 µg/mL, respectively (Smolarz et al., 2013).

Aloin A

*Weak antifungal (filamentous) anthraquinone:* Aloe-emodin inhibited the growth of *Aspergillus niger*, *Cladosporium herbarum*, and *Fusarium moniliforme* with MIC values of 250, 250 and 125 μg/mL, respectively (Ali et al., 1999).

*Strong broad-spectrum antiviral amphiphilic anthraquinones:* Chrysophanol (LogD = 2.5 at pH 7.4; molecular weight = 254.2 g/mol) inhibited the cytopathic effects of the Poliovirus types 2 and 3 viruses in Buffalo green monkey kidney cells with $IC_{50}$ values of 0.2 and 0.02 μg/mL, respectively). Aloe-emodin (LogD = 1.2 at pH 7.4; molecular weight = 270.2 g/mol) inhibited the cytopathic effects of the Poliovirus type 3 in Buffalo green monkey kidney cells with an $IC_{50}$ value of 0.5 μg/mL and a selectivity index of 3 (Semple et al., 2001). Aloe-emodin inhibited the replication of the Japanese encephalitis virus and Enterovirus 71 in TE-671 cells with $IC_{50}$ values of 1.5 and 0.1 μg/mL and selectivity indices of 1743 and 18,800 (Lin et al., 2008). In a subsequent study, aloe-emodin inhibited the replication of the Japanese encephalitis virus (T1P1 strain) in BHK-21 cells with an $IC_{50}$ value of 17.3 μg/mL and was virucidal with an $IC_{50}$ value of 0.4 μg/mL (Chang et al., 2014). Aloe-emodin, chrysophanol, and aloin B at a concentration of 10 μg/mL decreased HBsAg synthesis by 81.7, 65.5, and 62% in HepG2.2.15 cells (Parvez et al., 2019).

Chrysophanol

Aloe-emodin

*Antiviral enzyme inhibition by anthraquinone:* Aloe-emodin inhibited Severe acute respiratory-associated coronavirus 3C-like protease with an $EC_{50}$ value of 35.7 µg/mL (Lin et al., 2005).

Commentaries: (i) The plant brings to being the anthracene anthranol, the anthraquinone chrysophanic acid, aloin and aloe emodin, and the chromone *C*-glycoside aloesin (Chiang et al., 2012; Waller, 1978). (ii) Consider that $IC_{50}$ values as well as selectivity indices for compounds tested for antiviral activity *in vitro* depend on the type of host cell used. One could examine the anti-COVID-19 activities of aloe-emodin and chrysophanol.

### References

Ali, M.I., Shalaby, N.M., Elgamal, M.H. and Mousa, A.S., 1999. Antifungal effects of different plant extracts and their major components of selected Aloe species. *Phytotherapy Research*, *13*(5), pp. 401–407.

Chang, S.J., Huang, S.H., Lin, Y.J., Tsou, Y.Y. and Lin, C.W., 2014. Antiviral activity of *Rheum palmatum* methanol extract and chrysophanol against Japanese encephalitis virus. *Archives of Pharmacal Research*, *37*(9), pp. 1117–1123.

Chiang, H.M., Lin, Y.T., Hsiao, P.L., Su, Y.H., Tsao, H.T. and Wen, K.C., 2012. Determination of marked components—aloin and aloe-emodin—in Aloe vera before and after hydrolysis. *Journal of Food and Drug Analysis*, *20*(3), pp. 646–652.

Coopoosamy, R.M. and Magwa, M.L., 2006. Antibacterial activity of aloe emodin and aloin A isolated from *Aloe excelsa*. *African Journal of Biotechnology*, *5*(11).

Lin, C.W., Tsai, F.J., Tsai, C.H., Lai, C.C., Wan, L., Ho, T.Y., Hsieh, C.C. and Chao, P.D.L., 2005. Anti-SARS coronavirus 3C-like protease effects of *Isatis indigotica* root and plant-derived phenolic compounds. *Antiviral Research*, *68*(1), pp. 36–42.

Lin, C.W., Wu, C.F., Hsiao, N.W., Chang, C.Y., Li, S.W., Wan, L., Lin, Y.J. and Lin, W.Y., 2008. Aloe-emodin is an interferon-inducing agent with antiviral activity against Japanese encephalitis virus and enterovirus 71. *International Journal of Antimicrobial Agents*, *32*(4), pp. 355–359.

Pandey, R. and Mishra, A., 2010. Antibacterial activities of crude extract of *Aloe barbadensis* to clinically isolated bacterial pathogens. *Applied Biochemistry and Biotechnology*, *160*(5), pp. 1356–1361.

Parvez, M.K., Al-Dosari, M.S., Alam, P., Rehman, M., Alajmi, M.F. and Alqahtani, A.S., 2019. The anti-Hepatitis B virus therapeutic potential of anthraquinones derived from Aloe vera. *Phytotherapy Research*, *33*(11), pp. 2960–2970.

Semple, S.J., Pyke, S.M., Reynolds, G.D. and Flower, R.L., 2001. In vitro antiviral activity of the anthraquinone chrysophanic acid against poliovirus. *Antiviral Research*, *49*(3), pp. 169–178.

Smolarz, H.D., Swatko-Ossor, M., Ginalska, G. and Medyńska, E., 2013. Antimycobacterial effect of extract and its components from *Rheum rhaponticum*. *Journal of AOAC International*, *96*(1), pp. 155–160.

Waller, G.R., 1978. A Chemical Investigation of *Aloe barbadensis* Miller. *Proceedings of the Oklahoma Academy of Science* (Vol. 58, pp. 69–76).

### 3.2.1.2.2   *Dianella ensifolia* (L.) DC

Common names: Umbrella dracaena; Shan jian (China); milam (India); labeh labeh, jambaka (Indonesia); Kikyo ran (Japan); akar siak (Malaysia); bururl (Papua New Guinea); bariu-bariu (the Philippines); ha dia sua, lamphan (Thailand); Luoi dong (Vietnam)

Synonyms: *Anthericum adenanthera* G. Forst.; *Dianella javanica* Kunth; *Dianella mauritiana* Bl.; *Dianella montana* Bl.; *Dianella nemorosa* Lam.; *Dianella odorata* Bl.; *Dianella sandwicensis* Hook. & Arn.; *Dracaena ensata* Thunb. & Dallm.; *Dracaena ensifolia* L.; *Phalangium adenanthera* Poir.; *Rhuacophila javanica* Bl.; *Walleria paniculata* Fritsch

Habitat: Cultivated

Distribution: India, Sri Lanka, Nepal, Bangladesh, Bhutan, Myanmar, Southeast Asia, China, Taiwan, Japan, Papua New Guinea, Australia, and Pacific Islands

Botanical examination: This poisonous herb grows from a rhizome. The plant is used for protective rituals for rice farming in Indonesia. The leaves are ensiform, up to 80 cm×2.5 cm, serrate, and obtuse at apex. The inflorescence is a panicle at apex of a peduncle that can grow up to about 50 cm tall. The perianth includes 6 tepals, which are white, yellowish, or purplish, linear-lanceolate, and up to 7 mm long. The androecium includes 6 stamens, which are about the same length as the tepals and with oblong anthers. The style grows to about 6 mm long. The berries are blue, somewhat globose, about 5 mm across, and glossy.

Medicinal uses: Abscesses, carbuncles (China); herpes (Malaysia); colds, flu (Thailand); leucorrhea, blennorrhea (Cambodia, Laos, and Vietnam)

Phytochemical class: Phenolics

*Moderate antibacterial broad-spectrum essential oil:* Essential oil of aerial parts inhibited the growth of *Staphylococcus aureus* (ATCC 6538), *Bacillus subtilis* (ATCC 6633), *Escherichia coli* (ATCC 25922), and *Pseudomonas aeruginosa* (ATCC 27853) with MIC/MBC values of 160/160, 310/630, 160/310, and 310/630 µg/mL, respectively (He et al., 2019).

*Weak antibacterial anthraquinone:* Chrysophanol inhibited the growth of methicillin-resistant *Staphylococcus aureus* with an MIC of 256 µg/mL (Hatano et al., 1999). Chrysophanol inhibited the growth of *Staphylococcus epidermidis*, *Escherichia coli*, and *Proteus vulgaris* with MIC values of 31.2, 125, and 125 µg/mL (Coopoosamy & Magwa, 2006).

*Viral enzyme inhibition by mid-polar extract:* Ethyl acetate extract of fruits inhibited A(H1N1) Influenza virus neuraminidase by 31.5% at a concentration of 40 µg/mL (Liu et al., 2018).

Commentaries: (i) The plant generates series of chromanes and chromones (Nhung et al., 2019) as well as the anthraquinone chrysophanol (chrysophanic acid or 1,8-dihydroxy-3-methylanthraquinone) (Semple et al., 2001). Consider that anthraquinones are not uncommonly antibacterial, antifungal, and antiviral at least partially, on account of their planar structures, which favor DNA intercalation (hence risks of mutagenicity). For instance, emodin (LogD = 1.7 at pH 7.4; molecular mass = 270.3 g/mol) also known as 6-methyl-1,3,8-trihydroxyanthraquinone (which occurs in some members of the genus *Aloe* L. (Ombito et al., 2015) as well as plants belonging to the family Polygonaceae) inhibited the cytopathic effects of the Poliovirus type 3 in Buffalo green monkey kidney cells with an $IC_{50}$ value of 0.2 µg/mL and a selectivity index of 12.4 (Semple et al., 2001). Emodin inhibited Coxsackie B4 ($CVB_4$) plaque reduction in Hep-2 cells with an $EC_{50}$ of 12 µM and a selectivity index of 5. This anthraquinone given orally to BALB/c mice at a dose of 30 mg/Kg/day for 14 days increased lifetime as efficiently as Ribavirin at 10 mg/Kg/day. Emodin at a concentration of 50 µM inhibited the Severe acute respiratory-associated coronavirus S protein and angiotensin-converting enzyme 2 interaction by about 40% (Ho et al., 2007). Emodin at concentrations of 50 and 25 µg/mL reduced by about 50% the virus titer of Herpes simplex virus type-1 and Herpes simplex virus type-2, respectively. *In vivo*, emodin given orally at a dose of

6.7 g/Kg/day, three times daily for 7 days afforded a prolongation of the survival time of BALB/c mice intracerebrally infected with Herpes simplex virus type-1 and Herpes simplex virus type-2 superior to acyclovir given orally at the dose of 100 mg/Kg/day for 7 days (Xiong et al., 2011). Anthraquinones not uncommonly display strong antibacterial activities against Gram-positive bacteria. Emodin inhibited methicillin-resistant *Staphylococcus aureus* with an MIC of 25 µg/mL (Joung et al., 2012). One could examine the anti-COVID-19 properties of emodin.

Emodin

### References

Coopoosamy, R.M. and Magwa, M.L., 2006. Antibacterial activity of chrysophanol isolated from *Aloe excelsa* (Berger). *African Journal of Biotechnology*, 5(16).

Hatano, T., Uebayashi, H., Ito, H., Shiota, S., Tsuchiya, T. and Yoshida, T., 1999. Phenolic constituents of *Cassia* seed and antibacterial effect of some naphthalenes and antraquinones on Methicillin-Resistant *Staphylococcus aureus*. *Chemical Pharmacology Bulletin*, 47, pp. 1121–1127.

He, Z.Q., Shen, X.Y., Cheng, Z.Y., Wang, R.L., Lai, P.X. and Xing, X., 2019. Chemical composition, antibacterial, antioxidant and cytotoxic activities of the essential oil of *Dianella ensifolia*.

Ho, T.Y., Wu, S.L., Chen, J.C., Li, C.C. and Hsiang, C.Y., 2007. Emodin blocks the SARS coronavirus spike protein and angiotensin-converting enzyme 2 interaction. *Antiviral Research*, 74(2), pp. 92–101.

Joung, D.K., Joung, H., Yang, D.W., Kwon, D.Y., Choi, J.G., Woo, S., Shin, D.Y., Kweon, O.H., Kweon, K.T. and Shin, D.W., 2012. Synergistic effect of rhein in combination with ampicillin or oxacillin against methicillin-resistant *Staphylococcus aureus*. *Experimental and Therapeutic Medicine*, 3(4), pp. 608–612.

Liu, J., Zu, M., Chen, K., Gao, L., Min, H., Zhuo, W., Chen, W. and Liu, A., 2018. Screening of neuraminidase inhibitory activities of some medicinal plants traditionally used in Lingnan Chinese medicines. *BMC Complementary and Alternative Medicine*, 18(1), p. 102.

Nhung, L.T.H., Linh, N.T.T., Cham, B.T., Thuy, T.T., Tam, N.T., Thien, D.D., Huong, P.T.M., Tan, V.M., Tai, B.H. and Hoang Anh, N.T., 2019. New phenolics from *Dianella ensifolia*. *Natural Product Research*, pp. 1–8.

Ombito, J.O., Salano, E.N., Yegon, P.K., Ngetich, W.K., Mwangi, E.M., Koech, G.K.K. and Yegon, K., 2015. A review of the chemistry of some species of genus Aloe (*Xanthorrhoeaceae* family). *Journal of Scientific and Innovative Research*, 4(1), pp. 49–53.

Semple, S.J., Pyke, S.M., Reynolds, G.D. and Flower, R.L., 2001. In vitro antiviral activity of the anthraquinone chrysophanic acid against poliovirus. *Antiviral Research*, 49(3), pp. 169–178.

*3.2.1.2.3    Hemerocallis fulva* (L.) L.

Common names: fulvous daylily; xuan cao (China); Kuankai (India); Wonchuri (Korea)

Synonym: *Hemerocallis lilioasphodelus* var. *fulvus L.*

Habitat: Grassy lands, riverbanks, cultivated

Distribution: Himalaya, India, China, Taiwan, Japan, and Korea

Botanical observation: This ornamental herb grows up to a length of 1.5 m from tuberous roots. The plant is used for food by Buddhist monks. The leaves are simple, basal, and alternate. The blade is linear, up to 90 cm×3 cm, and acute at apex. The inflorescences are cymes at apex of a slender peduncle. The perianth includes 6 oblong and a somewhat dull, light reddish to orangish (papaya fruit flesh colored) or yellow tepals, which are recurved and about 10 cm long, wavy, and with a paler midrib. The androecium includes 6 slender stamens, which are about 5 cm long with purplish dorsifixed anthers, which are about 8 mm long. The style is slender and the stigma capitate. The capsule is oblong, trigonous, dehiscent, and up to about 2 cm long. The seeds are numerous and black.

Medicinal uses: Fever, jaundice (China)

*Antibacterial (Gram-positive) mid-polar extract:* Ethyl acetate extract (100 µg/disc) inhibited the growth of *Streptococcus mutans* (OMZ 176) with the inhibition zone diameter between about 13 and 15 mm (Do et al., 2002).

*Weak broad-spectrum antibacterial anthraquinones:* Chrysophanol inhibited the growth of *Bacillus subtilis, Staphylococcus epidermidis, Escherichia coli, Aeromonas hydrophila, Proteus vulgaris, Pseudomonas aeruginosa*, and *Vibrio harveyi* with MIC values of 250, 31.2, 125, 200, 125, 128, and 1000 µg/mL, respectively (Yusuf et al., 2019). Rhein inhibited the growth of *Escherichia coli* (K12), *Staphylococcus aureus* (209P) and methicillin-resistant *Staphylococcus aureus* (OM481) with MIC values of 128, 32, and 32 µg/mL, respectively (Hatano et al., 1999).

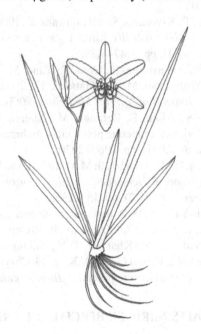

*Hemerocallis fulva* (L.) L.

*Strong broad-spectrum antifungal amphiphilic anthraquinone:* Chrysophanol (LogD = 2.5 at pH 7.4; molecular weight = 254.2 g/mol) inhibited the growth of *Blumeria graminis* f. sp. *hordei* with an IC$_{50}$ value of 4.7 µg/mL (Yusuf et al., 2019). At the concentration of 100 µg/mL it inhibited the growth of *Podosphaera xanthii* (Yusuf et al., 2019). Chrysophanol inhibited the growth of *Candida albicans, Cryptococcus neoformans, Trichophyton mentagrophytes*, and *Aspergillus fumigatus, Trichophyton rubrum,* and *Epidermophyton floccosum* with MIC values of 50, 50, 25, 50, 156, and 625 µg/mL, respectively (Yusuf et al., 2019). At a concentration of 500 µg/mL, chrysophanol inhibited the growth of *Botrytis cinerea* and *Rhizoctonia solani* by 21.2 and 22.5%, respectively (Yusuf et al., 2019).

*Strong antiviral (enveloped monopartite linear single-stranded (+) RNA) anthraquinone:* Chrysophanol inhibited the replication of the Japanese encephalitis virus with an $IC_{50}$ value of 15.8 µg/mL and was virucidal with an $IC_{50}$ value of 0.7 µg/mL (Chang et al., 2014).

Commentary: The plant synthetizes the anthraquinones: chrysophanol, methyl rhein, 1, 8-dihydroxy-3-methoxy-anthraquinone, and rhein (Sarg et al., 1990), as well as an interesting series of furans (Inoue et al., 1990), γ-lactam pyrrolidine alkaloids (Matsumoto et al., 2016), and diterpenes (Yang et al., 2003). Chrysophanol has anti-Hepatitis B activity *in vitro*, as presented earlier (Parvez et al., 2019).

### References

Chang, S.J., Huang, S.H., Lin, Y.J., Tsou, Y.Y. and Lin, C.W., 2014. Antiviral activity of Rheum palmatum methanol extract and chrysophanol against Japanese encephalitis virus. *Archives of Pharmacal Research*, 37(9), pp. 1117–1123.

Do, D.S., Lee, S.M., Na, M.K. and Bae, K.H., 2002. Antimicrobial activity of medicinal plant extracts against *Cariogenic Bacterium, Streptococcus* mutans OMZ 176. *Korean Journal of Pharmacognosy*, 33(4), pp. 319–323.

Hatano, T., Uebayashi, H., Ito, H., Shiota, S., Tsuchiya, T. and Yoshida, T., 1999. Phenolic constituents of Cassia seeds and antibacterial effect of some naphthalenes and anthraquinones on methicillin-resistant *Staphylococcus aureus*. *Chemical and Pharmaceutical Bulletin*, 47(8), pp. 1121–1127.

Inoue, T., Iwagoe, K., Konishi, T., Kiyosawa, S. and Fujiwara, Y., 1990. Novel 2, 5-dihydrofuryl-γ-lactam derivatives from *Hemerocallis fulva* L. var. kwanzo Regel. *Chemical and Pharmaceutical Bulletin*, 38(11), pp. 3187–3189.

Matsumoto, T., Nakamura, S., Nakashima, S., Ohta, T., Yano, M., Tsujihata, J., Tsukioka, J., Ogawa, K., Fukaya, M., Yoshikawa, M. and Matsuda, H., 2016. γ-Lactam alkaloids from the flower buds of daylily. *Journal of Natural Medicines*, 70(3), pp. 376–383.

Parvez, M.K., Al-Dosari, M.S., Alam, P., Rehman, M., Alajmi, M.F. and Alqahtani, A.S., 2019. The anti-Hepatitis B virus therapeutic potential of anthraquinones derived from Aloe vera. *Phytotherapy Research*, 33(11), pp. 2960–2970.

Sarg, T.M., Salem, S.A., Farrag, N.M., Abdel-Aal, M.M. and Ateya, A.M., 1990. Phytochemical and antimicrobial investigation of *Hemerocallis fulva* L. grown in Egypt. *International Journal of Crude Drug Research*, 28(2), pp. 153–156.

Yang, Z.D., Chen, H. and Li, Y.C., 2003. A new glycoside and a novel-type diterpene from *Hemerocallis fulva* (L.) L. *Helvetica Chimica Acta*, 86(10), pp. 3305–3309.

Yusuf, M.A., Singh, B.N., Sudheer, S., Kharwar, R.N., Siddiqui, S., Abdel-Azeem, A.M., Fernandes Fraceto, L., Dashora, K. and Gupta, V.K., 2019. Chrysophanol: a natural anthraquinone with multifaceted biotherapeutic potential. *Biomolecules*, 9(2), p. 68.

## 3.3 ORDER COMMELINALES MIRB. EX BERCHT. & J. PRESL. (1820)

### 3.3.1 Family Commelinaceae Mirbel (1804)

The family Commelinaceae consists of 50 genera and 700 species of tropical and often water loving herbs. The leaves are simple and spiral. The stems are often articulate, stoloniferous, and characterized by tubular bracts. The inflorescences are axillary clusters or terminal cymes or panicles, occasionally within a folded boat-shaped bract. The calyx comprises 3 sepals. The corolla consists of 3 petals. The androecium consists of 6 stamens, which are often hairy at base. The gynoecium consists of 3 carpels united into a compound, superior, and a 3-locular ovary. The style is terminal and simple and the stigma small. The fruits are loculicidally dehiscent capsules or succulent. The seeds are muricate, ridged, or reticulate.

### 3.3.1.1  *Commelina benghalensis* L.

Common names: Benghal Dayflower, tropical spiderwort; kanisira (Bangladesh); fan bao cao (China); kanchata, motishumliyu (India); tali korang (Indonesia); rumput mayiam (Malaysia); kabi-lau (Malaysia); myet-cho (Myanmar); kab pi (Laos); thài lài long (Vietnam)

Synonyms: *Commelina canescens* Vahl; *Commelina cavaleriei* H. Lév.; *Commelina delicatula* Schltdl.; *Commelina mollis* Jacq.; *Commelina nervosa* Burm. f.; *Commelina procurrens* Schltdl.; *Commelina turbinata* Vahl

Habitat: Watery and shady spots

Distribution: Tropical and subtropical Asia

Botanical examination: This discrete and creeping herb grows to a length of 70 cm. The leaves are simple, alternate, exstipulate. The petiole is sheathing at base and up to about 2 cm long. The blade is broadly lanceolate to elliptic, 3–7 × 1.5–3.5 cm, somewhat fleshy, wavy, and marked with about 3 pairs of longitudinal secondary nerves. The flowers are arranged in solitary and terminal cymes. The calyx includes 3 sepals, which are minute and membranous. The corolla include 3 petals of a peculiar plain blue, of which the upper 2 are about 5 mm long, membranous, clawed, somewhat kidney shaped, and wavy. The androecium includes 6 stamens: 3 fertile and 2 developing into sort of slender hooks. The capsule is oblong, about 5 mm long, trilobed, dehiscent, and sheltering numerous tiny and somewhat tuberculate seeds.

Medicinal use: Fever, leprosy (India)

*Broad-spectrum antibacterial halo developed by polar solvent:* Ethanol extract of leaves inhibited the growth of *Staphylococcus aureus*, *Streptococcus pyogenes*, *Shigella dysenteriae*, and *Salmonella typhi* with inhibition zone diameters of 10, 12, 10, and 10 mm, respectively (0.5 mg/disc) (Uddin et al., 2008).

Commentaries:

(i) Constituents in this plant are apparently unknown. Consider that members of the genus *Commelina* L. synthetize simple phenolics such as methyl gallate, *p*-hydroxybenzoic acid, and protocatechuic acid, as well as flavones and flavone glycosides and stilbenes, including rhaponticin (Yuan et al., 2013). Consider that aquatic plants not uncommonly produce simple phenolics, including methyl gallate.

Methyl gallate

(ii) Methyl gallate (also known as gallic acid methyl ester or methyl 3,4,5-trihydroxybenzoate, LogD = 1.1 at pH 7.4; molecular weight = 184.1 g/mol) inhibited the growth of *Shigella dysenteriae* 1 (NT4907), *Shigella flexneri* 2a (B294), *Shigella boydii* (BCH612), and *Shigella sonnei* (1DH00968SS) with MIC/MBC values of 128/256, 128/512, 128/256, and 256/512 μg/mL, respectively, via fatal insults of internal and external membrane inducing leaking of intracellular contents (Acharyya et al., 2015). Methyl gallate inhibited the growth of *Pseudomonas aeruginosa* (Z61), *Escherichia coli* (DC2), and *Staphylococcus aureus* (RN4220) with MIC values of 12.5, 25, and 200 μg/mL, respectively (Saxena et al., 1994). Methyl gallate inhibited the replication of the Herpes simplex virus type 2 (MS strain), Herpes simplex virus type 1 (MacIntyre strain), Cytomegalovirus, Influenza A, PR/8/34 (H1N1), and Vesicular stomatitis virus with $IC_{50}$ values of 0.2, 0.6, 6.9, 176, and 350 μg/mL (Kane et al., 1988). In Herpes simplex virus methyl gallate may interact with virus proteins and alter the adsorption and penetration of the virion (Kane et al., 1988). Note that methyl gallate at a concentration of 2.5 mg/mL inhibited the survival of intracellular *Mycobacterium fortuitum* and *Mycobacterium tuberculosis* in murine peritoneal macrophages by about 80% (Kim et al., 2009). Combinations of gallic acid and methyl gallate had stronger intracellular killing activity against *Mycobacterium fortuitum*, illustrating the phenomenon that compounds produced by plants have synergistic antimicrobial properties (Kim et al., 2009).

(iii) Rhaponticin inhibited the growth of *Mycobacterium tuberculosis* (H37Ra) and *M. bovis* with MIC/MBC values of 128/256 and 128/256 μg/mL, respectively (Smolarz et al., 2013). It may therefore have some level of activity against *Mycobacterium leprae*.

Rhaponticin

### References

Acharyya, S., Sarkar, P., Saha, D.R., Patra, A., Ramamurthy, T. and Bag, P.K., 2015. Intracellular and membrane-damaging activities of methyl gallate isolated from *Terminalia chebula* against multidrug-resistant Shigella spp. *Journal of Medical Microbiology*, 64(8), pp. 901–909.

Kane, C.J., Menna, J.H., Sung, C.C. and Yeh, Y.C., 1988. Methyl gallate, methyl-3, 4, 5-trihydroxybenzoate, is a potent and highly specific inhibitor of herpes simplex virus in vitro. II. Antiviral activity of methyl gallate and its derivatives. *Bioscience Reports*, 8(1), pp. 95–102.

Kim, C.E., Griffiths, W.J. and Taylor, P.W., 2009. Components derived from Pelargonium stimulate macrophage killing of *Mycobacterium* species. *Journal of Applied Microbiology*, 106(4), pp. 1184–1193.

Saxena, G., McCutcheon, A.R., Farmer, S., Towers, G.H.N. and Hancock, R.E.W., 1994. Antimicrobial constituents of *Rhus glabra*. *Journal of Ethnopharmacology*, 42(2), pp. 95–99.

Smolarz, H.D., Swatko-Ossor, M., Ginalska, G. and Medyńska, E., 2013. Antimycobacterial effect of extract and its components from *Rheum rhaponticum*. *Journal of AOAC International*, 96(1), pp. 155–160.

Uddin, S.J., Rouf, R., Shilpi, J.A., Alamgir, M., Nahar, L. and Sarker, S.D., 2008. Screening of some Bangladeshi medicinal plants for in vitro antibacterial activity. *Oriental Pharmacy and Experimental Medicine*, 8(3), pp. 316–321.

Yuan, H.E., Zhou, X.D., Meng, L.J., Qin, F.M. and Zhou, G.X., 2013. Chemical constituents from Commelina communis. *Zhongguo zhongyao zazhi = China Journal of Chinese Materia Medica*, 38(19), pp. 3304–3308.

### 3.3.1.2 *Commelina paludosa Bl.*

Common names: Swamp dayflower; Bat boitta shak, batbattye shak, janhi rachong, (Bangladesh); da bao ya zhi cao (China); kankowa, kena (India)

Synonyms: *Commelina obliqua* Buch.-Ham. ex D. Don; *Commelina obliqua* Vahl

Habitat: Riverbanks, wet and shady spots in forests

Distribution: Pakistan, India, Sri Lanka, Nepal, Sikkim, Bhutan, Myanmar, Cambodia, Laos, Vietnam, Thailand, Malaysia, China, Taiwan, Indonesia, and the Philippines

Botanical description: This herb grows up to 1 m long. The leaves are simple, sheathing at base, sessile, edible, and alternate. The sheath is hairy at the mouth. The blade is lanceolate to ovate, 6.5–20 cm×2–7 cm, hairy below, somewhat dark green and glossy, fleshy, with discrete longitudinal nerves, and acute at apex. The inflorescences are terminal or axillary. The calyx comprises 3 sepals, which are up to 6 mm long. The corolla includes 3 petals, 2 of which are spathulate, reniform, and acute at apex, with a very peculiar type of blue, and up to 8 mm long. The androecium includes 6 stamens of which 3 fertile and 2 developing into sort of slender hooks. The capsules are ovoid, trigonous, up to 4 mm long, and contain 3 seeds.

Medicinal use: Dysentery (Bangladesh); fever (Myanmar)

Pharmacology: Antibacterial

*Antibacterial (Gram-positive) halo developed by polar extract:* Methanol extract of leaves inhibited the growth of *Bacillus cereus* (clinical isolate), *Staphylococcus aureus* (MTCC 1144), and *Salmonella typhimurium* (MTCC 3216) (Panda et al., 2016).

Commentary: Apparently no one has ever isolated any constituents from this plant. Phenolics by means of mild astringent effects may, at least partially, account for the antidysenteric property.

### Reference

Panda, S.K., Mohanta, Y.K., Padhi, L., Park, Y.H., Mohanta, T.K. and Bae, H., 2016. Large scale screening of ethnomedicinal plants for identification of potential antibacterial compounds. *Molecules*, 21(3), p. 293.

### 3.3.1.3 *Tradescantia spathacea Sw.*

Common names: Oyster plant, Moses in a cradle, boat lily; zi bei wan nian qing (China); riri (Cook Islands); Bunga Adam dan Awa (Malaysia); Bangka bangkaan (the Philippines); mi guin gamoni (Myanmar); waan kap hoi yai (Thailand); faina (Tonga); so huyet (Vietnam)

Synonyms: *Ephemerum discolor* (L'Hér.) Moench; *Rhoeo discolor* (L'Hér.) Hance ex Walp.; *Rhoeo spathacea* (Sw.) Stearn; *Tradescantia discolor* L'Hér.

Habitat: Cultivated

Distribution: Tropical Asia and Pacific

Botanical examination: This ornamental herb grows up to about 25 cm tall. The leaves are simple, spiral, and exstipulate. The blade is narrowly lanceolate, coriaceous, purple below and dark green above, 20–40×3–6 cm, and sharply acuminate at apex. The flowers are immaculately white and enclosed in a purple spathe-like bract, which is broadly lanceolate, somewhat eery, purple, and up to about 4 cm long. The calyx includes 3 sepals. The corolla comprises 3 petals, which are broadly lanceolate and up to 8 mm long. The androecium includes 6 stamens. The capsules are trilobed, dehiscent, and contain a few rugose seeds.

Medicinal uses: Cough and dysentery (Myanmar)

Phytochemical class: Phenolics

Strong *Broad-spectrum antibacterial polar extract:* Ethanol extract of leaves inhibited the growth of *Escherichia coli* and *Listeria innocua* at concentrations of 4 and 4 µg/mL and was bactericidal (García-Varela et al., 2015). Ethanol extract of leaves inhibited the growth of *Streptococcus mustans* at concentrations of 4 µg/mL and was bactericidal as per flow cytometry analysis (García-Varela et al., 2015).

*Strong Broad-spectrum antibacterial polar extract:* Methanol extract of leaves inhibited the growth of *Candida albicans* at a concentration of 1 µg/mL and was fungicidal (García-Varela et al., 2015).

*Weak antimycobacterial polar extract:* Aqueous extract of leaves at a concentration of 1250 µg/mL inhibited the growth of *Mycobacterium tuberculosis* (H37Rv) and *Mycobacterium tuberculosis* (multidrug resistant clinical strain) by 32.1 and 28.6%, respectively (Radji et al., 2015).

*Strong antiviral (enveloped segmented linear single-stranded (−) RNA virus) flavonoid fraction:* Flavonoid fraction of leaves inhibited the replication of the Influenza virus strains A/Yucatan/2370 (H1N1) in Madin-Darby canine kidney cells with an $IC_{50}$ value of 0.2 µg/mL and a selectivity index of 4.5 post treatment (Sánchez-Roque et al., 2017).

Commentaries:

(i) The plant produces the flavonoid *C*-glucosides vitexin (apigenin 8-*C*-glucoside) and vicenin (apigenin 6,8-di-*C*-glucoside) (Martínez & Martinez, 1993). (ii) Vitexin has interesting antimicrobial properties *in vitro*. The formation of openings by a compound in the cytoplasmic membranes of bacteria and yeasts allowing the efflux of cytoplasmic macromolecules (molecular mass below 660 Da) and ions imposes an arrest in division (fungistatic or bacteriostatic effect). Once the compound is removed from the environment, the yeast or the bacterium continues its life cycle: this is the "phoenix effect" described by Jay in 2002. The effect might explain (at least partially) why inhibition zones are not uncommonly turbid in paper disc tests or agar well tests.

### References

García-Varela, R., García-García, R.M., Barba-Dávila, B.A., Fajardo-Ramírez, O.R., Serna-Saldívar, S.O. and Cardineau, G.A., 2015. Antimicrobial activity of Rhoeo discolor phenolic rich extracts determined by flow cytometry. *Molecules*, 20(10), pp. 18685–18703.

Jay, J., 2002. *Microbiologia Moderna de los Alimentos*, 4ª edn. Zaragoza, Espana: Editorial Acribia.

Martínez, M.A.D.P. and Martinez, A.J., 1993. Flavonoid distribution in Tradescantia. *Biochemical Systematics and Habitat*, 21(2), pp. 255–265.

Radji, M., Kurniati, M. and Kiranasari, A., 2015. Comparative antimycobacterial activity of some Indonesian medicinal plants against multi-drug resistant *Mycobacterium tuberculosis*. *Journal of Applied Pharmaceutical Science*, 5(1), pp. 019–022.

Sánchez-Roque, Y., Ayora-Talavera, G., Rincón-Rosales, R., Gutiérrez-Miceli, F.A., Meza-Gordillo, R., Winkler, R., Gamboa-Becerra, R., Ayora-Talavera, T.D. and Ruiz-Valdiviezo, V.M., 2017. The flavonoid fraction from rhoeo discolor leaves acting as antiviral against influenza a virus. *Records of Natural Products*, 11, pp. 532–546.

### 3.3.2    FAMILY PONTEDERIACEAE KUNTH (1816)

The family Pontederiaceae consists of 6 genera and 40 species of aquatic and fleshy herbs. The petiole is sheathing at base. The leaves are simple, basal, spiral, and exstipulate. The inflorescences are racemes or spikes in a spathe-like bract. The perianth includes 6 tepals, which are free or partially fused. The androecium presents 6 stamens. The gynoecium is made if 3 carpels merged in a trilocular ovary, each locule sheltering numerous ovules growing on parietal, axile placentas, or pendulous. The fruit is a 3-locular capsule or a nut. The seeds are minute and ribbed.

#### 3.3.2.1    *Eichhornia crassipes* (Mart.) Soms

Common names: Water hyacinth; Kochuripana (Bangladesh); kamplaok (Cambodia); feng yan lan (China); kelipuk (Indonesia); hotei aoi (Japan); bu re ok jam (Korea); tob pong (Laos); bunga jamban (Malaysia); beda-bin (Myanmar); jalkumbhi (Nepal); beo nat ban (Vietnam)

Synonyms: *Eichhornia speciosa* Kunth, *Heteranthera formosa* Miq., *Piaropus crassipes* (Mart.) Raf., *Piaropus mesomelas* Raf., *Pontederia azurea* Sw., *Pontederia crassipes* Mart., *Pontederia crassipes* Roem. & Schult., *Pontederia elongata* Balf.

Habitat: Ponds, lakes, slow rivers

Distribution: Tropical Asia and Pacific

Botanical examination: This graceful and rapidly multiplying aquatic floating herb grows from fibrous roots. In Indonesia, this plant is used as vegetable. The leaves are thick, fleshy, edible, somewhat spongy, smooth, and glossy (favorite food of dugongs which will disappear soon if no action is taken). The petiole is swollen at base and up to about 3.5 cm long. The blade is orbicular, glossy, somewhat light-glaucous and hydrophobic below, and 2.5–3.5 cm × 7.5–10 cm. The inflorescence is a spike, which bears up to about 15–flowers. The spathes are obovate and 4–10 cm long. The flowers open after sunrise and wilt by night. The perianth is heavenly light blue and develops 6 lobes, which are obovate and 1.5–3.5 cm long, the upper lobe presenting at base a yellow spot. The androecium comprises 2 sets of 3 stamens, one long and exerted and 3 shorter, all with dark bluish anthers. The style is trilobed. The capsules contain up to about 15 seeds, which are ribbed and minute.

Medicinal use: Skin infection (Malaysia)

*Broad-spectrum antibacterial polar extract:* Methanol extract (6 mm diameter paper disc impregnated with the extract at a concentration of 125 µg/mL) inhibited the growth of *Staphylococcus aureus* (ATCC 25923), *Staphylococcus aureus* (ATCC 29213), *Staphylococcus aureus* (ATCC 43300), methicillin-resistant *Staphylococcus*, oxacillin-sensitive *Staphylococcus aureus*, and coagulase-negative *Staphylococcus epidermidis* with inhibition zone diameters of 8.1, 9.7, 6.9, 7.6, 8.1, and 8.7 mm, respectively (Gutiérrez-Morales et al., 2017). Ethanol extract of leaves (50 µL/well, 25 µg/mL) inhibited the growth of *Escherichia coli* (MTCC-40), *Staphylococcus epidermidis* (MTCC10623), and *Bacillus subtilis* (MTCC-736) with inhibition zone diameters of 14, 10, and 10 mm, respectively (Joshi et al., 2013).

*Strong broad-spectrum antibacterial polar extract:* Ethanol extract of leaves inhibited the growth of *Escherichia coli*, *Bacillus subtilis*, *Bacillus cereus*, *Lactobacillus casei*, and *Pseudomonas aeruginosa* with MIC values of 16, 16, 32, 32, and 64 µg/mL, respectively (Haggag et al., 2017).

*Strong broad-spectrum antifungal polar extract:* Ethanol extract of leaves inhibited the growth of *Aspergillus flavus*, *Aspergillus niger*, *Alternaria alternata*, *Colletotrichum geosporioides*, *Fusarium solani*, and *Candida albicans* with MIC values of 16, 16, 16, 32, 32, and 64 µg/mL, respectively (Haggag et al., 2017).

*Strong anticandidal benzoindenone:* 2,5-dimethoxyl-4-phenyl-benzoindenone isolated from the leaves inhibited the growth of *Candida albicans* (Della Greca et al., 1991a, 1991b).

2,5-Dimethoxyl-4-phenyl-benzoindenone

Commentaries:

(i) The plant generates a bewildering array of phenalenes derivatives, of which 2,6-dimethoxy-9-phenylphenalenone (Della Greca et al., 1992) which is probably antimicrobial. Note that plants directly exposed to the sun protect themselves against UV light by synthetizing aromatic compounds with ketone groups conjugated with double bonds (often allowing electron resonance). To protect itself against the sun light, *Eichhornia crassipes* (Mart.) Solms also produces a benzyl amide alkaloid: $N^1$-acetyl-$N^2$-formyl-5-methoxykynuramine, which is cytoprotective and a human brain metabolite of melatonin (Hardeland et al., 2009). One could have the curiosity to examine the antimicrobial properties of $N^1$-acetyl-$N^2$-formyl-5-methoxykynuramine, which in the presence of free radicals is converted to $N^1$-acetyl-5-methoxykynuramine (Hardeland et al., 2009). One could wonder if some natural products from plants could be prodrugs, which upon penetration in bacterial or fungal cytoplasm and exposure to free radicals be converted into antibacterial or antifungal agents. Is it the case for $N^1$-acetyl-$N^2$-formyl-5-methoxykynuramine?

(ii) *Eichhornia crassipes* (Mart.) Solms synthetizes a ergostane-type steroids, including 4α-methyl-5α-ergosta8,14,24(28)-triene-3β, 4β-diol, 4α-methyl-5α-ergosta-8,24(28)- diene-3β, 4β-diol, and 4α-methyl-5α-ergosta-7,24(28)-diene3β, 4β-diol., with the ability to inhibit the proliferation of algae (Della Greca et al., 1991). Consider that antialgal principles are not uncommonly antibacterial or antifungal.

### References

Della Greca, M., Lanzetta, R., Mangoni, L., Monaco, P. and Previtera, L., 1991a. A bioactive benzoindenone from *Eichhornia crassipes* solms. *Bioorganic & Medicinal Chemistry Letters*, *1*(11), pp. 599–600.

Della Greca, M., Lanzetta, R., Molinaro, A., Monaco, P. and Previtera, L., 1992. Phenalenemetabolites from *Eichhornia crassipes*. *Bioorganic & Medicinal Chemistry Letters*, *2*(4), pp. 311–314.

Della Greca, M., Monaco, P. and Previtera, L., 1991b. New oxygenated sterols from the weed Eichhornia crassipes solms. *Tetrahedron*, *47*(34), pp. 7129–7134.

Gutiérrez-Morales, A., Velázquez-Ordoñez, V., Khusro, A., Salem, A.Z., Estrada-Zúñiga, M.E., Salem, M.Z., Valladares-Carranza, B. and Burrola-Aguilar, C., 2017. Antistaphylococcal properties of *Eichhornia crassipes*, *Pistacia vera*, and *Ziziphus amole* leaf extracts: isolates from cattle and rabbits. *Microbial Pathogenesis*, *113*, pp. 181–189.

Haggag, M.W., Abou El Ella, S.M. and Abouziena, H.F., 2017. Phytochemical analysis, antifungal, antimicrobial activities and application of *Eichhornia crassipes* against some plant pathogens. *Planta Daninha*, *35*.

Hardeland, R., Tan, D.X. and Reiter, R.J., 2009. Kynuramines, metabolites of melatonin and other indoles: the resurrection of an almost forgotten class of biogenic amines. *Journal of Pineal Research*, 47(2), pp. 109–126.

Joshi, M.A.H.A.V.I.R. and Kaur, S.A.N.D.E.E.P., 2013. In vitro evaluation of antimicrobial activity and phytochemical analysis of *Calotropis procera*, *Eichhornia crassipes* and *Datura innoxia* leaves. *Asian Journal of Pharmaceutical and Clinical Research*, 6(5), pp. 25–28.

### 3.3.2.2  *Monochoria hastata* (L.) Solms

Common names: Hastate-leaved pondweed arrowleaf Monochoria; chrach (Cambodia); jian ye yu jiu hua (China); bia bia (Indonesia); chacha ayer, kankon ayer (Malaysia); maoa (Papua New Guinea); payaw-payaw (the Philippines); phakpong (Thailand); rau mác (Vietnam)

Synonyms: *Monochoria dilatata* (Buch.–Ham.) Kunth, *Monochoria hastifolia* C. Presl; *Monochoria sagittata* (Roxb.) Kunth, *Pontederia dilatata* Buch–Ham.; *Pontederia hastata* L.; Pontederia sagittata Roxb.

Habitat: Pools, rice fields, and ditches

Distribution: India, Bhutan, Nepal, Sri Lanka, Cambodia, Myanmar, Vietnam, China, Malaysia, and Indonesia

Botanical observation: It is a somewhat dull-looking aquatic and rhizomatous herb. The leaves are simple, spiral, and to a certain degree araceous-like. The plant is eaten in India. The petiole is terete, smooth, and up to about 90 cm long. The blade is fleshy, hastate, glossy, and about 5 cm × 15 cm. The flowers are crowded on a short about 6 cm long raceme included in a leaf sheath. The perianth presents 6 lobes, which are bluish purple, about 1.5 cm long, and oblong. The androecium includes 6 stamens of inequal length, the anthers of which grow up to about 5 mm long and bright yellow. The style is slender and with the same color as the perianth. The capsules are globose, about 1 cm long, and contain numerous tiny and ribbed seeds.

Medicinal uses: Boils (the Philippines)

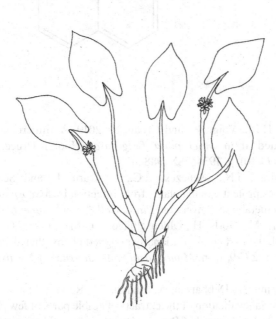

*Monochoria hastata* (L.) Solms

*Moderate broad-spectrum antibacterial polar extract:* Methanol extract of stems and leaves inhibited the growth of *Staphylococcus aureus* (ML -185), *Bacillus subtilis* (ATCC 39816), *Shigella dysenteriae* (ATCC 26591), *Escherichia coli* (ATCC 10536), and *Vibrio cholerae* (ATCC 3241) with MIC values of 900, 800, 900, 700, and 800 µg/mL, respectively (Ngomdir et al., 2007).

Commentary: Interesting phenylphenalenones, including anigorufone, 4-hydroxyanigorufone and lachnanthocarpone, are generated by *Monochoria elata* Ridl. (Hölscher et al., 2006) and are possibly to be found in *Monochoria hastata* (L.) Solms. These planar phenalen-1-one are probably antibacterial and antifungal or even antiviral by means, at least partially, of DNA intercalation. Consider that phenylphenalenones are phytoalexins in members of the genus *Musa* L. (family Musaceae Juss., order Zingiberales Griseb.), including anigorufone (Hidalgo et al., 2016). Anigorufone (also known as 2-hydroxy-9-phenylphenalen-1-one) isolated from *Macropidia fuliginosa* (Hook.) Druce (family Haemodoraceae R. Brown, order Commelinales) (50 µg/9 mm diameter paper well) inhibited the growth of *Bacillus subtilis* (6633 B1), *Staphylococcus aureus* (511 B3), methicillin-resistant *Staphylococcus aureus*, *Escherichia coli* (458 B4), *Pseudomonas aeruginosa* (SG 137 B7), and *Mycobacterium vaccae* (10 670 M4) with inhibition zone diameters of 13, 12, 10, 15, 17, and 17 mm, respectively (Brkljaca et al., 2019). The phenylphenalenone anigorufone (LogD =3.7 at pH 7.4; molecular weight = 272.2 g/mol) (50 µg/9 mm diameter paper disc) inhibited the growth of *Sporodobolomyces salmonicolor* (549 H4), *Candida albicans*, and *Penicillium notatum* (JP36 P1) with inhibition zone diameters of 21, 13, and 16 mm, respectively (Brkljaca et al., 2019). *In vivo* antibacterial or antifungal examination of anigorufone is warranted.

Anigorufone

### References

Brkljaca, R., Dahse, H.M., Voigt, K. and Urban, S., 2019. Antimicrobial evaluation of the constituents isolated from *Macropidia fuliginosa* (Hook.) Druce. *Natural Product Communications*, *14*(10), p. 1934578X19884411.

Hidalgo, W., Chandran, J.N., Menezes, R.C., Otálvaro, F. and Schneider, B., 2016. Phenylphenalenones protect banana plants from infection by *Mycosphaerella fijiensis* and are deactivated by metabolic conversion. *Plant, Cell & Environment*, *39*(3), pp. 492–513.

Hölscher, D., Reichert, M., Görls, H., Ohlenschläger, O., Bringmann, G. and Schneider, B., 2006. Monolaterol, the first configurationally assigned phenylphenalenone derivative with a stereogenic center at C-9, from *Monochoria elata*. *Journal of Natural Products*, *69*(11), pp. 1614–1617.

Ngomdir, M., Debbarma, B., Debbarma, A., Chanda, S., Raha, S., Saha, R., Pal, S. and De, B., 2007. Antibacterial evaluation of the extracts of edible parts of few plants used by tribal people of Tripura, India. *Journal of Pure and Applied Microbiology*, *1*(1), pp. 65–68.

### 3.3.2.3  *Monochoria vaginalis* (Burm. f.) C. Presl

Common names: Oval-leaved pondweed; chrach (Cambodia); ya she cao (China); indivarah (India); Eceng padi (Indonesia); ko nagi (Japan), mul dal gae bi (Korean); keladi agak (Malaysia); biga-bigaan (the Philippines); Diya habarala (Sri Lanka); khakhiat (Thailand); rau chóc (Vietnam)

Synonyms: *Gomphima vaginalis* (Burm. f.) Raf.; *Monochoria vaginalis* var. *pauciflora* (Bl.) Merr.; *Pontederia vaginalis* Burm. f.; *Pontederia plantaginea* (Roxb.) Kunth

Habitat: Pools, rice fields, and ditches

Distribution: India, Bhutan, Nepal, Sri Lanka, Cambodia, Myanmar, Vietnam, China, Malaysia, Indonesia, the Philippines

Botanical observation: It is an aquatic herb which grows up to about 50 cm tall. The leaves are simple, spiral, and exstipulate. The plant is eaten in India. The petiole is terete, smooth, and up to about 50 cm long. The blade is fleshy, somewhat narrowly cordate, glossy, 2–10 cm × 10–20 cm, and acute or acuminate at apex. The radical leaves have a broad sheath. The inflorescence is a raceme on an about 3 cm peduncle. The inflorescence is somewhat enclosed at base by a spathe-like leaf and sustains to about 12 flowers. The flowers are eaten in China. The perianth is made of 6 lobes, which are bluish purple, about 1.5 cm long, and oblong. The androecium includes 6 stamens of unequal length, the anthers of which grow up to about 4 mm long and bright yellow. The style is slender and with the same color as the tepals. The capsules are ovoid, about 1 cm long, and contain numerous tiny and ribbed seeds.

Medicinal uses: Boils, cough (India); cholera (Taiwan); fever (Indonesia); post-partum (Malaysia); toothache (Indonesia; Myanmar)

*Monochoria vaginalis* (Burm. f.) C. Presl

*Broad-spectrum antibacterial polar extract:* Aqueous extract of leaves inhibited the growth of *Staphylococcus aureus* (ATCC 25923), *Bacillus subtilis* (ATCC 6633), *Escherichia coli* (ATCC 25922), and *Pseudomonas aeruginosa* (ATCC 27853).

*Strong broad-spectrum antibacterial stigmastane-type steroidal saponin:* Stigmasterol 3-*O*-β-*D*-glucopyranoside (also known as stigmasta-5,22-dien-3-*O*-β-*D*-glucopyranoside) inhibited the growth of *Staphylococcus aureus* (MTCC 96), *Bacillus subtilis* (MTCC 441), *Enterococcus faecalis* (MTCC 439), *Escherichia coli* (MTCC 723), *Vibrio cholerae* (MTCC 3904), *Klebsiella pneumoniae* (MTC 932), *Proteus vulgaris* (MTCC 426), *Proteus mirabilis* (MTCC 425), *Shigella dysenteriae* (12/30), and *Pseudomonas aeruginosa* (MTCC 741) with MIC values below 30 µg/mL, and was bactericidal with 5–10 hours for all bacteria tested except *Staphylococcus aureus* (MTCC 96), *Klebsiella pneumoniae* (MTC 932), and *Proteus mirabilis* (MTCC 425).

Stigmasterol 3-*O*-β-*D*-glucopyranoside

*Antibacterial (Gram-negative) halo developed by megastigmane:* Vomifoliol (10 µg/disc) inhibited the growth of *Neisseria gonorrhoeae* (ATCC 49226) equivalently as 63.1% of ciprofloxacin (1 µg/disc) (Nair et al., 2013).

*Antifungal (filamentous) phenylphenalenone:* Methoxyanigofurone inhibited the growth of *Mycosphaerella fijiensis* (Otálvaro et al., 2007).

*Antibiotic potentiation by stigmastane-type steroidal saponin:* Stigmasterol 3-*O*-β-*D*-glucopyranoside (Zhou et al., 2007) increased the vulnerability *Pseudomonas aeruginosa* to methicillin, ciprofloxacin, gentamicin, and chloramphenicol and potentiated ciprofloxacin, gentamicin, and chloramphenicol against *Escherichia coli* (Subramaniam et al., 2014).

Commentary: Zheng et al. (2013) isolated from *Monochoria vaginalis* var *plantaginea* (Roxb.) Solms the steroids cholest-4-en-3,6-dione, stigmast-4-ene-3,6-dione, the methoxyanigorufone, the alkyl phenol (10Z)-1-(2,6-dihydroxyphenyl)octadec-10-en-1-one, 1-(4-methoxyphenyl)-7-phenyl-(6E)-6-hepten-3-one, the megastigmanes (+)-dehydrovomifoliol, and (3S, 5R, 6R, 7E,

9R)-5,6-epoxy-3,9-dihydroxy-7-megastigmene. The alkyl phenol (10Z)-1-(2,6-dihydroxyphenyl) octadec-10-en-1-one is most probably antibacterial and/or antifungal. One could have the curiosity to isolate antibacterial, antifungal, or antiviral principles from this plant.

### References

Nair, J.J., Mulaudzi, R.B., Chukwujekwu, J.C., Van Heerden, F.R. and Van Staden, J., 2013. Antigonococcal activity of *Ximenia caffra* Sond. (Olacaceae) and identification of the active principle. *South African Journal of Botany*, 86, pp. 111–115.

Otálvaro, F., Nanclares, J., Vásquez, L.E., Quinones, W., Echeverri, F., Arango, R. and Schneider, B., 2007. Phenalenone-type compounds from *Musa acuminata* var. "Yangambi km 5"(AAA) and their activity against *Mycosphaerella fijiensis*. *Journal of Natural Products*, 70(5), pp. 887–890.

Subramaniam, S., Keerthiraja, M. and Sivasubramanian, A., 2014. Synergistic antibacterial action of β-sitosterol-D-glucopyranoside isolated from *Desmostachya bipinnata* leaves with antibiotics against common human pathogens. *Revista Brasileira de Farmacognosia*, 24(1), pp. 44–50.

Zheng, H., Choi, S., Kang, S., Lee, D., Zee, O. and Kwak, J., 2013. Phytochemical constituents of *Monochoria vaginalis* var. plantaginea and their antioxidative and cytotoxic activities. *Planta Medica*, 79(13), p. PI113.

Zhou, Y.J., Xu, X.H., Qiao, F.Y., Zhang, J.P. and Yu, L.Q., 2007. Isolation and identification of an antioxidant from *Monochoria vaginalis*. *Chinese Journal of Applied Habitat*, 18(03), p. 509.

## 3.4   ORDER POALES SMALL (1903)

### 3.4.1   FAMILY CYPERACEAE A. L. de Jussieu (1789)

The family Cyperaceae consists of 70 genera and nearly 4000 species of rhizomatous sedges. The stems are often sharply angled and somewhat coriaceous, and bear at apex a few leaves which are simple, alternate, arranged in 3 whorls, and endowed with a closed sheath. The inflorescences are spikes or composed of spikelets. The perianth consists of 1 to several bristles. The androecium consists of 1–6 stamens. The gynoecium consists of 2–3 carpels forming a compound, unilocular, and superior ovary. The fruits are trigonous or lenticelled achenes. Members in this family produce an interesting series of sesquiterpenes, quinones, and stilbenes.

#### 3.4.1.1   *Cyperus cyperoides* (L.) Kuntze

Common names: Pacific island flat sedge, Bara guthubi, (Bangladesh); zhuan zi miao (China), suket lumbungan (Indonesia); kolpulu (India); inu kugu (Japan), bang dong sa ni a jae bi (Korea); rumput mesiyang (Malaysia); kode jhar (Nepal); kaiga, sap (Papua New Guinea); mangilang (the Philippines); yaa rang kaa (Thailand)

Synonyms: *Cyperus subumbellatus* Kük.; *Cyperus umbellatus* Benth.; *Kyllinga sumatrensis* Retz.; *Kyllinga umbellata* Rottb.; *Mariscus biglumis* Gaertn.; *Mariscus cyperoides* (L.) Urb.; *Mariscus nossibeensis* Steud.; *Mariscus philippensis* Steud.; *Mariscus sumatrensis* (Retz.) J. Raynal; *Mariscus umbellatus* Vahl; *Scirpus cyperoides* L.

Habitat: Moist soils

Distribution: Tropical Asia and Pacific

Botanical examination: It is a sedge with a pleasant aspect, which grows up to a height of about 50 cm. The rhizome is short, edible, and hard. The stems are trigonal, smooth, and swollen at base. The leaves are simple, terminal, and spiral. The blade is narrowly triangular, dull green, and up to 0.5 cm×3.5 cm. The involucral bracts are 5–8, leaflike, up to 25 cm long, and spreading. The inflorescences are spikes, which are cylindrical, up to 2.5×1 cm, and present many dense spikelets.

The spikelets are linear, up to 0.7 cm×7 mm, and 2- or 3-flowered. The glumes are yellow, oblong, and 3 mm long. The androecium includes 3 stamens, which are broadly linear and minute. The style is short with 3 stigmas. The nutlets are dark brown, narrowly oblong, trigonal, and minute.

Medicinal uses: Wounds (Nepal)

Commentary:

(i) The plant has apparently not been studied for its possible antimicrobial effects. Members of the genus *Cyperus* L. bring into being interesting series of furano 1,4-benzoquinones, such as cyperaquinones, scabequinones, and breviquinones (Allan et al., 1978; Morimoto et al., 1999), which so far have not been apparently studied for antimicrobial activity. 1,4-Benzoquinones from medicinal plants are most always antibacterial. The 1,4-benzoquinone framework inhibited the growth (bactericidal) of *Salmonella typhimurium*, *Escherichia coli*, *Staphylococcus aureus*, and *Bacillus cereus* with MIC values of 32, 64, 8, and 32 µg/mL, respectively (Kim et al., 2010). The precise antibacterial mode of action of 1,4-benzoquinones is apparently yet not fully understood. In eukaryotic cells, quinones produced by plants interfere, among other things, with mitochondrial ubiquinone or co-enzyme Q, complex I (NADH-ubiquinone oxidoreductase) (Ōmura et al., 2001) and might do the same in bacteria (Guénebaut et al., 1998). Nishina and Uchibori (1991) suggested that the antibacterial mechanism of benzoquinone derivatives is based on the inhibition of DNA synthesis.

(ii) Cyperaquinones, scabequinones, and breviquinones are purple, red to yellow pigments (Allan et al., 1978) and *Cyperus cyperoides* (L.) Kuntze used for the making of yellow face paint in Papua New Guinea (Hill, 2011) produces cyperaquinone and hydroxycyperaquinone (Allan et al., 1978).

### References

Allan, R.D., Wells, R.J., Correll, R.L. and MacLeod, J.K., 1978. The presence of quinones in the genus Cyperus as an aid to classification. *Phytochemistry*, *17*(2), pp. 263–266.

Guénebaut, V., Schlitt, A., Weiss, H., Leonard, K. and Friedrich, T., 1998. Consistent structure between bacterial and mitochondrial NADH: ubiquinone oxidoreductase (complex I). *Journal of Molecular Biology*, *276*(1), pp. 105–112.

Hill, R., 2011. Colour and ceremony: the role of paints among the Mendi and Sulka peoples of Papua New Guinea (Doctoral dissertation, Durham University).

Kim, M.H., Jo, S.H., Ha, K.S., Song, J.H., Jang, H.D. and Kwon, Y.I., 2010. Antimicrobial activities of 1, 4-benzoquinones and wheat germ extract. *Journal of Microbiology and Biotechnology*, *20*(8), pp. 1204–1209.

Morimoto, M., Fujii, Y. and Komai, K., 1999. Antifeedants in Cyperaceae: coumaran and quinones from Cyperus spp. *Phytochemistry*, *51*(5), pp. 605–608.

Nishina, A. and Uchibori, T., 1991. Antimicrobial activity of 2, 6-dimethoxy-p-benzoquinone, isolated from thick-stemmed bamboo, its analogs. *Agricultural and Biological Chemistry*, *55*(9), pp. 2395–2398.

Ōmura, S., Miyadera, H., Ui, H., Shiomi, K., Yamaguchi, Y., Masuma, R., Nagamitsu, T., Takano, D., Sunazuka, T., Harder, A. and Kölbl, H., 2001. An anthelmintic compound, nafuredin, shows selective inhibition of complex I in helminth mitochondria. *Proceedings of the National Academy of Sciences*, *98*(1), pp. 60–62.

### 3.4.1.2  *Cyperus rotundus* L.

Common names: Nut grass, coco grass, coco sedge; mutha, vadal (Bangladesh); kravanh chruk (Cambodia); tian tou xiang, xiang fu zi (China); ambuda, dilla, nagarmotha (India); mota (Indonesia); hewz hmu (Laos); rumput halia hitam, rumput teki (Malaysia); monhnyin bin (Myanmar); muther (Pakistan); boto-botones, mutha (the Philippines); co gau, sa thao (Vietnam); ya haeo mu (Thailand); topalak (Turkey)

Synonyms: *Chlorocyperus rotundus* (L.) *Palla*; *Cyperus agrestis* Willd. ex Spreng. & Link; *Cyperus* bicolor Vahl; *Cyperus hexastachyos* Rottb.; *Cyperus hydra* Michx.; *Cyperus rubicundus* Vahl; *Cyperus tetrastachyos* Desf.; *Cyperus tuberosus* Rottb.

Habitat: Paddy fields, marshes, riverbanks, and watery grassy lands

Distribution: Tropical Asia

Botanical observation: This sedge grows up to 50 cm tall. The rhizome is reddish brown, ovoid, about 1 cm long, edible, and aromatic. The stems are 15–60 cm long, trigonal, and slender. The leaves are simple, spiral, terminal, narrow, 6–20 cm long, sheathing at base, and single nerved. The inflorescences are 2.5–4.5 cm long umbels made of 4–6 dark red and 1.5–2.5 cm long spikes. The glumes are distichous, narrow, lanceolate, subacute, and imbricate. The rachis is winged. The style is ellipsoid angular, and greyish. The fruits are trigonous nuts achenes.

Medicinal uses: Cholera, fever, diarrhea, dysentery leprosy, and ulcers (India)

*Moderate broad-spectrum antibacterial polar extract:* Ethanol extract of rhizome inhibited the growth of *Staphylococcus epidermidis* (MTCC-3615), *Bacillus cereus* (MTCC-430), and *Pseudomonas aeruginosa* (MTCC-424) with MIC values of 250, 250, and 125 µg/mL, respectively.

*Antifungal (yeast) halo developed by polar extract:* Methanol extract of whole plant (500 µg/disc) inhibited the growth of *Trichosporon begelli* (NCIM 3404) with an inhibition zone diameter of 10 mm (Parekh & Chanda, 2008).

*Weak antiviral (enveloped circular double stranded DNA) patchoulane-type sesquiterpene:* $3\beta$-Hydroxycyperenoic acid isolated from the rhizome inhibited Hepatitis B virus surface antigen with an $IC_{50}$ value of 46.6 µM and the selectivity of 31 (Xu et al., 2015).

*Strong antiviral (enveloped circular double stranded DNA)* eudesmane-type sesquiterpene: 10-epieudesm-11-ene-$3\beta$, $5\alpha$-diol isolated from the rhizome inhibited Hepatitis B virus (Xu et al., 2015).

Commentaries: (i) Essential oil of rhizomes contains patchoulane-type sesquiterpenes, such as α-cyperone, cyperene, and α-selinene (Zhang et al., 2017). (ii) The plant is used for the treatment of COVID-19 in China (Wang et al., 2020). Siddha healers in South India use the plant to treat COVID-19 (Sasikumar et al., 2020). Are patchoulane-type sesquiterpenes active against Severe acute respiratory syndrome-associated coronavirus type-2?

### References

Parekh, J. and Chanda, S., 2008. In vitro antifungal activity of methanol extracts of some Indian medicinal plants against pathogenic yeast and moulds. *African Journal of Biotechnology*, 7(23).

Sasikumar, R., Priya, S.D. and Jeganathan, C., 2020. A case study on domestics spread of SARS-CoV-2 pandemic in India. *International Journal of Advanced Science and Technology*, 29(7), pp. 2570–2574.

Wang, S.X., Wang, Y., Lu, Y.B., Li, J.Y., Song, Y.J., Nyamgerelt, M. and Wang, X.X., 2020. Diagnosis and treatment of novel coronavirus pneumonia based on the theory of traditional Chinese medicine. *Journal of Integrative Medicine*, 18(4), pp. 275–283.

Xu, H.B., Ma, Y.B., Huang, X.Y., Geng, C.A., Wang, H., Zhao, Y., Yang, T.H., Chen, X.L., Yang, C.Y., Zhang, X.M. and Chen, J.J., 2015. Bioactivity-guided isolation of anti-Hepatitis B virus active sesquiterpenoids from the traditional Chinese medicine: rhizomes of *Cyperus rotundus*. *Journal of Ethnopharmacology*, 171, pp. 131–140.

Zhang, L.L., Zhang, L.F., Hu, Q.P., Hao, D.L. and Xu, J.G., 2017. Chemical composition, antibacterial activity of Cyperus rotundus rhizomes essential oil against *Staphylococcus aureus* via membrane disruption and apoptosis pathway. *Food Control*, 80, pp. 290–296.

### 3.4.1.3   *Scirpus ternatanus* Reinw. ex Miq.

Common name: Bai sui biao cao (China)

Synonym: *Scirpus chinensis* Munro

Habitat: Wet grassy soils

Distribution: India, Bhutan, Myanmar, Vietnam, Thailand, China, Taiwan, Japan, Indonesia, Philippines, Papua New Guinea, and Pacific islands

Botanical examination: It is a sedge that grows up to about 1 m tall. The stems are trigonal. The leaves are simple, basal, and cauline. The blade is grass-like, up to 1.5×100 cm wide, somewhat coriaceous, dark green, and glossy. The involucral bracts are 5 or 6 and leaflike. The inflorescence is a cyme of oblong, and about 8 mm long spikelets. The glumes are spirally arranged, brown, membranous, and minute. The perianth comprises 2 or 3 tepals. The androecium includes 2–3 stamens. The style is slender. The nutlet is yellowish, somewhat ovoid, and minute.

Medicinal uses: Wounds (Taiwan)

Commentaries:

(i) Methanol extract of roots *Scirpus fluviatilis* (Torr.) A. Gray (Local name: *Kei san ryo* (Japan) and *Ching sang leng* (China) at the concentration of 100 µg/mL inhibited RNA-dependent DNA polymerase and ribonuclease H activity of Human immunodeficiency virus type-1 reverse transcriptase and Human immunodeficiency virus type-1 protease by 9.5, 0.1, and 8.4%, respectively (Min et al., 2001). The plant produces the hydroxystilbene dimers scirpusin A and B as well as the stilbene resveratrol and 3,3′, 4,5′-tetrahydroxystilbene (Nakajima et al., 1978). Scirpusin A (LogD = 3.7 at pH 7.4; molecular weight = 470.4 g/mol) inhibited the replication of Human immunodeficiency virus type-1 (strain IIIB) with an $EC_{50}$ value of 10 µg/mL (Yang et al., 2005). Consider that at a concentration of 200 µM resveratrol protected Vero E6 cells against Middle East respiratory syndrome coronavirus (HCoV-EMC/2012) with a decrease of viral RNA levels in resveratrol-treated cells (Lin et al., 2017).

Scirpusin A

(ii) One could have the curiosity to examine the antimicrobial properties of *Scirpus ternatanus* Reinw. ex Miq. and look for antiviral principles. The medicinal use of the plant is most probably owed to stilbenes and/or stilbenes oligomer.

### References
Lin, S.C., Ho, C.T., Chuo, W.H., Li, S., Wang, T.T. and Lin, C.C., 2017. Effective inhibition of MERS-CoV infection by resveratrol. *BMC Infectious Diseases*, *17*(1), pp. 1–10.

Min, B.S., Kim, Y.H., Tomiyama, M., Nakamura, N., Miyashiro, H., Otake, T. and Hattori, M., 2001. Inhibitory effects of Korean plants on HIV -1 activities. *Phytotherapy Research*, *15*(6), pp. 481–486.

Nakajima, K., Taguchi, H., Endo, T. and Yosioka, I., 1978. The constituents of *Scirpus fluviatilis* (Torr.) A. Gray. I.: the structures of two new hydroxystilbene dimers, scirpusin A and B. *Chemical and Pharmaceutical Bulletin*, *26*(10), pp. 3050–3057.

Yang, G.X., Zhou, J.T., Li, Y.Z. and Hu, C.Q., 2005. Anti-HIV bioactive stilbene dimers of *Caragana rosea*. *Planta Medica*, *71*(06), pp. 569–571.

### 3.4.1.4  *Scleria levis* Retz.

Common names: Mao guo zhen zhu mao (China); teteles (Indonesia); sialit dudok (Malaysia); shinju gaya (Japan); daat (Philippines); goda karawu (Sri Lanka); yaa saam khom (Thailand); dưng láng (Vietnam)

Synonyms: *Scleria hebecarpa* Nees; *Scleria pubescens* Steud.

Habitat: Sun-exposed grassy spots

Distribution: India, Sri Lanka, Myanmar, Cambodia, Laos, Vietnam, China, Japan, Thailand, Malaysia, Indonesia, the Philippines, Australia, and Pacific Islands.

Botanical examination: It is a sedge growing from a woody rhizome and covered with purple scales. The stems grow up to about 90 cm tall, trigonal, scabrous, and hairy. The leaves are simple and spiral. The blade is linear and up to about 30 cm × 1 cm. The involucral bracts are leaflike and up to 15 cm long. The bractlets are hairy, auriculate at base, and present barbate auricles. The inflorescence is paniculate with 1–2 lateral branches that are up to about 10 cm long and bear spikelets, which are brown, ovoid, sessile, and 3 mm long. The glumes are 3 mm long and membranous. The male flowers comprise 3 stamens. The female flower presents 3 stigmas. The nutlets are white, spherical, 2 mm in diameter, trigonal, and smooth.

Medicinal use: Dysentery (India); cough (Malaysia)

Commentary: Consider that the sesquiterpene endoperoxide okundoperoxide was isolated from an African member of the genus *Scleria* L. (Efange et al., 2009). Sesquiterpene endoperoxides are not uncommonly antimycobacterial and one could enquire if these type of compounds are present in *Scleria levis* Retz and account for its medicinal uses (patients suffering from tuberculosis have persistent bloody cough), as well as other members in the genus *Scleria* L. *Scleria sumatrensis* Retz. is traditionally used to treat gonorrhea in Malaysia and in Indonesia, *Scleria purpurascens* Steud. (local name: Sialit tajam) is traditionally used to treat cough. Okundoperoxide inhibited the growth of *Staphylococcus aureus* (ATCC 33862) and *Staphylococcus aureus* (clinical stains) (Mbah et al., 2012). Is okundoperoxide antimycobacterial? Why are non-hydrophilic endoperoxides antimycobacterial? The endoperoxide sesquiterpene lactone artemisin (antiplasmodial agent from *Artemisia annua* L., family Asteraceae) inhibited the growth of *Mycobacterium bovis* with the MIC value of 200 µg/mL (Patel et al., 2019). Artemisinin increased the antimycobacterial potencies of rifampicin by generating free radicals, which can target membrane lipids, thereby inducing lipid peroxidation (Patel et al., 2019). The lipophilic triterpene ergosterol-5,8-endoperoxide (LogD = 6.9 at pH 7.4; molecular weight = 412.6 g/mol) isolated from *Ajuga remota* Wall. ex Benth. (family Lamiaceae) inhibited the growth *Mycobacterium tuberculosis* (H37Rv) with an MIC value of 1 µg/mL (Cantrel et al., 1999).

Ergosterol-5,8-endoperoxide

### References

Cantrell, C.L., Rajab, M.S., Franzblau, S.G., Fronczek, F.R. and Fischer, N.H., 1999. Antimycobacterial ergosterol-5, 8-endoperoxide from *Ajuga remota*. *Planta Medica*, 65(08), pp. 732–734.

Efange, S.M., Brun, R., Wittlin, S., Connolly, J.D., Hoye, T.R., McAkam, T., Makolo, F.L., Mbah, J.A., Nelson, D.P., Nyongbela, K.D. and Wirmum, C.K., 2009. Okundoperoxide, a bicyclic cyclofarnesylsesquiterpene endoperoxide from *Scleria striatinux* with antiplasmodial activity. *Journal of Natural Products*, 72(2), pp. 280–283.

Mbah, J.A., Ngemenya, M.N., Abawah, A.L., Babiaka, S.B., Nubed, L.N., Nyongbela, K.D., Lemuh, N.D. and Efange, S.M., 2012. Bioassay-guided discovery of antibacterial agents: in vitro screening of *Peperomia vulcanica*, *Peperomia fernandopoioana* and *Scleria striatinux*. *Annals of Clinical Microbiology and Antimicrobials*, 11(1), p. 10.

Patel, Y.S., Mistry, N. and Mehra, S., 2019. Repurposing artemisinin as an anti-mycobacterial agent in synergy with rifampicin. *Tuberculosis*, 115, pp. 146–153.

### 3.4.2  FAMILY POACEAE BARNHART (1895)

The family Poaceae is a vast taxon that consists of 700 genera and nearly 11,000 species of grasses. The stems are often soft, and bear simple and alternate leaves. The flowers are arranged in spikes or spikelets. The perianth consists of 1 to several bristles. The androecium consists of 1–6 stamens. The gynoecium consists of 1–3 carpels with plumose stigma. Plants in this taxon are anemophilous: they are pollinated by wind. The fruit is a caryopsis. The word Poaceae comes from the genus *Poa* L. from the Greek *poa* = fodder. Members in this family are known to elaborate *C*-glycosyl flavonoids and essential oils.

### 3.4.2.1  *Chrysopogon aciculatus* (Retz.) Trin.

Synonyms: *Andropogon acicularis* Retz. ex Roem. & Schult.; *Andropogon acicularis* Willd.; *Andropogon aciculatus* Retz.; *Andropogon gryllus* L.; *Andropogon javanicus* Steud.; *Andropogon subulatus* J. Presl; *Centrophorum chinense* Trin.; *Chrysopogon acicularis* Duthie; *Chrysopogon subulatus* (J. Presl) Trin. ex Steud.; *Chrysopogon trivialis* Arn. & Nees; *Holcus aciculatus* (Retz.) R. Brown; *Rhaphis acicularis* (Retz. ex Roem. & Schult.) Desv.; *Rhaphis aciculatus* (Retz.) Honda; *Rhaphis javanica* Nees; *Rhaphis javanica* Nees; *Rhaphis javanica* Nees; *Rhaphis javanica* Nees; *Rhaphis javanica* Nees; *Rhaphis trivialis* Lour.

Common names: Golden false beard grass, Love grass; Shashashwar (Bangladesh); smau kântraëy (Cambodia); zhu jie cao (China); Chorkanta (India); *Salohot* (Indonesia); rumput jarum, kemuncup (Malaysia); Kuro ghaans (Nepal); Knarbru (Papua New Guinea); Amorseko (Philippines); ya chao-chu (Thailand); co bong (Vietnam)

Habitat: Roadsides and grassy areas

Distribution: Tropical and Subtropical Asia and Pacific

Botanical examination: It is a beautiful grass that grows to a height of 40 cm tall. The stems are glabrous, smooth, terete, and about 1 mm in diameter. The leaves are simple and spiral. The blade is glabrous, 5–2.3 cm×3 mm, lanceolate, and shows 7 distinct nerves. The margin is ciliate. The inflorescences are terminal and somewhat purplish 4–6 cm long spikes.

Medicinal uses: chronic fever (India); dysentery (Bangladesh); diarrhea, urinary tract infections (the Philippines)

*Antibacterial (Gram-positive) halo developed by polar extract:* Ethanol extract (100 mg/mL, agar well) of whole plant inhibited the growth of *Staphylococcus aureus* (SJTUF 20745), *Staphylococcus aureus* (SJTUF 20746), *Staphylococcus aureus* (SJTUF 20758), *Staphylococcus aureus* (SJTUF 20978), and *Staphylococcus aureus* (SJTUF 20991) with inhibition zone diameters of 11.4, 12.6, 12.3, 25, and 14.2 mm, respectively (Kim et al., 2020).

Commentary: The plant is used medicinally to treat microbial infections in various countries pointing to the presence of potent antimicrobial principles. It brings into being the *C*-glycosyl flavonol aciculatin (Lai et al., 2012), which has the ability to bind to DNA (Carte et al., 1991) and as such may have antimicrobial effects. Consider that *C*-glycosyl flavonols are not uncommonly antibacterial, antifungal, or antiviral.

### References

Carte, B.K., Carr, S., DeBrosse, C., Hemling, M.E., MacKenzie, L., Offen, P. and Berry, D.E., 1991. Aciculatin, a novel flavone-C-glycoside with DNA binding activity from *Chrysopogon aciculatis*. *Tetrahedron*, 47(10–11), pp. 1815–1822.

Kim, G., Gan, R.Y., Zhang, D., Farha, A.K., Habimana, O., Mavumengwana, V., Li, H.B., Wang, X.H. and Corke, H., 2020. Large-Scale Screening of 239 traditional Chinese medicinal plant extracts for their antibacterial activities against multidrug-resistant *Staphylococcus aureus* and cytotoxic activities. *Pathogens*, 9(3), p. 185.

Lai, C.Y., Tsai, A.C., Chen, M.C., Chang, L.H., Sun, H.L., Chang, Y.L., Chen, C.C., Teng, C.M. and Pan, S.L., 2012. Aciculatin induces p53-dependent apoptosis via MDM2 depletion in human cancer cells in vitro and in vivo. *PLoS One*, 7(8), p. e42192.

### 3.4.2.2   *Coix lacryma-jobi* L.

Common names: Job's tears; sada kunch, sada hongai (Bangladesh); skuoy (Cambodia); guy a diu, yi yi (China); gargaria, gavedhukah, oshito (India); jail (Indonesia); duay (Laos, Thailand); jelai batu (Malaysia); ka leik (Myanmar); adlay (the Philippines); bo bo (Vietnam)

Synonyms: *Coix agrestis* Lour.; *Coix arundinacea* J. Koenig ex Willd.; *Coix arundinacea* Lam.; *Coix exaltata* J. Jacq.; *Coix exaltata* Jacq. ex Spreng.; *Coix lacryma* L.; *Coix ovata* Stokes; *Coix pendula* Salisb.; *Lithagrostis lacryma-jobi* (L.) Gaertn.; *Sphaerium lacryma* (L.) Kuntze

Habitat: River banks, watery soils, forests, and cultivated in villages

Distribution: Tropical Asia and Pacific

Botanical examination: It is a stout herb that grows up to 3 m tall. The leaves are simple, alternate, and cauline. The ligule is minute. The blade is linear-lanceolate, 10–45×1.5–5 cm, somewhat rounded at base, and acute at apex. The inflorescence is a raceme that grows up to 4 cm long and presents a few spikelets arranged in pairs or in threes, oblong, and up to 8 mm long. The androecium includes 3 stamens, which are up to 5 mm long, with a pair of stigma. The nuts are magnificent (used to make prayer beds, rosaries, and necklaces) somewhat bluish, pyriform and glossy, grow up to 1.5 cm long, and contain an edible caryopsis.

Medicinal uses: Diseases of the lung, ulcers, and wounds (India)

*Weak antibacterial (Gram-positive) polar extract:* Methanol extract of germinating seeds inhibited the growth of *Bacillus subtilis* (IFO 3009) and *Staphylococcus aureus* (IFO 12732) with MIC values of 2500 and 1000 µg/mL, respectively (Ishiguro et al., 1993).

*Antifungal (filamentous) polar extract:* Methanol extract (10 mg/mL) of seeds evoked a decrease in the diameter of colonies of *Alicyclobacillus sacchari* (M001) and *Chaetomium funicola* (M002) (Sato et al., 2000).

*Antifungal (filamentous) halo developed by benzoxazolinone:* Coixol (also known as 6-methoxy-2-benzoxazolinone) (LogP = 1.4 at pH = 7.4; molecular weight = 165.1 g/mol) (1 μmol/6 mm diameter paper disc) inhibited the mycelial growth of *Coprinus comatus* and at 2.5 μmol/disc inhibited the growth of *Fusarium oxysporum and Rhizoctonia solani* (Wang et al., 2001).

Coixol

*Viral enzyme inhibition by polar extract:* Aqueous extract of seeds at a concentration of 250 μg/mL inhibited Human immunodeficiency virus type-1 protease by 40.1% (Xu et al., 1996).

*Viral inhibition by benzoxazolinone:* Coixol at the concentration of 80 μM inhibited Human immunodeficiency virus type-1 reverse transcriptase by 19.5% (Wang et al., 2001).

Commentary: The plant brings to being trans-feruloyl phytosterol and trans-feruloyl campestanol (Kondo et al., 1988). Consider that 2(3H)-benzoxazolinone inhibited the replication of the Influenza virus (A/Puerto Rico/8/34 (H1N1), PR/8) with an $IC_{50}$ value of 46 μM and selectivity index of 7.6 (Gu et al., 2015). The plant is used for the treatment of COVID-19 in China (Wang et al., 2020), and one could have the curiosity to examine the anti-Severe acute respiratory syndrome-associated coronavirus 2 properties of coixol.

### References

Gu, W., Wang, W., Li, X.N., Zhang, Y., Wang, L.P., Yuan, C.M., Huang, L.J. and Hao, X.J., 2015. A novel isocoumarin with anti-influenza virus activity from *Strobilanthes cusia*. *Fitoterapia*, *107*, pp. 60–62.

Ishiguro, Y., Okamoto, K. and Sonoda, Y., 1993. Antimicrobial activity in etiolated seedlings of adlay. *Nippon Shokuhin Kogyo Gakkaishi*, *40*(5), pp. 353–356.

Kondo, Y., Nakajima, K., Nozoe, S. and Suzuki S., 1988. Isolation of ovulatory-active substances from crops of Job's tears (*Coix lacryma-jobi* L. var. ma-yuen STAPF). *Chemical and Pharmaceutical Bulletin*, *36*(8), pp. 3147–3152.

Sato, J., Goto, K., Nanjo, F., Kawai, S. and Murata, K., 2000. Antifungal activity of plant extracts against *Arthrinium sacchari* and *Chaetomium funicola*. *Journal of Bioscience and Bioengineering*, *90*(4), pp. 442–446.

Wang, S.X., Wang, Y., Lu, Y.B., Li, J.Y., Song, Y.J., Nyamgerelt, M. and Wang, X.X., 2020. Diagnosis and treatment of novel coronavirus pneumonia based on the theory of traditional Chinese medicine. *Journal of Integrative Medicine*, *18*(4), pp. 275–283.

Xu, H.X., Wan, M., Loh, B.N., Kon, O.L., Chow, P.W. and Sim, K.Y., 1996. Screening of traditional medicines for their inhibitory activity against HIV -1 protease. *Phytotherapy Research*, *10*(3), pp. 207–210.

### 3.4.2.3 *Cymbopogon citratus* (DC.) Stapf.

Common names: citronella grass, fever grass, lemon grass; sapalin, hkum-bang-pan, wine-baing (Myanmar); gondho ghas (Bangladesh); slek krey sabou (Cambodia); mao xiang cao (China); serai, serah (Indonesia, Malaysia); agya ghas, haona, sungandhitrna, vashna pulla (India); sing khai

(Laos); sapalin (Myanmar); balioko, tangyad (the Philippines); chakai; khrai (Thailand); hurong mao, sa Chan (Vietnam)

Synonyms: *Andropogon cerifer* Hack.; *Andropogon citratus* DC.; *Andropogon citratus* DC. ex Nees; *Andropogon citriodorum* hort. ex Desf.; *Andropogon roxburghii* Nees ex Wight & Arn.; *Andropogon schoenanthus* L.; *Cymbopogon nardus* (L.) Rendle

Habitat: Cultivated

Distribution: Tropical Asia and the Pacific

Botanical description: It is a tufted herb that grows to a height of 2m from a rhizome. The leaves at base form a sort of elongated bulb, which is light yellow to somewhat pinkish, deliciously aromatic (somewhat lemony), and edible (used in wonderful Thai dishes). The blade is narrow, dull green, somewhat fibrous, and tapering at apex. The Malays believe that precious stones dwell under these plants. The blade is lanceolate, dull light green, linear, and 15–60 cm × 1–2 cm. The inflorescences are racemes of spikelets, which are up to 2 cm long.

Medicinal uses: Cholera (Myanmar); diarrhea (China); fungal infection (India); tonsilitis (Bangladesh); fever (India; Myanmar)

*Strong antibacterial (Gram-positive) essential oil:* Essential oil inhibited the growth of *Staphylococcus aureus* with an MIC of 0.6 μL/mL (Miller et al., 2015).

*Broad-spectrum antibacterial halo developed by essential oil:* Essential oil of leaves (containing mainly myrcene, α-citral (geranial) and β-citral (neral)), inhibited the growth of *Staphylococcus aureus* (NCTC 5671), *Bacillus subtilis* (NCTC 8236), *Escherichia coli* (NCTC 9001), and *Pseudomonas aeruginosa* (NCTC 6750), with inhibition zone diameters ranging from 15 to 32 mm (6 mm diameter paper disc impregnated with 20 μL) (Onawunmi et al., 1984).

*Antibacterial (Gram-positive) halo developed by amphiphilic aliphatic monoterpenes:* Myrcene (LogD = 4.3 at pH 7.4; molecular weight = 136.2 g/mol) inhibited the growth of *Staphylococcus aureus* (NCTC 5671) with an inhibition zone diameter of 11 mm (6 mm diameter paper disc impregnated with 20 μL) (Onawunmi et al., 1984).

*Broad-spectrum antibacterial (Gram-positive) halo developed by amphiphilic aliphatic monoterpenes:* α-Citral (also known as geranial, LogD = 3.1; molecular weight = 152.2 g/mol) inhibited the growth of *Staphylococcus aureus* (NCTC 5671), *Bacillus subtilis* (NCTC 8236), and *Escherichia coli* (NCTC 9001) with inhibition zone diameters of 30, 40, and 15 mm, respectively (6 mm diameter paper disc impregnated with 20 μL), and this effect against *Bacillus subtilis* (NCTC 8236) was increased in the presence of myrcene (Onawunmi et al., 1984). β-Citral (also known as neral, LogD = 3.1 at pH 7.4; molecular weight = 152.2 g/mol) inhibited the growth *Staphylococcus aureus* (NCTC 5671), *Bacillus subtilis* (NCTC 8236), and *Escherichia coli* (NCTC 9001) with inhibition zone diameters of 17, 20, and 12 mm, respectively (6 mm diameter paper disc impregnated with 20 μL) (Onawunmi et al., 1984). Citronellal (LogD = 3.4 at pH 7.4; molecular weight = 154.2 g/mol) inhibited the growth of *Escherichia coli* (NCTC 9001), *Bacillus subtilis* (NCTC 8236), and *Staphylococcus aureus* (NCTC 567) with inhibition zone diameters of 7, 13, and 10 mm, respectively (6 mm diameter paper disc impregnated with 20 μL) (Onawunmi et al., 1984). Citronellol (LogD = 3.5 at pH 7.4; molecular weight = 156.2 g/mol) inhibited the growth of *Escherichia coli* (NCTC 9001), *Bacillus subtilis* (NCTC 8236), and *Staphylococcus aureus* (NCTC 567) with inhibition zone diameters of 7, 13, and 11 mm, respectively (6 mm diameter paper disc impregnated with 20 μL) (Onawunmi et al., 1984). Geraniol (also known as nerol, LogD = 3.4 at pH 7.4; molecular weight = 154.2 g/mol) inhibited the growth of *Escherichia coli* (NCTC 9001), *Bacillus subtilis* (NCTC 8236), and *Staphylococcus aureus* (NCTC 567) with inhibition zone diameters of 9, 14, and 12 mm, respectively (6 mm diameter paper disc impregnated with 20 μL) (Onawunmi et al., 1984).

*Anticandidal halo developed by essential oil:* Essential oil (containing mainly α-citral and β-citral; 2 μL/6 mm diameter paper disc) inhibited the growth of *Candida albicans* (ATCC 10231), *Candida albicans* (ATCC 18804), *Candida glabrata* (ATCC 2001), *Candida krusei* (ATCC 6258), *Candida parapsilosis* (ATCC 22019), and *Candida tropicalis* (ATCC 750) with inhibition zone diameters ranging from 12.8 to 19.3 mm, respectively (Silva et al., 2008).

*Strong broad-spectrum antifungal essential oil:* Essential oil inhibited the mycelial growth of *Candida albicans* by 85% at a concentration of 10 µg/mL (Abe et al., 2003). Essential oil inhibited the growth of *Alternaria alternata, Aspergillus niger, Fusarium oxysporum, Penicillium roquefortii, Candida albicans, Candida oeophila, Hansenula anomala, Metschnikowia fructicola, Saccharomyces cerevisiae, Saccharomyces uvarum,* and *Schizosaccharomyces pombe* with the MIC/MCF values of 0.06/0.06, 0.06/0.06, 0.06/0.3, 0.3/1.2, 0.3/0.3, 0.1/0.1, 1.2/1.2, 1.2/2.5, 5/10, 2.5/2.5, and 0.3/1.2 µL/mL, respectively (Irkin & Korukluoglu, 2009).

*Anticandidal halo developed by aliphatic monoterpenes:* Mixture of α-citral and β-citral (2 µL/6 mm diameter paper disc) inhibited the growth of *Candida albicans* (ATCC 10231), *Candida albicans* (ATCC 18804), *Candida glabrata* (ATCC 2001), *Candida krusei* (ATCC 6258), *Candida parapsilosis* (ATCC 22019), and *Candida tropicalis* (ATCC 750) with inhibition zone diameters ranging from 12.4 to 18.84 mm, (Silva et al., 2008).

*Strong anticandidal aliphatic monoterpene:* α-Citral inhibited the mycelial growth of *Candida albicans* with the $IC_{50}$ value of about 30 µg/mL (Abe et al., 2003).

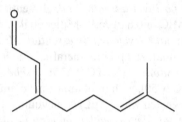

α-Citral

*Strong antiviral (non-enveloped monopartite linear double stranded DNA) polar extract:* Ethanol extract of aerial parts inhibited the replication of the Human adenovirus C serotype 5 (HAdV-5) in A549 cells with an $IC_{50}$ of 90 µg/mL (Chiamenti et al., 2019).

*Strong antiviral (enveloped monopartite linear single stranded (+) RNA) polar extract:* Methanol extract of roots inhibited the replication of the Dengue virus type-2 (strain New Guinea C) in Huh7it-1 cells with an $IC_{50}$ value of 29.3 µg/mL and a selectivity index of 6.2 (Rosmalena et al., 2019).

*Strong antiviral (enveloped linear monopartite single stranded (-) RNA virus) C-glycosyl flavonol:* Isoorientin (LogD = −1.2 at pH = 7.4; molecular mass = 448.3 g/mol) inhibited the replication of the Respiratory syncytial virus (A2 strain) and Respiratory syncytial virus (Long strain) in HEp-2 cells with $IC_{50}$ values of 3.6 and 2.4 µg/mL, respectively (Zhu et al., 2015).

*Weak antiviral (enveloped monopartite single-stranded (+) RNA) hydroxycinnamic acid:* Chlorogenic acid inhibited the replication of the Human coronavirus (HCoV) NL63 with an $IC_{50}$ value of 43.4 µM and plaque formation in LLC-MK2 cells with an $IC_{50}$ value of 43.3 µM (Weng et al., 2019).

*Strong antiviral (enveloped monopartite linear single-stranded (+) RNA) hydroxycinnamic acid:* Caffeic acid inhibited Human coronavirus-NL63 yield, plaque formation, and virus attachment with an $IC_{50}$ value of 3.5, 5.4, and 8.1 µM, respectively (Weng et al., 2019).

*In vivo antiviral (enveloped linear monopartite single-stranded (-) RNA virus) C-glycosyl flavonol:* Isoorientin given for 3 days to BALB/c mice experimentally infected with Respiratory syncytial virus (Long strain) evoked a decrease in pulmonary viral load and inflammation (Zhu et al., 2015).

Commentary: (i) Consider that the increased antibacterial potencies of α-citral in the presence of myrcene provides evidence of synergy among components of essential oils. (ii) The Human adenovirus C serotype 5 is a non-enveloped DNA virus belonging to the genus Mastedovirus in the family Adenoviridae responsible for endemic pneumonias (Chen & Tian, 2018). (iii) Kanyinda (2020) reported improvement in the health of patients testing positive for COVID-19 taking infusions of teas containing *Cymbopogon citratus* (DC.) Stapf. (iv) The plant produces the *C*-glucosyl flavones isoorientin, swertiajaponin, and orientin, and the hydroxycinammic acid derivatives chlorogenic acid and caffeic acid (Cheel et al., 2005). (v) One could endeavor to examine the anti-Influenza and anti-Severe acute respiratory syndrome-associated coronavirus 2 properties of this herb.

(vi) The lipophilic cyclic monoterpenes (+)-limonene and (-)-limonene were inactive against *Escherichia coli* (NCTC 9001), *Bacillus subtilis* (NCTC 8236), and *Staphylococcus aureus* (NCTC 567) and all these linear monoterpenes were inactive against *Pseudomonas aeruginosa* (NCTC 6750) in the experiment conducted by Onawunmi et al. (1984). *Pseudomonas aeruginosa* and Gram-negative bacteria in general resist a broad spectrum of xenobiotics (or molecules stranger to bacterial physiology; from Greek *xeno* = foreign) because of their outer membrane and porin, but also because they have an extra class of efflux pumps compared to Gram-positive bacteria. Efflux pumps can be categorized so far into five classes (Kumar et al., 2013; Nishino & Yamaguchi, 2008)

1. ABC (ATP binding cassette): efflux pumps located in the cytoplasmic membrane of both Gram-positive and Gram-negative bacteria, which use the energy derived from ATP hydrolysis to expel xenobiotics.
2. Resistance nodulation cell-division (RND): efflux pumps located in the cytoplasmic and outer membrane of Gram-negative bacteria (specific to Gram-negative bacteria), which expel xenobiotics using the $H^+$ gradient (antiporters) (Nishino & Yamaguchi, 2008).
3. Major facilitator (MF): efflux pumps located in the cytoplasmic membrane of both Gram-positive and Gram-negative bacteria, which expel xenobiotics using the $H^+$ gradient (antiporters).
4. Small multidrug resistance (SMR): efflux pumps located in the cytoplasmic membrane of both Gram-positive and Gram-negative bacteria, which expel xenobiotics using the $H^+$ gradient (antiporters)
5. Multidrug and toxic compound extrusion (MATE): efflux pump located in the cytoplasmic membrane and are antiporters (the exit of the xenobiotic coincides with the entry of a $Na^+$)

## References

Abe, S., Sato, Y., Inoue, S., Ishibashi, H., Maruyama, N., Takizawa, T., Oshima, H. and Yamaguchi, H., 2003. Anti-*Candida albicans* activity of essential oils including lemongrass (*Cymbopogon citratus*) oil and its component, citral. *Nippon Ishinkin Gakkai Zasshi*, 44(4), pp. 285–291.

Cheel, J., Theoduloz, C., Rodríguez, J. and Schmeda-Hirschmann, G., 2005. Free radical scavengers and antioxidants from Lemongrass (*Cymbopogon citratus* (DC.) Stapf.). *Journal of Agricultural and Food Chemistry*, 53(7), pp. 2511–2517.

Chen, S. and Tian, X., 2018. Vaccine development for human mastadenovirus. *Journal of Thoracic Disease*, 10(Suppl 19), p. S2280.

Chiamenti, L., Silva, F.P.D., Schallemberger, K., Demoliner, M., Rigotto, C. and Fleck, J.D., 2019. Cytotoxicity and antiviral activity evaluation of Cymbopogon spp hydroethanolic extracts. *Brazilian Journal of Pharmaceutical Sciences*, 55.

Irkin, R. and Korukluoglu, M., 2009. Effectiveness of *Cymbopogon citratus* L. essential oil to inhibit the growth of some filamentous fungi and yeasts. *Journal of Medicinal Food*, 12(1), pp. 193–197.

Kanyinda, J.N.M., 2020. Coronavirus (COVID-19): a protocol for prevention and treatment (Covalyse®). *European Journal of Medical and Health Sciences*, 2, pp. 1–4.

Kumar, S., Mukherjee, M.M. and Varela, M.F., 2013. Modulation of bacterial multidrug resistance efflux pumps of the major facilitator superfamily. *International Journal of Bacteriology*.

Miller, A.B., Cates, R.G., Lawrence, M., Soria, J.A.F., Espinoza, L.V., Martinez, J.V. and
    Arbizú, D.A., 2015. The antibacterial and antifungal activity of essential oils extracted
    from Guatemalan medicinal plants. *Pharmaceutical Biology*, *53*(4), pp. 548–554.

Nishino, K. and Yamaguchi, A., 2008. Role of xenobiotic transporters in bacterial drug resis-
    tance and virulence. *IUBMB Life*, *60*(9), pp. 569–574.

Onawunmi, G.O., Yisak, W.A. and Ogunlana, E.O., 1984. Antibacterial constituents in the
    essential oil of *Cymbopogon citratus* (DC.) Stapf. *Journal of Ethnopharmacology*, *12*(3),
    pp. 279–286.

Rosmalena, R., Elya, B., Dewi, B.E., Fithriyah, F., Desti, H., Angelina, M., Hanafi, M.,
    Lotulung, P.D., Prasasty, V.D. and Seto, D., 2019. The antiviral effect of indonesian medici-
    nal plant extracts against dengue virus in vitro and in silico. *Pathogens*, *8*(2), p. 85.

Silva, C.D.B.D., Guterres, S.S., Weisheimer, V. and Schapoval, E.E., 2008. Antifungal activ-
    ity of the lemongrass oil and citral against Candida spp. *Brazilian Journal of Infectious
    Diseases*, *12*(1), pp. 63–66.

Weng, J.R., Lin, C.S., Lai, H.C., Lin, Y.P., Wang, C.Y., Tsai, Y.C., Wu, K.C., Huang, S.H.
    and Lin, C.W., 2019. Antiviral activity of Sambucus FormosanaNakai ethanol extract and
    related phenolic acid constituents against human coronavirus NL63. *Virus Research*, *273*,
    p. 197767.

Zhu, X.Z., Shen, W.W., Gong, C.Y., Wang, Y., Ye, W.C., Li, Y.L. and Li, M.M., 2015. Antiviral
    activity of isoorientin against respiratory syncytial virus in vitro and in vivo. *Journal of
    Sun Yat-sen University (Medical Sciences)*, (3), p. 5.

### 3.4.2.4  *Cynodon dactylon* (L.) Pers.

Common names: Bermuda grass, Bahama grass; dupa (Bangladesh); smao anchien (Cambodia);
gou ya gen (China); durva (India); jukut kakawatan (Indonesia); hnaz fed (Laos); rumput minyak
(Malaysia); mie sa miet (Myanmar); kapot kapot (the Philippines); ya-phraek (Thailand); co chi cog
(Vietnam)

Habitat: Roadsides, grassy spots, vacant urban areas

Distribution: Tropical Asia and Pacific

Botanical observation: It a discrete grass that grows up to about 30cm tall from a rhizome. The
plant is sacred in India. The stem is somewhat purplish and stoloniferous. The leaves are simple,
alternate, exstipulate, and form sheaths bearded at mouth. A ligule is present and displays a line
of hairs. The blade is linear, 1–16 cm × 1–4 mm, linear, and dull green. The umbels of spikes are
terminal and up to 6 cm long.

Medicinal uses: Cough, cold, cuts, wounds, and syphilis (India)

*Broad-spectrum antibacterial halo developed by polar extract:* Aqueous extract (4 mm diameter
well filled with 20 µL of extract at a concentration of 50 mg/mL) inhibited the growth of *Escherichia
coli*, *Staphylococcus aureus*, and *Pseudomonas aeruginosa* with inhibition zone diameters of 8, 5,
and 5 mm, respectively (Rao et al., 2011).

*Moderate broad-spectrum antibacterial polar extract:* Ethanol extract inhibited the growth
of extended-spectrum β-lactamase-producing *Escherichia coli*, *Escherichia coli* (ATCC 25922),
*Vibrio cholerae*, multidrug-resistant *Enterobacter cloacae*, imipenem-resistant *Pseudomonas
aeruginosa*, *Salmonella typhimurium*, *Salmonella enteritidis*, extended-spectrum β-lactamase-
producing *Klebsiella pneumoniae*, methicillin-resistant *Staphylococcus aureus*, *Staphylococcus
aureus* (ATCC 25923), *Streptococcus pyogenes*, and *Streptococcus agalactiae* with MIC values
ranging from 195 to 3125 µg/mL (Marasini et al., 2015).

*Strong antiviral (enveloped monopartite linear single-stranded (+) RNA) polar extract:* Ethanol
extract at a concentration of 50 µg/mL inhibited the replication of the Chikungunya virus in Vero
cells by 98% (Murali et al., 2015).

*Antiviral (enveloped monopartite linear (+)RNA) polar extract:* Ethanol extract abrogated the replication of the Porcine reproductive and respiratory syndrome virus at a concentration of 780 μg/ mL in MARC-145 cells.

*In vivo antiviral (enveloped DNA virus) polar extract:* Aqueous extract mixed at 2% of pellet food protected prawns against the White spot syndrome virus (Balasubramanian et al., 2007).

Commentaries:

(i) Consider that flavonoids with a 5,7-dihydroxyflavone framework are often anti-Chikungunya virus *in vitro* (Pohjala et al., 2011).

(ii) The White spot virus responsible for loss in prawn farms is an enveloped DNA virus that belongs to the genus Whispovirus in the family Nimaviridae (Mahy & Van Regenmortel, 2008).

(iii) The plant produces a series of flavonoids, including the C-glycosyl flavonols vitexin and orientin (Ashokkumar et al., 2013). Is it active against COVID-19?

### References

Ashokkumar, K., Selvaraj, K. and Muthukrishnan, S.D., 2013. *Cynodon dactylon* (L.) Pers.: an updated review of its phytochemistry and pharmacology. *Journal of Medicinal Plants Research*, 7(48), pp. 3477–3483.

Balasubramanian, G., Sarathi, M., Kumar, S.R. and Hameed, A.S., 2007. Screening the antiviral activity of Indian medicinal plants against white spot syndrome virus in shrimp. *Aquaculture*, 263(1–4), pp. 15–19.

Mahy, B.W. and Van Regenmortel, M.H., 2008. *Encyclopedia of Virology*. Academic Press, USA.

Marasini, B.P., Baral, P., Aryal, P., Ghimire, K.R., Neupane, S., Dahal, N., Singh, A., Ghimire, L. and Shrestha, K., 2015. Evaluation of antibacterial activity of some traditionally used medicinal plants against human pathogenic bacteria. *BioMed Research International*.

Murali, K.S., Sivasubramanian, S., Vincent, S., Murugan, S.B., Giridaran, B., Dinesh, S., Gunasekaran, P., Krishnasamy, K. and Sathishkumar, R., 2015. Anti—chikungunya activity of luteolin and apigenin rich fraction from *Cynodon dactylon*. *Asian Pacific Journal of Tropical Medicine*, 8(5), pp. 352–358.

Pohjala, L., Utt, A., Varjak, M., Lulla, A., Merits, A., Ahola, T. and Tammela, P., 2011. Inhibitors of alphavirus entry and replication identified with a stable Chikungunya replicon cell line and virus-based assays. *PLoS One*, 6(12), p. e28923.

Pringproa, K., Khonghiran, O., Kunanoppadol, S., Potha, T. and Chuammitri, P., 2014. In vitro virucidal and virustatic properties of the crude extract of *Cynodon dactylon* against porcine reproductive and respiratory syndrome virus. *Veterinary Medicine International*.

Rao, A.S., Nayanatara, A.K., Kaup, R., Sharma, A., Kumar, A., Vaghasiya, B.D., Kishan, K. and Pai, S.R., 2011. Potential antibacterial and antinfungal activity of aqueous extract of *Cynodon dactylon*. *International Journal of Pharmaceutical Research and Development*, 2(11), pp. 2889–2893.

### 3.4.2.5  *Eleusine indica* (L.) Gaertn.

Synonyms: *Cynodon indicus* (L.) Raspail; *Cynosurus indicus* L.

Common name: Crowfoot grass, goose grass, Indian goosegrass; sursuri ghas, malkantari (Bangladesh); choeung kras (Cambodia); niu jin cao (China); rumput sambau (Malaysia); singno-myet (Myanmar); nandimuki (India); bila bila (Philippines); co man trau, thanh tam (Vietnam)

Habitat: Roadsides, grassy spots, and vacant urban areas

Distribution: Tropical Asia and Pacific

Botanical observation: This discrete grass grows up to about 90 cm tall. The stems are flattened and glabrous. The plant is sacred in India. The leaves are simple, alternate, exstipulate, and form sheaths, which are glabrous or pilose. The ligule is minute, membranous, and with few hairs.

The blade is linear, 1–16 cm×1–4 mm, linear, and dull green. The inflorescences are terminal umbels of 2–7 spikes, which grow up to 10 cm long. The spikelets are elliptic and about 5 mm long.

Medicinal uses: Dysentery (Malaysia, Philippines); fever (Cambodia, Laos, Vietnam)

*Broad-antibacterial halo developed by mid-polar extract:* Chloroform extract inhibited the growth of *Staphylococcus aureus* (NCTC 6571), *Escherichia coli* (NCTC 1048), *Enterobacter aerogenes*, *Proteus vulgaris*, *Streptococcus* sp., *Bacillus* sp., *Pseudomonas aerogenes*, and *Klebsiella aerogenes* at the dilution of 0.125 (Alaekwe et al., 2015).

*Anticandidal halo developed by polar extract:* Methanol extract (1 mg/6 mm paper disc) inhibited the growth of *Candida albicans* with an inhibition zone diameter of 7 mm (Wiart et al., 2004).

*Strong antiviral (enveloped monopartite linear double-stranded DNA) polar extract:* Ethanol extract inhibited the replication of the Herpes simplex virus type-1 in HeLa cells with an $IC_{50}$ value of 100 µg/mL (Hamidi et al., 1996).

Commentary: (i) Consider that ethyl acetate extract of leaves inhibited angiotensin-converting enzyme (Tutor & Chichioco-Hernandez, 2018), which is an anti-Severe acute respiratory syndrome-associated coronavirus 2 target (Tutor & Chichioco-Hernandez, 2018). (ii) The plant brings to being the *C-glucoside* flavonols schaftoside (6-*C*-β-glucoside-8-*C*-α-arabinopyranosyl apigenin) and vitexin (8-*C*-β-glucosyl apigenin) which are anti-inflammatory and probably antipyretic (De Melo et al., 2005). Isovitexin, orientin, and saponanin (also known as isovitexin-7-*O*-glucoside) most probably synergistically account for the medicinal uses. One could have the curiosity to enquire anti-Severe acute respiratory syndrome-associated coronavirus 2 activity of this common Asian grass.

### References

Alaekwe, I.O., Ajiwe, V.I.E., Ajiwe, A.C. and Aningo, G.N., 2015. Phytochemical and antimicrobial screening of the aerial parts of *Eleusine indica*. *International Journal of Pure & Applied Bioscience*, 3(1), pp. 257–264.

De Melo, G.O., Muzitano, M.F., Legora-Machado, A., Almeida, T.A., De Oliveira, D.B., Kaiser, C.R., Koatz, V.L.G. and Costa, S.S., 2005. C-glycosylflavones from the aerial parts of *Eleusine indica* inhibit LPS-induced mouse lung inflammation. *Planta Medica*, 71(04), pp. 362–363.

Hamidi, J.A., Ismaili, N.H., Ahmadi, F.B. and Lajisi, N.H., 1996. Antiviral and cytotoxic activities of some plants used in Malaysian indigenous medicine. *Pertanika Journal of Tropical Agricultural Science*, 19(2/3), pp. 129–136.

Tutor, J.T. and Chichioco-Hernandez, C.L., 2018. Angiotensin-converting enzyme inhibition of fractions from *Eleusine indica* leaf extracts. *Pharmacognosy Journal*, 10(1), pp. 25–28

Wiart, C., Mogana, S., Khalifah, S., Mahan, M., Ismail, S., Buckle, M., Narayana, A.K. and Sulaiman, M., 2004. Antimicrobial screening of plants used for traditional medicine in the state of Perak, Peninsular Malaysia. *Fitoterapia*, 75(1), pp. 68–73.

### 3.4.2.6  *Imperata cylindrica* (L.) Raeusch.

Common names: Thatch grass, cogongrass; sbaw (Cambodia); bai mao (China); darha (India); alang-alang (Indonesia); nazkah (Laos); lalang (Malaysia); kyet mei (Myanmar); kuani (Papua New Guinea); kugon (Philippines); ya kha (Thailand); co thanh (Vietnam)

Synonyms: *Imperata allang* Jungh.; *Imperata arundinacea* Cirillo; *Lagurus cylindricus* L.; *Saccharum cylindricum* (L.) Lam.; *Saccharum koenigii* Retz.; *Saccharum thunbergii* Retz.

Habitat: Grassy areas, roadsides

Distribution: Afghanistan, Kazakhstan, Kyrgyzstan, Turkmenistan, Uzbekistan, Pakistan, India, Sri Lanka, Nepal, Bhutan, Myanmar, Thailand, Vietnam, China, Korea, Japan, Malaysia, Indonesia, the Philippines, Papua New Guinea, Australia, and Pacific Islands

Botanical examination: This rhizomatous herb grows up to about 1.2 m tall. The leaves are simple and spiral. The leaf sheath is glabrous or pilose at margin and mouth. The ligule is minute. The blade is linear, up to 1 m×2 cm, and long acuminate at apex. The inflorescence is a spike, which is somewhat cylindrical, up to 20 cm long, whitish, of an heavenly grace under the sunlight at dawn or sunrise. The glume is 5–9-veined and presents long silky hairs. The androecium comprises a pair of stamens, the anthers are elongated, yellow, and reach 4 mm long. The gynoecium presents a pair of dark purplish and plumose stigmas.

Medicinal uses: Urinary tract infection (the Philippines)

*Antiviral (non-enveloped, monopartite, linear, single stranded (+) RNA)) polar extract:* Ethanol extract of leaves inhibited the replication of the Foot and Mouth Diseases virus type O (FMDV) (KPS/005/2545) in BHK-21 cells at a concentration of 195 µg/mL (Chungsamarnyart et al., 2007).

Commentary: The plant produces the megastigmane tabanone which is most probably antibacterial as it comprises a cyclohexenone framework as well as the simple phenol vinyl guaiacol (Cerdeira et al., 2012), chromones, and flavones (Xuan et al., 2013). The plant is used for the treatment of Severe acute respiratory syndrome-associated coronavirus 2 in China (Liu et al., 2012).

### References

Cerdeira, A.L., Cantrell, C.L., Dayan, F.E., Byrd, J.D. and Duke, S.O., 2012. Tabanone, a new phytotoxic constituent of cogongrass (*Imperata cylindrica*). *Weed Science*, 60(2), pp. 212–218.

Chungsamarnyart, N., Sirinarumitr, T., Chumsing, W. and Wajjawalku, W., 2007. In vitro study of antiviral activity of plant crude-extracts against the foot and mouth disease virus. *Agriculture and Natural Resources*, 41(5), pp. 97–103.

Liu, X., Zhang, M., He, L. and Li, Y., 2012. Chinese herbs combined with Western medicine for severe acute respiratory syndrome (SARS). *Cochrane Database of Systematic Reviews*, (10).

Xuan, L.I.U., Zhang, B.F., Li, Y.A.N.G., Gui-Xin, C.H.O.U. and Zheng-Tao, W.A.N.G., 2013. Two new chromones and a new flavone glycoside from *Imperata cylindrica*. *Chinese Journal of Natural Medicines*, 11(1), pp. 77–80.

### 3.4.2.7  *Lophatherum gracile* Brongn.

Common names: Bamboo-leaf; dan zhu ye, pa lao, shan ji mi (China); rumput bambu (Indonesia); tan chiku yo (Japan); dam juk yeap (Korea); cekrek, ruput bulu (Malaysia); phai pen lek (Thailand)

Synonyms: *Acroelytrum japonicum* Steud.; *Allelotheca urvillei* Steud.; *Lophatherum annulatum* Franch. & Sav.; *Lophatherum dubium* Steud.; *Lophatherum elatum* Zoll. & Moritzi; *Lophatherum geminatum* Baker; *Lophatherum humile* Miq.; *Lophatherum japonicum* (Steud.) Steud.; *Lophatherum lehmannii* Nees; *Lophatherum multiflorum* Steud.; *Lophatherum pilosulum* Steud.; *Lophatherum zeylanicum* Hook. f.

Habitat: Shady and moist soils, roadsides, and villages

Distribution: India, Sri Lanka, Nepal, Cambodia, Myanmar, Vietnam, Thailand, China, Taiwan, Japan, Korea, Malaysia, Indonesia, the Philippines, Papua New Guinea, Australia, and Pacific Islands

Botanical examination: This herb somewhat resembles a small bamboo and grows up to about 1.5 m tall from elongated tuberous roots. The leaves are simple and spiral. The ligule is brown and hairy, and the pseudopetiole about 1 cm long is present. The blade is lanceolate, 5–30×2–5 cm, round at base, and somewhat dark dull green. The spikelets are ovate and about 1 cm long. The inflorescence is a lax terminal raceme of spikes, which grows up to a length of 25 cm. The glumes are ovate, 5-veined, glabrous, and about 5 mm long.

Medicinal use: Chancre (Malaysia); pharyngitis (China); cold, fever, sore throat (Vietnam); mouth sores (China)

*Strong antiviral (enveloped linear monopartite single-stranded (-) RNA) polar extract:* Ethanol extract inhibited the replication of the Respiratory syncytial virus (Long strain), and Respiratory syncytial virus A2 (ATCC VR-1540) strain in the HEp-2 cell line, with $IC_{50}$ values of 20 and 25 µg/mL, respectively (Chen et al., 2019).

*Strong antiviral (enveloped linear monopartite single-stranded (-) RNA) hydrophilic C-glycosyl flavonols:* Isoorientin (LogD = −1.2 at pH = 7.4; molecular mass = 448.3 g/mol) inhibited the replication of the Respiratory syncytial virus (Long strain) and Respiratory syncytial virus (A2 strain) in the HEp-2 cell line with $IC_{50}$ values of 3.1 and 3.1 µg/mL and selectivity indices of 138.7 and 138.7, respectively (Chen et al., 2019). Swertiajaponin inhibited the replication of the Respiratory syncytial virus (Long strain) and Respiratory syncytial virus (A2 strain) in the HEp-2 cell line with $IC_{50}$ values of 6.3 and 6.3 µg/mL and selectivity indices of 39 and 397, respectively (Chen et al., 2019).

Isoorientin

*Moderate antiviral (enveloped linear monopartite single-stranded (-) RNA) stilbene glycoside:* Piceatannol-3′-O-β-D-glucopyranoside inhibited the replication of Respiratory syncytial virus (Long strain) and Respiratory syncytial virus (A2 strain) in the HEp-2 cell line with $IC_{50}$ values of 100 and 100 µg/mL, respectively, and a selectivity index >4 (Chen et al., 2019). Rhaponticin inhibited the replication of the Respiratory syncytial virus (Long strain) and Respiratory syncytial virus (A2 strain) in HEp-2 cell line with $IC_{50}$ values of 60 and 60 µg/mL and selectivity indices of >6.7 and >6.7, respectively (Chen et al., 2019).

*Moderate antiviral (enveloped linear monopartite single-stranded (-)RNA virus) hydroxycinnamic acid derivatives:* 3,5-di-caffeoyl quinic acid inhibited the replication of the Respiratory syncytial virus (Long strain) and Respiratory syncytial virus (A2 strain) in the HEp-2 cell line with $IC_{50}$ values of 2.5 and 2.2 µg/mL and selectivity indices of >80 and >90.9, respectively (Chen et al., 2019). 3,4-di-caffeoyl quinic acid inhibited the replication of the Respiratory syncytial virus (Long strain) and Respiratory syncytial virus (A2 strain) in the HEp-2 cell line with $IC_{50}$ values of 2 and 1.8 µg/mL and selectivity indices of >100 and >111.1, respectively (Chen et al., 2019).

*In vivo antiviral (enveloped monopartite linear (−) RNA) polar extract:* BALB/c mice given an ethanol extract orally at a dose of 2000 mg/Kg/day for 4 days had a reduction of viral load in lung tissues equivalent to ribavirin given orally at 50 mg/Kg/day (Chen et al., 2019).

Commentary: The plant generates a series of *C*-glycosyl flavonols: orientin, homoorientin, vitexin, and isovitexin (Xue et al., 2009) and is used for the treatment of COVID-19 in China (Wang et al., 2020).

### References

Chen, L.F., Zhong, Y.L., Luo, D., Liu, Z., Tang, W., Cheng, W., Xiong, S., Li, Y.L. and Li, M.M., 2019. Antiviral activity of ethanol extract of *Lophatherum gracile* against respiratory syncytial virus infection. *Journal of Ethnopharmacology*, 242, p. 111575.

Wang, S.X., Wang, Y., Lu, Y.B., Li, J.Y., Song, Y.J., Nyamgerelt, M. and Wang, X.X., 2020. Diagnosis and treatment of novel coronavirus pneumonia based on the theory of traditional Chinese medicine. *Journal of Integrative Medicine*, 18(4), pp. 275–283.

Xue, Y.Q., Song, J., Ye, S.P. and Yuan, K., 2009. Separation, identification and its antibacterial activity of glycosylflavones in *Lophatherum gracile* Brongn. *West China Journal of Pharmaceutical Sciences*, 24(3), pp. 218–220.

### 3.4.2.8  *Milium effusum* L.

Synonyms: *Agrostis effusa* (L.) Lam., *Decandolia effusa* (L.) Bastard, *Melica effusa* (L.) Salisb., *Miliarium effusum* (L.) Moench, *Milium confertum* L., *Milium schmidtianum* K. Koch, *Milium transsilvanicum* Schur, *Paspalum effusum* (L.) Raspail

Common names: Millet grass; nol twin (Bangladesh); su cao (China)

Habitat: Forest edges, scrublands, roadsides, and moist soils

Distribution: Iran, Afghanistan, Kazakhstan, Tajikistan, Pakistan, India, Bhutan, Bangladesh, China, Taiwan, and Japan

Botanical examination: This rhizomatous herb grows up to a height of 1.8 m. The leaves are simple, spiral, and exstipulate. The sheaths are somewhat inflated, lax, and glabrous. The ligule is lanceolate and up to 1 cm long. The blade is linear, thin, soft, flat, 10–30 cm × 5–15 mm, glabrous, and acute at apex. The panicle is lax and 10–30 cm long. The spikelets are about 4 mm long. The glumes are elliptic-ovate and membranous.

Medicinal uses: Wounds (Bangladesh)

Commentary: The plant has apparently not been examined for possible antimicrobial properties. It generates quercetin 3-*O*-rutinoside, quercetin 3-*O*-glucoside, kaempferol 3-*O*-rutinoside, and kaempferol 3-*O*-glucoside (Moulton & Whittle, 1989).

#### Reference

Moulton, R.J. and Whittle, S.J., 1989. The major flavonoids in the leaves of *Milium effusum* L. *Biochemical Systematics and Habitat*, 17(3), pp. 197–198.

### 3.4.2.9  *Phragmites australis* (Cav.) Trin. Ex Steud.

Synonyms: *Arundo australis* Cay.; *Arundo phragmites* L.; *Phragmites communis* Trin.; *Phragmites maxima* (Forssk.) Blatter & McCann; *Trichoon phragmites (Li.)* Rendle

Common names: Common reed grass; lu wei (China); dila (Pakistan)

Habitat: River banks, lakes, and marshes

Distribution: Temperate Asia

Botanical examination: It is a rhizomatous herb that grows up to about 3 m tall. The leaves are simple, spiral, and exstipulate. The blade is linear, grows up to about 60 cm ×1–3 cm, and tapering at apex. The panicle is fluffy, whitish, and grows up to about 50 cm ×10 cm. The spikelets are up to about 2 cm long. The glume is lanceolate, up to 1 cm long, and apiculate. The flower presents 3 stamens and 2 plumose stigmas.

Medicinal uses: Typhoid (Bangladesh); sore throat (Australia)

Commentaries: (i) The rhizome accumulates gallotannins and gallic acid (Weidenhamer et al., 2013), and releases gallic acid in the soil it grows (Weidenhamer et al., 2013). The flowers contain the *C*-glycosylflavones swertiajaponin, isoswertiajaponin, 3'-*O*-gentiobioside, and 3'-*O*-glucoside of swertiajaponin, as well as the flavone glycosides rhamnetin 3-*O*-rutinoside and rhamnetin 3-*O*-glucoside (Nawwar et al., 1980). (ii) The plant is used in China for the treatment of COVID-19 Wang et al., 2020). Once could have the curiosity to isolate anti-Severe acute respiratory syndrome-associated coronavirus 2 principle from this herb. Consider that Poaceae are not uncommonly used for the treatment of COVID-19 in China.

#### References

Nawwar, M.A., El Sissi, H.I. and Baracat, H.H., 1980. The flavonoids of Phragmites australis flowers. *Phytochemistry*, 19(8), pp. 1854–1856.

Wang, S.X., Wang, Y., Lu, Y.B., Li, J.Y., Song, Y.J., Nyamgerelt, M. and Wang, X.X., 2020. Diagnosis and treatment of novel coronavirus pneumonia based on the theory of traditional Chinese medicine. *Journal of Integrative Medicine, 18*(4), pp. 275–283.

Weidenhamer, J.D., Li, M., Allman, J., Bergosh, R.G. and Posner, M., 2013. Evidence does not support a role for gallic acid in *Phragmites australis* invasion success. *Journal of Chemical Habitat, 39*(2), pp. 323–332.

### 3.4.2.10  *Phyllostachys nigra* (Lodd. ex Lindl.) Munro

Common name: Black bamboo; zi zhu (China); somdae (Korea)

Habitat: Open forests, valley, roadsides, or ornamental

Distribution: China and Korea

Botanical examination: This bamboo grows up to about 8 m tall. The ligule is purple and ciliate. The leaves are simple, exstipulate, and alternate. The blade is lanceolate and about 7–10 cm × 1.2 cm. The young shoots are edible. The spikelets are lanceolate, up to 2 cm. The glumes is hairy. The androecium includes 3 stamens, which are about 1 cm long. The ovary develops 3 stigmas.

Medicinal use: Cough (Korea)

*Viral enzyme inhibition by polar extract:* Methanol extract of leaves at a concentration of 100 μg/mL inhibited RNA-dependent DNA polymerase and ribonuclease H activities of Human immunodeficiency virus type-1 reverse transcriptase and Human immunodeficiency virus type-1 protease by 14.6, 1.1, and 18.4%, respectively (Min et al., 2001).

Commentary: Consider that methanol extract inhibited angiotensin-converting enzyme (Park & Jhon, 2010) and as such may have some activity against the Severe acute respiratory syndrome-associated coronavirus 2. The leaves contain the flavonols tricin, the flavonol glycosides tricin-7-*O*-β-D-glucopyranoside, tricin-7-O-neohesperidoside, and vittariflavone; the flavones *C*-glucosides vitexin (Sun et al., 2008), isoorientin, orientin, luteolin 6-*C*-(6″-*O*-*trans*-caffeoylglucoside); and the hydroxycinnamic acid derivatives *cis*-coumaric acid and *p*-coumaric acid (Lee et al., 2010). Note that flavones *C*-glucosides are not uncommon in members of the Poaceae, and such compounds are often antiviral.

### References

Lee, H.J., Kim, K.A., Kang, K.D., Lee, E.H., Kim, C.Y., Um, B.H. and Jung, S.H., 2010. The compound isolated from the leaves of *Phyllostachys nigra* protects oxidative stress-induced retinal ganglion cells death. *Food and Chemical Toxicology, 48*(6), pp. 1721–1727.

Min, B.S., Kim, Y.H., Tomiyama, M., Nakamura, N., Miyashiro, H., Otake, T. and Hattori, M., 2001. Inhibitory effects of Korean plants on HIV-1 activities. *Phytotherapy Research, 15*(6), pp. 481–486.

Park, E.J. and Jhon, D.Y., 2010. The antioxidant, angiotensin converting enzyme inhibition activity, and phenolic compounds of bamboo shoot extracts. *LWT-Food Science and Technology, 43*(4), pp. 655–659.

Sun, W.X., Li, X., Li, N. and Meng, D.L., 2008. Chemical constituents of the extraction of bamboo leaves from *Phyllostachys nigra* (Loddex Lindl) *Munro varhenonis* (Mitf) Stepfex Rendle. *Journal of Shenyang Pharmaceutical University, 25*(1), pp. 39–43.

### 3.4.2.11  *Vetiveria zizanioides* (L.) Nash

Synonyms: *Agrostis verticillata* Lam.; *Anatherum muricatum* (Retz.) P. Beauv.; *Andropogon muricatus* Retz.; *Andropogon squarrosus* Hook. f.; *Phalaris zizanioides* Linn.; *Sorghum zizanioides* (Linn.) O. Ktze.; *Vetiveria muricata* (Retz.) Griseb.; *Vetiveria odoratissima* Lem.-Lisanc.

Common names: Vetiver, sirmou, khas khas (Bangladesh, India); ranapriya (India); akar wangi (Indonesia, Malaysia); amora (Philippines); ya faekhom, huong bai (Thailand)

Habitat: Swamps or cultivated

Distribution: India, Bangladesh, Myanmar

Botanical observation: This massive and aromatic tufted grass grows to about 4m tall from fibrous roots (somewhat remotely leek-like). The leaves form sheaths, which are lax and up to about 20cm long. A minute and ciliate ligule is present. The blade is linear, and 30–60 cm×4–10mm. The panicles are terminal, oblong, dark purplish and about 20cm long. Sessile spikelets are about 5mm long and linear lanceolate. The androecium includes 3 stamens with orange anthers, which are about 2mm long. The gynoecium includes 2 plumose and purple stigmas. The caryopsis is elongated.

Medicinal uses: Boils, fever, sores in the mouth, urinary tract infection (India)

*Strong broad-spectrum antibacterial non-polar extract:* Hexane extract of roots inhibited the growth of *Mycobacterium smegmatis* (wild type MC2 155), *Mycobacterium smegmatis* (resistant to ciprofloxacin, lomofloxacin, norfloxacin MSR 101), and *Escherichia coli* (resistant to nalidixic acid NK5819) with MIC values of 62.5, 62.5, and 1000 µg/mL, respectively (Luqman et al., 2005). Essential oil inhibited the growth of *Streptococcus mutans* with an MIC of 0.4 µg/mL (Miller et al., 2015).

*Strong antifungal (filamentous) essential oil:* Essential oil inhibited the growth of *Aspergillus niger* (ATCC 16888), *Candida albicans* (ATCC 18804), and *Cryptococcus neoformans* (ATCC 24607) with MIC values of 78, 313, and 20 µg/mL, respectively (Powers et al., 2018).

*Strong antiviral (enveloped segmented linear single-stranded (−) RNA virus) amphiphilic flavonol:* Tricin (also known as 5,7,4'-trihydroxy-3', 5'-dimethoxyflavone, LogD = 1.1 at pH 7.4; molecular mass = 330.2 g/mol) inhibited the replication of the Influenza A/Solomon Islands (H1N1), Influenza A/Hiroshima/52/2005 (H3N2), Influenza A/Naritra/1/2009 (H1N1 pdm), Influenza A/California/07/2009 (H1N1 pdm), and Influenza B/Malaysia/2506/2004 with $IC_{50}$ values of 4.6, 3.4, 8.2, 10.2, and 4.9µM, respectively (Yazawa et al., 2011).

Tricin

*Strong antiviral (enveloped monopartite linear dimeric single-stranded (+) RNA) amphiphilic flavonol:* Tricin inhibited the replication of the Human immunodeficiency virus type-1 (strain IIIB) with an $IC_{50}$ value of 14.4 µg/mL and a selectivity index of 40 (Matsuta et al., 2011).

*In vivo antiviral (enveloped segmented linear single-stranded (−) RNA virus) amphiphilic flavonol:* Mice (DBA/2 Cr) intranasally infected with Influenza A/PR/8/34 and orally administered with tricin at 100 µg/Kg survived by 83% (Yazawa et al., 2011).

Commentary: Essential oil contains mainly the bicyclic sesquiterpenes β-vetispirene, khusimol, vetiselinenol, and α-vetivone (Champagnat et al., 2006). The plant generates a series of flavonoids: carlinoside, neocarlinoside, 6,8-di-C-arabinopyranosyl luteolin, isoorientin, and tricin-5-O-glucoside (Champagnat et al., 2008). Is tricin active against the Severe acute respiratory syndrome-associated coronavirus?

**References**

Champagnat, P., Figueredo, G., Chalchat, J.C., Carnat, A.P. and Bessiere, J.M., 2006. A study on the composition of commercial *Vetiveria zizanioides* oils from different geographical origins. *Journal of Essential Oil Research, 18*(4), pp. 416–422.

Champagnat, P., Heitz, A., Carnat, A., Fraisse, D., Carnat, A.P. and Lamaison, J.L., 2008. Flavonoids from *Vetiveria zizanioides* and *Vetiveria nigritana* (Poaceae). *Biochemical Systematics and Habitat, 1*(36), pp. 68–70.

Luqman, S., Srivastava, S., Darokar, M.P. and Khanuja, S.P., 2005. Detection of antibacterial activity in spent roots of two genotypes of aromatic grass *Vetiveria zizanioides*. *Pharmaceutical Biology, 43*(8), pp. 732–736.

Matsuta, T., Sakagami, H., Satoh, K., Kanamoto, T., Terakubo, S., Nakashima, H., Kitajima, M., Oizumi, H. and Oizumi, T., 2011. Biological activity of luteolin glycosides and tricin from *Sasa senanensis* Rehder. *In Vivo, 25*(5), pp. 757–762.

Miller, A.B., Cates, R.G., Lawrence, M., Soria, J.A.F., Espinoza, L.V., Martinez, J.V. and Arbizú, D.A., 2015. The antibacterial and antifungal activity of essential oils extracted from Guatemalan medicinal plants. *Pharmaceutical Biology, 53*(4), pp. 548–554.

Powers, C.N., Osier, J.L., McFeeters, R.L., Brazell, C.B., Olsen, E.L., Moriarity, D.M., Satyal, P. and Setzer, W.N., 2018. Antifungal and cytotoxic activities of sixty commercially-available essential oils. *Molecules, 23*(7), p. 1549.

Yazawa, K., Kurokawa, M., Obuchi, M., Li, Y., Yamada, R., Sadanari, H., Matsubara, K., Watanabe, K., Koketsu, M., Tuchida, Y. and Murayama, T., 2011. Anti-influenza virus activity of tricin, 4′, 5, 7-trihydroxy-3′, 5′-dimethoxyflavone. *Antiviral Chemistry and Chemotherapy, 22*(1), pp. 1–11.

## 3.5 ORDER ZINGIBERALES GRISEB. (1854)

### 3.5.1 FAMILY CANNACEAE A.L. de Jussieu (1789)

The family Cannaceae consists of the genus *Canna* L.

#### 3.5.1.1 *Canna indica* L.

Synonyms: *Canna chinensis* Willd.; *Canna edulis* Ker Gawl.

Common names: Canna, Queensland arrowroot; Indian shot; kalabati, surbo jaya (Bangladesh); ché:k té:hs (Cambodia); hen mie, mei ren jiao (China); devakuli, kulvaleimuni, Krishna tamarah (India); minbuah tasbeh (Indonesia); daun tasbeh (Malaysia); adakut, butsarana (Myanmar); hakik (Pakistan); balunsaying (the Philippines); phuttharaksa (Thailand); khoai dao (Vietnam)

Habitat: Sun-exposed watery soils, cultivated ornamental

Distribution: Native to South America, tropical Asia, and Pacific

Botanical examination: This magnificent rhizomatous herb grows up to a height of 3 m. The rhizome yields an edible starch. The leaves are simple, spiral, and exstipulate. The leaf sheath is green. A pseudopetiole is present. The blade is ovate to oblong 30–60 cm × 10–20 cm, edible when young, somewhat pale green, and with some sort of secondary nerves. The inflorescence is a terminal spike-like raceme. The calyx consists of 3 sepals, which are purplish green, lanceolate, and about 1.5 cm long. The corolla is tubular, 3 lobed, the lobes linear, about 4 cm long, and yellowish green. The androecium includes 3 staminodes as well as a labellum (made of 2 fused staminodes), which are oblong to spathulate, about 5 cm long, of an heavenly light pink, plain red, or yellow, and a single, about 6 cm long fertile anther. The ovary is globose, 6 mm across, rugose, and develops a 6 cm long style. The capsule is ovoid to obovoid, trilocular, somewhat spiny, of a peculiar dull light green, about 2 cm long, and opening to release beautiful, globose to ovoid, glossy, and blackish-brown seeds (used to make rosaries, necklaces, and beads). The dry capsule is golden brownish, rough to the touch, papery, and glossy.

Medicinal use: Fever (India, Myanmar); diarrhea, sores (Myanmar)

*Viral enzyme inhibition by polar extract:* Water extract of rhizomes inhibited the Human immunodeficiency virus type-1 reverse transcriptase by 92.9% at a concentration of 200 μg/mL (Woradulayapinij et al., 2005).

Commentary: One could have the curiosity to look for antimicrobial and particularly antiviral principles in this plant.

### Reference

Woradulayapinij, W., Soonthornchareonnon, N. and Wiwat, C., 2005. In vitro HIV reverse transcriptase inhibitory activities of Thai medicinal plants and *Canna indica* L. rhizomes. *Journal of Ethnopharmacology, 101*(1–3), pp. 84–89.

### 3.5.2 FAMILY COSTACEAE NAKAI (1941)

The family Costaceae consists of 4 genera and nearly 120 species of herbs growing from rhizomes. The stems are often fleshy, even spiral, and bear simple and spirally arranged leaves. The flowers are arranged in terminal spikes. The calyx is tubular and 2–3 lobed. The corolla is tubular with 3 irregular lobes. The androecium consists of 1 stamen. The gynoecium consists of 2–3 carpels forming a 2–3 locular ovary with ovules on axil placentas. The fruit is a capsule containing numerous black and arillate seeds.

### 3.5.2.1 *Costus speciosus* (J. Koenig ex Retz.) Sm.

Common names: Spiral ginger; keowa, ketoki (Bangladesh); trâthôk (Cambodia); bi qiao jiang (China); canda, kembuka (India); setawar (Indonesia); uangs (Laos) tawar (Malaysia), tubong usa (the Philippines); uang-yai (Thailand); mía do (Vietnam)

Synonyms: *Banksia speciosa* König, *Costus formosanus* Nakai, *Costus speciosus* var. *hirsutus* Blume, *Costus speciosus* var. *leocalyx* (K. Schum.) Nakai, *Costus spicatus* var. *pubescens* Griseb., *Hellenia grandiflora* Retz.

Habitat: Damp, open spots, cultivated ornamental

Distribution: India, Bangladesh, Malaysia, Cambodia, Laos, Vietnam, Thailand, Sri Lanka, Philippines, Australia, and Indonesia.

Botanical description: This eery herb grows up to 2 m tall from a rhizome. The stem is reddish, and characteristically spiral. The leaves are shortly petiolate, 18 cm × 10 cm, hairy below, and oblong acuminate. The spikes are terminal, bullet shaped, reddish, glossy, scaly and up to about 15 cm long. The calyx is short and red. The corolla is 6 cm long, white, somewhat eery, membranaceous, showy, and often pinkish and with up to 12 cm across, white and yellow lip. The stamen is 6 cm long, hairy, and orange. The flowers are used by the malays for exorcism.

Medicinal uses: dysentery (Bangladesh, India); bronchitis, fever, leprosy (India)

*Antibacterial halo developed by polar extract:* Aqueous extract of rhizome (4 mg/6 mm paper disc) inhibited the growth of *Staphylococcus aureus.*

*Moderate antibacterial* (Gram-positive) *developed by mid-polar extract:* Chloroform extract of rhizomes inhibited the growth of *Staphylococcus aureus*, *Staphylococcus epidermidis*, and *Bacillus subtilis* with MIC values of 1250, 625, and 1250 μg/mL, respectively ().

*Moderate antifungal (filamentous) non-polar extract:* Hexane extract of rhizomes inhibited the growth of *Trichophyton mentagrophytes*, *Epidermophyton floccosum*, *Trichophyton simii*, *Trichophytum rubrum*, and *Magnaporthe grisea* with the MIC values of 620, 620, 620, 125, and 125 μg/mL, respectively (Duraipandiyan et al., 2012).

*Moderate antifungal (yeast) sesquiterpene lactones:* The germacrane-type sesquiterpene lactone costunolide inhibited the growth of *Trichophyton mentagrophytes*, *Trichophyton simii*, *Trichophyton rubrum*, *Epidermophytum floccosum*, *Scopulariopsis* sp., *Aspergillus niger*, *Curvularia lunata*, and *Magnaporthe grisea* with the MIC values of 62.5, 31.2, 62.5, 125, 250, 250,

125, and 250 μg/mL, respectively (Duraipandiyan et al., 2012). Eremanthin of *Trichophyton men-tagrophytes*, *Trichophyton simii*, *Trichophyton rubrum*, *Epidermophyton floccosum*, *Aspergillus niger*, *Curvularia lunata*, and *Magnaporthe grisea* with MIC values of 125, 62.5, 250, 125, 125, 250, and 250 μg/mL, respectively (Duraipandiyan et al., 2012).

*Strong antiviral (enveloped monopartite linear double-stranded DNA):* Ethanol extract inhibited the replication of the Herpes simplex virus type-1 with an MIC value of 100 μg/mL (Hamidi et al., 1996).

*Strong antiviral (enveloped non-segmented (-) RNA) polar extract:* Ethanol extract inhibited the replication of the Vesicular stomatitis virus with an MIC value of 20 μg/mL (Hamidi et al., 1996).

*Strong antiviral (enveloped circular double-stranded DNA) amphiphilic sesquiterpene lactone:* Costunolide (LogD = 3.3 at pH 7.4; molecular weight = 232.3 g/mol) inhibited the replication of the Hepatitis B virus with an $IC_{50}$ value of 1 μM (Chen et al., 1995).

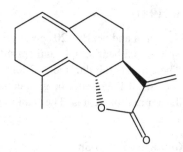

Costunolide

Commentaries: (i) The plants brings into being the steroidal diosgenin as well as a series of aliphatic hydroxyketones, including 24-hydroxyhentriacontan-27-one and 24-hydroxytriacontan-26-one (Gupta et al., 1981). (ii) The plant is mentioned in the *Atharva Veda* (1000 BC) and was imported from India by Romans for the worship of Apollo (Cobb, 2018). (iii) Siddha healers in South India use the plant to treat COVID-19 (Sasikumar et al., 2020). (iv) The plant generates the antiviral spirostane-type steroid diosgenin (Rathore & Khanna, 1979), which is not uncommonly found in medicinal plants used to treat COVID-19 in Asia. One could examine the Severe acute respiratory syndrome-associated coronavirus type-2 activity of diosgenin.

### References

Chen, H.C., Chou, C.K., Lee, S.D., Wang, J.C. and Yeh, S.F., 1995. Active compounds from *Saussurea lappa* Clarks that suppress Hepatitis B virus surface antigen gene expression in human hepatoma cells. *Antiviral Research*, 27(1–2), pp. 99–109.

Cobb, M.A., 2018. *Rome and the Indian Ocean trade from Augustus to the early third century CE*. Brill, USA.

Duraipandiyan, V., Al-Harbi, N.A., Ignacimuthu, S. and Muthukumar, C., 2012. Antimicrobial activity of sesquiterpene lactones isolated from traditional medicinal plant, *Costus specio-sus* (Koen ex. Retz.) Sm. *BMC Complementary and Alternative Medicine*, 12(1), pp. 1–6.

Gupta, M.M., Lal, R.N. and Shukla, Y.N., 1981. Aliphatic hydroxyketones and diosgenin from *Costus speciosus* roots. *Phytochemistry*, 20(11), pp. 2553–2555.

Hamidi, J.A., Ismaili, N.H., Ahmadi, F.B. and Lajisi, N.H., 1996. Antiviral and cytotoxic activities of some plants used in Malaysian indigenous medicine. *Pertanika Journal of Tropical Agricultural Science*, 19(2/3), pp. 129–136.

Rathore, A.K. and Khanna, P., 1979. Steroidal constituents of *Costus speciosus* (Koen) SM. callus cultures. *Planta Medica*, 35(03), pp. 289–290.

Saraf, A., 2010. Phytochemical and antimicrobial studies of medicinal plant *Costus speciosus* (Koen.). *E-Journal of Chemistry*, 7, pp. 413–455.

Sasikumar, R., Priya, S.D. and Jeganathan, C., 2020. A case study on domestics spread of SARS-CoV-2 pandemic in India. *International Journal of Advanced Science and Technology*, 29(7), pp. 2570–2574.

### 3.5.3 FAMILY MUSACEAE A. L. de JUSSIEU (1789)

The family Musaceae consists of 3 genera and about 40 species of large herbs. The stems are often fleshy, fibrous, and consisting of leave sheaths. The leaves are simple, spiral, and petiolate. The blade is fleshy, not uncommonly very large, and dull green. The flowers are terminal. The perianth comprises 6 tepals. The androecium consists of 5 stamens. The gynoecium consists of 3 carpels forming a 3 locular ovary with ovules on axil placentas. The fruit is a berry.

#### 3.5.3.1 *Musa paradisiaca* L.

Synonyms: *Musa sapientum* L.

Common names: Banana; plantain; kala, kola, kela, Lai fung (Bangladesh); chek (Cambodia); da jiao (China); dairakala, anshumatiphal (India); kwàyz khauz (Laos); pisang (Indonesia, Malaysia); saging (Philippines); kluai (Thailand); chuoi (Vietnam); murgueb (Yap)

Habitat: Vacant grassy and moist spots of land, roadsides, villages, cultivated

Distrribution: Tropical Asia and Pacific

Botanical description: This tree looks like a giant ginger and grows up to about 10 m tall. The plant is sacred in India and of extraordinary usefulness for animals and humans. The pseudostem is massive, fibrous, fleshy, green, and consists of somewhat sappy and glossy light yellow leaf sheaths. The leaves are simple and alternate. The blade can grow up to about 3 m long, it is oblong, somewhat rectangular, light green, glossy, smooth, fleshy, and with a deeply sunken midrib. The spikes are stout and about 1 m long and present burgundy red bracts, which are 15–20 cm long, ovate, concave, fleshy, open in succession. The perianth is about 2 cm long and consists of 5 fused tepals forming a golden yellow lamina. The androecium comprises 5 stamens. The ovary develops a style with a trilobed stigma. The berries are yellow, slightly falcate, oblong, fleshy, 5–15 cm long containing a few seeds (in a white and delicious pulp) which are black, glossy and about 4 mm across.

Medicinal uses: Diabetic foot ulcers (Malaysia); tonsillitis, typhoid, and tooth infection in (Bangladesh); dysentery (India, Bangladesh, Yap); diarrhea, tuberculosis (India); blennorhea, fever (the Philippines)

*Broad halo developed by polar extracts:* Methanol extract of peel (2 mg/agar well (8.5 mm)) inhibited *Enterobacter aerogenes* (ATCC13048), *Klebsiella pneumoniae* (NCIM2719), and *Salmonella typhimurium* (ATCC2364) with inhibition zone diameters 10, 15.5, and 9 mm, respectively (Rakholiya et al., 2014). Exsudate from the pseudostems (100 μL/well) inhibited the growth of *Lactophilus acidophilus*, *Staphylococcus aureus*, *Escherichia coli*, and *Pseudomonas aeruginosa* with inhibition zone diameters of 11, 19, 17, and 12 mm, respectively (Ghany et al., 2019).

*Strong broad-spectrum antibacterial polar extract:* Ethanol/Aqueous (50:50) extract of fresh flowers inhibited the growth of *Staphylococcus aureus* (ATCC 25293), *Escherichia coli* (MTCC 443), *Pseudomonas aeruginosa* (ATCC 9027), *Bacillus subtilis* (MTCC 121), *Bacillus cereus* (MTCC 109), *Klebsiella Pneumoniae I(MTCC 109)*, *P. mirabilis* (MTCC 1429), *Salmonella typhimurium* (MTCC 2672), and *Streptococcus pneumonia* (MTCC 2672) (Jawla et al., 2012).

*Strong anticandidal antibacterial polar extract:* Ethanol/aqueous (50:50) extract of fresh flowers inhibited the growth of *Candida albicans* (MTCC 183) and *Candida albidus* (MTCC 2661) with MIC values of 9.8 and 7.6 μg/mL, respectively (Jawla et al., 2012).

*Antiviral (non-enveloped monopartite linear (+) RNA) polar extract:* Exudate from the pseudostem inhibited the replication of the Hepatitis A virus (HAV-10) in Vero cells (Ghany et al., (2019).

*Strong antiviral (enveloped monopartite linear single-stranded (+) RNA) polar extract:* Ethanol extract of leaves inhibited the replication of the Chikungunya virus (899 strain) in Vero cells with an $IC_{50}$ value of 10.8 µg/mL ($CC_{50}$ value for Vero cells about 60.9 µg/mL, SI = 5.6) (Panda et al., 2020). Ethanol extract of leaves inhibited the replication of the Yellow fever virus (YFV) (17D Stamari l strain) cultured in human liver (Huh) cells with an $IC_{50}$ value of 16.3 µg/mL and selectivity index > 100 (Panda et al., 2020).

Commentaries: (i) The antibacterial properties of the plant was reported by Scott et al. (1949). It generates phenalenone-type phytoalexins, irenolone and emenolone (Luis et al., 1993), which are planar and bear quinone moieties and as such are probably, at least partially, DNA-intercalating antimicrobial agents. Other secondary metabolites synthetized are the cycloartane-type triterpenes cycloeucalenone, 31-norcyclolaudenone, and 24-methylene-cycloartanol (Accioly et al., 2012). 31-Norcyclolaudenone and 24-methylene-cycloartanol are antileishmanial (Accioly et al., 2012) and as such probably antifungal (amphotericin is antileishmanial) (Sundar et al., 2004). One could have some interest to look for antiviral principles in this plant. (ii) The medicinal properties of banana fruit are found in the *The Canon of Medicine* of our ancient peer Ibn Sina also known as Avicenna (980–1037).

### References

Accioly, M.P., Bevilaqua, C.M.L., Rondon, F.C., de Morais, S.M., Machado, L.K., Almeida, C.A., de Andrade Jr, H.F. and Cardoso, R.P., 2012. Leishmanicidal activity in vitro of *Musa paradisiaca* L. and *Spondias mombin* L. fractions. *Veterinary Parasitology*, 187(1–2), pp. 79–84.

Ghany, T.A., Ganash, M., Alawlaqi, M.M. and Al-Rajhi, A.M., 2019. Antioxidant, antitumor, antimicrobial activities evaluation of *Musa paradisiaca* L. Pseudostem exudate cultivated in Saudi Arabia. *BioNanoScience*, 9(1), pp. 172–178.

Jawla, S., Kumar, Y. and Khan, M.S.Y., 2012. Antimicrobial and antihyperglycemic activities of *Musa paradisiaca* flowers. *Asian Pacific Journal of Tropical Biomedicine*, 2(2), pp. S914–S918.

Luis, J.G., Echeverri, F., Quinones, W., Brito, I., Lopez, M., Torres, F., Cardona, G., Aguiar, Z., Pelaez, C. and Rojas, M., 1993. Irenolone and emenolone: two new types of phytoalexin from *Musa paradisiaca*. *The Journal of Organic Chemistry*, 58(16), pp. 4306–4308.

Panda, S.K., Castro, A.H.F., Jouneghani, R.S., Leyssen, P., Neyts, J., Swennen, R. and Luyten, W., 2020. Antiviral and cytotoxic activity of different plant parts of banana (Musa spp.). *Viruses*, 12(5), p. 549.

Rakholiya, K., Kaneria, M. and Chanda, S., 2014. Inhibition of microbial pathogens using fruit and vegetable peel extracts. *International Journal of Food Sciences and Nutrition*, 65(6), pp. 733–739.

Scott, W.E., McKay, H.H., Schaffer, P.S. and Fontaine, T.D., 1949. The partial purification and properties of antibiotic substances from the banana (*Musa sapientum*). *The Journal of Clinical Investigation*, 28(5), pp. 899–902.

Sundar, S., Mehta, H., Suresh, A.V., Singh, S.P., Madhukar, R. and Murray, H.W., 2004. Amphotericin B treatment for Indian visceral leishmaniasis: conventional versus lipid formulations. *Clinical Infectious Diseases*, 38(3), pp. 377–383.

## 3.5.4 FAMILY ZINGIBERACEAE MARTYNOV (1820)

The family Zingiberaceae consists of about 45 genera and 700 species of gingers. The precise identification of members of this family is difficult. The leaves are simple and spiral. The blade presents a prominent midrib and numerous lateral nerves. The inflorescence is a cyme or a spike. The calyx consists of 3 sepals, which are greenish and united below to form a tube. The corolla is tubular with 3 linear or oblong lobes. The lip is broad, entire, or trilobed. The androecium comprises a single stamen. The gynoecium consists of 3 carpels united into a trilocular ovary and develops a papillate stigma. The fruits are capsules. The seeds are numerous, often aromatic, and arillate. Members of the family Zingiberaceae bring to being antimicrobial flavonoids and a series of monoterpenes, sesquiterpenes, and diterpenes.

### 3.5.4.1  *Alpinia conchigera* Griff.

Synonym: *Languas conchigera* (Griff.) Burkill

Common names: Lesser alpinia; khetranga; ketranga, (Bangladesh); jie bian shan jiang (China); lengkuas kencil, lengkuas padang (Malaysia); khaa ling (Thailand)

Habitat: Forests

Distribution: India, Bangladesh, Myanmar, Cambodia, Laos, Thailand, Vietnam, Malaysia, and Indonesia

Botanical description: This ginger grows to about 2 m tall from a rhizome, which is light yellow to somewhat pinkish. The leaves are simple, sheathing, and with an entire ligule about 5 mm long. The pseudopetiole is up to 1 cm long. The blade is elliptic, glossy, somewhat dull dark green, 20–30 cm × 10–15 cm, obtuse at base, and acute at apex. The inflorescences are terminal panicles, which are about 25 cm long. The calyx is tubular and 3-lobed. The corolla is heavenly, about 1 cm long, with an obovate and emarginate labellum penciled with orangish-red lines. The androecium comprises 1 conspicuous stamen bent downward. The ovary is minute. The capsules are ovoid, glossy, green turning bright red, about 1 cm long, and marked at apex with vestiges of perianth.

Medicinal uses: Dysentery (Bangladesh)

*Moderate antibacterial (Gram-negative) polar extract:* Methanol extract of rhizomes inhibited *Neisseria gonorrhea* with an MIC of 513.4 µg/mL (Chomnawang et al., 2009).

*Strong broad-spectrum antibacterial essential oil:* Essential oil of leaves inhibited the growth of *Pseudomonas aeruginosa* (UI 60690), *Pseudomonas cepacia*, and *Staphylococcus epidermidis* (ATCC 1228) with MIC values of 25.9, 25.9, and 25.9 µg/mL, respectively (Ibrahim et al., 2009).

*Strong antibacterial (Gram-positive) amphiphilic phenylpropanoid:* Chavicol acetate (also known as 4-allylphenyl acetate, LogD = 2.7 at pH = 7.4; molecular weight = 176.2 g/mol) inhibited the growth of *Staphylococcus aureus* (ATCC 29213), methicillin-resistant *Staphylococcus aureus* (ATCC 33591), and vancomycin intermediate-resistant *Staphylococcus aureus* (ATCC 700699) with the MIC values of 35.7, 17.8, and 35.7 µg/mL, respectively (Aziz et al., 2013). 1′-Hydroxychavicol acetate (Log P = 1.5 at pH 7.4; molecular weight = 192.2 g/mol) inhibited the growth of *Staphylococcus aureus* (ATCC 29213), methicillin-resistant *Staphylococcus aureus* (ATCC 33591), and ancomycin intermediate-resistant *Staphylococcus aureus* (ATCC 700699), (VISA 24), and (VRSA 156) with MIC values of 1250, 625, 1250, 156, and 625 µg/mL, respectively (Aziz et al., 2013).

Chavicol acetate

1′-Hydroxychavicol acetate

*Moderate antibacterial (Gram-positive) hydroxycinnamic acid derivatives:* p-Hydroxy Cinnamaldehyde inhibited the growth of *Staphylococcus aureus* (ATCC 29213), methicillin-resistant *Staphylococcus aureus* (ATCC 33591), and vancomycin intermediate-resistant *Staphylococcus aureus* (ATCC 700699) with MIC values of 625, 313, and 156 μg/mL, respectively (Aziz et al., 2013). Trans-p-coumaryl diacetate (4-hydroxy cinnamyl alcohol diacetate) isolated from the rhizomes inhibited the growth of *Staphylococcus aureus* (ATCC 29213), methicillin-resistant *Staphylococcus aureus* (ATCC 33591), and vancomycin intermediate-resistant *Staphylococcus aureus* (ATCC 700699), (VISA 24), and (VRSA 156) with MIC values of 625, 625, 625, 156, and 1250 μg/mL, respectively (Aziz et al., 2013). p-hydroxycinnamyl acetate inhibited the growth of *Staphylococcus aureus* (ATCC 29213), methicillin-resistant *Staphylococcus aureus* (ATCC 33591), vancomycin intermediate-resistant *Staphylococcus aureus* (ATCC 700699), methicillin-resistant *Staphylococcus aureus* with intermediate resistance to vancomycin (VISA 24), and methicillin-resistant *Staphylococcus aureus* with complete resistance to vancomycin (VRSA 156) with MIC values of 1250, 1250, 2500, 39, and 625 μg/mL, respectively (Aziz et al., 2013).

*Moderate antibacterial (Gram-positive) phenylpropanoid:* 1′S-1′-Acetoxyeugenol acetate isolated from the rhizomes inhibited the growth of *Staphylococcus aureus* (ATCC 29213), methicillin-resistant *Staphylococcus aureus* (ATCC 33591), vancomycin intermediate-resistant *Staphylococcus aureus* (ATCC 700699), methicillin-resistant *Staphylococcus aureus* with intermediate resistance to vancomycin (VISA 24), and methicillin-resistant *Staphylococcus aureus* with complete resistance to vancomycin (VRSA 156) with MIC values of 625, 625, 625, 313, and 313 μg/mL, respectively (Aziz et al., 2013).

*Weak antibacterial (gram-positive) simple phenol:* 4-Hydroxybenzaldehyde (para-hydroxybenzaldehyde) isolated from the rhizomes inhibited the growth of *Staphylococcus aureus* (ATCC 29213), methicillin-resistant *Staphylococcus aureus* (ATCC 33591), and vancomycin intermediate-resistant *Staphylococcus aureus* (ATCC 700699), methicillin-resistant *Staphylococcus aureus* with intermediate resistance to vancomycin (VISA 24), and methicillin-resistant *Staphylococcus aureus* with complete resistance to vancomycin (VRSA 156) (Aziz et al., 2013).

*Strong antifungal (filamentous) essential oil:* Essential oil of leaves inhibited the growth of *Microsporium canis* (ATCC 36299), *Trichophyton mentagrophytes* (ATCC 18748), and *Trichophyton rubrum* (ATCC 28188) (Ibrahim et al., 2009).

*Moderate broad-spectrum antifungal mid-polar extract:* Dichloromethane extract of rhizomes inhibited the growth of *Candida albicans* (ATCC 10231), *Microsporium canis* (ATCC 3629), and *Trichophyton rubrum* (ATCC 28188) with MIC values of 625, 156, and 156 μg/mL, respectively (Aziz et al., 2013).

*Very weak broad-spectrum antifungal hydroxycinnamic acid derivative:* Trans-p-coumary diacetate (4-hydroxy cinnamyl alcohol diacetate) isolated from the rhizomes inhibited the growth of *Candida albicans* (ATCC 10231), *Microsporium canis* (ATCC 3629), and *Trichophyton rubrum* (ATCC 28188) with MIC values of 625, 625, and 625 μg/mL, respectively (Aziz et al., 2013). p-Hydroxycinnamyl acetate isolated from the rhizomes inhibited the growth of *Candida albicans* (ATCC 10231), *Microsporium canis* (ATCC 3629), and *Trichophyton rubrum* (ATCC 28188) with MIC values of 1000, 2000, and 2000 μg/mL, respectively (Aziz et al., 2013).

*Moderate weak broad-spectrum antifungal phenylpropanoid:* 1′-Acetoxyeugenol acetate isolated from the rhizomes inhibited the growth of *Candida albicans* (ATCC 10231), *Microsporium canis* (ATCC 3629), and *Trichophyton rubrum* (ATCC 28188) with MIC values of 1250, 313, and 312 μg/mL, respectively (Aziz et al., 2013).

*Moderate broad-spectrum antifungal simple phenol:* 4-Hydroxybenzaldehyde (para-hydroxybenzaldehyde) isolated from the rhizomes inhibited the growth of *Candida albicans* (ATCC 10231), *Microsporium canis* (ATCC 3629), and *Trichophyton rubrum* (ATCC 28188) (Aziz et al., 2013).

*Strong antiviral (enveloped monopartite linear dimeric single stranded (+) RNA) phenylpropanoid:* 1′-Acetoxychavicol acetate at a concentration of 4 μM inhibited the replication of the Human immunodeficiency virus type-1 by more than 80% (Ye & Li, 2006).

1'-Acetoxychavicol acetate

Commentary: The major constituents in the leaf essential oil are Chavicol, β-Bisabolene, β-emelene, β-Pinene, and β-Sesquiphellandrene (Ibrahim et al., 2009). Being active against the Human immunodeficiency virus type-1 1'-acetoxychavicol acetate could be of value against COVID-19.

### References

Aziz, A.N., Ibrahim, H., Syamsir, D.R., Mohtar, M., Vejayan, J. and Awang, K., 2013. Antimicrobial compounds from *Alpinia conchigera*. *Journal of Ethnopharmacology*, *145*(3), pp. 798–802.

Chomnawang, M.T., Trinapakul, C. and Gritsanapan, W., 2009. In vitro antigonococcal activity of *Coscinium fenestratum* stem extract. *Journal of Ethnopharmacology*, *122*(3), pp. 445–449.

Ibrahim, H., Aziz, A.N., Syamsir, D.R., Ali, N.A.M., Mohtar, M., Ali, R.M. and Awang, K., 2009. Essential oils of *Alpinia conchigera* Griff. and their antimicrobial activities. *Food Chemistry*, *113*(2), pp. 575–577.

Ye, Y. and Li, B., 2006. 1' S-1'-acetoxychavicol acetate isolated from *Alpinia galanga* inhibits human immunodeficiency virus type 1 replication by blocking Rev transport. *Journal of General Virology*, *87*(7), pp. 2047–2053.

### 3.5.4.2 *Alpinia galanga* (L.) Willd.

Synonym: *Amomum galanga* (L.) Lour.; *Languas galanga* (L.) Stuntz; *Maranta galanga* L.; *Zingiber galanga* (L.) Stokes

Common names: Galanga, greater galangal, Siamese ginger; pras (Cambodia); hong dou kou (China); hoimboti-boch; mohavori-boch (Bangladesh); langkuas (Indonesia); chittharattai, kanghoo (India); kelawas (Indonesia); khaksta deng (Laos); lankuas (Malaysia), padaykogi; (Myanmar); dai koryokyo (Japan); lengkuas (Malaysia); langkawas (the Philippines); kha taa daeng (Thailand); sơn nại (Vietnam)

Habitat: Forests, cultivated

Distribution: India, Bangladesh, Myanmar, Thailand, Vietnam, China, and Taiwan

Botanical examination: This ginger dwells in the forests of China, Taiwan, India, Indonesia, Malaysia, Myanmar, Thailand, and Vietnam. The rhizome is tuberous, pinkish, thick, edible, and aromatic. The stems can reach 2 m tall. The leaves are simple and alternate. The ligule is suborbicular and about 0.5 cm long. The petiole is 0.6 cm long. The blade is oblong, 25 cm×6–35 cm×10 cm, hairy, attenuate at base and acute at apex. The inflorescence is a panicle, which is 20 cm×30 cm. The flowers are green-white and fragrant. The calyx is tubular, 0.6–10 cm long, and persistent. The corolla tube is 0.6–1 cm long. The corolla lobes are oblong and 1.6–1.8 cm long. The labellum is white with red lines, obovate-spathulate, 2 cm long, and bifid at apex. The androecium includes a stamen, which is about 1.7 cm long. The capsules are red, oblong, slightly contracted at the middle, 1–1.5 cm×0.7 cm, thin, glabrous, and contain 3–6 seeds.

Medicinal uses: Cough (Bangladesh, India); dysentery (Cambodia, India, Myanmar, Laos, and Vietnam); eye infection (Bangladesh), fever (India); herpes (Indonesia), post-partum (Malaysia), sore throat (India); tuberculosis (India)

*Strong broad-spectrum antibacterial polar extract:* Methanol extract of rhizomes inhibited the growth of *Bacillus subtilis* (MTCC 2391), *Enterococcus faecalis, Staphylococcus aureus, Staphylococcus epidermis, Enterobacter aerogene, Enterobacter cloacae, Escherichia coli* (MTCC 1563), *Klebsiella pneumoniae, Pseudomonas aeruginosa* (MTCC 6642), and *Salmonella typhimurium* with MIC/MBC values of 40/160, 40/320, 40/320, 40/160, 160/320, 160/640, 40/80, 320/1280, 640/2560, and 40/160 µg/mL, respectively (Rao et al., 2010).

*Moderate antibacterial (Gram-positive) flavonol:* Galangin inhibited the growth of penicillin-resistant *Staphylococcus aureus* DMST 20651 and *Staphylococcus aureus* (ATCC 29213) with MIC values of 300 and 100 µg/mL, respectively (Eumkeb et al., 2010).

*Strong antimycobacterial phenylpropanoid:* 1′-Acetoxychavicol acetate inhibited the growth of *Mycobacterium tuberculosis* (H37Ra ATCC 25177) and *Mycobacterium tuberculosis* (H37 Rv ATCC 27294) with MIC values of 0.2 and 0.7 µg/mL, respectively (Warit et al., 2017).

*Antibiotic potentiator flavonol:* Galangin (also known as 3,5,7-trihydroxyflavone) potentiated ceftazidime against penicillin-resistant *Staphylococcus aureus* DMST 20651 with a FICI value below 0.03 via probable inhibition penicillinase and β-lactamase (Eumkeb et al., 2010).

*Strong antifungal (filamentous) mid-polar extract:* Chloroform extract of rhizomes inhibited the growth of *Cryptococcus neoformans* (clinical strain) and *Microsporium gypseum* (clinical strain) with MIC values of 128 and 16 µg/mL, respectively (Phongpaichit et al., 2005).

*Strong antiviral (enveloped segmented linear single-stranded (−) RNA virus) phenylpropanoid:* 1′-Acetoxychavicol acetate inhibited the replication of the Influenza virus A/WSN/33(H1N1) with the $IC_{50}$ value of 2 µM and selectivity index of 2.8 (Watanabe et al., 2011).

Commentary: Essential oil of rhizome shelters 1,8-cineole, chavicol acetate, 1′-acetoxychavicol acetate, β-sesquiphellandrene, terpinen-4-ol, and α-terpineol (Rana et al., 2010). Terpinen-4-ol in the essential oil of rhizomes displayed antimicrobial effects (Janssen & Scheffer, 1985). Acetoxychavicol acetate from the rhizome inhibited the growth of fungi with MIC values ranging from 50 to 250 µg/mL (Janssen & Scheffer, 1985). One could examine the anti-Severe acute respiratory syndrome-associated coronavirus 2 properties of this ginger. Consider that 1′-acetoxychavicol acetate inhibited the Human immunodeficiency virus type 1 replication (Ye and Li, 2006) and since anti-HIV principles are often active against coronaviruses. One couls examine the Severe acute respiratory syndrome-associated coronavirus 2 activity of 1′-acetoxychavicol acetate.

## References

Eumkeb, G., Sakdarat, S. and Siriwong, S., 2010. Reversing β-lactam antibiotic resistance of *Staphylococcus aureus* with galangin from *Alpinia officinarum* Hance and synergism with ceftazidime. *Phytomedicine, 18*(1), pp. 40–45.

Janssen, A.M. and Scheffer, J.J.C., 1985. Acetoxychavicol acetate, an antifungal component of *Alpinia galangal. Planta Medica, 51*(06), pp. 507–511.

Phongpaichit, S., Subhadhirasakul, S. and Wattanapiromsakul, C., 2005. Antifungal activities of extracts from Thai medicinal plants against opportunistic fungal pathogens associated with AIDS patients. *Mycoses, 48*(5), pp. 333–338.

Rana, V.S., Verdeguer, M. and Blazquez, M.A., 2010. GC and GC/MS analysis of the volatile constituents of the oils of *Alpinia galanga* (L.) Willd and A. officinarum Hance rhizomes. *Journal of Essential Oil Research, 22*(6), pp. 521–524.

Rao, K., Ch, B., Narasu, L.M. and Giri, A., 2010. Antibacterial activity of *Alpinia galanga* (L) Willd crude extracts. *Applied Biochemistry and Biotechnology, 162*(3), pp. 871–884.

Warit, S., Rukseree, K., Prammananan, T., Hongmanee, P., Billamas, P., Jaitrong, S., Chaiprasert, A., Jaki, B.U., Pauli, G.F., Franzblau, S.G. and Palittapongarnpim, P., 2017. In vitro activities of enantiopure and racemic 1′-acetoxychavicol acetate against clinical isolates of mycobacterium tuberculosis. *Scientia Pharmaceutica, 85*(3), p. 32.

Watanabe, K., Takatsuki, H., Sonoda, M., Tamura, S., Murakami, N. and Kobayashi, N., 2011. Anti-influenza viral effects of novel nuclear export inhibitors from Valerianae Radix and *Alpinia galanga*. *Drug Discoveries & Therapeutics*, 5(1), pp. 26–31.

Ye, Y. and Li, B., 2006. 1′ S-1′-acetoxychavicol acetate isolated from *Alpinia galanga* inhibits human immunodeficiency virus type 1 replication by blocking Rev transport. *Journal of General Virology*, 87(7), pp. 2047–2053.

### 3.5.4.3 *Alpinia nigra* (Gaertn.) B. L. Burtt.

Synonyms: *Alpinia allughas* (Retz.) Roscoe; *Alpinia aquatica* (J. Koenig) Roscoe; *Heritiera allughas* Retz.; *Languas allughas* (Retz.) Burkill; *Languas aquatica* J. Koenig; *Zingiber nigrum* Gaertn.

Common names: Black-fruited galangal; bhulchengi, khetranga, tara, taruka (Bangladesh); pullei (India)

Habitat: Forests

Distribution: India, Sri Lanka, Bhutan, Bangladesh, and Thailand

Botanical examination: This herb grows to about 2.5 m tall from an aromatic rhizome. The leaves are simple, with a ligule, which is about 5 mm and a short petiole. The young shoots are edible. The blade is elliptic, glossy, 25–35×6–10 cm, base and apex acute. The inflorescences are terminal panicles which are about 30 cm long. The calyx is tubular, 3-lobed, and about 1.5 cm long. The corolla is tubular, heavenly pinkish, with a white labellum, which is about 1.5 cm long and bifid. The androecium comprises 1 stamen, which is about 1.5 cm long. The ovary is minute and develops a slender style. The capsule is globose, black, and up to 1.5 cm in diameter.

Medicinal uses: Cough, jaundice (Bangladesh)

*Antifungal halo developed by essential oil:* Essential oil of rhizome (400 µg/5 mm disc) inhibited the growth of *Malassezia furfur* and *Microsporium gypseum* with inhibition zone diameters of 210 and 240 mm, respectively (Rajapaksha et al., 2017).

*Strong broad-spectrum antibacterial essential oil:* Essential oil of rhizomes inhibited the growth of *Staphylococcus aureus* (ATCC 6538), *Bacillus cereus* (ATCC 11778), *Listeria monocytogenes* (ATCC 19115), *Escherichia coli* (ATCC 25922), *Salmonella paratyphi* (MTCC 735), *Escherichia coli* (enterotoxic MTCC 723), and *Yersinia enterolitica* (MTCC) with MIC/MBC values of 3.1/6.2, 3.1/6.2, 3.1/3.1, 3.1/3.1, 6.2/6.2, 3.1/3.1, and 1.5/3.1 µg/mL, respectively (Gosh et al., 2014).

*Strong broad-spectrum antibacterial labdane diterpenes:* (E)-labda-8(17), 12-diene-15, 16-dial isolated from the seeds inhibited the growth of *Staphylococcus aureus* (ATCC 6538), *Bacillus cereus* (ATCC 11778), *Listeria monocytogenes* (ATCC 19115), *Escherichia coli* (ATCC 25922), *Salmonella paratyphi* (MTCC 735), *Escherichia coli* (enterotoxic MTCC 723), and *Yersinia enterolitica* (MTCC) with MIC/MBC values of 12.5/25, 12.5/12.5, 25/25, 25/50, 12.5/12.5, 25/25, and 12.5/25 µg/mL, respectively (Gosh et al., 2013). (E)-8β, 17-Epoxylabd-12-ene-15, 16-dial isolated from the seeds inhibited the growth of *Staphylococcus aureus* (ATCC 6538), *Bacillus cereus* (ATCC 11778), *Listeria monocytogenes* (ATCC 19115), *Escherichia coli* (ATCC 25922), *Salmonella paratyphi* (MTCC 735), *Escherichia coli* (enterotoxic MTCC 723), and *Yersinia enterolitica* (MTCC) with MIC/MBC values of 3.3/6.7, 6.7/6.7, 12.5/12.5, 12.5/12.5, 6.7/12.5, 6.7/6.7, and 3.3/3.3 µg/mL, respectively (Gosh et al., 2013). Both compounds evoked membrane insults and cytoplasmic constituents leakage in *Staphylococcus aureus* and *Yersinia enterocolitica* (Gosh et al., 2013).

(E)-8β, 17-Epoxylabd-12-ene-15, 16-dial

*Anticandidal lipophilic labdane diterpenes*: (E)-8β, 17-Epoxylabd-12-ene-15, 16-dial inhibited the growth of *Candida albicans* via a mechanism involving disturbance in cytoplasmic membrane permeability (Haraguchi et al., 1996).

Commentaries: (i) Essential oil of rhizomes shelters principally β-pinene, β-caryophyllene, and α-humulene (Gosh et al., 2014). The plant has apparently not been studied for possible antiviral properties. Consider that aqueous extract of rhizomes of *Alpinia officinarum* L. at a concentration of 250 μg/mL inhibited Human immunodeficiency virus type-1 protease by 65.8% (Xu et al., 1996). (ii) (E)-8β, 17-Epoxylabd-12-ene-15, 16-dial exemplifies the concept that to lipophilic secondary metabolite from medicinal targets the cytoplasmic membrane of bacteria. In general lipophilic or amphiphilic natural products have affinity and remain in the lipophilic core of fungal and bacterial cytoplasmic membranes.

### References

Ghosh, S., Indukuri, K., Bondalapati, S., Saikia, A.K. and Rangan, L., 2013. Unveiling the mode of action of antibacterial labdane diterpenes from *Alpinia nigra* (Gaertn.) BL Burtt seeds. *European Journal of Medicinal Chemistry*, *66*, pp. 101–105.

Ghosh, S., Ozek, T., Tabanca, N., Ali, A., ur Rehman, J., Khan, I.A. and Rangan, L., 2014. Chemical composition and bioactivity studies of *Alpinia nigra* essential oils. *Industrial Crops and Products*, *53*, pp. 111–119.

Haraguchi, H., Kuwata, Y., Inada, K., Shingu, K., Miyahara, K., Nagao, M. and Yagi, A., 1996. Antifungal activity from *Alpinia galanga* and the competition for incorporation of unsaturated fatty acids in cell growth. *Planta Medica*, *62*(04), pp. 308–313.

Rajapaksha, R.S.C.G., Wickramarachchi, W.J. and Hansini, K.G.D.M., 2017. Evaluation of the antifungal activity of the extracts of the rhizome of *Alpinia Nigra*. *International Journal of Advances in Agricultural & Environmental Engineering*, *4*, pp. 86–88.

Xu, H.X., Wan, M., Loh, B.N., Kon, O.L., Chow, P.W. and Sim, K.Y., 1996. Screening of traditional medicines for their inhibitory activity against HIV -1 protease. *Phytotherapy Research*, *10*(3), pp. 207–210.

### 3.5.4.4 *Amomum dealbatum* Roxb.

Synonym: *Cardamomum dealbatum* (Roxb.) Kuntze

Common names: Long-fruited amomum; palachengay (Bangladesh); aidu, alach (India); chang guo sha ren (China)

Habitat: Forests

Distribution: India, Nepal, Sikkim, Bangladesh, and Thailand

Botanical examination: It is a herb that grows to about 2.5 m tall from a rhizome. The leaves are simple, sheathing, and with an orbicular, bifid, ligule, which is about 1.5 cm long. The pseudopetiole is up to 3 cm long. The blade is broadly lanceolate, 50–70×6–15 cm, hairy below, cuneate at base, and acuminate at apex. The inflorescences are basal and sessile somewhat globose spikes, which are about 5 cm in diameter. The calyx is tubular and 3-lobed. The corolla is tubular, pure white, membranous, about 2.5 cm long, and develops a labellum, which is elliptic, marked with a yellow patch, and emarginate. The androecium contains a single stamen. The capsules are ribbed and about 3 cm long. The infructescence at first glance resembles a cactus.

Medicinal uses: Abscess (Bangladesh, India); childbirth (Indonesia)

Commentaries:

(i) Members in the genus *Amomum* Roxb. often generate the chalcone cardamonin (Rao et al., 1976), which inhibits Human immunodeficiency virus type-1 protease and Dengue virus type-2 (DV2) NS3 protease (Gonçalves et al., 2014).

(ii) This ginger has apparently not been examined for possible antimicrobial activities. Consider that the inflorescence before flowers open has somewhat the shape and form of a spiked virus, and following the "Theory of signatures" it could be an indication that the plant could be used for the treatment of viral diseases (?). So far no one has managed to explain why plants have so many forms, shapes, and colors and it could be inferred that plants send messages to humans for the treatment of their diseases. Malays of Perak consider *Mitragyna speciosa* (Korth.) Havil. has a cure for the treatment of COVID-19 because its globose inflorescence looks very much like the Severe acute respiratory syndrome-associated coronavirus type-2. From this Malaysian tree, the amphiphilic yohimbane-type indole alkaloid hirsutine (LogD = 2.6 at pH 7.4; molecular 368.5 g/mol) (which is not uncommon in members of the genus *Mitragyna* Korth) inhibited the replication of Influenza A (subtype H3N2) with an $IC_{50}$ value of 0.4 µg/mL and a selectivity index of 58.8 (Moradi et al., 2018), as well as Dengue virus (Hishiki et al., 2017). Is hirsutine active against COVID-19? *Amomum kravanh* Pierre ex Gagnep. and *Amomum villosum* Lour. are used for the treatment of COVID-19 in China (Luo, 2020; Wang et al., 2020).

Hirsutine

### References

Gonçalves, L.M., Valente, I.M. and Rodrigues, J.A., 2014. An overview on cardamonin. *Journal of Medicinal Food*, 17(6), pp. 633–640.

Hishiki, T., Kato, F., Tajima, S., Toume, K., Umezaki, M., Takasaki, T. and Miura, T., 2017. Hirsutine, an indole alkaloid of *Uncaria rhynchophylla*, inhibits late step in dengue virus lifecycle. *Frontiers in Microbiology*, 8, p. 1674.

Luo, A., 2020. Positive SARS-Cov-2 test in a woman with COVID-19 at 22 days after hospital discharge: a case report. *Journal of Traditional Chinese Medical Sciences, 7*(4), pp. 413–417.

Moradi, M.T., Karimi, A. and Lorigooini, Z., 2018. Alkaloids as the natural anti-influenza virus agents: a systematic review. *Toxin Reviews, 37*(1), pp. 11–18.

Rao, C.B., Rao, T.N. and Suryaprakasam, S., 1976. Cardamonin and alpinetin from the seeds of *Amomum subulatum. Planta Medica, 29*(04), pp. 391–392.

Wang, S.X., Wang, Y., Lu, Y.B., Li, J.Y., Song, Y.J., Nyamgerelt, M. and Wang, X.X., 2020. Diagnosis and treatment of novel coronavirus pneumonia based on the theory of traditional Chinese medicine. *Journal of Integrative Medicine, 18*(4), pp. 275–283.

### 3.5.4.5 *Boesenbergia rotunda* (L.) Manfs.

Synonyms: *Boesenbergia pandurata* (Roxb.) Schltr.; *Curcuma rotunda* L.; *Gastrochilus rotundus* (L.) Alston; *Kaempferia pandurata* Roxb.

Common names: Fingerroot, Chinese ginger, Thai ginseng; ao chun jiang (China), gajutu, oban-gajutsu (Japan); gajai, neng kieng (Laos); temu kunci (Malaysia); nue (Nicobar); kra chai (Thailand), cu ngai (Vietnam)

Habitat: Forests, cultivated

Distribution: India, and Sri Lanka, Bangladesh, Myanmar, Cambodia, Laos, Vietnam, Thailand, Malaysia, and Indonesia

Botanical description: This ginger is cultivated for its aromatic and medicinal rhizomes. It grows to 50 cm tall and the rhizome is odd, very peculiarly and somewhat monstrously hand shaped, edible, yellowish beige, and strongly aromatic. The leaves are simple and somewhat basal. The leaf sheath is red and the ligule bifid and 0.5 cm long. The petiole is 7–16 cm long and channeled. The blade is ovate-oblong, 25 cm×7–50 cm×12 cm, round at base and apiculate at apex. The inflorescence is subsessile, aromatic, and 3–7 cm long. The bracts are lanceolate and 4–5 cm long. The calyx is 1.5–2 cm long and bifid at apex. The corolla tube is 4.5–5.5 cm long and light pink, the labellum marked with whitish pink stripe, 2.5–3.5 cm long, concave, and crisped. The androecium includes a single stamen, which is up to about 1.3 cm long. The fruit is a capsule

Medicinal uses: Dysentery, leucorrhea (Thailand)

*Moderate antibacterial (Gram-positive) polar extract:* Ethanol extract of rhizomes inhibited the growth of *Staphylococcus aureus* (ATCC 25923), *Staphylococcus epidermidis*, and *Bacillus subtilis* (ATCC 6633), with MIC values of 310, 160, and 40 µg/mL, respectively (Jitvaropas et al., 2012).

*Strong antibacterial (Gram-positive) polar extract:* Chloroform extract inhibited the growth of *Bacillus cereus* (ATCC 1778), *Staphylococcus aureus* (25923), methicillin-resistant *Staphylococcus aureus* (PSU039), with MIC values of 10, 10, and 10 µg/mL (Voravuthikunchai et al., 2006). The extract was bactericidal for methicillin-resistant *Staphylococcus aureus* (PSU039) (Voravuthikunchai et al., 2006).

*Weak antifungal (yeasts) polar extract:* Ethanol extract of rhizomes inhibited the growth of *Candida albicans* (ATCC 10231) and *Saccahromyces cerevisiae* with the MIC values of 2500 and 2500 µg/mL, respectively (Jitvaropas et al., 2012).

*Strong antiviral (enveloped monopartite linear single-stranded (+) RNA) polar extract:* Ethanol extract of rhizomes inhibited the replication of the Severe acute respiratory syndrome-associated coronavirus type-2 virus (SARS-CoV-2/01/human/Jan2020/Thailand) in Vero E6 cells post-infection with an $IC_{50}$ value of 3.6 µg/mL ($CC_{50}$ = 28 µg/mL).

*Strong antiviral (enveloped linear single stranded (+)RNA) prenylated phenolic:* The prenylated chalcone panduratin A (molecular weight 406.5 g/mol) isolated from the rhizome inhibited the replication of the Severe acute respiratory syndrome -associated coronavirus 2 (SARS-CoV-2/01/human/Jan2020/Thailand) in Vero E6 cells post-infection with an $IC_{50}$ value of 0.8 µM ($CC_{50}$ = 14.71 µM) (Kanjanasirirat et al., 2020). Panduratin A at the pre-entry phase inhibited Severe acute respiratory syndrome-associated coronavirus 2 infection with an $IC_{50}$ of 5.3 µM ($CC_{50}$ = 43.4 µM) (Kanjanasirirat et al., 2020).

Panduratin A

*Viral enzyme inhibition by prenylated phenolic:* Panduratin A and 4-hydroxypanduratin A at 80 ppm inhibited dengue 2 NS2B/3 virus protease by 66.7 and 78.1%, respectively (Kiat et al., 2006). (-) Panduratin A and (-)-isopanduratin A at a concentration of 10 μM inhibited aminopeptidase N by 62.1 and 44.5% (Morikawa et al., 2008).

*Viral enzyme inhibition by simple phenolic:* Geranyl-2,4-dihydroxy-6-phenylbenzoate at the concentration of 10 μM inhibited aminopeptidase N by 56.4% (Morikawa et al., 2008).

Commentaries:

(i) The rhizomes generate a fascinating array of flavonoids and phenolics, including panduratin A, 4-hydroxypanduratin A. and isopanduratin A, the flavanone pinocembrin (see earlier), and the chalcone cardamonin (Morikawa et al., 2008).

(ii) Clinical examination of the rhizome of this plant or panduratin A for the treatment or prevention of COVID-19 (and forthcoming viral pandemics) is warranted.

### References

Jitvaropas, R., Saenthaweesuk, S., Somparn, N., Thuppia, A., Sireeratawong, S. and Phoolcharoen, W., 2012. Antioxidant, antimicrobial and wound healing activities of *Boesenbergia rotunda*. *Natural Product Communications*, 7(7), p. 1934578X1200700727.

Kanjanasirirat, P., Suksatu, A., Manopwisedjaroen, S., Munyoo, B., Tuchinda, P., Jearawuttanakul, K., Seemakhan, S., Charoensutarakul, S., Wongtrakoongate, P., Rangkasenee, N. and Pitiporn, S., 2020. High-Content Screening of Thai Medicinal Plants Reveals *Boesenbergia rotunda* Extract and its Component Panduratin A as Anti-SARS-CoV-2 Agents. *Scientific Report*, 10(1), pp. 1–12.

Kiat, T.S., Pippen, R., Yusof, R., Ibrahim, H., Khalid, N. and Abd Rahman, N., 2006. Inhibitory activity of cyclohexenyl chalcone derivatives and flavonoids of fingerroot, *Boesenbergia rotunda* (L.), towards dengue-2 virus NS3 protease. *Bioorganic & Medicinal Chemistry Letters*, 16(12), pp. 3337–3340.

Morikawa, T., Funakoshi, K., Ninomiya, K., Yasuda, D., Miyagawa, K., Matsuda, H. and Yoshikawa, M., 2008. Medicinal foodstuffs. XXXIV. Structures of new prenylchalcones and prenylflavanones with TNF-α and aminopeptidase N inhibitory activities from *Boesenbergia rotunda*. *Chemical and Pharmaceutical Bulletin*, 56(7), pp. 956–962.

Voravuthikunchai, S.P., Limsuwan, S., Supapol, O. and Subhadhirasakul, S., 2006. Antibacterial activity of extracts from family *Zingiberaceae* against foodborne pathogens. *Journal of Food Safety*, 26(4), pp. 325–334.

### 3.5.4.6  *Kaempferia marginata* Carey ex Roscoe

Synonyms: *Alpinia sessilis* J. Koenig, *Kaempferia galanga* L.; *Kaempferia humilis* Salisb., *Kaempferia latifolia* Donn ex Hornem, *Kaempferia marginata* Carey ex Roscoe, *Kaempferia plantaginifolia* Salisb., *Kaempferia procumbens* Noronha

Common names: East Indian galangal, resurrection lily; ku shan nai, shan nai (China); kacholam (India); cekur (Indonesia); van horn (Laos); cekur (Malaysia); kosol (the Philippines); pro hom, tup mup (Thailand); tam nai (Vietnam)

Habitat: Grassy and shady spots

Distribution: India, Myanmar, and Thailand

Botanical examination: This ginger grows from a tuberous rhizome. The leaves are spread on ground and sessile. The ligule is triangular. The blade is pale green above, purplish below, suborbicular, 8–11 cm×6–11 cm, and cordate at base. The inflorescence emerges from the rhizome before the leaves. The flowers are fragrant. The bracts are light whitish green. The calyx is 2.5 long and bifid. The corolla lobes are spreading, linear and 2.5 cm long and develop a lip, which is labellum purplish and up to about 3 cm long. The androecium comprises a single 5 mm long stamen.

Medicinal use: Cough (Thailand)

*Broad-spectrum antibacterial essential oil:* Essential oil of rhizome (15 µL/disc, 6 mm paper disc) inhibited the growth of *Salmonella typhimurium* (TISTR 292) and *Streptococcus enteriditis* (DMST 17368) with inhibition zone diameters of 11.5 and 8 mm, respectively (Wannissorn et al., 2005). Essential oil of rhizome (10 µL/disc, 6 mm paper disc) inhibited the growth of *Staphylococcus aureus* (ATCC 25923), *Streptococcus faecalis*, *Bacillus subtilis*, *Salmonella typhi*, *Shigella flexneri*, and *Escherichia coli* (ATCC 25922) with inhibition zone diameters of 12, 14, 16, 9, 12, and 8 mm, respectively (Tewtrakul et al., 2005).

*Anticandidal essential oil:* Essential oil of rhizome (15 µL/disc, 6 mm paper disc) inhibited the growth of *Candida albicans* with an inhibition zone diameter of 31 mm (Tewtrakul et al., 2005).

*Broad-spectrum antifungal polar extract:* Ethanol extract of rhizomes (5 mg, 6 mm discs) inhibited the growth of *Aspergillus niger* (MTCC 2612), *Aspergillus flavus* (MTCC 2813), *A. fumigatus* (MTCC 2584), and *Candida albicans* (MTCC 1637) with the inhibition zone diameters of 16.3, 15.3, 14, and 12.2 mm, respectively (Kochuthressia et al., 2012).

*Viral enzyme inhibition by polar extract:* Methanol extract of rhizomes inhibited Human immunodeficiency virus type-1 protease activity by more than 80% (Sookkongwaree et al., 2006).

*Viral enzyme inhibition by flavone:* 5-Hydroxy-3,7-dimethoxyflavone (also known as Galangin 3,7-dimethyl ether) isolated from the rhizome inhibited Human immunodeficiency virus type-1 protease, Hepatitis C virus NS3 protease, and human Cytomegalovirus protease activities with $IC_{50}$ values of 66.1, 192.9, and 248.5 µg/mL, respectively (Sookkongwaree et al., 2006).

Commentary: Essential oil of rhizome shelters principally ethyl-p-methoxycinnamate, methylcinnamate, carvone, and eucalyptol (Tewtrakul et al., 2005). The plant generates kaempferol (Kanjanapothi et al., 2004), which inhibited 3a channel protein of Severe acute respiratory syndrome-associated coronavirus at a concentration of 20 µM (Schwarz et al., 2014).

### References

Kanjanapothi, D., Panthong, A., Lertprasertsuke, N., Taesotikul, T., Rujjanawate, C., Kaewpinit, D., Sudthayakorn, R., Choochote, W., Chaithong, U., Jitpakdi, A. and Pitasawat, B., 2004. Toxicity of crude rhizome extract of *Kaempferia galanga* L. (Proh Hom). *Journal of Ethnopharmacology*, 90(2–3), pp. 359–365.

Kochuthressia, K.P., Britto, S.J., Jaseentha, M.O. and Raphael, R., 2012. In vitro antimicrobial evaluation of *Kaempferia galanga* L. rhizome extract. *American Journal of Biotechnology and Molecular Sciences*, 2(1), pp. 1–5.

Schwarz, S., Sauter, D., Wang, K., Zhang, R., Sun, B., Karioti, A., Bilia, A.R., Efferth, T. and Schwarz, W., 2014. Kaempferol derivatives as antiviral drugs against the 3a channel protein of coronavirus. *Planta Medica*, 80(02–03), p. 177.

Sookkongwaree, K., Geitmann, M., Roengsumran, S., Petsom, A. and Danielson, U.H., 2006. Inhibition of viral proteases by Zingiberaceae extracts and flavones isolated from *Kaempferia parviflora. Die Pharmazie-An International Journal of Pharmaceutical Sciences*, *61*(8), pp. 717–721.

Tewtrakul, S., Yuenyongsawad, S., Kummee, S. and Atsawajaruwan, L., 2005. Chemical components and biological activities of volatile oil of *Kaempferia galanga* Linn. *Songklanakarin Journal of Science and Technology*, *27*(2), pp. 503–507.

Wannissorn, B., Jarikasem, S., Siriwangchai, T. and Thubthimthed, S., 2005. Antibacterial properties of essential oils from Thai medicinal plants. *Fitoterapia*, *76*(2), pp. 233–236.

### 3.5.4.7 *Curcuma wenyujin* Y.H. Chen & C. Ling

Common names: Wen yu jin (China)

   Habitat: Grassy lands, cultivated

   Distribution: China

Botanical examination: This magnificent ginger grows up to a height of 1.5 m tall. The rhizome is yellow inside, whitish outside, and ovoid, fusiform and tuberous roots are present. The leaves are simple, spiral, and basal. The petiole is up to 30 cm tall. The blade is oblong, 35–75 × 15–20 cm, cuneate at base, and acute to somewhat caudate at apex. The spikes are 20–30 cm long at apex of 15–20 cm tall peduncles. The spike presents oblong bracts, which are a heavenly pink, and up to 8 cm long. The calyx is tubular, white, splitting on 1 side, and up to about 1.2 cm long. The corolla is tubular, white, and about 2.5 cm long. The lateral staminodes are yellow, oblong, and 1.5 cm long. The labellum is reflexed, yellowish, ovate, emarginate, and 2 cm long. The androecium includes a single stamen with a spurred and versatile anther. The gynoecium is trilocular and hairy. The capsule is trilobed and dehiscent.

   Medicinal use: COVID-19 (China)

*Moderate broad-spectrum essential oil:* Essential oil of rhizomes inhibited the growth of *Propionibacterium acnes* (ATCC 11827) and *Staphylococcus aureus* with MIC values of 250 and 250 µg/mL, respectively (Zhu et al., 2013).

*Very weakly broad-spectrum antibacterial sesquiterpene:* Germacrone inhibited the growth of *Escherichia coli* (clinical isolate), *Escherichia coli* (ATCC 25922), and *Bacillus subtilis* (ATCC 6633 (Radulović et al., 2010).

*Moderate antifungal (dimorphic) essential oil:* Essential oil inhibited the growth of *Malassezia furfur* (ATCC 44344) with MIC values of 125 µg/mL (Zhu et al., 2013).

*Moderate antifungal (dimorphic) monoterpene:* 1,8-Cineole isolated from the rhizome inhibited the growth of *Malassezia furfur* (ATCC 44344) with an MIC value of 62.5 µg/mL (Zhu et al., 2013).

*Moderate antifungal (dimorphic) monoterpenes:* Isoborneol and camphor isolated from the rhizome inhibited the growth of *Malassezia furfur* (ATCC 44344) with an MIC value of 250 µg/mL (Zhu et al., 2013).

*Weak antifungal (dimorphic) sesquiterpenes:* Germacrone, curdione, and β-elemene isolated from the rhizome inhibited the growth of *Malassezia furfur* (ATCC 44344) with MIC values of 250, 250, and 125 µg/mL (Zhu et al., 2013).

*Strong antiviral (enveloped monopartite linear single-stranded (+) RNA) amphiphilic diarylheptanoid:* Curcumin (LogD = 2.4; molecular mass = 368.3 g/mol) inhibited the replication of Zika (HD78788 strain) and Chikungunya virus, with $IC_{50}$ values of 1.9 and 3.8 µM, respectively (Mounce et al., 2017).

*Strong antiviral (enveloped non-segmented (-)RNA) amphiphilic diarylheptanoid:* Curcumin inhibited the replication of Vesicular stomatitis virus (Indiana strain), with an $IC_{50}$ value of 4.9 µM (Mounce et al., 2017).

*Weak antiviral (non-enveloped linear monopartite (+) RNA virus) amphiphilic diarylheptanoid:* Curcumin inhibited the replication of Coxsackie B3 (Nancy strain) with an $IC_{50}$ value of 94.5 µM (Mounce et al., 2017).

Curcumin

*Viral enzyme inhibition by diarylheptanoids:* Curcumin inhibited the Severe acute respiratory syndrome-associated coronavirus chymotrypsin-like protease activity with an $IC_{50}$ value of $40\,\mu M$ (Wen et al., 2007). Curcumin and demethoxycurcumin inhibited aminopeptidase N with $IC_{50}$ values of 10 and $20\,\mu M$, respectively (Bauvois & Dauzonne, 2006).

Commentary: The roots contain the sesquiterpenes 4-epi-curcumenol, curcumenol, curcuma-diol, and gweicurculactone, as well as the curcuminoids curcumin, demethoxy-curcumin, and bis-demethoxycurcumin (Wang et al., 2008). Consider that curcumin inhibited NorA efflux pump in *Staphylococcus aureus* (Sharma et al., 2019). NorA belongs to the major facilitator efflux pumps group.

### References

Bauvois, B. and Dauzonne, D., 2006. Aminopeptidase-N/CD13 (EC 3.4. 11.2) inhibitors: chemistry, biological evaluations, and therapeutic prospects. *Medicinal Research Reviews*, 26(1), pp. 88–130.

Mounce, B.C., Cesaro, T., Carrau, L., Vallet, T. and Vignuzzi, M., 2017. Curcumin inhibits Zika and chikungunya virus infection by inhibiting cell binding. *Antiviral Research*, *142*, pp. 148–157.

Radulović, N.S., Dekić, M.S., Stojanović-Radić, Z.Z. and Zoranić, S.K., 2010. *Geranium macrorhizum* L. (Geraniaceae) essential oil: a potent agent against Bacillus subtilis. *Chemistry & Biodiversity*, 7(11), pp. 2783–2800.

Sharma, A., Gupta, V.K. and Pathania, R., 2019. Efflux pump inhibitors for bacterial pathogens: from bench to bedside. *The Indian Journal of Medical Research*, 149(2), p. 129.

Wang, D., Huang, W., Shi, Q., Hong, C., Cheng, Y., Ma, Z. and Qu, H., 2008. Isolation and cytotoxic activity of compounds from the root tuber of *Curcuma wenyujin*. *Natural Product Communications*, 3(6), p. 1934578X0800300606.

Wen, C.C., Kuo, Y.H., Jan, J.T., Liang, P.H., Wang, S.Y., Liu, H.G., Lee, C.K., Chang, S.T., Kuo, C.J., Lee, S.S. and Hou, C.C., 2007. Specific plant terpenoids and lignoids possess potent antiviral activities against severe acute respiratory syndrome coronavirus. *Journal of Medicinal Chemistry*, 50(17), pp. 4087–4095.

Zhu, J., Lower-Nedza, A.D., Hong, M., Jiec, S., Wang, Z., Yingmao, D., Tschiggerl, C., Bucar, F. and Brantner, A.H., 2013. Chemical composition and antimicrobial activity of three essential oils from *Curcuma wenyujin*. *Natural Product Communications*, 8(4), p. 1934578X1300800430.

### 3.5.4.8　*Zingiber cassumunar* Roxb.

Synonyms: *Amomum cassumunar* (Roxb.) Donn; *Cassumunar roxburghii* (Roxb.) Colla; *Amomum montanum* J. Koenig ex Retz.; *Zingiber montanum* (J. Koenig ex Retz.) Link ex A. Dietr.

Common names: Bengal ginger, cassumunar ginger; bun ada (Bangadesh); ardika, banada, tekhao yaikhu (India); banglai (Indonesia); bongelai (Malaysia); meik tha lin, hta nah (Myanmar); plai, wan fai (Thailand)

Habitat: Forests, villages, cultivated

Distribution: India, Bangladesh, Myanmar, Thailand, Malaysia, and Indonesia

Botanical examination: This ginger grows to about 2 m tall from an aromatic (somewhat camphor-like odor), bright yellow inside, terete, whitish, and somewhat fleshy rhizome. The leaves are simple and alternate. The blade is dull green, with a minute ligule, subsessile, 20–35 cm × 7.5–15 cm, narrowly oblong, acute or round at base, acuminate at apex, glossy, and sheathing. The inflorescence is a terminal, fusiform, dark reddish, somewhat scaly, about 15 cm long, fleshy spike on a 20 cm tall stem. The calyx is tubular, white, and about 2 cm long. The corolla is tubular, white and develops a broad labellum, which is about 5 cm long and tinted yellow. The stamen is about 1 cm long and slender. The ovary is about 5 mm long across and develops a slender style and a obconic stigma. The capsule is globose, about 1.5 cm and contains numerous purplish little seeds.

Medicinal uses: Diarrhea (Bangladesh, India, Malaysia); post-partum (Malaysia)

*Broad-spectrum antibacterial halo developed by essential oil*: Essential oil of rhizomes (15 µL/disc) inhibited the growth of *Salmonella typhi*, *Streptococcus enteriditis*, *Escherichia coli*, *Clostridium pefringens*, and *Campilobacter jejuni* with inhibition zone diameters of 27.5, 24.5, 28.5, 21, and 32 mm, respectively.

*Moderate broad-spectrum antibacterial essential oil:* Essential oil of rhizomes (mainly containing) sabinene and terpinen-4-ol) inhibited the growth of *Staphylococcus aureus* (MTCC 96), *Staphylococcus epidermidis* (MTCC 435), *Streptococcus mutans* (MTCC 890), *Klebsiella pneumoniae* (MTCC 109), *Pseudomonas aeruginosa* (MTCC 741), *Escherichia coli* (MTCC 723), and *Salmonella typhimurium* (MTCC 98) with MIC values of 500, 250, 500, 500, 500, 250, and 125 µg/mL, respectively (Verma et al., 2018).

*Broad-spectrum antibacterial sesquiterpene*: The monocyclic sesquiterpene zerumbone inhibited the growth of *Bacillus cereus* (F 4810), *Staphylococcus aureus* (FRI 722), *Escherichia coli* (MTCC 108), and *Yersinia enterolitica* (MTCC 851) (Kumar et al., 2013).

*Moderate antibacterial (Gram-positive) sesquiterpene:* Zerumbone was bactericidal for *Streptococcus mutans* (ATCC 35668) with an MIC/MBC value of 250/500 µg/mL (da Silva et al., 2018).

*Moderate anticandidal essential oil:* Essential oil of rhizomes inhibited the growth of *Candida albicans* (ATCC 14053) and *Candida albicans* (MTCC 1637) (Verma et al., 2018).

*Moderate anticandidal sesquiterpene:* Zerumbone (Log P = 3.9) inhibited the growth of *Candida albicans* (ATCC 14053), *Candida albicans* (GNU 1656) and *Candida albicans* (GNU 3967) with $IC_{80}$ values of 64, 128, and 64 µg/mL, respectively (Shin & Eom, 2019). Zerumbone inhibited hyphal growth and biofilm secretion (Shin & Eom, 2019).

*Moderate antiviral (enveloped segmented linear single-stranded (−) RNA virus) non-polar extract:* Hexane extract inhibited 52.6% of Avian Influenza virus H5N1 infection at a concentration of 200 µg/mL (Klaywong et al., 2014).

*Antiviral (enveloped linear monopartite dimeric single-stranded (+)RNA) sesquiterpene:* Zerumbone inhibited the replication the Human immunodeficiency virus (Dai et al., 1997).

Zerumbone

Commentaries: Members of the genus *Zingiber* Mill. are used to treat viral and bacterial infections in Asia Pacific: Siddha healers in South India use *Zingiber officinale* Roscoe to treat COVID-19 (Sasikumar et al., 2020). In Bangladesh *Zingiber officinale* Roscoe affords a traditional remedy for dysentery (local name: *Ada*) and *Zingiber purpureum* Roxb. (Local name: *Bao ada*) is used to assuage leucorrhea. *Zingiber chrysostachys* Ridl. is used to treat fever in Malaysia. In China, *Zingiber officinale* Roscoe is used to treat COVID-19 (local name: *Jiang*) (Wang et al., 2020; Luo, 2020; Yang et al., 2020) in combination with oseltamivir phosphate (combination of extracts or teas of medicinal plants and conventional drugs is a sensible and economical therapeutic strategy. For instance, bitter gourd also known as *Momordica charantia* L. is combined with antidiabetic drugs in Bangladesh, allowing significant delay in starting insulin therapy in type 2 diabetics). *Zingiber officinale* Roscoe is used in India for the treatment of Chikungunya virus infection. Ethanol extract of *Zingiber officinale* Roscoe inhibited the replication of the Influenza virus (A/Chile/1/83) with an $EC_{50}$ value of 78 μg/mL and reduced plaque formation in Madin-Darby canine kidney cells with an $IC_{50}$ value of 39 μg/mL (Shin et al., 2010). Isolation and characterization of antiviral agents for COVID-19 and other future pandemics from members of the genus *Zingiber* Mill. is warranted.

### References

da Silva, T.M., Pinheiro, C.D., Orlandi, P.P., Pinheiro, C.C. and Pontes, G.S., 2018. Zerumbone from *Zingiber zerumbet* (L.) smith: a potential prophylactic and therapeutic agent against the cariogenic bacterium *Streptococcus mutans*. *BMC Complementary and Alternative Medicine, 18*(1), pp. 1–9.

Dai, J.R., Cardellina, J.H., Mahon, J.B.M. and Boyd, M.R., 1997. Zerumbone, an human HIV-inhibitory and cytotoxic sesquiterpene of *Zingiber aromaticum* and Z. *zerumbet*. *Natural Product Letters, 10*(2), pp. 115–118.

Klaywong, K., Khutrakul, G., Choowongkomon, K., Lekcharoensuk, C., Petcharat, N., Leckcharoensuk, P. and Ramasoota, P., 2014. Screening for lead compounds and herbal extracts with potential anti-influenza viral activity. *Southeast Asian Journal of Tropical Medicine and Public Health, 45*(1), p. 62.

Kumar, S.S., Srinivas, P., Negi, P.S. and Bettadaiah, B.K., 2013. Antibacterial and antimutagenic activities of novel zerumbone analogues. *Food Chemistry, 141*(2), pp. 1097–1103.

Luo, A., 2020. Positive SARS-Cov-2 test in a woman with COVID-19 at 22 days after hospital discharge: a case report. *Journal of Traditional Chinese Medical Sciences, 7*(4), pp. 413–417.

Sasikumar, R., Priya, S.D. and Jeganathan, C., 2020. A case study on domestics spread of SARS-CoV-2 pandemic in India. *International Journal of Advanced Science and Technology, 29*(7), pp. 2570–2574.

Shin, D.S. and Eom, Y.B., 2019. Zerumbone inhibits *Candida albicans* biofilm formation and hyphal growth. *Canadian Journal of Microbiology, 65*(10), pp. 713–721.

Wang, S.X., Wang, Y., Lu, Y.B., Li, J.Y., Song, Y.J., Nyamgerelt, M. and Wang, X.X., 2020. Diagnosis and treatment of novel coronavirus pneumonia based on the theory of traditional Chinese medicine. *Journal of Integrative Medicine, 18*(4), pp. 275–283.

Yang, Q.X., Zhao, T.H., Sun, C.Z., Wu, L.M., Dai, Q., Wang, S.D. and Tian, H., 2020. New thinking in the treatment of 2019 novel coronavirus pneumonia. *Complementary Therapies in Clinical Practice, 39*, p. 101131.

# 4 The Clade Ranunculids

## 4.1 ORDER RANUNCULALES JUSS. EX BERCHT. & J. PRESL. (1820)

### 4.1.1 FAMILY BERBERIDACEAE A.L. de. Jussieu (1789)

The family Berberidaceae consists of 2 genera and 650 species of herbs or shrubs. The wood of shrubs is often yellow owed to alkaloids. The leaves are simple or compound, opposite, alternate, and stipulate or exstipulate. The petiole is enlarged at base and apex. The calyx comprises 6–9 sepals. The corolla includes mostly 6 petals. The androecium includes often 6 stamens. Sepals and petals have often the same color and shape. The gynoecium comprises 1 carpel containing numerous ovules. The fruit is somewhat baccate and contain 1 to numerous seeds. Members of this family produce antimicrobial protoberberines, bisbenzylisoquinolines, and protopine alkaloids as well as aryltetralin lignans. Is is an archetype of alkaloid producing plant family. Isoquinolines are plant defenses in phylogenetically basal families.

#### 4.1.1.1 *Berberis aristata* Sims

Common names: Indian barberry; daru haldi, daruharidra (India); chutro, komme, tisy (Nepal)

Habitat: Open, moist places

Distribution: West Himalaya and Nepal

Botanical examination: This graceful shrub grows to 3 m tall. The stems are terete, yellowish, somewhat zigzag shaped, and thorned. The wood is bright yellow. The thorns are sharp, straight, woody, and about 2 cm long. The leaves are simple, subsessile, and exstipulate, and gathered on short peduncles. The blade is obovate-elliptic, 3–7.5 cm×1–3 cm, coriaceous, elliptic to spathulate and aristate. The inflorescences are 2–4 cm long, pendulous, and bear numerous flowers. The corolla is yellow and consists of 6 petals, which are about 7 mm×4.5 mm obovate and retuse at apex. The 6 stamens are about 5 mm long. The berries are elliptic, edible, up to about 1 cm long, beautifully colored from some sort of purple to red, and contain 3–4 seeds.

Medicinal uses: Fever, dysentery, sore throat (Nepal)

*Strong broad-spectrum antibacterial polar extract:* Hydroalcoholic extract of roots inhibited the growth of *Micrococcus luteus* (MTCC-106), *Bacillus subtilis* (MTCC-121), *Bacillus cereus* (MTCC-430), *Enterobacter aerogenus* (MTCC-111), *Escherichia coli* (MTCC-443), *Klebsiella pneumoniae* (109), *Proteus mirabilis* (MTCC-1429), *Pseudomonas aeruginosa* (MTCC-429), *Staphylococcus aureus* (MTCC-96), and *Streptococcus pneumoniae* (MTCC-2672) with MIC values of 0.6, 0.3, 0.6, 2.5, 0.3, 0.6, 0.6, 0.6, 0.3, and 0.3 µg/mL, respectively (Singh et al., 2009). Ethanol extract inhibited the growth of clinical carbapenem-resistant NDM-1, producing *Escherichia coli* (Strain -09) with the MIC value of 25 µg/mL (Thakur et al., 2016).

*Strong antibacterial (Gram-positive) hydrophilic protoberberine alkaloid:* Berberine (LogD=−1.3 at pH 7.4; molecular weight=336.4 g/mol) inhibited 20 strains of *Propionibacterium acnes* with MIC between 5 and 25 µg/mL (Slobodníková et al., 2004). Berberine inhibited 14 strains of *Staphylococcus epidermidis* with MIC between 25 and 500 µg/mL (Slobodníková et al., 2004). Berberine inhibited the growth of *Staphylococcus aureus, Bacillus subtilis, Escherichia coli,* and *Pseudomonas aeruginosa* with MIC values of 250, 497, 230, and 226 µg/mL, respectively (Čerňáková & Košťálová, 2002). Berberine inhibited the growth of several methicillin-resistant *Staphylococcus aureus* strains with MIC values ranging from 32 to 128 µg/mL (Yu et al., 2005). In a subsequent study, berberine inhibited *Staphylococcus aureus, Bacillus subtillis, Escherichia coli,* and *Shigella dysenteriae* with MIC values ranging from 115 to 500 µg/mL, respectively (Wang et al., 2008). Berberine inhibited the growth of *Staphylococcus capitis* (ATCC 35661), *Staphylococcus epidermidis* (ATCC 12228), *Staphylococcus intermedius* (ATCC 29663), *Staphylococcus lentus* (ATCC 700403), and *Staphylococcus lugdunensis* (ATCC 49576) with MIC values of 16, 32, 64, 64, and 64 µg/mL, respectively (Wojtyczka et al., 2014). Berberine inhibited the growth of *Streptococcus agalactiae* (CVCC 1886) with an MIC value of 78 µg/mL via a mechanism involving cytoplasmic insults, protein synthesis inhibition, and DNA synthesis inhibition (Peng et al., 2015). Berberine inhibited the growth of *Actinobacillus pleuropneumoniae* with an MIC value of 312.5 µg/mL via DNA synthesis inhibition, inhibition of protein synthesis, decreased cytoplasm, lost cytoplasm, edged nuclear area, plasmolysis, unequal division, cell vacuolization (Kang et al., 2015).

Berberine

*Antibiotic potentiator hydrophilic protoberberine alkaloid:* Berberine decreased the MICs of ampicillin and oxacillin against methicillin-resistant *Staphylococcus aureus* (Yu et al., 2005). Berberine inhibited the MexAB antibiotic efflux pump in *Pseudomonas aeruginosa* (Blanco et al., 2018).

*Antibiotic potentiator polar extract:* Ethanol extract of stem bark increased the sensitivity of clinical carbapenem-resistant NDM-1 producing *Escherichia coli* (Strain -09) toward colistin, tygecycline, and augmentin (Thakur et al., 2016).

*Strong broad-spectrum antifungal hydrophilic protoberberine:* Berberine inhibited the growth of fluconazole resistant clinical strains of *Candida tropicalis*, *Candida albicans*, *Candida parapsilosis*, *Cryptococcus neoformans*, as well as *Candida krusei* (ATCC 6258), and *Candida parapsilosis* (ATCC 22019) (da Silva et al., 2016). This protoberberine alkaloid evoked plasma and mitochondrial membrane insults as well as DNA damages (da Silva et al., 2016).

*Antifungal potentiator hydrophilic protoberberine alkaloid:* Berberine at the concentration of 1.9 µg/mL decreased the MIC of fluconazole from 1.9 to 0.4 µg/mL (Iwazaki et al., 2010).

*Strong antiviral (enveloped linear monopartite double-stranded DNA) hydrophilic protoberberine alkaloid:* Berberine inhibited the growth of human Cytomegalovirus with an $IC_{50}$ value of 0.6 µM via a mechanism involving interference with intracellular events after virus penetration into the host cells and before viral DNA synthesis (Hayashi et al., 2007).

*Strong antiviral (enveloped, segmented, single-stranded (-)RNA) )-hydrophilic protoberberine alkaloid:* Berberine inhibited the replication of the Influenza virus A/FM/1/47 (H1N1) with an $IC_{50}$ value of 25 µg/mL and a selectivity index of 9.6 (Shao et al., 2020).

*Strong antiviral (enveloped monopartite linear single-stranded (+) RNA) hydrophilic protoberberine:* Berberine inhibited the replication of the Chikungunya virus with an $EC_{50}$ value of 4.5 µM and a selectivity index of 45 in HEK-293T cells (Varghese et al., 2016).

*Strong antiviral (non-enveloped segmented (+) RNA virus) protoberberine:* Berberine inhibited the replication of the Coxsackievirus strain B1 and B2 with an $EC_{50}$ value of 7.3 µM and a selectivity index of 16.3. Berberine inhibited the replication of Coxsackievirus strain B3, B4, B5, and B6 viruses with $IC_{50}$/selectivity indices of 6.4/18.7, 5.8/20.5, 13.8/8.7, and 9.4 $\mu$M/12.7, respectively (Zeng et al., 2020).

*Strong antiviral (enveloped monopartite linear dimeric single-stranded (+) RNA) hydrophilic protoberberine:* Berberine inhibited the Human immunodeficiency virus type-1 syncytia formation in a dose-dependent manner with an $IC_{50}$ value of 5.7 $\mu$g/mL via impairment of viral fusion (Shao et al., 2020).

*In vivo antiviral (enveloped, segmented, single-stranded (-)RNA) hydrophilic protoberberine:* Berberine given intraperitoneally at a dose of 5 mg/Kg/day for 7 days to mice intranasally infected with Influenza virus decreased the mortality rate from 90 to 55% with reduction in lung viral titer, reduced lung inflammation, and improved lung histology (Wu et al., 2011).

Commentaries:

(i) In Nepal, the root and stem bark afford a yellow dye used to paint doors and windows. The yellow dye consists of prototoberberine alkaloids such as berberine (Chandra & Purohit, 1980).

(ii) Collyres containing berberine are used in therapeutic treatment of eye infections.

(iii) Amin and coworkers (1969) observed that berberine was bactericidal to *Vibrio cholerae* and bacteriostatic to *Staphylococcus aureus*, at concentrations of 35 and 50 $\mu$g/mL via almost immediate inhibition of protein synthesis. Subsequently it was ascertained that berberine binds to DNA (Bandyopadhyay et al., 2013) and being planar it might be able to intercalate into DNA and block the replication machinery.

(iv) Berberine has a low oral bioavailability.

(v) Being a broad-spectrum antiviral principle, berberine could be active against COVID-19. Further studies in this direction are warranted.

## References

Bandyopadhyay, S., Patra, P.H., Mahanti, A., Mondal, D.K., Dandapat, P., Bandyopadhyay, S., Samanta, I., Lodh, C., Bera, A.K., Bhattacharyya, D. and Sarkar, M., 2013. Potential antibacterial activity of berberine against multi drug resistant enterovirulent *Escherichia coli* isolated from yaks (*Poephagus grunniens*) with haemorrhagic diarrhoea. *Asian Pacific Journal of Tropical Medicine*, 6(4), pp. 315–319.

Blanco, P., Sanz-García, F., Hernando-Amado, S., Martínez, J.L. and Alcalde-Rico, M., 2018. The development of efflux pump inhibitors to treat Gram-negative infections. *Expert Opinion on Drug Discovery*, 13(10), pp. 919–931.

Čerňáková, M. and Košťálová, D., 2002. Antimicrobial activity of berberine—a constituent of *Mahonia aquifolium*. *Folia Microbiologica*, 47(4), pp. 375–378.

Chandra, P. and Purohit, A.N., 1980. Berberine contents and alkaloid profile of Berberis species from different altitudes. *Biochemical Systematics and Habitat*, 8(4), pp. 379–380.

da Silva, A.R., de Andrade Neto, J.B., da Silva, C.R., de Sousa Campos, R., Silva, R.A.C., Freitas, D.D., do Nascimento, F.B.S.A., de Andrade, L.N.D., Sampaio, L.S., Grangeiro, T.B. and Magalhães, H.I.F., 2016. Berberine antifungal activity in fluconazole-resistant pathogenic yeasts: action mechanism evaluated by flow cytometry and biofilm growth inhibition in Candida spp. *Antimicrobial Agents and Chemotherapy*, 60(6), pp. 3551–3557.

Hayashi, K., Minoda, K., Nagaoka, Y., Hayashi, T. and Uesato, S., 2007. Antiviral activity of berberine and related compounds against human cytomegalovirus. *Bioorganic & Medicinal Chemistry Letters*, 17(6), pp. 1562–1564.

Iwazaki, R.S., Endo, E.H., Ueda-Nakamura, T., Nakamura, C.V., Garcia, L.B. and Dias Filho, B.P., 2010. In vitro antifungal activity of the berberine and its synergism with fluconazole. *Antonie Van Leeuwenhoek*, 97(2), p. 201.

Kang, S., Li, Z., Yin, Z., Jia, R., Song, X., Li, L., Chen, Z., Peng, L., Qu, J., Hu, Z. and Lai, X., 2015. The antibacterial mechanism of berberine against *Actinobacillus pleuropneumoniae*. *Natural Product Research*, 29(23), pp. 2203–2206.

Peng, L., Kang, S., Yin, Z., Jia, R., Song, X., Li, L., Li, Z., Zou, Y., Liang, X., Li, L. and He, C., 2015. Antibacterial activity and mechanism of berberine against *Streptococcus agalactiae*. *International Journal of Clinical and Experimental Pathology*, 8(5), p. 5217.

Shao, J., Zeng, D., Tian, S., Liu, G. and Fu, J., 2020. Identification of the natural product berberine as an antiviral drug. *AMB Express*, 10(1), pp. 1–11.

Singh, M., Srivastava, S. and Rawat, A.K.S., 2009. Antimicrobial studies of stem of different Berberis species. *Natural Product Sciences*, 15(2), pp. 60–65.

Thakur, P., Chawla, R., Goel, R., Narula, A., Arora, R. and Sharma, R.K., 2016. Augmenting the potency of third-line antibiotics with *Berberis aristata*: in vitro synergistic activity against carbapenem-resistant *Escherichia coli*. *Journal of Global Antimicrobial Resistance*, 6, pp. 10–16.

Varghese, F.S., Thaa, B., Amrun, S.N., Simarmata, D., Rausalu, K., Nyman, T.A., Merits, A., McInerney, G.M., Ng, L.F. and Ahola, T., 2016. The antiviral alkaloid berberine reduces chikungunya virus-induced mitogen-activated protein kinase signaling. *Journal of Virology*, 90(21), pp. 9743–9757.

Wang, L.J., Ye, X.L., Li, X.G., Sun, Q.L., Yu, G., Cao, X.G., Liang, Y.T. and Zhou, J.Z., 2008. Synthesis and antimicrobial activity of 3-alkoxyjatrorrhizine derivatives. *Planta Medica*, 74(03), pp. 290–292.

Wojtyczka, R.D., Dziedzic, A., Kępa, M., Kubina, R., Kabała-Dzik, A., Mularz, T. and Idzik, D., 2014. Berberine enhances the antibacterial activity of selected antibiotics against coagulase-negative *Staphylococcus* strains in vitro. *Molecules*, 19(5), pp. 6583–6596.

Yu, H.H., Kim, K.J., Cha, J.D., Kim, H.K., Lee, Y.E., Choi, N.Y. and You, Y.O., 2005. Antimicrobial activity of berberine alone and in combination with ampicillin or oxacillin against methicillin-resistant *Staphylococcus aureus*. *Journal of Medicinal Food*, 8(4), pp. 454–461.

Zeng, Q.X., Wang, H.Q., Wei, W., Guo, T.T., Yu, L., Wang, Y.X., Li, Y.H. and Song, D.Q., 2020. Synthesis and biological evaluation of berberine derivatives as a new class of broad-spectrum antiviral agents against Coxsackievirus B. *Bioorganic Chemistry*, 95, p. 103490.

### 4.1.1.2 *Berberis asiatica* Roxb. ex DC.

Common names: Tree turmeric; kilmora, daruharidra (India); chutro, chotr, chutra (Nepal)

Habitat: Open, sunny and dry lands

Distribution: Himalayas, Nepal, Bhutan, Assam, and China

Botanical examination: It is a shrub that grows to about 3 m tall. The wood is bright yellow. The stems are light yellowish, somewhat zigzag shaped, and thorny. The thorns are straight, woody, and sharp. The leaves are simple, exstipulate, and gathered on short peduncles. The petiole is up to 1 cm long. The blade is oblong, elliptic, or broadly obovate, 1.5–9 cm×0.5–3.8 cm, coriaceous, dark green above, glaucous below, and laxly spinose-dentate. The inflorescences are fascicle-like racemes, which are up to 3 cm long and pendulous. The corolla is yellow and consists of 6 yellow and minute petals. The androecium includes 6 stamens. The fruits are berries which are ovoid, edible, blueberry purple, up to about 1 cm long.

Medicinal use: Boils, fever, wounds (India); dysentery, conjunctivitis (Nepal)

*Strong broad-spectrum antibacterial polar extract*: Hydroalcoholic extract of stem inhibited *Micrococcus luteus* (MTCC-106), *Bacillus subtilis* (MTCC-121), *Bacillus cereus* (MTCC-430), *Enterobacter aerogenus* (MTCC-111), *Escherichia coli* (MTCC-443), *Klebsiella pneumoniae* (109), *Proteus mirabilis* (MTCC-1429), *Pseudomonas aeruginosa* (MTCC-429), *Staphylococcus aureus*

(MTCC-96), and *Streptococcus pneumoniae* (MTCC-2672) with MIC values ranging from 0.3 to 0.6 µg/mL, respectively (Singh et al., 2009).

*Moderate broad-spectrum antibacterial alkaloid fraction:* Alkaloid fraction of stem bark inhibited the growth of *Staphylococcus aureus*, *Enterococcus facecalis*, *Shigella dysenteriae* 1, *Vibrio cholerae*, *Vibrio parahaemolyticus*, and *Plesiomonas shigellioides* with MIC values of 312.5, 39, 78.1, 78.1, 156.2, 156.2, and 156.2 µg/mL respectively (Bhandari et al., 2000).

*Strong antifungal polar extract:* Hydroalcoholic extract of roots inhibited *Candida albicans*, *Aspergillus spinulosus*, and *Aspergillus flavus* (Singh et al., 2009).

*Strong anticandidal alkaloid fraction:* Alkaloid fraction of stem bark inhibited the growth of *Candida albicans* with an MIC value of 39 µg/mL, respectively (Bhandari et al., 2000).

Commentary: The plant, like most members of the genus *Berberis* L., produces the broad-spectrum antimicrobial protoberberine alkaloid berberine (Srivastava et al., 2004).

### References

Bhandari, D.K., Nath, G., Ray, A.B. and Tewari, P.V., 2000. Antimicrobial activity of crude extracts from *Berberis asiatica* stem bark. *Pharmaceutical Biology*, *38*(4), pp. 254–257.

Singh, M., Srivastava, S. and Rawat, A.K.S., 2009. Antimicrobial studies of stem of different Berberis species. *Natural Product Sciences*, *15*(2), pp. 60–65.

Srivastava, S.K., Singh Rawat, A.K. and Mehrotra, S., 2004. Pharmacognostic evaluation of the root of *Berberis asiatica*. *Pharmaceutical Biology*, *42*(6), pp. 467–473.

### 4.1.1.3 *Berberis chitria* Lindl.+6

Common name: Kindotaa (India); chutro (Nepal)

Habitat: Open lands

Distribution: Himalayas, India, Nepal, and Bhutan

Botanical examination: It is a shrub, which grows to 3 m tall. The stem are terete, reddish brown, and thorned. The wood is bright yellow. The thorns are sharp, woody and about 2 cm long. The leaves are simple, subsessile, and exstipulate, and gathered on short peduncles. The blade is obovate-elliptic, 2–6 cm × 1.5–2.5 cm, coriaceous, dull green, with 4–6 pairs of secondary nerves, and laxly spinose-dentate. The inflorescences are 8–12 cm long panicles and bear numerous flowers, which are about 1.5 cm across and yellow. The calyx includes 6 sepals, which are obovate and up to 1 cm long. The corolla is yellow and consists of 6 petals, which are about 8 mm long, and oblong. The androecium includes 6 stamens, which are about 7 mm long. The berries are narrowly ovoid, up to about 1.5 cm long, dark red-brown to purple, and glossy.

Medicinal uses: Fever and wounds (India)

*Strong broad-spectrum antibacterial polar extract:* hydroalcoholic extract of stems inhibited *Micrococcus luteus* (MTCC-106), *Bacillus subtilis* (MTCC-121), *Bacillus cereus* (MTCC-430), *Enterobacter aerogenus* (MTCC-111), *Escherichia coli* (MTCC-443), *Klebsiella pneumoniae* (109), *Proteus mirabilis* (MTCC-1429), *Pseudomonas aeruginosa* (MTCC-429), *Staphylococcus aureus* (MTCC-96), and *Streptococcus pneumoniae* (MTCC-2672) with MIC values of 0.6, 1.2, 0.6, 2.5, 2.5, 0.6, 0.6, 0.6, 0.6, and 0.3 µg/mL, respectively (Singh et al., 2009).

*Strong antifungal (filamentous) polar extract:* Hydroalcoholic extract of roots inhibited *Aspergillus nidulans*, *Cryptococcus albidus*, and *Aspergillus flavus* (Singh et al., 2009).

*Strong antibacterial (Gram-positive) hydrophilic protoberberine alkaloid:* Jatrorrhizine (LogD = −1.5 at pH 7.4; molecular weight = 338.3 g/mol) inhibited 20 strains of *Propionibacterium acnes* with MIC between 5 and 50 µg/mL, respectively (Slobodníková et al., 2004). Jatrorrhizine inhibited 14 strains of *Staphylococcus epidermidis* with MIC between 100 and 500 µg/mL (Slobodníková et al., 2004). In a subsequent study, jatrorrhizine inhibited the growth of *Staphylococcus aureus*, *Bacillus subtilis*, *Escherichia coli*, and *Shigella dysenteriae* with MIC values of 320, 640, 320, and 640 µg/mL, respectively (Wang et al., 2008).

Jatrorrhizine

*Moderate broad-spectrum antifungal protoberberine alkaloid:* Jatrorrhizine inhibited the growth of clinical isolates of *Epidermophyton floccosum, Trichophyton rubrum, Trichophyton interdigitale, Trichophyton violaceum, Trichophyton mentagrophytes* v. *granulosum, Trichophyton equinum, Microsporium canis, Microsporium gypseum, Candida albicans,* and *Candida tropicalis* with the MIC values of 62.5, 62.5, 62.5, 62.5, 125, 125, 62.5, 125, 500, and 250 µg/mL, respectively (Volleková et al., 2003).

*Antibiotic potentiator protoberberine alkaloid:* Palmatine (Log P=3.7) inhibited the MexAB antibiotic efflux pump in *Pseudomonas aeruginosa* (Blanco et al., 2018).

*Strong antiviral (enveloped monopartite linear single-stranded (+) RNA) hydrophilic protoberberine alkaloid:* Palmatine (LogD=−0.9 at pH 7.4; molecular weight=352.4 g/mol) inhibited the replication of the West Nile virus, Dengue 2 virus, and Yellow fever virus with $IC_{50}$ values of 3.6, 26.4, and 7.3 µM and selectivity indices of 286, 39, and 141, respectively (Jia et al., 2010). Palmatine inhibited the replication of the Zika virus in Vero cells with an $IC_{50}$ value of about 30 µM via inhibition of the virus entry in host cell (Ho et al., 2019).

Palmatine

*Strong antiviral (enveloped monopartite linear double-stranded DNA) hydrophilic protoberberine alkaloid:* Palmatine inhibited the replication of the Herpes simplex virus type-1 (strain 7401H) with $IC_{50}$ values of 34 µg/mL in Vero cells (Nakamura et al., 1999).

*Viral enzyme inhibition by protoberberine alkaloids:* Jatrorrhizine and berberine inhibited the Human immunodeficiency virus reverse transcriptase with the $IC_{50}$ values of 71 and 100 µg/mL, respectively (Tan et al., 1991). Palmatine inhibited the West Nile virus NS2B-NS3 protease with an $IC_{50}$ value of 96 μM (Ho et al., 2019).

Commentary: (i) The plant constructs protoberberine alkaloids, including berberine, palmatine, and jatrorrhizine (Hussaini & Shoeb, 1985). Natural products from members of the family Berberidaceae are not uncommonly able to inhibit viral proteases: extract of leaves of *Epimedium sagittatum* (Sieb. et Zucc.) Maxim. Used medicinally in China (*san zhi jiu ye cao*) at a concentration of 250 μg/mL inhibited Human immunodeficiency virus type-1 protease by 100% (Xu et al., 1996). (ii) The MexAB-OprM efflux pump is responsible for the resistance of *Pseudomonas aeruginosa* to antibiotics and other xenobiotics including natural products from plants (Pan et al., 2016). Natural products having effects on the central nervous system of mammals are often able to inhibit bacterial efflux pumps. Why? This is the case for palmatine (Dhingra & Bhankher, A., 2014). Inhibitors of bacterial efflux pumps from medicinal plants (Prasch & Bucar, 2015) might be the solution for global increase in bacterial resistance.

## References

Blanco, P., Sanz-García, F., Hernando-Amado, S., Martínez, J.L. and Alcalde-Rico, M., 2018. The development of efflux pump inhibitors to treat Gram-negative infections. *Expert opinion on drug discovery*, 13(10), pp. 919–931.

Dhingra, D. and Bhankher, A., 2014. Behavioral and biochemical evidences for antidepressant-like activity of palmatine in mice subjected to chronic unpredictable mild stress. *Pharmacological Reports*, 66(1), pp. 1–9.

Ho, Y.J., Lu, J.W., Huang, Y.L. and Lai, Z.Z., 2019. Palmatine inhibits Zika virus infection by disrupting virus binding, entry, and stability. *Biochemical and Biophysical Research Communications*, 518(4), pp. 732–738.

Hussaini, F.A. and Shoeb, A., 1985. Isoquinoline derived alkaloids from *Berberis chitria*. *Phytochemistry*, 24(3), p. 633.

Jia, F., Zou, G., Fan, J. and Yuan, Z., 2010. Identification of palmatine as an inhibitor of West Nile virus. *Archives of Virology*, 155(8), pp. 1325–1329.

Nakamura, N., Hattori, M., Kurokawa, M., Shiraki, K., Kashiwaba, N. and Ono, M., 1999. Anti-herpes simplex virus activity of alkaloids isolated from *Stephania cepharantha*. *Biological and Pharmaceutical Bulletin*, 22(3), pp. 268–274.

Pan, Y.P., Xu, Y.H., Wang, Z.X., Fang, Y.P. and Shen, J.L., 2016. Overexpression of MexAB-OprM efflux pump in carbapenem-resistant Pseudomonas aeruginosa. *Archives of microbiology*, 198(6), pp. 565–571.

Prasch, S. and Bucar, F., 2015. Plant derived inhibitors of bacterial efflux pumps: an update. *Phytochemistry Reviews*, 14(6), pp. 961–974.

Singh, M., Srivastava, S. and Rawat, A.K.S., 2009. Antimicrobial studies of stem of different *Berberis* species. *Natural Product Sciences*, 15(2), pp. 60–65.

Tan, G.T., Pezzuto, J.M., Kinghorn, A.D. and Hughes, S.H., 1991. Evaluation of natural products as inhibitors of human immunodeficiency virus type 1 (HIV -1) reverse transcriptase. *Journal of Natural products*, 54(1), pp. 143–154.

Volleková, A., Košt'álová, D., Kettmann, V. and Tóth, J., 2003. Antifungal activity of Mahonia aquifolium extract and its major protoberberine alkaloids. *Phytotherapy Research*, 17(7), pp. 834–837.

Wang, L.J., Ye, X.L., Li, X.G., Sun, Q.L., Yu, G., Cao, X.G., Liang, Y.T. and Zhou, J.Z., 2008. Synthesis and antimicrobial activity of 3-alkoxyjatrorrhizine derivatives. *Planta Medica*, 74(03), pp. 290–292.

Xu, H.X., Wan, M., Loh, B.N., Kon, O.L., Chow, P.W. and Sim, K.Y., 1996. Screening of traditional medicines for their inhibitory activity against HIV -1 protease. *Phytotherapy Research*, 10(3), pp. 207–210.

#### 4.1.1.4 *Berberis lycium* Royle

Common names: Zarch, sumbalu (Pakistan)
  Habitat: Open lands
  Distribution: Kashmir, Pakistan, and West Himalayas
  Botanical examination: This handsome shrub grows to 3 m tall. The wood is bright yellow. The stem are terete, light grey, and thorned. The thorns are sharp, woody, and about 2 cm long. The leaves are simple, subsessile, and exstipulate, and gathered on short peduncles. The blade is oblong-ovate, 3–6 cm×0.6–1.2 cm, coriaceous, whitish below, and laxly spinose-dentate. The panicles are 3–6 cm long and many flowered. The 6 sepals are obovate and about 5 mm long. The 6 petals are minute, yellow, about 4 mm long, and oblong. The androecium includes 6 stamens. The berries are narrowly ovoid, up to about 8 mm long, and somewhat blueberry purple.
  Medicinal uses: Wounds, fever, jaundice (India)
  *Moderate broad-spectrum antibacterial polar extract:* Hydroalcoholic extract of roots inhibited the growth of *Mycobacterium luteus* (MTCC-106), *Bacillus subtilis* (MTCC-121), *Bacillus cereus* (MTCC-430), *Enterobacter aerogenus* (MTCC-111), *Escherichia coli* (MTCC-443), *Klebsiella pneumoniae* (109), *Proteus mirabilis* (MTCC-1429), *Pseudomonas aeruginosa* (MTCC-429), *Staphylococcus aureus* (MTCC-96), *Salmonella typhimurium* (MTCC-98), and *Streptococcus pneumoniae* (MTCC-2672) with the MIC values of 0.3, 0.3, 2.5, 1.2, 0.6, 0.6, 0.3, 0.3, 0.3, 0.6, and 1.2 mg/mL, respectively (Singh et al., 2009).
  *Moderate antifungal polar extract:* Hydroalcoholic extract of roots inhibited *Aspergillus terreus* with an MIC value of 0.6 mg/mL (Singh et al., 2009).
  *Strong antiviral (enveloped monopartite linear single-stranded (+) RNA) polar extract:* Methanol extract at the concentration of 20 µg/mL decreased Hepatitis C virus (clinical isolate) count in HepG2 cells culture by 30% (Yousaf et al., 2018).
  Commentary: The plant produces broad-spectrum antibacterial alkaloids of which berberine (Chandra & Purohit, 1980).

#### References

Chandra, P. and Purohit, A.N., 1980. Berberine contents and alkaloid profile of Berberis species from different altitudes. *Biochemical Systematics and Habitat*, 8(4), pp. 379–380.
Singh, M., Srivastava, S. and Rawat, A.K.S., 2009. Antimicrobial studies of stem of different Berberis species. *Natural Product Sciences*, 15(2), pp. 60–65.
Yousaf, T., Rafique, S., Wahid, F., Rehman, S., Nazir, A., Rafique, J., Aslam, K., Shabir, G. and Shah, S.M., 2018. Phytochemical profiling and antiviral activity of *Ajuga bracteosa*, *Ajuga parviflora*, *Berberis lycium* and *Citrus lemon* against Hepatitis C virus. *Microbial Pathogenesis*, 118, pp. 154–158.

#### 4.1.1.5 *Berberis thunbergii* DC

Common names: Japanese barberry; ri ben xiao bo (China)
  Habitat: Forests, grasslands, ornamental
  Distribution: Native to Japan, North America, Europe, China, and Russia
  Botanical examination: This shrub grows to 2 m tall. The stems are angled, reddish purple, and thorned. The thorns are sharp, woody, and about 1.5 cm long. The leaves are simple, exstipulate, and gathered on short peduncles. The petiole is up to about 8 mm long. The blade is spathulate, 1–2 cm×0.5–1.2 cm, green or more or less tinted in red, attenuate at base, and emarginate or obtuse at apex. The inflorescences are about 2 cm long fascicles of light yellow to whitish and 8 mm diameter flowers.The 6 sepals are obovate and about 5 mm long. The 6 petals are about 5 mm long and oblong.

The androecium includes 6 stamens. The berries are ellipsoid, up to about 8 mm long, red, glossy, and contain 1–2 seeds.

Medicinal uses: Fever, antiseptic (China); dental caries (Vietnam)

*Moderate antibacterial (Gram-positive) polar extract:* Ethanol extract of stems inhibited *Bacillus subtilis, Staphylococcus aureus,* and *Bacillus thuringensis* with MIC of 2300, 500, and 500 µg/mL, respectively (Li et al., 2007).

*Moderate broad-spectrum antibacterial hydrophilic protoberberine:* Berberine (LogD = −1.3 at pH 7.4; molecular weight = 336.4 g/mol) inhibited the growth of *Escherichia coli* (ATCC 8739), *Staphylococcus aureus* (ATCC 6538), *Streptococcus mutans* (ATCC 25175), and *Streptococcus pyogenes* (ATCC 19615) with MIC values of 250, 125, 125, and 31 µg/mL, respectively (Villinski et al., 2003).

Commentary: The roots contain the broad-spectrum antimicrobial protoberberine alkaloids berberine, palmatine, and jatrorrhizine (Villinski et al., 2003). Consider that when a natural product is able to inhibit the growth of bacteria, fungi and viruses, it points to the possibility that this product targets DNA.

### References

Li, A.R., Zhu, Y., Li, X.N. and Tian, X.J., 2007. Antimicrobial activity of four species of Berberidaceae. *Fitoterapia, 78*(5), pp. 379–381.

Villinski, J., Dumas, E., Chai, H.B., Pezzuto, J., Angerhofer, C. and Gafner, S., 2003. Antibacterial activity and alkaloid content of *Berberis thunbergii, Berberis vulgaris* and *Hydrastis canadensis. Pharmaceutical Biology, 41*(8), pp. 551–557.

### 4.1.1.6 *Mahonia bealei* (Fortune) Carrière

Synonym: *Berberis bealei* Fortune

Common names: Beale's barberry; kuo ye shi da gong lao (China)

Habitat: Forests, streamside, roadside

Distribution: China

Botanical examination: This shrub grows to about 3 m tall. The stems are terete. The leaves are compound, spiral, exstipulate, and up to about 50 cm long. The leaves comprise 4–10 pairs of folioles, which are 0.5–10 cm × 1.5–13 cm, oblong to rhombic, rounded to cordate at base, laxly spinose at margin, and dark glossy green above. The rachis is reddish brown and somewhat glossy. The racemes are erect, showy, up to about 25 cm long, and supports numerous yellow flowers. The calyx comprises 9 yellow sepals, the outer sepals are about 5 mm long. The corolla comprises 6 petals, which are about 7 mm long. The androecium includes 6 stamens. The ovary is oblong, minute, with a short style. The berries are ovoid and about 1.5 cm long.

Medicinal use: Colds, dysentery, diarrhea, jaundice, tuberculosis (China)

*Weak antibacterial (Gram-positive) polar extract:* Ethanol extract of stems inhibited the growth of *Bacillus subtilis, Staphylococcus aureus,* and *Bacillus thuringiensis* with the MIC of 4500, 500, and 2300 µg/mL, respectively (Li et al., 2007).

*Antiviral (enveloped, segmented, single-stranded (-) RNA)) alkaloid fraction:* Alkaloid fraction of roots (containing the bisbenzylisoquinoline alkaloid isotetrandrine) inhibited the replication of the Influenza virus A (Zeng et al., 2006).

*Moderate anticandidal antifungal hydrophilic aporphine:* Magnoflorine (LogD = −1.2 at pH 7.4; molecular weight = 342.4 g/mol) inhibited the growth of *Candida albicans* (KCTC7965), *Candida albicans* (KACC30071), and *Candida parapsilosis* var. *parapsilosis* (KACC45480) with MIC values of 50, 100, and 100 µg/mL, respectively (Kim et al., 2018).

Magnoflorine

Commentary: The plant brings to being the antimicrobial protoberberine alkaloids berberine, palmatine, jatrorrhizine, and the aporphine alkaloid magnoflorine (Ji et al., 2000), which most probably account for the medicinal uses.

### References

Ji, X., Li, Y., Liu, H., Yan, Y. and Li, J., 2000. Determination of the alkaloid content in different parts of some Mahonia plants by HPCE. *Pharmaceutica Acta Helvetiae*, 74(4), pp. 387–391.

Kim, J., Bao, T.H.Q., Shin, Y.K. and Kim, K.Y., 2018. Antifungal activity of magnoflorine against Candida strains. *World Journal of Microbiology and Biotechnology*, 34(11), p. 167.

Li, A.R., Zhu, Y., Li, X.N. and Tian, X.J., 2007. Antimicrobial activity of four species of Berberidaceae. *Fitoterapia*, 78(5), pp. 379–381.

Zeng, X., Dong, Y., Sheng, G., Dong, X., Sun, X. and Fu, J., 2006. Isolation and structure determination of anti-influenza component from *Mahonia bealei*. *Journal of Ethnopharmacology*, 108(3), pp. 317–319.

### 4.1.1.7 *Mahonia fortunei* (Lindl.) Fedde

Synonym: *Berberis fortunei* Lindl.
   Common names: Chinese Mahonia; shi da gong lao (China)
   Habitat: Forests, roadsides, open lands
   Distribution: China and Taiwan
   Botanical observation: This shrub grows to about 2 m tall. The stems are terete and yield a yellow pigment. The leaves are compound, exstipulate, and up to about 20 cm long. The leaves comprise 2–5 pairs of folioles, which are 4.5–15 cm×0.9–2.5 cm, narrowly elliptic, cuneate at base, spinose at margin, coriaceous, acute at apex, and dull green above. The racemes are erect, showy, up to about 7 cm long, and present numerous yellow flowers. The 9 sepals are up to about 8 mm long. The 6 petals are oblong and about 4 mm long. The androecium includes 6 stamens. The ovary is without style. The berries are purplish, ovoid, and about 6 mm long.
   Medicinal use: Colds, fever, Influenza, jaundice (China)
   *Weak antibacterial (Gram-positive) polar extract:* Ethanol extract of stems inhibited *Bacillus subtilis*, *Staphylococcus aureus*, and *Bacillus thuringiensis* with the MIC of 1.1, 0.2, and 1.1 mg/mL, respectively, (Li et al., 2007).
   *Moderate antibacterial (Gram-positive) polar extract:* Ethanol extract of stems inhibited *Bacillus subtilis*, *Staphylococcus aureus*, and *Bacillus thuringiensis* with MIC of 1150, 290, and 1150 µg/mL, respectively (Li et al., 2007).

Commentary: The plant produces the antimicrobial protobeberine alkaloids columbamine, jatrorrhizine, palmatine, and berberine (Liu et al., 2020).

### References

Li, A.R., Zhu, Y., Li, X.N. and Tian, X.J., 2007. Antimicrobial activity of four species of Berberidaceae. *Fitoterapia*, *78*(5), pp. 379–381.

Liu, L., Cui, Z.X., Yang, X.W., Xu, W., Zhang, Y.B., Li, F.J., Gong, Y., Liu, N.F., Peng, K.F. and Zhang, P., 2020. Simultaneous characterisation of multiple *Mahonia fortunei* bioactive compounds in rat plasma by UPLC–MS/MS for application in pharmacokinetic studies and anti-inflammatory activity in vitro. *Journal of Pharmaceutical and Biomedical Analysis*, *179*, p. 113013.

### 4.1.1.8 *Mahonia napaulensis* DC

Synonyms: *Berberis acanthifolia* (Wall. ex G. Don) Wall. ex Walp.; *Berberis gautamae* Laferr.; *Berberis griffithii* (Takeda) Laferr.; *Berberis leschenaultii* Wall. ex Wight & Arn.; *Berberis longlinensis* (Y.S. Wang & P.G. Xiao) Laferr.; *Berberis manipurensis* (Takeda) Laferr.; *Berberis miccia* Buch.-Ham. ex D. Don; *Berberis napaulensis* (DC.) Laferr.; *Berberis pomensis* (Ahrendt) Laferr.; *Berberis salweenensis* (Ahrendt) Laferr.; *Mahonia acanthifolia* Wall. ex G. Don; *Mahonia griffithii* Takeda; *Mahonia leschenaultii* (Wall. ex Wight & Arn.) Takeda ex Dunn; *Mahonia longlinensis* Y.S. Wang & P.G. Xiao; *Mahonia manipurensis* Takeda; *Mahonia miccia* Buch.-Ham. ex D. Don; *Mahonia nepalensis* DC.; *Mahonia pomensis* Ahrendt; *Mahonia salweenensis* Ahrendt; *Mahonia sikkimensis* Takeda

Common names: Ni bo er shi da gong lao (China); taming (India); jamanemandro, komo (Nepal)

Habitat: Forests, forest margins, and thickets

Distribution: India, Bhutan, Nepal, Myanmar, Vietnam, and China

Botanical observation: This treelet grows to about 5 m tall. The stems are terete. The leaves are compound, exstipulate, and up to about 60 cm long. The leaves comprise 5–12 pairs of folioles, which are 0.5–9.5 cm × 1.2–5 cm, oblong-ovate, rounded at base, spinose at margin, coriaceous, and acute at apex, dark green and glossy above. The racemes are erect, showy, up to about 25 cm long and present yellow flowers. The calyx comprises 6 sepals, which are up to about 8 mm long. The corolla comprises 6 oblong petals which are elliptic and about 7 mm long. The androecium includes 6 stamens. The ovary develops a short style. The berries are purplish black, oblong, and about 7 mm long.

Medicinal uses: Dysentery (China, Nepal); diarrhea (Nepal)

*Moderate broad-spectrum antibacterial alkaloid fraction*: Alkaloid fraction of stem bark inhibited *Bacillus cereus*, *Enterococcus faecalis*, and *Shigella flexneri* with MIC equal to 256, 256, and 512 µg/mL (Pfoze et al., 2011).

*Antifungal (filamentous) polar extract:* Methanol extract of leaves at 80 ppm inhibited the growth of phytopathogenic *Colletotrichum capsici* (MTCC 2071), *Leptosphaerulina trifolii* (MTCC 2328), *Alternaria brassicicola* (MTCC 2102), and *Helminthosporium solani* (MTCC2075) by 70, 80, 83.3, and 83.3%, respectively (Bajpai & Vankar, 2007).

*Strong antiviral (enveloped monopartite linear double-stranded DNA) amphiphilic bisbenzylisoquinoline alkaloids:* Homoaromoline (LogD = 3.5 at pH 7.4; molecular weight = 608.7 g/mol) and isotetrandrine (LogD = 4.1 at pH 7.4; molecular mass = 622.7 g/mol) inhibited the replication of the Herpes simplex virus type-1 (strain 7401H) with $IC_{50}$ values of 15.1 and 17.3 µg/mL in Vero cells ($CC_{50}$ = 30 µg/mL). Homoaromoline inhibited the replication of the Herpes simplex virus type-1 (TK- B2006) and Herpes simplex virus type-2 (Ito-1262) with $IC_{50}$ values of 17.4 and 23.8 µg/mL in Vero cells (Nakamura et al., 1999).

Homoaromoline

Isotetrandrine

Commentary: The plant is used for dyeing textile (Bajpai & Vankar, 2007) on account of alkaloids such as the bisbenzylisoquinolines homoaromoline and isotetrandrine (Mai et al., 2009).

### References

Bajpai, D. and Vankar, P.S., 2007. Antifungal textile dyeing with mahonia napaulensis dc leaves extract based on its antifungal activity. *Fibers and Polymers, 8*(5), p. 487.

Mai, N.T., Tuan, T.A., Huong, H.T., Van Minh, C., Ban, N.K. and Van Kiem, P., 2009. Bisbenzylisoquinoline alkaloids from *Mahonia nepalensis. Vietnam Journal of Chemistry, 47*(3), p. 368.

Nakamura, N., Hattori, M., Kurokawa, M., Shiraki, K., Kashiwaba, N. and Ono, M., 1999. Anti-herpes simplex virus activity of alkaloids isolated from *Stephania cepharantha*. *Biological and Pharmaceutical Bulletin*, 22(3), pp. 268–274.

Pfoze, N.L., Kumar, Y., Myrboh, B., Bhagobaty, R.K. and Joshi, S.R., 2011. In vitro antibacterial activity of alkaloid extract from stem bark of *Mahonia manipurensis* Takeda. *Journal of Medicinal Plants Research*, 5(5), pp. 859–861.

### 4.1.1.9  *Nandina domestica* Thunb.

Common names: Heavenly bamboo, sacred bamboo; nan tian zhu, tein chok (China)

Habitat: Forests, roadsides, thickets

Distribution: India, China, and Japan

Botanical observation: This dense shrub grows to about 3 m tall. The plant is sacred in China, where it is used to decorate altars. The stems are reddish at apex. The leaves are alternate, bi- or tripinnate, up to about 30cm long, and exstipulate. The petioles are enlarged at base. The folioles are elliptic-lanceolate, 2–10×0.5–2.5 cm, green or reddish, cuneate at base, and acuminate at apex. The panicles are many flowered and axillary. The calyx includes numerous sepals, which are about 4mm long. The 6 petals are oblong, pure white, somewhat recurved, and about 4mm long. The androecium is crown-like and includes oblong-elliptic stamens, which are 3.5mm long. The ovary is ellipsoid and develops a short style. The berries are red, globose, about 7mm in diameter, and contain 1–3 seeds.

Medicinal use: Tracheitis (China)

*Weak antibacterial (Gram-positive) polar extract:* Ethanol extract inhibited *Staphylococcus aureus*, *Streptococcus faecalis*, and *Bacillus thuringensis* with MIC values of 2300, 4500, and 4500 µg/mL, respectively (Li et al., 2007). Aqueous extract of leaves inhibited the growth of *Streptococcus pyogenes* (CMCC(B)32175) with the MIC value of 3000 µg/mL via with condensation of DNA, cytoplasmic membrane insults, and leakage of cytoplasm (Guo et al., 2018).

*Moderate Antifungal (filamentous) essential oil:* Essential oil of flowers inhibited the growth of phytopathogenic *Fusarium oxysporum* (KACC 41083), *Fusarium solani* (KACC 41092), *Phytophthora capsici* (KACC 41078), *Colletotrichum capsici* (KACC 41078), *Sclerotina sclerotiorum* (KACC 41065), *Botrytis cinerea* (KACC 40573), and *Rhizoctoma solani* (KACC 40111) with MIC values ranging from 125 to 1000 µg/mL, respectively (Bajpai et al., 2009).

*Strong broad spectrum antibacterial amphiphilic protopine alkaloid:* Protopine (LogD=2.3 at pH 7.4; molecular weight=353.3 g/mol) inhibited the growth of *Escherichia coli*, *Pseudomonas aeruginosa*, *Proteus mirabilis*, *Klebsiella pneumoniae*, *Acinetobacter baumannii*, *Staphylococcus aureus*, and *Bacillus subtilis*, with MIC values of 32, 64, 32, 8, 8, 64, and 128µg/mL, respectively (Orhan et al., 2007).

Protopine

*Strong anticandidal amphiphilic protopine alkaloid:* Protopine inhibited the growth of *Candida albicans* with an MIC value of 4 μg/mL (Orhan et al., 2007).

*Viral enzyme inhibition by polar extract:* Methanol extract of leaves at 100 μg/mL inhibited Human immunodeficiency virus protease by 27% (Park et al., 2002). Aqueous extract of leaves at a concentration of 250 μg/mL inhibited Human immunodeficiency virus type-1 protease by 65.8% (Xu et al., 1996).

Commentaries: The plant produces the protoberberine nandinine, the aporphine alkaloid nandazurine, the protopine-type alkaloid protopine, berberine, and jatrorrhizine (Ikuta & Itokawa, 1988).

### References

Bajpai, V.K., Lee, T.J. and Kang, S.C., 2009. Chemical composition and in vitro control of agricultural plant pathogens by the essential oil and various extracts of *Nandina domestica* Thunb. *Journal of the Science of Food and Agriculture, 89*(1), pp. 109–116.

Guo, Z.Y., Zhang, Z.Y., Xiao, J.Q., Qin, J.H. and Zhao, W., 2018. Antibacterial effects of leaf extract of *Nandina domestica* and the underlined mechanism. *Evidence-Based Complementary and Alternative Medicine.*

Ikuta, A. and Itokawa, H., 1988. Alkaloids of tissue cultures of *Nandina domestica. Phytochemistry, 27*(7), pp. 2143–2145.

Li, A.R., Zhu, Y., Li, X.N. and Tian, X.J., 2007. Antimicrobial activity of four species of Berberidaceae. *Fitoterapia, 78*(5), pp. 379–381.

Orhan, I., Özçelik, B., Karaoğlu, T. and Şener, B., 2007. Antiviral and antimicrobial profiles of selected isoquinoline alkaloids from Fumaria and Corydalis species. *Zeitschrift für Naturforschung, 62c*, pp. 19–26.

Park, J.C., Hur, J.M., Park, J.G., Hatano, T., Yoshida, T., Miyashiro, H., Min, B.S. and Hattori, M., 2002. Inhibitory effects of Korean medicinal plants and camelliatannin H from *Camellia japonica* on human immunodeficiency virus type 1 protease. *Phytotherapy Research, 16*(5), pp. 422–426.

Slobodníková, L., KoSt'álová, D., Labudová, D., Kotulová, D. and Kettmann, V., 2004. Antimicrobial activity of *Mahonia aquifolium* crude extract and its major isolated alkaloids. *Phytotherapy Research, 18*(8), pp. 674–676.

Xu, H.X., Wan, M., Loh, B.N., Kon, O.L., Chow, P.W. and Sim, K.Y., 1996. Screening of traditional medicines for their inhibitory activity against HIV -1 protease. *Phytotherapy Research, 10*(3), pp. 207–210.

### 4.1.1.10 *Podophyllum hexandrum* Royle

Synonyms: *Podophyllum emodi* Wall. ex Hook. f. & Thomson; *Sinopodophyllum hexandrum* (Royle) T.S. Ying

Common names: Indian podophyllum; tao er qi (China); bankakri, giriparpata (India); langhupatra, shinmendo (Nepal)

Habitat: Forest margins, meadows, grassy and wet soils

Distribution: Afghanistan, Pakistan, Kashmir, India, Himalayas, Nepal, Bhutan, Tibet, and China

Botanical examination: This graceful herb grows up to about 80 cm in height from a rhizome with tuberous yellowish roots exuding some sort of acrid sap. The stems are angled, and develops 2 leaves, which are simple. At very first glance the plant has a somewhat Ranunculaceous look. The petiole is up to about 25 cm long. The blade is orbicular, cordate at base, 11–20×18–30 cm, serrate or not, 3–5-lobed, (apex of lobes acute or acuminate) glossy, margin entire or coarsely dentate. The inflorescence is solitary and cauliflorous or terminal. The calyx includes 3 sepals. The corolla presents 6 pure white or light pink petals, which are obovate or obovate-oblong, and

2.5–3.5×1.5–1.8 cm. The androecium includes 6 stamens, which are about 2 cm long with elongated anthers. The ovary is 1.2 cm, oblong, light green, smooth, and develops a minute style. The berries are red, pendulous, edible (Nepal) smooth, ovoid-globose, up to 7 mm×4 cm, fleshy, and contain numerous ovoid-triangular, 3 mm long, and brownish seeds.

Medicinal use: Cold, cough (Nepal); cuts, wounds (India)

*Strong broad-spectrum antibacterial polar extract:* Ethanol extract of leaves inhibited the growth of *Bacillus subtilis, Staphylococcus aureus, Escherichia coli,* and *Pseudomonas aeruginosa* with MIC values of 8, 64, 16, and 32 µg/mL, respectively (Kumar et al., 2010).

*Broad-spectrum antifungal aryltetralin lignans:* 4′-*O*-Demethyldehydropodophyllotoxin and picropodophyllone isolated from the leaves at a concentration of 200 µg/mL inhibited the growth of *Epidermophyton floccosum, Curvularia lunata, Nigrospora oryzae, Microsporium canis, Allescheria boydii,* and *Pleurotus ostreatus* (Ashraf et al., 1995).

*Strong antiviral (enveloped monopartite linear double-stranded DNA) amphiphilic aryltetralin lignan:* Podophyllotoxin (LogD = 1.9 at pH 7.4; molecular weight = 414.4 g/mol) abrogated the formation of plaques by Herpes simplex virus type-1 (strain SC9) in Vero cells at a concentration of $10^3$ nM via a mechanism involving early stages of virus replication (Hammonds et al., 1996).

Podophyllotoxin

Commentaries: (i) One could suggest that members of the genus *Podophyllum* L. which abound with lignans would be better at home in the family Podophyllaceae rather than in the Berberidaceae. (ii) The roots contain the aryltetralin lignans, mainly podophyllotoxin and 4′-*O*-demethylpodophyllotoxin (Jackson & Dewick, 1984; Giri & Narasu, 2000). Podophyllotoxin is a cytotoxic agent that inhibits tubulin incorporation into microtubules in mammalian cells, and at high concentrations interacts directly with the microtubule (Hammonds et al., 1996). Consider that microtubules may act as an "anchor" for Herpes simplex virus type-1 viral capsid proteins to allow processes such as transport of naked Herpesvirus to the nucleus, mRNA transcription, DNA replication, and DNA packaging prior to virion release (Hammonds et al., 1996). Podophyllotoxin has antiviral activity at early stages of replication against the DNA virus murine Cytomegalovirus but not against the RNA Sindbis virus (Hammonds et al., 1996). Hence, natural products inhibiting microtubule machinery (including colchicine and vincristine) have potential as antiviral agents. The infection of host cells by coronaviruses involves the interaction of the spike protein with host cell cytoskeletal proteins, and microtubules are involved in the transport and assembly of spike proteins into virions during the replication (Schlesinger et al., 2020). Hence natural products that disrupt microtubule machinery (including colchicine or podophyllotoxin) need to be examined for their clinical values for the treatment of COVID-19 and other zoonotic viruses that will follow. Furthermore, natural products interfering with the microtubule system of white blood cells are theoretically able to mitigate pulmonary inflammatory cascades during respiratory virus infection.

### References

Ashraf, M., Choudhary, M.I. and Kazmi, M.H., 1995. Antifungal aryltetralin lignans from leaves of *Podophyllum hexandrum*. *Phytochemistry*, *40*(2), pp. 427–431.

Giri, A. and Narasu, M.L., 2000. Production of podophyllotoxin from *Podophyllum hexandrum*: a potential natural product for clinically useful anticancer drugs. *Cytotechnology*, *34*(1–2), pp. 17–26.

Hammonds, T.R., Denyer, S.P., Jackson, D.E. and Irving, W.L., 1996. Studies to show that with podophyllotoxin the early replicative stages of herpes simplex virus type 1 depend upon functional cytoplasmic microtubules. *Journal of Medical Microbiology*, *45*(3), pp. 167–172.

Jackson, D.E. and Dewick, P.M., 1984. Aryltetralin lignans from *Podophyllum hexandrum* and *Podophyllum peltatum*. *Phytochemistry*, *23*(5), pp. 1147–1152.

Kumar, R., Badere, R. and Singh, S.B., 2010. Antibacterial and antioxidant activities of ethanol extracts from trans Himalayan medicinal plants. *Pharmacognosy Journal*, *2*(17), pp. 66–69.

Schlesinger, N., Firestein, B.L. and Brunetti, L., 2020. Colchicine in COVID-19: an old drug, new use. *Current Pharmacology Reports*, *6*(4), pp. 137–145.

### 4.1.2   FAMILY LARDIZABALACEAE R. BROWN (1821)

The family Lardizabalaceae consists of 8 genera and about 30 species of twining woody climbers or shrubs. The leaves are alternate, digitately compound or rarely pinnate. The flowers are racemose, unisexual, and develop with the leaves from perulate buds. The calyx consists of 3 or 6 imbricate sepals. The corolla consists of 6 petals, which are smaller than the sepals. The androecium is made of 6 stamens, which are free or connate. The gynoecium consists of 3 carpels, each containing 1 to numerous ovules. The fruits are succulent, colored, and, indehiscent ripe carpels. Members of this family are known so far to produce lignans, triterpene saponins, and phenylpropanoids.

#### 4.1.2.1   *Akebia quinata* (Houtt.) Decne.

Synonyms: *Akebia micrantha* Nakai; *Rajania quinata* Houtt.

Common names: Chocolate vine, five-leaf akebia; mu tong (China); moku-tsu (Japan)

Habitat: Forest margin and mountain thickets

Distribution: China, Korea, and Japan

Botanical examination: This graceful climber has slender, twining, cylindric, and woody stems. The leaves are palmately compound, spiral, and exstipulate. The petiole is slender, and up to 10 cm long and the petiolules can reach up to 2 cm long. The folioles are 5, obovate-elliptic, glaucous below, acute at base, rounded and emarginate at apex, and 2–5 × 1.5–2.5 cm. The racemes are axillary and pendulous and grow to about 12 cm long. The calyx includes 3 sepals which are light to dark purple or green to white, hooded, broadly ovoid, 6 mm–2 cm × 4 mm–1.5 cm, and rounded at apex. The androecium includes 6 subsessile and purplish stamens with elongated and curved anthers. The gynoecium includes 3–6 free, cylindrical, and purplish carpels moist at apex. The follicle is fleshy (annonaceous-like at first glance) purplish, oblong, slightly curved, 5–8 × 3–4 cm, dehiscent, and shelters numerous seeds embedded in a white pulp.

Medicinal uses: Fever, colds, wounds (China)

*Viral enzyme inhibition by polar extract:* Methanol extract of aerial parts at a concentration of 100 μg/mL inhibited RNA-dependent DNA polymerase activity of Human immunodeficiency virus type-1 reverse transcriptase and Human immunodeficiency virus type-1 protease by 23.2 and 20.1%, respectively (Min et al., 2001).

Commentary: The plant has apparently not been examined for its potential antimicrobial effects. Fruits abounds with noroleanane-type triterpene saponins (Iwanaga et al., 2012; Wang et al., 2014; Xu et al., 2016). Consider that oleanane aglycones often inhibit the enzymatic activity of Human immunodeficiency virus reverse transcriptase, oleanolic acid, β-amyrin, and erythrodiol inhibited the enzymatic activity of Human immunodeficiency virus reverse transcriptase with an $IC_{50}$ value of 3.1, 4.7, and 5 µM, respectively (Akihisa et al., 2001).

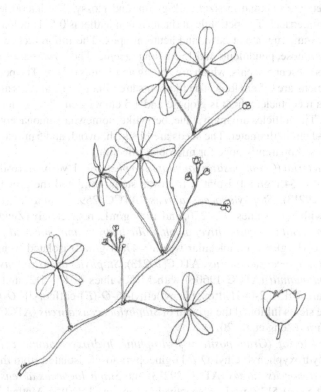

*Akebia quinata* (Houtt.) Decne.

### References

Akihisa, T., Ogihara, J., Kato, J., Yasukawa, K., Ukiya, M., Yamanouchi, S. and Oishi, K., 2001. Inhibitory effects of triterpenoids and sterols on human immunodeficiency virus-1 reverse transcriptase. *Lipids*, *36*(5), pp. 507–512.

Iwanaga, S., Warashina, T. and Miyase, T., 2012. Triterpene saponins from the pericarps of *Akebia trifoliata*. *Chemical and Pharmaceutical Bulletin*, *60*(10), pp. 1264–1274.

Min, B.S., Kim, Y.H., Tomiyama, M., Nakamura, N., Miyashiro, H., Otake, T. and Hattori, M., 2001. Inhibitory effects of Korean plants on HIV -1 activities. *Phytotherapy Research*, *15*(6), pp. 481–486.

Wang, J., Xu, Q.L., Zheng, M.F., Ren, H., Lei, T., Wu, P., Zhou, Z.Y., Wei, X.Y. and Tan, J.W., 2014. Bioactive 30-noroleanane triterpenes from the pericarps of *Akebia trifoliata*. *Molecules*, *19*(4), pp. 4301–4312.

Xu, Q.L., Wang, J., Dong, L.M., Zhang, Q., Luo, B., Jia, Y.X., Wang, H.F. and Tan, J.W., 2016. Two new pentacyclic triterpene saponins from the leaves of *Akebia trifoliata*. *Molecules*, *21*(7), p. 962.

### 4.1.2.2  *Sargentodoxa cuneata* (Oliv.) Rehder & E.H. Wilson

Synonym: *Holboellia cuneata* Oliv. *Sargentodoxa simplicifolia* S. Z. Qu & C. L. Min
   Common names: Daxue teng (China)
   Habitat: Forests, cultivated
   Distribution: Vietnam, Laos, and China
   Botanical examination: This handsome woody climber grows up to about 6 m long. The bark is fissured longitudinally. The stems are red brown, terete, exude a red sap, and present showy medullary rays in cross section. The leaves are 3-foliolate. The petiole is 5–10 cm long, slender, and channeled. The leaflets are unequal in size, dark green, and glossy. The lateral leaflets are subsessile, deltoid, and asymmetrical. The petiolule of the terminal leaflet is 0.5–1 cm long. The blade is rhomboid, 5–15 cm×3 cm, cuneate at base, and acute at apex. The inflorescence is a solitary, densely flowered, loose, globose, pendulous, and 15 cm long raceme. The flowers are fragrant and yellowish. The male flowers present 5 sepals, which are oblong and 1 cm×0.5 cm. The petals are rhomboid and minute. The stamens are 0.5 cm long. The filaments are minute. The anthers are oblong. The female flower presents a receptacle, which is globose and 1.5 cm×1.2 cm. The gynoecium present numerous free carpels. The follicles are many, blue, berrylike, somewhat annonaceous-like at first glance, stipitate and 0.7–1 cm in diameter. The seeds are blackish, ovoid, and 5 mm long.
   Medicinal uses: Apparently none for microbial infections
   *Strong antibacterial (Gram-positive) amphiphilic phenolic:* Hydroxytyrosol (LogP=0.3 at pH 4; molecular weight=154.1 g/mol) isolated from the stems inhibited the growth of *Staphylococcus aureus* (ATCC 29213), *Staphylococcus aureus* (ATCC 25923), and *Acinetobacter baumannii* (ATCC 19606) with MIC values of 2, 256, and 128 µg/mL, respectively (Zeng et al., 2015).
   *Moderate antibacterial (Gram-positive) amphiphilic hydroxycinnamic acid glycoside derivatives:* Calceolarioside B (LogP=1.0; molecular weight=478.4 g/mol) isolated from the stems inhibited the growth of *Staphylococcus aureus* (ATCC 29213), *Staphylococcus aureus* (ATCC 25923), and *Acinetobacter baumannii* (ATCC 19606) with MIC values of 64, 512, and 512 µg/mL, respectively (Zeng et al., 2015). 2-(4-Hydroxyphenyl)ethyl-[6-*O*-(*E*)-caffeoryl]- *O*-β-D-glucopyranoside isolated from the stems inhibited the growth of *Staphylococcus aureus* (ATCC 29213) with an MIC value of 128 µg/mL (Zeng et al., 2015).
   *Strong antibacterial (Gram-positive) hydrophilic hydroxycinnamic acid glycoside derivatives:* 2-(3,4-Dihydroxyphenyl) ethyl-*O*-β-D-glucopyranoside isolated from the stems inhibited the growth of *Staphylococcus aureus* (ATCC 29213) and *Staphylococcus aureus* (ATCC 25923) with an MIC value of 8 and 512 µg/mL, respectively (Zeng et al., 2015). Methyl 3-*O*- caffeoylquinate (LogP=−0.1) isolated from the stems inhibited the growth of *Staphylococcus aureus* (ATCC 29213) with an MIC value of 32 µg/mL (Zeng et al., 2015).
   *Moderate antibacterial (Gram-positive) amphiphilic flavanols:* Catechin, epicatechin (LogD=0.5; molecular weight=290.2 g/mol), and dulcisflavan isolated from the stems inhibited the growth of *Staphylococcus aureus* (ATCC 29213) with an MIC value of 256 µg/mL (Zeng et al., 2015).
   *Strong antiviral (enveloped monopartite linear dimeric single-stranded (+) RNA) amphiphilic phenylpropanoid:* Hydroxytyrosol (LogD=0.3 at pH 4; molecular weight=154.1 g/mol) inhibited the formation of syncytium and p24 production by Human immunodeficiency virus type-1 (strain IIIB) in H9 cells with an $IC_{50}$ value of 61 and 68 nM, respectively (CC50> 10,000) (Lee-Huang et al., 2007).

Hydroxytyrosol

*Antiviral (enveloped, segmented, single-stranded (-)RNA) phenolic:* Hydroxytyrosol at a concentration of 10 μg/mL decreased Avian Influenza viruses A/chicken/Yokohama/aq55/01 (H9N2) titer in Madin-Darby Canine Kidney cells from about 5.5 to below 1.2 $Log_{10}TCID_{50}$/mL (Yamada et al., 2009).

*Antiviral (enveloped monopartite single-stranded (-) RNA) hydrophilic phenolic:* Hydroxytyrosol at a concentration of 100 μg/mL decreased Newcastle disease virus (Ibaraki/85 strain) titer in Madin-Darby Canine Kidney cells from about 6.5 to below 1.2 $Log_{10}TCID_{50}$/mL (Yamada et al., 2009).

*Viral enzyme inhibition by hydroxycinnamic acid derivative:* Calceolarioside B exhibited binding affinity on Human immunodeficiency virus gp41 with $IC_{50}$ values of 100 μg/mL (Kim et al., 2002).

Commentary: The plant abounds with antiviral phenolics and hydroxycinnamic acid glycosides, which are inviting *in vitro* and *in vivo* examinations for possible anti-COVID-19 (and other pathogenic viruses). Consider that phenolics and hydroxycinnamic acid glycosides are most often antiviral.

### References

Kim, H.J., Yu, Y.G., Park, H. and Lee, Y.S., 2002. HIV gp41 binding phenolic components from *Fraxinus sieboldiana* var. angustata. *Planta Medica*, 68(11), pp. 1034–1036.

Lee-Huang, S., Huang, P.L., Zhang, D., Lee, J.W., Bao, J., Sun, Y., Chang, Y.T., Zhang, J. and Huang, P.L., 2007. Discovery of small-molecule HIV fusion and integrase inhibitors oleuropein and hydroxytyrosol: part I. Integrase inhibition. *Biochemical and Biophysical Research Communications*, 354(4), pp. 872–878.

Yamada, K., Ogawa, H., Hara, A., Yoshida, Y., Yonezawa, Y., Karibe, K., Nghia, V.B., Yoshimura, H., Yamamoto, Y., Yamada, M. and Nakamura, K., 2009. Mechanism of the antiviral effect of hydroxytyrosol on influenza virus appears to involve morphological change of the virus. *Antiviral Research*, 83(1), pp. 35–44.

Zeng, X., Wang, H., Gong, Z., Huang, J., Pei, W., Wang, X., Zhang, J. and Tang, X., 2015. Antimicrobial and cytotoxic phenolics and phenolic glycosides from *Sargentodoxa cuneata. Fitoterapia*, 101, pp. 153–161.

## 4.1.3 FAMILY PAPAVERACEAE A.L. de Jussieu (1789)

The family Papaveraceae comprises about 40 genera and 800 species (including the Fumariaceae) of often laticiferous or sappy herbs, which are often graceful, fleshy, and somewhat light glaucous. The stems are terete. The leaves are spiral, often in rosette, simple, exstipulate. The inflorescences are racemes, panicles, cymes, or solitary flowers. The flowers are often showy. The calyx comprises 2–4 sepals, which are caducous and enclose the corolla in flower buds. The corolla comprises up to 4–6 membranous petals. The androecium is often showy and includes numerous free stamens. The gynoecium consists of 2 to many fused carpels forming an ovary with parietal placentation. The fruits are follicles, achenes, capsules, or berries containing minute seeds. Papaveraceae produce antimicrobial isoquinoline alkaloids.

### 4.1.3.1 *Argemone mexicana* L.

Synonyms: *Argemone leiocarpa* Greene; *Argemone mucronata* Dum. Cours. ex Steud.; *Argemone ochroleuca* Sweet; *Argemone sexvalis* Stokes; *Argemone spinosa* Moench; *Argemone versicolor* Salisb.; *Argemone vulgaris* Spach; *Echtrus mexicanus* (L.) Nieuwl.; *Echtrus mexicanus* Nieuwl.; *Echtrus trivialis* Lour.

Common names: Mexican prickly poppy; shial kanta (Bangladesh, India); piramathandu (India); khaya (Myanmar); ji ying su (China); druju (Indonesia); celang keringan (Malaysia); kasumbang aso (the Philippines); fin naam (Thailand); gai cua (Vietnam)

Habitat: Dry, sunny roadsides, waste places, desolate lands, and disturbed areas

Distribution: Turkey, Iran, Afghanistan, Turkmenistan, Tajikistan, Pakistan, India, Bangladesh, Bhutan, Cambodia, Thailand, Vietnam, China, Indonesia, Nepal, the Philippines, and Taiwan

Botanical observation: This herb grows to about 1 m tall. The stem is terete, somewhat glaucous, and fleshy. The leaves are simple, spiral, sessile, and exstipulate. The blade is somewhat glaucous, pinnatifid, spiny, and 5–20 cm×2–8 cm. The flowers are solitary, terminal, and showy. The calyx comprises 3, 1.2 cm long and spiny sepals. The corolla comprises 4–6 obovate, 2.5–3.5 cm×2–2.5 cm, bright yellow, and membranous petals. The androecium includes numerous slender stamens, which are up to about 1.2 cm long. The gynoecium consists of numerous carpels fused into an ovate ovary, which is 0.8–1 cm long, spiny, and 3–6-lobed. The capsules are spiny, dehiscent, oblong, 2.5–4 cm×1.2–2 cm, 3–6-lobed, and contain numerous minute seeds.

Medicinal uses: Ulcers (Indonesia, India, Bangladesh); fungal infection (Bangladesh); skin diseases (India; Myanmar)

*Strong antibacterial polar extract:* Methanol extract of seeds (50 μL of 2000 μg/mL extract, per agar well) inhibited the growth of *Pseudomonas aeruginosa*, *Staphylococcus aureus*, *Escherichia coli*, and *Bacillus cereus* with inhibition zone diameters of 22, 18, 21, and 20 mm, respectively (Bhattacharjee et al., 2006). Aqueous extract of leaves inhibited the growth of *Klebsiella oxytoca*, *Vibrio damsella*, *Enterobacter aerogenes*, and *Escherichia coli* with the MIC values of 10, 10, 0.5, and 10 μg/mL (Vivek & Bhat, 2010).

*Moderate antibacterial mid-polar extract:* Chloroform fraction (containing N-demethyloxysanguinarine) of seeds inhibited the growth of *Staphylococcus aureus*, *Escherichia coli*, *Klebsiella pneumonia*, and *Pseudomonas aeruginosa* (Bhattacharjee et al., 2010).

*Antifungal polar extract:* Methanol extract of stems inhibited the growth of *Mucor indicus*, *Aspergillus flavus*, *Aspergillus niger*, and *Penicillum notatum* (More & Kharat, 2016).

*Strong antiviral (enveloped monopartite linear dimeric single-stranded (+) RNA) amphiphilic protopine-type alkaloid:* Protopine (LogD=2.3 at pH 7.4; molecular weight=353.3 g/mol) from the leaves inhibited the replication of the Human immunodeficiency virus (strain NL4.3) virus in Human CD4+ T cells by 39.3% with a maximum non-cytotoxic concentration of 5 μg/mL (Sabde et al., 2011).

*Strong antiviral (enveloped monopartite linear dimeric single-stranded (+) RNA) hydrophilic protoberberine:* Berberine (LogD=−1.3 at pH 7.4; molecular weight=336.4 g/mol) inhibited the replication of the Human immunodeficiency virus type-1 (strain NL4.3) in Human CD4+ T cells by 66.6% with a maximum noncytotoxic concentration of 0.1 μg/mL (Sabde et al., 2011; Tomar et al., 2015).

Commentary: The plant produces isoquinoline alkaloids such as dehydrocorydalmine, jatrorrhizine, and berberine (Singh et al., 2010), which are antibacterial and antifungal and account, at least partially, for the medicinal uses. The plant is also poisonous (Sanghvi et al., 1960; Tomar et al., 2015).

### References

Bhattacharjee, I., Chatterjee, S.K. and Chandra, G., 2010. Isolation and identification of antibacterial components in seed extracts of *Argemone mexicana* L. (Papaveraceae). *Asian Pacific Journal of Tropical Medicine*, 3(7), pp. 547–551.

Bhattacharjee, I., Chatterjee, S.K., Chatterjee, S. and Chandra, G., 2006. Antibacterial potentiality of *Argemone mexicana* solvent extracts against some pathogenic bacteria. *Memórias do Instituto Oswaldo Cruz*, 101(6), pp. 645–648.

More, N. and Kharat, A., 2016. Antifungal and anticancer potential of *Argemone mexicana* L. *Medicines*, 3(4), p. 28.

More, N.V., Kharat, K.R. and Kharat, A.S., 2017. Berberine from *Argemone mexicana* L exhibits broadspectrum antibacterial activity. *Acta Biochimica Polonica*, 64(4), 653–660.

Sabde, S., Bodiwala, H.S., Karmase, A., Deshpande, P.J., Kaur, A., Ahmed, N., Chauthe, S.K., Brahmbhatt, K.G., Phadke, R.U., Mitra, D. and Bhutani, K.K., 2011. Anti-HIV activity of Indian medicinal plants. *Journal of Natural Medicines*, 65(3–4), pp. 662–669.

Sanghvi, L.M., Misra, S.N. and Bose, T.K., 1960. Cardiovascular manifestations in *Argemone mexicana* poisoning (epidemic dropsy). *Circulation*, 21(6), pp. 1096–1106.

Singh, S., Singh, T.D., Singh, V.P. and Pandey, V.B., 2010. Quaternary alkaloids of *Argemone mexicana*. *Pharmaceutical Biology*, 48(2), pp. 158–160.

Tomar, L.R., Raizada, A., Yadav, A. and Agarwal, S., 2015. Epidemic dropsy 2013: case series. *Tropical Doctor*, 45(2), pp. 137–139.

Vivek, K. and Bhat, S.K., 2010. Bacteriostatic potential of *Argemone mexicana* Linn. against enteropathogenic bacteria. *Indian Journal of Natural Products and Resources*, 1(3), pp. 338–341.

### 4.1.3.2   *Chelidonium majus* L.

Synonym: *Chelidonium sinense* DC

Common names: Greater celandine; bai qu cai (China); kusano o (Japan); kirlangiç out, sarılık otu (Turkey)

Habitat: Forests margins, grasslands, roadsides, riverbanks, and rocky slopes

Distribution: Turkey, Iran, Kazakhstan, China, Japan, and Korea

Botanical observation: This gracious herb grows to about 60 cm tall from fibrous roots. The stems are terete, somewhat hairy, and yield a yellow acrid latex upon incision. The leaves are mostly basal, pinnatisect, and exstipulate. The petiole is 2–5 cm long. The blade is up to 20 cm long, with 2–4 pairs of lobes, which are oblong, crenate, and 2–10 × 1–5 cm. The inflorescences are umbels on 2–8 cm long pedicles. The calyx comprises a pair of ovoid, 5–8 mm long, and caducous sepals. The 4 petals are obovate, about 1 cm long membranous, and yellow. The numerous stamens are about 8 mm long. The ovary is linear, about 8 mm long, glabrous, 1-loculed, and consists of 2 fused carpels and 2-lobed stigmas. The capsules are terete, up to about 5 cm long, dehiscent, and contain numerous minute glossy seeds.

Medicinal use: Hepatitis (Turkey)

*Strong broad-spectrum antibacterial benzophenanthridine alkaloid:* 8-Hydroxydihydrosanguinarine (molecular weight = 349.3 g/mol) inhibited the growth is 20 methicillin-resistant *Staphylococcus aureus* clinical isolates with MIC values ranging from 0.4 to 7.8 μg/mL and MBC values ranging from 1.9 to 31.2 μg/mL (Zuo et al., 2009). 8-Hydroxydihydrosanguinarine inhibited the growth of extended-spectrum β-lactamase (ESBL)-producing strains of *Escherichia coli* with MIC values ranging from 15.6 to 250 μg/mL and MBC values ranging from 62.5 to 500 μg/mL (Zuo et al., 2011). 8-Hydroxydihydrosanguinarine inhibited the growth of *Escherichia coli* with the MIC values of 15.6 μg/mL and the MBC values of 31.2 μg/mL (Zuo et al., 2011). 8-Hydroxydihydrosanguinarine inhibited the growth of extended-spectrum β-lactamase (ESBL)-producing strains of *Klebsiella pneumoniae* with MIC ranging from 93.8 to 375 μg/mL and MBC ranging from 375 to 1500 μg/mL (Zuo et al., 2011).

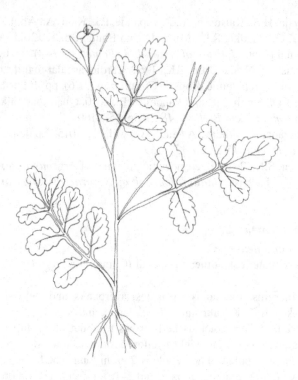

*Chelidonium majus* L.

8-Hydroxydihydrosanguinarine

*Strong antibacterial (Gram-positive) benzophenanthridine alkaloid:* 8-Hydroxydihydro-chelerythrine (molecular weight = 365.4 g/mol) inhibited the growth of 20 methicillin-resistant *Staphylococcus aureus* clinical isolates with MIC ranging from 0.9 to 15.6 μg/mL and MBC ranging from 7.81 to 62.5 μg/mL (Zuo et al., 2009). 8-Hydroxydihydrochelerythrine inhibited the growth of *Escherichia coli* with the MIC value of 125 μg/mL and MBC of 250 μg/mL (Zuo et al., 2011). 8-Hydroxydihydrochelerythrine inhibited the growth of extended-spectrum β-lactamase (ESBL)-producing strains of *Escherichia coli* with MIC ranging from 62.5 to 500 μg/mL and MBC ranging from 125 to 1000 μg/mL (Zuo et al., 2018). 8-Hydroxydihydrochelerythrine inhibited the growth of extended-spectrum β-lactamase (ESBL)-producing strains of *Klebsiella pneumoniae* with MIC ranging from 187.5 to 750 μg/mL and MBC ranging from 375 to 1500 μg/mL (Zuo et al., 2011).

8-Hydroxydihydrochelerythrine

*Strong broad-spectrum antibacterial amphiphilic benzophenanthridine alkaloid:* Dihydrochelerythrine (LogD=4.3 at pH 7.4; molecular weight=349.4 g/mol) inhibited methicillin-resistant *Staphylococcus aureus* (SK1), *Escherichia coli* (TISTR 780), and *Salmonella typhimurium* (TISTR 292) with MIC values of 8, 16, and 128 μg/mL (Tantapakul et al., 2012). Dihydrochelerythrine inhibited the growth of *Staphylococcus epidermidis* (ATCC 12228), *Staphylococcus aureus* (ATCC 6538), *Streptococcus pyogenes* (ATCC 19615), *Bacillus subtilis* (ATCC 6633), *Klebsiella pneumoniae* (ATCC 13883), and *Escherichia coli* (ATCC 25922) with MIC/MBC values of 6.2/12.5, 12.5/50, 12.5/50, 25/50, 12.5/25, and 25/25 μg/mL, respectively (Tavares et al., 2014).

Dihydrochelerythrine

*Moderate anticandidal benzophenanthridine*: 8-Hydroxydihydrosanguinarine inhibited clinical strains of *Candida parapsilosis, Candida tropicalis, Candida krusei, Candida glabrata*, and *Candida neoformans* with MIC/MFC values of 40/80, 80/160, 50/100, 80/160, and 15/30 μg/mL, respectively (Meng et al., 2009).

*Strong antiviral (enveloped monopartite linear double-stranded DNA) lipophilic benzophenanthridine alkaloid:* Chelidimerine (LogD=6.9 at pH 7.4; molecular weight=720.7 g/mol) inhibited Herpes simplex virus replication in Madin-Darby Canine Kidney cells with maximum nontoxic concentration of 64 μg/mL (Orhan et al., 2007).

Chelidimerine

*Strong antiviral (enveloped monopartite linear single-stranded (-) RNA) benzophenanthridine alkaloid:* Chelidimerine inhibited Parainfluenza-3 virus in Vero cells with maximum nontoxic concentration of 32 μg/mL (Orhan et al., 2007).

*In vivo antiviral (enveloped monopartite linear dimeric single-stranded (+) RNA) polar extract in vivo:* Aqueous fraction extracted from the aerial parts prevented Human immunodeficiency virus type-1 infection and virus-induced syncytium formation in human CD4+ cells (5 μg/well) (Gerenčer et al., 2006). Pre-treatment of $2.5 \times 10^5$ TCID$_{50}$ of Human immunodeficiency virus type-1 (strain IIIB) with 25 μg of the fraction prevented infection of cells as determined by a decrease in reverse transcriptase activity and p24 levels in the cell culture (Gerenčer et al., 2006). The fraction given intraperitoneally at a dose of 1 mg five times a week for 2 weeks starting 4 weeks following infection to C57Bl/6 mouse infected by intraperitoneal injection of the stock virus pool of defective murine leukemia retroviruses (<uLVs) LP-BM5 normalized the weight of spleen and cervical lymph nodes (Gerenčer et al., 2006).

*Viral enzyme inhibition by polar extract:* Methanol extract of aerial parts at a concentration of 100 μg/mL inhibited Human immunodeficiency virus type-1 protease by 19.9%, respectively (Min et al., 2001).

Commentaries: (i) Extended-spectrum β-lactamase (ESBL) are responsible for the resistance of *Klebsiella pneumoniae* and *Escherichia coli* to broad-spectrum third-generation cephalosporins (cefotaxim) and monobactams (aztreonam) (Mandell et al., 2009). Benzophenanthridines derive from protopine-type alkaloids, which derive from tetrahydroprotoberberines, which themselves derive from benzylisoquinolines (Phillips & Castle, 1981). These alkaloids are planar and intercalate into bacterial or fungal (also mammalian) cells and block topoisomerase (Makhey et al., 2003), therefore eliciting antibacterial, antifungal, or antiviral (also cytotoxic) activities.

### References

Gerenčer, M., Turecek, P.L., Kistner, O., Mitterer, A., Savidis-Dacho, H. and Barrett, N.P., 2006. In vitro and in vivo anti-retroviral activity of the substance purified from the aqueous extract of *Chelidonium majus* L. *Antiviral Research*, 72(2), pp. 153–156.

Makhey, D., Li, D., Zhao, B., Sim, S.P., Li, T.K., Liu, A., Liu, L.F. and LaVoie, E.J., 2003. Substituted benzo [i] phenanthridines as mammalian topoisomerase-targeting agents. *Bioorganic & Medicinal Chemistry*, 11(8), pp. 1809–1820.

Mandell, G., Dolin, R. and Bennett, J., 2009. *Mandell, Douglas, and Bennett's Principles and Practice of Infectious Diseases* (No. Ed. 7). Elsevier, USA.

Meng, F., Zuo, G., Hao, X., Wang, G., Xiao, H., Zhang, J. and Xu, G., 2009. Antifungal activity of the benzo [c] phenanthridine alkaloids from *Chelidonium majus* Linn against resistant clinical yeast isolates. *Journal of Ethnopharmacology*, 125(3), pp. 494–496.

Min, B.S., Kim, Y.H., Tomiyama, M., Nakamura, N., Miyashiro, H., Otake, T. and Hattori, M., 2001. Inhibitory effects of Korean plants on HIV-1 activities. *Phytotherapy Research*, 15(6), pp. 481–486.

Orhan, I., Ozcelik, B., Karaoğlu, T. and Şener, B., 2007. Antiviral and antimicrobial profiles of selected isoquinoline alkaloids from Fumaria and Corydalis species. *Zeitschrift für Naturforschung C*, 62(1–2), pp. 19–26.

Phillips, S.D. and Castle, R.N., 1981. A review of the chemistry of the antitumor benzo [c] phenanthridine alkaloids nitidine and fagaronine and of the related antitumor alkaloid coralyne. *Journal of Heterocyclic Chemistry*, 18(2), pp. 223–232.

Tantapakul, C., Phakhodee, W., Ritthiwigrom, T., Yossathera, K., Deachathai, S. and Laphookhieo, S., 2012. Antibacterial compounds from *Zanthoxylum rhetsa*. *Archives of Pharmacal Research*, 35(7), pp. 1139–1142.

Tavares, L.D.C., Zanon, G., Weber, A.D., Neto, A.T., Mostardeiro, C.P., Da Cruz, I.B., Oliveira, R.M., Ilha, V., Dalcol, I.I. and Morel, A.F., 2014. Structure-activity relationship of benzophenanthridine alkaloids from *Zanthoxylum rhoifolium* having antimicrobial activity. *PLoS One*, 9(5), p. e97000.

Zuo, G.Y., Meng, F.Y., Han, J., Hao, X.Y., Wang, G.C., Zhang, Y.L. and Zhang, Q., 2011. In vitro activity of plant extracts and alkaloids against clinical isolates of extended-spectrum b-lactamase (ESBL)-producing strains. *Molecules*, 16(7), pp. 5453–5459.

Zuo, G.Y., Meng, F.Y., Hao, X.Y., Zhang, Y.L., Wang, G.C. and Xu, G.L., 2009. Antibacterial alkaloids from *Chelidonium majus* Linn (Papaveraceae) against clinical isolates of methicillin-resistant *Staphylococcus aureus*. *Journal of Pharmacy & Pharmaceutical Sciences*, 11(4), pp. 90–94.

### 4.1.3.3 *Corydalis bulbosa* DC.

Synonyms: *Capnoides solida* (L.) Moench; *Corydalis bulbosa* var. *solida* L.; *Corydalis halleri* Willd.; *Corydalis solida* (L.) Clairv.; *Fumaria bulbosa* var. *cava* L.; *Fumaria solida* L.; *Pistolochia bulbosa* Soják; *Pistolochia solida* Bernh.

Common names: Solid-rooted Corydalis, bird-in-a-bush

Habitat: Old walls, rocky soil

Distribution: Siberia, Mongolia, and China

Botanical observation: This herb grows to about 25 cm tall from a short and globose tuber. The stems are terete and slender. The leaves are simple, spiral, and exstipulate. The blade is bi- or triternate. The racemes are terminal and support 6–20 flowers. The 2 sepals are lanceolate and minute. The corolla is up to about 2.2–3 cm long, purplish, with an elongated and curved spur. The androecium includes 2 trifid stamens. The ovary develops a filiform style. The capsule is linear about 1.5 cm long, and contain minute black and glossy and papillose seeds.

Medicinal use: Boils (China)

*Strong broad-spectrum antibacterial amphiphilic phthalide isoquinoline-type alkaloid:* Bicuculline (LogD = 2.4; molecular weight = 367.3 g/mol) inhibited the growth of *Escherichia coli*, *Pseudomonas aeruginosa*, *Proteus mirabilis*, *Klebsiella pneumonia*, *Acinetobacter baumannii*, *Staphylococcus aureus* and *Bacillus subtilis* with the MIC values of 32, 32, 32, 32, 32, 64, and 64 μg/mL respectively (Orhan et al., 2007).

Bicuculline

*Strong broad-spectrum antibacterial amphiphilic aporphine alkaloids:* Isoboldine (LogD=2.4 at pH 7.4; molecular weight=327.4 g/mol) inhibited the growth of *Escherichia coli, Pseudomonas aeruginosa, Proteus mirabilis, Klebsiella pneumonia, Acinetobacter baumannii, Staphylococcus aureus,* and *Bacillus subtilis* with the MIC values of 32, 32, 32, 32, 32, 64, and 64 µg/mL, respectively (Orhan et al., 2007). Bulbocapnine (Log P=3.1 at pH 7.4; molecular weight=325.4 g/mol) inhibited the growth of *Escherichia coli, Pseudomonas aeruginosa, Proteus mirabilis, Klebsiella pneumonia, Acinetobacter baumannii, Staphylococcus aureus,* and *Bacillus subtilis* with the MIC values of 32, 64, 32, 8, 8, 64, and 128 µg/mL, respectively (Orhan et al., 2007).

Bulbocapnine

*Strong antiviral (enveloped monopartite linear double-stranded DNA) aporphine alkaloid:* Bulbocapnine inhibited Herpes simplex virus replication in Madin-Darby canine kidney cells with a maximum nontoxic concentration of 64 µg/mL (Orhan et al., 2007).

　　*Moderate antiviral (non-enveloped single-stranded linear (+) RNA) aporphine alkaloid:* Isoboldine inhibited cytopathic effects in Vero cells of the Poliomyelitis virus type 2 (vaccinal strain Sabin II) with an $IC_{50}$ value of 15 µM and a selectivity index of 14.5 (Boustie et al., 1998).

　　*Strong antiviral (enveloped linear double-stranded DNA) aporphine alkaloid:* Bulbocapnine inhibited the Herpes simplex virus replication in Madin-Darby canine kidney cells with a maximum nontoxic concentration of 64 µg/mL (Orhan et al., 2007).

　　*Strong antiviral (enveloped monopartite linear single-stranded (-)RNA) aporphine alkaloid:* Bulbocapnine inhibited Parainfluenza-3 virus in Vero cells with a maximum nontoxic concentration of 32 µg/mL (Orhan et al., 2007).

Commentaries: (i) The plant produces the protoberberine-type alkaloids stylopine and bicuculline, the aporphine-type alkaloids bulbocapnine, protopine, isoboldine, glaucine and thaliporphine (Kiryakov et al., 1981), which are antibacterial and antifungal (Udvardy et al., 2014). (ii) Bicuculline is a nightmarish neurotoxin (Rodrigues et al., 2004). Consider that neuroactive substances are not uncommonly able to inhibit bacteria efflux pumps (for instance reserpine and palmatine). (iii) The aporphine framework often inhibit the replication of Polio virus (Boustie et al., 1998). One could examine the antibacterial or antiviral potential of isoboldine *in vivo*. Consider that the aporphine boldine inhibits the Human immunodeficiency (Tietjjen et al., 2015) and is being used in therapeutic for the improvement of liver function. Anti-HIV principles are often active against coronaviruses and thus one could endeaviour to examine its anti-COVID-19 properties of boldine. Boldine could be added to the list of first line natural products to test against coming pandemic zoonotic microbes.

### References

Boustie, J., Stigliani, J.L., Montanha, J., Amoros, M., Payard, M. and Girre, L., 1998. Antipoliovirus structure– activity relationships of some aporphine alkaloids. *Journal of Natural Products*, *61*(4), pp. 480–484.

Kiryakov, H.G., Iskrenova, E., Kuzmanov, B. and Evstatieva, L., 1981. Alkaloids from *Corydalis bulbosa*. *Planta Medica*, *43*(09), pp. 51–55.

Orhan, I., Ozcelik, B., Karaoğlu, T. and Şener, B., 2007. Antiviral and antimicrobial profiles of selected isoquinoline alkaloids from Fumaria and Corydalis species. Zeitschrift für Naturforschung C, 62(1–2), pp. 19–26.

Rodrigues, M.C.A., de Oliveira Beleboni, R., Coutinho-Netto, J., dos Santos, W.F. and Garcia-Cairasco, N., 2004. Behavioral effects of bicuculline microinjection in the dorsal versus ventral hippocampal formation of rats, and control of seizures by nigral muscimol. *Epilepsy Research*, *58*(2–3), pp. 155–165.

Tietjen, Ian, Ntie-Kang, F., Mwimanzi, P., et al., 2015. Screening of the pan-African Natural Product Library identifies ixoratannin A-2 and boldine as novel HIV-1 inhibitors. *PLoS One*, *10*(4), p. e0121099.

Udvardy, A., Miskovics, A. and Sipos, A., 2014. Antibacterial aporphinoids–progress and perspectives based on structure-activity analysis. *International Bulletin of Drug Research*, *4*(6), pp. 1–34.

### 4.1.3.4  *Corydalis govaniana* Wall.

Synonym: *Corydalis swatensis* Kitam.

Common name: Govan's Corydalis; bhootke, cheri pawa (Pakistan); Bhootakeshi, bhutkes (India); ku mang huang jin (China)

Habitat: Mountain forests, grassy slopes

Distribution: Pakistan, Himalayas; Kashmir, and Nepal

Botanical observation: This herb grows to about 35 cm tall from fibrous roots. The stems are terete and slender. In India, the plant is used to ward off evil spirits. The leaves are simple, spiral, and exstipulate. The petiole is up to 10 cm long and sheathing at base. The blade is incised, bi- or tripinnate, with 4–5 pairs of pinnae, and 3–15 × 1–7 cm. The inflorescences are 15 cm long racemes of up to about 30 flowers. The 2 sepals are whitish, dentate, and minute. The corolla is up to about 2.5 cm long, golden yellow tinged green, with an elongated and slightly curved or straight spur. The androecium includes a pair of trifid stamens. The gynoecium develops a filiform style. The capsule is obovoid about 1.5 cm long, and contain minute black, glossy, and smooth seeds.

Medicinal use: Antiseptic, syphilitic, scrofulous, cutaneous infections (India)

Commentaries:

(i) The plant has not been studied apparently for its antimicrobial properties. The roots produce protopine, corlumine, bicuculline, and isocorydine (Edwards & Handa, 1961). The leaves and stems contain govadine and govanine (Mukhopadhyay et al., 1987). These isoquinolines may participate in the traditional use of the plant in India. The plant needs to be assessed for its antimicrobial properties.

(ii) In traditional Asian medicines, the treatment of microbial infections not uncommonly encompasses some rituals to repel evil spirits (!). In fact the concept of evil spirit that are responsible for microbial infections is a common belief in South and Southeast Asia. In the Philippines (Visayas) the herbalists burn incense sticks or aromatic herbs or resins and invoke the dead or invisible people for assistance (sometime it works). In Indonesia, Malaysia, Thailand and in fact in most of rural Asia, such beliefs are not uncommon.

### References
Edwards, O.E. and Handa, K.L., 1961. The alkaloids of *Corydalis govaniana*. *Canadian Journal of Chemistry*, *39*(9), pp. 1801–1804.

Mukhopadhyay, S., Banerjee, S.K., Atal, C.K., Lin, L.J. and Cordell, G.A., 1987. Alkaloids of *Corydalis govaniana*. *Journal of Natural Products*, *50*(2), pp. 270–272.

### 4.1.3.5   *Corydalis incisa* (Thunb.) Pers.

Synonym: *Fumaria incisa* Thunb.

Common name: Ke ye zi jin (China)

Habitat: Forests, roadsides, waste places, and riverbanks

Distribution: China, Korea, and Japan

Botanical observation: This herb grows to about 50 cm tall from a tuber. The stems are angular and slender. The leaves are simple, spiral, and exstipulate. The petiole is up to 15 cm long and sheathing at base. The blade is deeply incised, bi- or tripinnate, with 4–5 pairs of pinnae, and 3–10×3–10 cm. The racemes are 3–12 cm long and support about 20 flowers. The 2 sepals are whitish, dentate, and minute. The corolla is up to about 1.5 cm long, light purplish with darker lobes and develops a straight spur. The androecium includes 2 trifid stamens. The ovary develops a filiform style. The capsules are oblong, dehiscent, about 1.2–1.8 cm long and contain minute seeds.

Medicinal uses: Abscess, skin diseases (China); skin diseases (Japan)

*Antifungal (filamentous) amphiphilic benzophenanthridine alkaloids:* Corynoline (LogD = 2.6 at pH 7.4; molecular weight = 367.4 g/mol) and acetylcorynoline (LogD = 3.5; molecular weight = 409.4 g/mol) from the aerial parts inhibited the growth of *Cladosporium herbarum* with a minimum amount of 3 μg/TLC spot (Guang Ma et al., 1999).

Corynoline

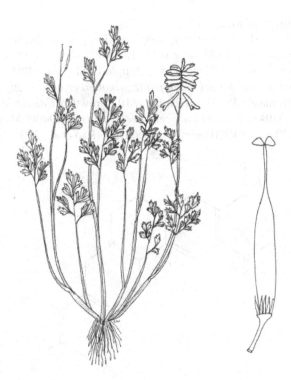

*Corydalis incisa* (Thunb.) Pers.

Commentary: Corynoline is cytotoxic for mammalian cells (Choi et al., 2007). In general, cytotoxic compounds for mammalian cells have very often antifungal activities. Benzophenanthridines are planar and tend to intercalate into DNA. Corynoline and acetylcorynoline are most probable strong broad-spectrum antibacterial alkaloids accounting, at least, for the anti-abscess property of the plant.

### References
Choi, S.U., Baek, N.L., Kim, S.H., Yang, J.H., Eun, J.S., Shin, T.Y., Lim, J.P., Lee, J.H., Jeon, H., Yun, M.Y. and Leem, K.H., 2007. Cytotoxic isoquinoline alkaloids from the aerial parts of *Corydalis incisa. Archives of Pharmacal Research, 30*(2), pp. 151–154.
Guang Ma, W., Fukushi, Y. and Tahara, S., 1999. Fungitoxic alkaloids from Hokkaido Corydalis species. *Fitoterapia, 70*(3), pp. 258–265.

### 4.1.3.6 *Corydalis racemosa* (Thunb.) Pers.
Synonyms: *Corydalis fumaria* H. Lév. & Vaniot; *Corydalis handel-mazzettii* Fedde; *Fumaria racemosa* Thunb.
Common name: Xiao hua huang jin (China)
Habitat: Forests margins, wastelands, and roadsides
Distribution: China, Japan, and Taiwan
Botanical observation: This herb grows to about 50 cm tall. The stems are ridged and slender. The leaves are simple, spiral, and exstipulate. The blade is incised, lobed, bipinnate. The racemes support about 15 flowers. The 2 sepals are ovate and minute. The corolla is up to about 1 cm long, yellow, and develops a minute upward curved lobed spur. The androecium includes a pair of trifid stamens. The ovary develops a bifid style. The capsule is linear about 3 cm long and contain minute black and tuberculate seeds.

Medicinal use: Abscess (China)

*Weak antibacterial (Gram-positive) amphiphilic protoberberine:* Tetrahydropalmatine (LogD=3.7; molecular weight=355.4 g/mol) inhibited the growth of methicillin-resistant *Staphylococcus aureus* with an MIC value of 3120 µg/mL (Shi et al., 2015).

*Strong broad-spectrum antifungal amphiphilic protoberberine:* Tetrahydropalmatine (Log D = 3.7 at pH 7.4; molecular mass=355.4 g/mol) inhibited the growth of *Candida albicans* (ATCCY0109), *Candida albicans* (KM943), and *Cryptococcus neoformans* (SM9406204) with MIC/MFC values of 320/>640, 320/>640, and 80/320 µg/mL, respectively (Rao et al., 2009).

Tetrahydropalmatine

Commentaries: (i) In general, filamentous fungi are more resistant to antifungal molecules than yeasts on account (at least partially) of the structure of their walls, which are thicker and more complex than in yeasts (Riquelme et al., 2018). Both type of fungi express efflux pumps (Espinel-Ingroff, 2008). (ii) Tetrahydropalmatine is neuroactive (Leung et al., 2003) and as such is most probably able to inhibit bacterial or fungal efflux pumps.

### References

Chu, L.H., Hsu, F.L., Chueh, F.Y., Niu, C.S. and Cheng, J.T., 1996. Antihypertensive activity of dl-tetrahydropalmatine, an active alkaloid isolated from the tubers of *Corydalis racemosa*. *Journal of the Chinese Chemical Society*, 43(6), pp. 489–492.

Espinel-Ingroff, A., 2008. Mechanisms of resistance to antifungal agents: yeasts and filamentous fungi. *Revista iberoamericana de micología*, 25(2), p. 101.

Leung, W.C., Zheng, H., Huen, M., Law, S.L. and Xue, H., 2003. Anxiolytic-like action of orally administered dl-tetrahydropalmatine in elevated plus-maze. *Progress in Neuro-Psychopharmacology and Biological Psychiatry*, 27(5), pp. 775–779.

Rao, G.X., Zhang, S., Wang, H.M., Li, Z.M., Gao, S. and Xu, G.L., 2009. Antifungal alkaloids from the fresh rattan stem of *Fibraurea recisa* Pierre. *Journal of Ethnopharmacology*, 123(1), pp. 1–5.

Riquelme, M., Aguirre, J., Bartnicki-García, S., Braus, G.H., Feldbrügge, M., Fleig, U., Hansberg, W., Herrera-Estrella, A., Kämper, J., Kück, U. and Mouriño-Pérez, R.R., 2018. Fungal morphogenesis, from the polarized growth of hyphae to complex reproduction and infection structures. *Microbiology and Molecular Biology Reviews*, 82(2), 1–47.

Shi, X., Li, X. and Zou, M., 2015. Chemical constituents and biological activities of *Stephania yunnanensis* HS Lo. *Biomedical Research*, 26, pp. 715–720.

### 4.1.3.7   *Corydalis saxicola* Bunting

Synonym: *Corydalis thalictrifolia* Franch.

   Common name: Yan huang lian (China)

   Habitat: Rocky soil

   Distribution: China

   Botanical observation: This herb grows to about 40 cm tall from a curved taperoot. The stems are terete and slender. The leaves are simple, spiral, and exstipulate. The petiole is up to 15 cm long. The blade is incised, pinnate, or bipinnate, the lobes somewhat rhombic or elliptic, 2–4×2–3 cm, and incised. The inflorescences are many-flowered racemes, which are of up to about 18 cm long. The calyx comprises 2 sepals, which are orbicular and minute. The corolla is up to about 3 cm long, golden yellow, and develops a lobed spur. The androecium includes 2 trifid stamens. The gynoecium comprises a 1-celled ovary and a slender style. The fruit is a dehiscent capsule, which is linear about 2.5 cm long and containing minute seeds.

   Medicinal use: Hepatitis (China)

   *Moderate antiviral (enveloped circular double-stranded DNA) protoberberines:* Dehydrocavidine, dehydroapocavidine, and dehydroisoapocavidine at a concentration of 250 µg/mL inhibited Hepatitis B surface antigen by 53, 59, and 54%, respectively, from Hepatitis B virus-producing human hepatoblastoma cell line (Li et al., 2008). Dehydrocavidive, dehydroapocavidine, and dehydroisoapocavidine at a concentration of 250 µg/mL inhibited Hepatitis B virus e-antigen by 41, 43, and 43%, respectively, from Hepatitis B virus producing human hepatoblastoma cell line (Li et al., 2008). Dehydrocheilanthifoline inhibited the secretions of Hepatitis B surface antigen and Hepatitis B virus e-antigen in HepG2.2.15 cells with $IC_{50}$ values of 15.8 and 17.1 µM, and with a selectivity index of 7.3 and 6.7, respectively (Zeng et al., 2013). Dihydrochelerythrine inhibited Hepatitis B surface antigen and Hepatitis B virus e-antigen secretions with $IC_{50}$ value below 0.05 µM and selectivity index > 3.5 (Wu et al., 2007).

   *Moderate antiviral (non-enveloped linear single-stranded DNA) protoberberine:* 2,9-Dihydroxy-3,11-dimethoxy-1,10-dinitrotetrahydroprotoberberine protected F81 cells against the Canine parvovirus with an $EC_{50}$ value of 182 µM and a selectivity index of 2.2 (Huang et al., 2012).

   Commentaries: The plant contains the morphinandienone alkaloid pallidine and the tetrahydroberberine alkaloid discretamine, which inhibit topoisomerase I as potently as camptothecin (Cheng et al., 2008). Note that inhibitors of topoisomerase I are antibacterial (Wu et al., 2013). The quaternary ammonium ion of protoberberine alkaloids and their polyaromatic planar framework account for topoisomerase I inhibition (Cheng et al., 2008) or DNA damages (Xue & Warshawsky, 2005). These compounds might have antifungal properties.

### References

Cheng, X., Wang, D., Jiang, L. and Yang, D., 2008. DNA topoisomerase I inhibitory alkaloids from *Corydalis saxicola*. *Chemistry & Biodiversity*, 5(7), pp. 1335–1344.

Huang, Q.Q., Bi, J.L., Sun, Q.Y., Yang, F.M., Wang, Y.H., Tang, G.H., Zhao, F.W., Wang, H., Xu, J.J., Kennelly, E.J. and Long, C.L., 2012. Bioactive isoquinoline alkaloids from *Corydalis saxicola*. *Planta Medica*, 78(01), pp. 65–70.

Li, H.L., Han, T., Liu, R.H., Zhang, C., Chen, H.S. and Zhang, W.D., 2008. Alkaloids from *Corydalis saxicola* and their anti-Hepatitis B virus activity. *Chemistry & Biodiversity*, 5(5), pp. 777–783.

Wu, J.Y., Chang, M.C., Chen, C.S., Lin, H.C., Tsai, H.P., Yang, C.C., Yang, C.H. and Lin, C.M., 2013. Topoisomerase I inhibitor evodiamine acts as an antibacterial agent against drug-resistant *Klebsiella pneumoniae*. *Planta Medica*, 79(01), pp. 27–29.

Wu, Y.R., Ma, Y.B., Zhao, Y.X., Yao, S.Y., Zhou, J., Zhou, Y. and Chen, J.J., 2007. Two new quaternary alkaloids and anti-Hepatitis B virus active constituents from *Corydalis saxicola*. *Planta Medica*, 73(08), pp. 787–791.

Xue, W. and Warshawsky, D., 2005. Metabolic activation of polycyclic and heterocyclic aromatic hydrocarbons and DNA damage: a review. *Toxicology and Applied Pharmacology*, *206*(1), pp. 73–93.

Zeng, F.L., Xiang, Y.F., Liang, Z.R., Wang, X., Huang, D.E., Zhu, S.N., Li, M.M., Yang, D.P., Wang, D.M. and Wang, Y.F., 2013. Anti-Hepatitis B virus effects of dehydrocheilanthifoline from *Corydalis saxicola*. *The American Journal of Chinese Medicine*, *41*(01), pp. 119–130.

### 4.1.3.8   *Fumaria indica* Pugsley

Synonym: *Fumaria parviflora* Lam.

Common names: Indian fumitory; parpata, pitpapra, pitpopdo (India); shaturuj (Iran)

Habitat: Roadsides, field margins, and wastelands

Distribution: Iran, Afghanistan, Central Asia, Himalayas, Pakistan, and India

Botanical observation: This herb grows to about 50cm tall. The stems are terete, slender, and glaucous. The leaves are simple, spiral, and exstipulate. The blade is glaucous, 3–7 cm×2–3cm, incised, bi-or tripinnate, the lobes 0.5–1.5cm long and linear. The inflorescences are many flowered racemes which are opposite to leaves or terminal, many-flowered, and about 2–5cm long. The 2 sepals are lanceolate, membranous, and minute. The corolla is up to about 8mm long, light pinkish and darker at apex, and develops a lobed spur. The androecium includes a pair of trifid stamens. The ovary develops a slender style. The capsules are globose, minute, dehiscent and contain a single seed. The plant is a vegetable in India.

Medicinal use: Fever (India)

*Weak antifungal (filamentous) protoberberine:* Berberine at a concentration of 500ppm inhibited the germination of spores of *Curvularia lunata*, *Erysiphe cichoracearum*, *Fusarium udum*, and *Penicillium* sp. by more than 80% (Sarma et al., 1999).

*Weak antifungal (filamentous) tetrahydrobenzyl isoquinoline:* Fuyuziphine at a concentration of 500ppm inhibited the germination of spores of *Alternaria brassicicola*, *Alternaria solani*, *Alternaria melongenae*, *Curvularia maculans*, *Erysiphe cichoracearum*, and *Helminthosporium pennisetti* by more than 80% (Pandey et al., 2007).

*Strong broad-spectrum antibacterial amphiphilic benzophenanthridine alkaloid*: Norsanguinarine (LogD=1.7 at pH 7.4; molecular weight=313.4 g/mol) inhibited the growth of *Escherichia coli*, *Pseudomonas aeruginosa*, *Proteus mirabilis*, *Klebsiella pneumoniae*, *Acinetobacter baumannii*, *Staphylococcus aureus*, and *Bacillus subtilis* with MIC values of 32, 32, 32, 32, 32, 64, and 64 µg/mL, respectively (Orhan et al., 2007).

Norsanguinarine

*Strong broad-spectrum antibacterial amphiphilic spirobenzylisoquinoline-type alkaloid:* (+)-parfumine (Log P=2.7 at pH 7.4; molecular weight=353.4 g/mol) inhibited the growth of *Escherichia coli*, *Pseudomonas aeruginosa*, *Proteus mirabilis*, *Klebsiella pneumonia*, *Acinetobacter baumannii*, *Staphylococcus aureus*, and *Bacillus subtilis* with MIC values of 32, 32, 32, 32, 32, 64, and 64 µg/mL respectively (Orhan et al., 2007).

*Strong antiviral (enveloped monopartite linear single-stranded (+) RNA) polar extract:* Methanol extract of whole plant was antiviral against Dengue virus-1, -2, -3, and -4 with $IC_{50}$ equal to 12.5, 7, 11.3, and 12.5 µg/mL, respectively (Sood et al., 2015).

*Strong antiviral (enveloped monopartite linear double-stranded DNA) amphiphilic benzophenanthridine:* Norsanguinarine inhibited Herpes simplex virus replication in Madin-Darby canine kidney cells with a maximum nontoxic concentration of 32 µg/mL (Orhan et al., 2007).

*Strong antiviral (enveloped monopartite linear single-stranded (-) RNA) amphiphilic benzophenanthridine:* Norsanguinarine inhibited Parainfluenza-3 virus replication in Vero cells with a maximum nontoxic concentration of 32 µg/mL, respectively (Orhan et al., 2007).

*Strong antiviral (enveloped monopartite linear double-stranded DNA) amphiphilic spirobenzylisoquinoline-type alkaloid:* (+)-Parfumine inhibited Herpes simplex virus replication in Madin-Darby canine kidney cells with a maximum nontoxic concentration of 32 µg/mL (Orhan et al., 2007).

*Strong antiviral (enveloped monopartite linear single-stranded (-)RNA virus) amphiphilic spirobenzylisoquinoline-type alkaloid:* (+)-Parfumine inhibited Parainfluenza-3 virus with a maximum nontoxic concentration of 64 µg/mL (Orhan et al., 2007).

Commentaries: (i) Tripathi et al. (1994) observed changes in antibacterial alkaloid concentrations over a period of 90 days in this plant and found that the amount of protopine decreased from 200 to 0 mg/120 gm. Also, norsanguinarine contents increased from 1.8 to 5 mg/120 gm. Therefore, the antimicrobial properties of plant extract depend on the age of the plant. (ii) Protopine, tetrahydrocoptisine, and norsanguinarine from this plant evoked some levels of antifungal effects (Singh et al., 1997). (iii) The plant appears to have no toxicity in animals (Singh et al., 2011) and could potentially undergo preclinical trials as antiviral (including anti-COVID-19) phytomedication. (iii) Epidemic diseases or *Amraz-e-Waba* were known and quite well understood by our peer Ibn Sina (or Avicenna) in his medical treatise *Al-Qanoon Fit Tib* (Canon of Medicine) and a formulation called *Joshandah Sual* in Unani medicine is used to manage flu-like pandemics (Alam et al., 2020).

## References

Alam, M.A., Quamri, M.A., Sofi, G., Ayman, U., Ansari, S. and Ahad, M., 2020. Understanding COVID-19 in the light of epidemic disease described in Unani medicine. *Drug Metabolism and Personalized Therapy, 1*(ahead-of-print).

Orhan, I., Ozcelik, B., Karaoğlu, T. and Şener, B., 2007. Antiviral and antimicrobial profiles of selected isoquinoline alkaloids from Fumaria and Corydalis species. *Zeitschrift für Naturforschung C, 62*(1–2), pp. 19–26.

Pandey, M.B., Singh, A.K., Singh, A.K. and Singh, U.P., 2007. Inhibitive effect of fuyuziphine isolated from plant (Pittapapra) (*Fumaria indica*) on spore germination of some fungi. *Mycobiology, 35*(3), pp. 157–158.

Sarma, B.K., Pandey, V.B., Mishra, G.D. and Singh, U.P. 1999. Antifungal activity of berberine iodide, a constituent of *Fumaria indica*. *Folia Microbiologica, 44*, pp. 164–166.

Singh, G.K., Chauhan, S.K., Rai, G. and Kumar, V., 2011. Fumaria indica is safe during chronic toxicity and cytotoxicity: a preclinical study. *Journal of Pharmacology & Pharmacotherapeutics, 2*(3), p. 191.

Singh, R.A., Singh, U.P., Tripathi, V.K., Roy, R. and Pandey, V.B. 1997. Effect of *Fumaria indica* alkaloids on conidial germination of some fungi. *Oriental Journal of Chemistry, 13*, pp. 177–180.

Sood, R., Raut, R., Tyagi, P., Pareek, P.K., Barman, T.K., Singhal, S., Shirumalla, R.K., Kanoje, V., Subbarayan, R., Rajerethinam, R. and Sharma, N., 2015. Cissampelos pareira Linn: natural source of potent antiviral activity against all four dengue virus serotypes. *PLoS Neglected Tropical Diseases, 9*(12), p. e0004255.

Tripathi, Y.C., Rathore, M. and Kumar, H., 1994. On the variation of alkaloidal contents of *Fumaria indica* at different stages of life span. *Ancient Science of Life, 13*(3–4), p. 271.

### 4.1.3.9   *Fumaria officinalis* L.

Synonyms: *Fumaria officinalis* Chaub.; *Fumaria officinalis* Hohen.

Common names: Common fumitory; Şahtere (Turkey); yan jin (China); Shatra (Iran); shaahtara, pitapapra (India)

Habitat: Waste places, forests, and roadsides

Distribution: Turkey, Iran, Afghanistan, Central Asia, India, China, Taiwan, and Japan

Botanical observation: This herb grows to about 50 cm tall. The stems are terete, slender, and glaucous. The leaves are simple, spiral, and exstipulate. The blade is glaucous, 3–7 cm×2–3 cm, incised, bi- or tripinnate, the lobes 0.5–1.5 cm long and linear. The inflorescences are up to 40-flowered racemes, which are opposite to leaves or terminal, and about 3–7 cm long. The calyx comprises 2 sepals, which dentate and minute. The corolla is up to about 1 cm long, purplish pink or whitish, and darker at apex, and develops a lobed spur. The androecium includes 2 trifid stamens. The ovary develops a slender style. The capsule is a somewhat reniform, rugose, minute, and dehiscent containing a single seed.

Medicinal uses: Fever (India); infections (Turkey)

*Moderate broad-spectrum antibacterial alkaloidal fraction:* Total alkaloid fraction extracted from the aerial parts inhibited the growth of *Propionibacterium acnes*, *Corynebacterium xerosis* and *Acinetobacter calcoaceticus* with MIC values of 0.33, 2.7, and 0.9 mg/mL, respectively (Khamtache-Abderrahim et al., 2016).

*Strong broad-spectrum antibacterial amphiphilic protopine-type alkaloid*: Protopine (LogD = 2.3 at pH 7.4; molecular weight = 353.3 g/mol) inhibited the growth of *Escherichia coli*, *Pseudomonas aeruginosa*, *Proteus mirabilis*, *Klebsiella pneumonia*, *Acinetobacter baumannii*, *Staphylococcus aureus*, and *Bacillus subtilis* with MIC values of 32, 64, 32, 8, 8, 64, and 128 μg/mL, respectively (Orhan et al., 2007).

*Strong broad-spectrum antibacterial benzophenanthridine alkaloids:* Sanguinarine (molecular weight = 332.3 g/mol) inhibited the growth of *Escherichia coli*, *Pseudomonas aeruginosa*, *Proteus mirabilis*, *Klebsiella pneumonia*, *Acinetobacter baumannii*, *Staphylococcus aureus*, and *Bacillus subtilis* with MIC values of 32, 32, 32, 32, 32, 64, and 64 μg/mL, respectively (Orhan et al., 2007). Sanguinarine inhibited the growth of *Streptococcus mutans* (ATCC 25175) with an MIC value 32 μg/mL (Park et al., 2003).

*Strong broad-spectrum antibacterial amphiphilic spirobenzylisoquinoline-type alkaloid*: Fumarophycine (LogD = 2.7 g/mol; molecular weight = 397.4 g/mol) inhibited the growth of *Escherichia coli*, *Pseudomonas aeruginosa*, *Proteus mirabilis*, *Klebsiella pneumoniae*, *Acinetobacter baumannii*, *Staphylococcus aureus*, and *Bacillus subtilis* with MIC values of 32, 64, 32, 8, 8, 64, and 128 μg/mL, respectively (Orhan et al., 2007). The spirobenzylisoquinoline-type alkaloid (+)-fumariline inhibited the growth of *Escherichia coli*, *Pseudomonas aeruginosa*, *Proteus mirabilis*, *Klebsiella pneumoniae*, *Acinetobacter baumanii*, *Staphylococcus aureus*, and *Bacillus subtilis* with MIC values of 32, 32, 32, 32, 32, 64, and 64 μg/mL, respectively (Orhan et al., 2007).

Fumarophycine

*Strong broad-spectrum antibacterial amphiphilic phthalide isoquinoline-type alkaloid:* Adlumidine (LogD=2.4 at pH 7.4; molecular weight=367.3 g/mol) inhibited the growth of *Escherichia coli, Pseudomonas aeruginosa, Proteus mirabilis, Klebsiella pneumoniae, Acinetobacter baumannii, Staphylococcus aureus*, and *Bacillus subtilis* with MIC values of 32, 32, 32, 32, 32, 64, and 64 µg/mL, respectively (Orhan et al., 2007).

Adlumidine

*Strong broad-spectrum antibacterial amphiphilic protoberberines:* Stylopine (also known as tetrahydrocoptisine; LogP=3.7 at pH 7.4; molecular weight=323.3 g/mol) inhibited the growth of *Escherichia coli, Pseudomonas aeruginosa, Proteus mirabilis, Klebsiella pneumoniae, Acinetobacter baumannii, Staphylococcus aureus*, and *Bacillus subtilis* with MIC values of 32, 32, 32, 32, 32, 64, and 64 µg/mL, respectively (Orhan et al., 2007). Sinactine (also known as tetrahydroepiberberine; LogD=3.6 at pH 7.4; molecular weight=339.4 g/mol) inhibited the growth of *Escherichia coli, Pseudomonas aeruginosa, Proteus mirabilis, Klebsiella pneumoniae, Acinetobacter baumannii, Staphylococcus aureus*, and *Bacillus subtilis* with MIC values of 32, 32, 32, 8, 8, 64, and 64 µg/mL, respectively (Orhan et al., 2007).

*Strong antiviral (enveloped monopartite linear double-stranded DNA) amphiphilic protopine-type alkaloid*: Protopine (LogD=2.3 at pH 7.4; molecular weight=353.3 g/mol) inhibited Herpes simplex virus replication in Madin-Darby canine kidney cells with a maximum nontoxic concentration of 64 μg/mL (Orhan et al., 2007).

*Strong antiviral (enveloped monopartite linear double-stranded DNA) hydrophilic benzophenanthridine alkaloid:* Sanguinarine inhibited Herpes simplex virus replication in Madin-Darby canine kidney cells with a maximum nontoxic concentration of 32 μg/mL (Orhan et al., 2007).

*Strong antiviral (enveloped monopartite linear double-stranded DNA) amphiphilic spirobenzylisoquinoline-type alkaloid*: Fumariline (LogD=3.3 at pH 7.4; molecular weight=351.3 g/mol) inhibited Herpes simplex virus replication in Madin-Darby canine kidney cells with a maximum nontoxic concentration of 32 μg/mL (Orhan et al., 2007).

*Strong antiviral (enveloped monopartite linear double-stranded DNA) amphiphilic phthalide isoquinoline-type alkaloid:* Bicuculline (LogD=2.4; molecular weight=367.3 g/mol) inhibited Herpes simplex virus replication in Madin-Darby canine kidney cells with a maximum nontoxic concentration of 32 μg/mL (Orhan et al., 2007).

*Strong antiviral (enveloped monopartite linear double-stranded DNA) amphiphilic phthalide isoquinoline-type alkaloid:* Adlumidine (LogD=2.4 at pH 7.4; molecular weight=367.3 g/mol) inhibited Herpes simplex virus replication in Madin-Darby canine kidney cells with a maximum nontoxic concentration of 32 μg/mL (Orhan et al., 2007).

*Strong antiviral (enveloped monopartite linear single-stranded (-) RNA virus) isoquinoline alkaloids*: Protopine inhibited Parainfluenza-3 virus in Vero cells with a maximum nontoxic concentration of 32 μg/mL (Orhan et al., 2007).

*Strong antiviral (enveloped monopartite linear single-stranded (–) RNA) benzophenanthridine alkaloids:* Sanguinarine inhibited Parainfluenza-3 virus in Vero cells with a maximum nontoxic concentration of 32 μg/mL (Orhan et al., 2007).

*Strong antiviral (enveloped monopartite linear single-stranded (–) RNA) spirobenzylisoquinoline-type alkaloid*: Fumariline inhibited Parainfluenza-3 virus in Vero cells with a maximum nontoxic concentration of 64 μg/mL, respectively.

*Strong antiviral (enveloped monopartite linear single-stranded (–) RNA) amphiphilic phthalide isoquinoline-type alkaloid:* Bicuculline inhibited Parainfluenza-3 virus in Vero cells with a maximum nontoxic concentration of 64 μg/mL, respectively (Orhan et al., 2007).

*Strong antiviral (enveloped monopartite linear single-stranded (–) RNA) phthalide isoquinoline-type alkaloid:* Adlumidine inhibited Parainfluenza-3 virus in Vero cells with a maximum nontoxic concentration of 64 μg/mL (Orhan et al., 2007).

Commentary: The plant appears to have no human toxicity (Al-Snafi, 2020), it may have activity against the Severe acute respiratory syndrome-acquired coronavirus.

**References**

Al-Snafi, A.E., 2020. Constituents and pharmacology of *Fumaria officinalis*–a review. *IOSR Journal of Pharmacy*, *10*(1), pp. 17–25.

Khamtache-Abderrahim, S., Lequart-Pillon, M., Gontier, E., Gaillard, I., Pilard, S., Mathiron, D., Djoudad-Kadji, H. and Maiza-Benabdesselam, F., 2016. Isoquinoline alkaloid fractions of *Fumaria officinalis*: characterization and evaluation of their antioxidant and antibacterial activities. *Industrial Crops and Products*, *94*, pp. 1001–1008.

Orhan, I., Özçelik, B., Karaoğlu, T. and Şener, B., 2007. Antiviral and antimicrobial profiles of selected isoquinoline alkaloids from *Fumaria* and *Corydalis* species. *Zeitschrift für Naturforschung C*, *62*(1–2), pp. 19–26.

Park, K.M., You, J.S., Lee, H.Y., Baek, N.I. and Hwang, J.K., 2003. Kuwanon G: an antibacterial agent from the root bark of Morus alba against oral pathogens. *Journal of Ethnopharmacology*, *84*(2–3), pp. 181–185.

### 4.1.3.10 *Hylomecon japonica* (Thunb.) Prantl & Kündig

Synonyms: *Chelidonium japonicum* Thunb.; *Hylomecon hylomeconoides* (Nakai) Y.N.Lee; *Stylophorum japonicum* (Thunb.) Miq.

 Common names: Japanese woodland poppy; he qing hua (China)
 Habitat: Forests, ditch sides, and shaded grassy spots
 Distribution: Korea, Japan, China, Siberia, and Russia
 Botanical observation: This herb grows to about 40 cm tall. The stems are terete, green to pur-plish and yields a yellow latex upon incision. The leaves are mostly basal, pinnatisect, and exstip-ulate. The petiole is elongated. The blade is 10–15 cm long, with 2–3 pairs of lobes, which are broadly obovate-rhombic, crenate, 3–8.5 × 1–5 cm, cuneate at base, and acuminate at apex. The inflorescences are few flowered terminal or axillary clusters. The flower peduncles are 3.5–7 cm long. The 2 sepals are ovate, 1–1.5 cm long, and caducous. The 4 petals are obovate, 1.5–2 cm long, membranous, and light yellow. The stamens are about 6 mm long. The ovary is oblong, about 7 mm long, glabrous, and consists of 2 fused carpels and minute 2-lobed stigmas. The cap-sules are linear, up to about 8 cm long, dehiscent, and contain numerous minute seeds.

 Medicinal use: Apparently none for the treatment of microbial infections

 *Antibacterial (Gram-positive) halo developed benzophenanthridine alkaloid:* 6-Methoxydihy-drosanguinarine (10 µg/paper disc) inhibited the growth of *Staphylococcus aureus* and methicillin-resistant *Staphylococcus aureus* with an inhibition zone diameter of about 17 mm (Choi et al., 2010).

 *Strong broad-spectrum antibacterial benzophenantridine alkaloid:* 6-Methoxydihydrosanguinarine from the roots inhibited the growth of *Enterococcus faecalis, Staphylococcus aureus* (ATCC 25925), and *Escherichia coli* (ATCC 25922) with MIC/MBC of 5/10, 2.5/5, and 20/160 µg/mL 6-Methoxydihydrosanguinarine induced *Staphylococcus aureus* cell wall and cytoplasm insults (Xue et al., 2017).

6-Methoxydihydrosanguinarine

*Strong anticandidal benzophenanthridine:* 6-Methoxydihydrosanguinarine from the roots inhib-ited the growth of *Candida albicans* (CMCC 85021) for MIC/MBC 20/80 µg/mL (Xue et al., 2017). In this experiment, berberine inhibited the growth of *Candida albicans* (CMCC 85021) for MIC/MBC with MIC/MBC of 160/640 µg/mL (Xue et al., 2017).

 Commentaries: (i) Prior to isolating antimicrobial compounds or testing extract from a plant, it is of utmost importance to have that plant identified properly. If the teaching of pharmacy in current dysfunctional neoliberal universities (Corlet, 1988) was done, *"secundum praecepta artis"* students would have proper botanical training and well equipped for drug discovery. (ii) Sanguinarine is planar and most probably intercalate in bacterial and fungal DNA.

**References**

Choi, J.G., Kang, O.H., Chae, H.S., Obiang-Obounou, B., Lee, Y.S., Oh, Y.C., Kim, M.S., Shin, D.W., Kim, J.A., Kim, Y.H. and Kwon, D.Y., 2010. Antibacterial activity of Hylomecon hylomeconoides against methicillin-resistant *Staphylococcus aureus*. *Applied Biochemistry and Biotechnology*, *160*(8), pp. 2467–2474.

Corlett, J. Angelo., 1988. Alienation in capitalist society. *Journal of Business Ethics*, *7*(9), 699–701.

Xue, X., Zhang, H., Zhang, X., Liu, X., Xi, K., Han, Y. and Guo, Z., 2017. TLC bioautography-guided isolation and antimicrobial, antifungal effects of 12 alkaloids from *Hylomecon japonica* roots. *Natural Product Communications*, *12*(9), p. 1934578X1701200914.

### 4.1.3.11  *Macleaya cordata* (Willd.) R. Brown

Synonyms: *Bocconia cordata* Willd.; *Macleaya yedoensis* André

Common names: Plume poppy; bo luo hui (China); takeni gusa (Japan)

Distribution: China, Taiwan, and Japan

Habitat: Forests, riverbanks, or cultivated

Botanical observation: This erect herb grows to about 3 m tall. The stem is terete, smooth, glaucous, and yields a yellow and poisonous latex. The leaves are simple, spiral, and exstipulate. The petiole is 1–12 cm long and channeled above. The blade is somewhat dull glaucous, cordate at base, broadly ovate, 5–25×5–25 cm, deeply lobed to undulate at margin, rounded at apex and with about 4 pairs of secondary nerves. The inflorescences are showy terminal or axillary and somewhat plumose yellowish panicles, which are 15–40 cm. The calyx comprises 4 sepals, which are whitish, narrowly oblong, and about 1 cm long. The androecium includes 24–30 stamens with linear and about 5 mm long anthers. The ovary is obovoid, about 4 mm long, and with 2-lobed stigmas. The capsules are spathulate, up to 3 cm long, rounded at apex and contain a few seeds, which are minute and ovoid.

Medicinal uses: Sores, abscesses, ringworms (China)

*Strong broad-spectrum antibacterial extract:* Alkaloidal extract evoked the MIC values of 16 µg/mL for *Staphylococcus aureus*, 32 µg/mL for *Enterococcus faecalis*, and 64 µg/mL for *Staphylococcus hyicus* and *Escherichia coli* (Opletal et al., 2014).

*Strong antibacterial (Gram-positive) benzophenanthridine alkaloids:* Sanguinarine inhibited the growth of *Staphylococcus aureus*, *Pseudomonas aeruginosa*, *Escherichia coli*, and *Streptococcus agalactiae* with MIC of 31.3, 250, 62.5, and 15.6 µg/mL, respectively (Kosina et al., 2010). Chelerythrine (also known as toddalin inhibited the growth of *Staphylococcus aureus*, *Pseudomonas aeruginosa*, *Escherichia coli*, and *Streptococcus agalactiae* with MIC of 31.3, 500, 125, and 7.8 µg/mL, respectively (Kosina et al., 2010). Chelerythrine inhibited the growth of *Staphylococcus epidermidis* (ATCC 12228), *Staphylococcus aureus* (ATCC 6538), *Streptococcus pyogenes* (ATCC 19615), *Bacillus subtilis* (ATCC 6633), *Klebsiella pneumoniae* (ATCC 13883), and *Escherichia coli* (ATCC 25922) with the MIC/MBC values of 1.5/12.5, 1.5/3.1, 1.5/6.2, 1.5/50, 1.5/50, and 1.5/25 µg/mL, respectively (Tavares et al., 2014).

Sanguinarine

Chelerythrine

*Moderate broad-spectrum antibacterial protopine-type alkaloid:* Protopine (also known as fumarine or corydinine) inhibited the growth of *Staphylococcus aureus, Pseudomonas aeruginosa, Escherichia coli,* and *Streptococcus agalactiae* with MIC of 250, 125, 125, and 125 µg/mL, respectively (Kosina et al., 2010). Allocryptopine inhibited the growth of *Staphylococcus aureus, Pseudomonas aeruginosa, Escherichia coli,* and *Streptococcus agalactiae* with MIC of 250, 125, 125, and 125 µg/mL, respectively (Kosina et al., 2010).

*Very weakly antibacterial (Gram-negative) benzophenanthridine alkaloid:* Dihydrosanguinarine inhibited the growth of *Escherichia coli* with MIC of 500 µg/mL (Kosina et al., 2010).

*Strong broad spectrum antifungal benzophenanthridine alkaloid:* Chelerythrine inhibited the growth of *Candida albicans* (ATCC 10231), *Saccharomyces cerevisiae* (ATCC 2601), and *C. neoformans* (ATCC 28952) with MIC/MBC values of 3.1/3.1, 6.2/6.2, and 3.1/6.2 µg/mL, respectively (Tavares et al., 2014).

*Moderate antiviral (enveloped monopartite linear double-stranded DNA) protopine-type alkaloid:* Allocryptopine inhibited Herpes simplex virus replication in Madin-Darby canine kidney cells in Vero cells with a maximum nontoxic concentration of 64 µg/mL, respectively (Orhan et al., 2007).

*Moderate antiviral (enveloped monopartite linear single-stranded (-) RNA) amphiphilic protopine-type alkaloid:* Allocryptopine inhibited replication in Madin-Darby canine kidney cells of the Parainfluenza-3 virus with a maximum nontoxic concentration of 32 µg/mL, respectively (Orhan et al., 2007).

*Viral enzyme inhibition by benzophenanthridine alkaloids:* Chelerythrine, sanguinarine, and fagaronine at a concentration of 50 µM inhibited Aminopeptidase N by about 30, 50, and 30%, respectively (Bauvois & Dauzonne, 2006).

Commentaries: Being able to inhibit the growth of bacteria, fungi, and viruses, and being planar, benzophenanthridines probably target DNA. Consider that sanguinarine is toxic to humans (Dalvi, 1985).

### References

Bauvois, B. and Dauzonne, D., 2006. Aminopeptidase-N/CD13 (EC 3.4. 11.2) inhibitors: chemistry, biological evaluations, and therapeutic prospects. *Medicinal Research Reviews,* *26*(1), pp. 88–130.

Dalvi, R.R., 1985. Sanguinarine: its potential, as a liver toxic alkaloid present in the seeds of *Argemone mexicana. Experientia,* *41*(1), pp. 77–78.

Kosina, P., Gregorova, J., Gruz, J., Vacek, J., Kolar, M., Vogel, M., Roos, W., Naumann, K., Simanek, V. and Ulrichova, J., 2010. Phytochemical and antimicrobial characterization of *Macleaya cordata* herb. *Fitoterapia,* *81*(8), pp. 1006–1012.

Opletal, L.,Ločárek, M., Fraňková, A., Chlebek, J., Smid, J., Hošt'álková, A., Safratova, M., Hulcová, D., Klouček, P., Rozkot, M. and Cahlíková, L., 2014. Antimicrobial activity of extracts and isoquinoline alkaloids of selected papaveraceae plants. *Natural Product Communications*, 9(12), pp. 1709–1712.

Orhan, I., Özçelik, B., Karaoğlu, T. and Şener, B., 2007. Antiviral and antimicrobial profiles of selected isoquinoline alkaloids from Fumaria and Corydalis species. *Zeitschrift für Naturforschung C*, 62(1–2), pp. 19–26.

Tavares, L.D.C., Zanon, G., Weber, A.D., Neto, A.T., Mostardeiro, C.P., Da Cruz, I.B., Oliveira, R.M., Ilha, V., Dalcol, I.I. and Morel, A.F., 2014. Structure-activity relationship of benzophenanthridine alkaloids from *Zanthoxylum rhoifolium* having antimicrobial activity. *PLoS One*, 9(5), p. e97000.

### 4.1.3.12   *Meconopsis aculeata* Royle

Common names: Blue poppy; gul-e-nilam, kanderi, kalihaari, tser-non, achatsarmum (India); pi ci lu rong hao (China)

Habitat: Rocky soils, streamsides, and mountains

Distribution: Pakistan, Himalaya, India, and China

Botanical observation: This graceful erect herb grows up to about 70 cm tall from a taproot. The whole plant is covered with small prickles and yields a yellow latex. The leaves are simple, basal, in rosette, and exstipulate. The petiole is 10–15 cm long and sheathing at base. The blade is 10–20 cm×2.5–5 cm, tapering at base, pinnatilobed with obtuse to acute lobes, which are about 0.5–3 cm×0.3–1 cm. The racemes are terminal and present a few showy, slightly nodding flowers which are 5–7 in diameter on 2–13.5 cm long peduncles. The calyx comprises 2 sepals, which are caducous, 2–2.5 cm long, and ovate. The 6 petals are bluish, wavy, somewhat translucent, 2–3 cm×4–4 cm, and membranous. The androecium is showy and comprises numerous yellow stamens, which are linear and about 1 cm long. The gynoecium is about 1 cm long and consists of a few carpels fused to form an ovoid ovary, which develops columnar style and 4 capitate stigmas. The capsules are obconic, about 1.5 cm long, 5–7 valved, elongated, and contain numerous minute reniform seeds.

Medicinal uses: Blood purifier, ulcers (India)

Commentary: The plant has apparently not been investigated for its possible antimicrobial properties. Protopine and sanguinarine, which are known antibacterial and antifungal alkaloids, have been reported in this plant (Hemingway et al., 1981).

### Reference

Hemingway, S.R., Phillipson, J.D. and Verpoorte, R., 1981. Meconopsis cambrica alkaloids. *Journal of Natural Products*, 44(1), pp. 67–74.

### 4.1.3.13   *Papaver nudicaule* L.

Synonym: *Papaver nudicaule* var. *chinense* (L.) Fedde

Common names: Icelanding poppy; ye ying su (China); kham scur (Pakistan)

Habitat: Forests margins, grasslands, grassy steppes, rocky streamsides, and mountains

Distribution: Pakistan, Afghanistan, Kazakhstan, Kyrgyzstan, Tajikistan, Uzbekistan, Korea, Mongolia, Russia, and China.

Botanical observation: This graceful herb grows to about 60 cm tall from a taproot. The leaves are simple, in rosette, and exstipulate. The petiole is 5–12 cm long and sheathing at base. The blade is somewhat glaucous, ovate to lanceolate, 3–8 cm long, pinnatilobate with pairs of lobes, and hairy. The flowers are showy, slightly aromatic, solitary, and terminal on a scape and about 4–6 cm in diameter. The calyx comprises 2 sepals, which are elliptic. The corolla includes 4 petals, which are yellow, whitish, orange or red, 2–3 cm long, membranous, and minutely crenate. The androecium includes numerous stamens, which are yellow and up to about 1 cm long. The ovary is obovoid, 5–10 mm long, hairy, and develops 4–8 stigmas. The capsule is obovoid, up to about 1.5 cm long, hairy, ribbed, poricidal, and contain numerous tiny seeds.

Medicinal uses: Wounds (China)

Commentary: This plant does not appear to have been studied for its possible antimicrobial activities. It is known to produce the alkaloid allocryptopine (Istatkova et al., 2008). The yellow color of the petals is owed the indole alkaloid nudicaulin (Tatsis et al., 2012), which could be antibacterial, antifungal, or antiviral (Bondarenko & Frasinyuk, 2019).

**References**

Bondarenko, S.P. and Frasinyuk, M.S., 2019. Chromone alkaloids: structural features, distribution in nature, and biological activity. *Chemistry of Natural Compounds*, pp. 1–34.

Istatkova, R., Philipov, S., Yadamsurenghiin, G.O., Samdan, J. and Dangaa, S., 2008. Alkaloids from *Papaver nudicaule* L. *Natural Product Research*, 22(7), pp. 607–611.

Tatsis, E.C., Schaumlöffel, A., Warskulat, A.C., Massiot, G., Schneider, B. and Bringmann, G., 2012. Nudicaulins, yellow flower pigments of *Papaver nudicaule*: revised constitution and assignment of absolute configuration. *Organic letters*, 15(1), pp. 156–159.

### 4.1.3.14 *Papaver macrostomum* Boiss. & A. Huet

Synonym: *Papaver dalechianum* Fedde; *Papaver divergens* Fedde; *Papaver kurdistanicum* Fedde; *Papaver piptostigma* Bien. ex. Fedde; *Papaver tubuliferum* Fedde

Common names: Large-mouth poppy; gelincik, lala, kulilkasor (Turkey)

Habitat: grasslands, rocky soil, and mountains

Distribution: Turkey, Afghanistan, India, Himalaya, Iran, Pakistan, Russia, and Turkey

Botanical observation: This laticiferous herb grows to about 40 cm tall. The stems are terete and hairy. The leaves are simple, in rosette, sessile, and exstipulate. The blade is pinnatisect, serrate, and up to about 15 cm long. The flowers are showy, 3–6 cm diameter, solitary and terminal on a scape, which is 10–20 cm long. The calyx comprises 2 sepals, which are elliptic. The corolla includes 4 petals, which are bright red, 2–3 cm long, obovate, membranous, and marked with a blackish spot at base. The androecium includes numerous stamens, which are purplish and minute. The ovary is oblong. The capsule is ellipsoid, up to about 2.5 cm glaucous, ribbed, poricidal, and contain numerous tiny, reticulated seeds.

Medicinal use: Cough (Turkey)

*Weak antibacterial polar extract:* Ethanol extract of aerial parts inhibited the growth of *Staphylococcus aureus* (ATCC 6538), *Staphylococcus epidermidis* (ATCC 12228), and *Pseudomonas aeruginosa* (ATCC 1539) (Ünsal et al., 2007).

*Weak anticandidal polar extract:* Ethanol extract of aerial parts inhibited the growth of *Candida albicans* (ATCC 10231) and *Candida guilliermondii* (KUEN 998) with the MIC values of 2031.2 and 4062.5 µg/mL, respectively (Ünsal et al., 2007).

Commentary: Ünsal et al. (2007) evaluated the antibacterial activities of ethanol extract of aerial parts of *Papaver macrostomum* Boiss. & A. Huet collected from 2 different locations at 1 year interval and found different antimicrobial strengths. The sample collected from Malatya in June 2001 demonstrated an MIC value of 4062.5 µg/mL against *Pseudomonas aeruginosa* and the sample from Van collected in June 2002 evoked an MIC value of 3487.5 µg/mL. This is due to a difference in, at least, alkaloid contents. Mecambrine was isolated from the sample collected at Malatya, whereas a sample collected at Van yielded cheilantifoline and laudanosine. Protopine and rhoeadine were the major alkaloids found in a sample collected in Central Europe (Mnatsakanyan et al., 1977). Thus, the antimicrobial properties of medicinal plants depend on the time and place of collection, which explains the differences in results observed in literature (in addition to different types of antimicrobial methods used). It also gives rationale to the importance given by Asian healers and shamans (not uncommonly by mean of dreams (!)) to the location and time of collection of a specific plant to treat illnesses.

### References

Mnatsakanyan, V.A., Preininger, V., Šimánek, V., Juřina, J., Klasek, A., Dolejš, L. and Šantavý, F., 1977. Isolation and chemistry of the alkaloids from *Papaver macrostomum* BOISS. et HUET. *Collection of Czechoslovak Chemical Communications*, *42*(4), pp. 1421–1430.

Ünsal, Ç., Sarıyar, G., Akarsu, B.G. and Çevikbaş, A., 2007. Antimicrobial activity and phytochemical studies on Turkish samples of *Papaver macrostomum*. *Pharmaceutical Biology*, *45*(8), pp. 626–630.

### 4.1.3.15 *Papaver dubium* L.

Synonyms: *Papaver dubium* var. *laevigatum* (M. Bieb.) Elk.; *Papaver laevigatum* M. Bieb.; *Papaver litwinowii* Fedde ex Bornm. Engler; *Papaver rhoeas* var. *dubium* (L.) Schmalh.; *Papaver turbinatum* DC.

Habitat: Stream banks, roadsides, rocky slopes

Common names: Long-headed Poppy, blindeyes; kanguni (India); kopekyagi, lala, kulilkasor (Turkey)

Distribution: Turkey, Iran, Iraq, Afghanistan, Turkey, Pakistan, Himalayas, India, and Nepal

Botanical observation: This laticiferous herb grows to about 60 cm tall. The stems are terete and hairy and yield a milky latex upon incision. The leaves are simple, in rosette, sessile, and exstipulate. The blade is pinnatifid and 7–15 cm×2–3 cm. The flowers are showy, 3–5 cm diameter, solitary, and terminal on a scape, which is 10–20 cm long. The calyx comprises 2 sepals, which are ovoid. The corolla includes 4 petals, which are bright red or pinkish, 2–4 cm long, obovate, and membranous. The numerous stamens minute. The ovary is obconic. The capsule is somewhat cylindrical or obconic, up to about 2.5 cm long, glaucous, ribbed, poricidal, with 7–9 stigmatic rays at apex, and contain numerous tiny reniform seeds.

Medicinal use: Cough (Turkey)

*Strong broad-spectrum antibacterial mid-polar extract:* Diethyl ether fraction of aerial parts inhibited the growth of *Staphylococcus aureus* (ATCC 65538), *Staphylococcus epidermidis* (ATCC 12228), *Klebsiella pneumoniae* (ATCC 4352), *Proteus mirabilis* (ATCC 14153), *Escherichia coli* (ATCC 25922), and *Pseudomonas aeruginosa* (ATCC) with the MIC values of 19.5, 312.5, 1250, 625, 1250, and 625 µg/mL, respectively (Ünsal et al., 2009).

Commentary: The plant is known to produce berberine, stylopine, and allocryptopine (Mat et al., 2000) which may, at least, partially account for the antibacterial property in the medicinal use.

### References

Mat, A., Sariyar, G., Ünsal, Ç., Deliorman, A., Atay, M. and Özhatay, N., 2000. Alkaloids and bioactivity of *Papaver dubium* subsp. dubium and *P. dubium* subsp. laevigatum. *Natural Product Letters*, *14*(3), pp. 205–210.

Ünsal, Ç., Özbek, B., Sarıyar, G. and Mat, A., 2009. Antimicrobial activity of four annual *Papaver* species growing in Turkey. *Pharmaceutical Biology*, *47*(1), pp. 4–6.

### 4.1.3.16 *Papaver rhoeas* L.

Synonyms: *Papaver rhoeas* var. *strigosum* Boenn.; *Papaver strigosum* Schur

Common names: Shirley's poppy, corn poppy, common poppy; yu mei ren (China); laal posta, rakta posta, laal kaskas (India); ceybuhaten, şişık, gelincik (Turkey)

Habitats: Grassy open lands, waste places, and roadsides

Distribution: South West Asia, cultivated

Botanical observation: This laticiferous herb grows to about 90 cm tall. The stems are terete and hairy and yield a milky latex upon incision. The leaves are simple, in rosette, and exstipulate. The blade is lanceolate, pinnatifid, and 3–15×1–6 cm. The flowers are showy, 6–9 cm diameter, solitary and terminal on a scape, which is 10–15 cm long. The calyx comprises 2 sepals, which are elliptic and about 1.5 cm long. The corolla includes 4 petals, which are bright red, up to 4.5 cm long, obovate, and membranous. The numerous stamens are about 1 cm long and purplish. The gynoecium includes an obovoid ovary with 8–12 stigmas. The capsule is obovoid, up to about 1.8 cm long, glaucous, inconspicuously ribbed, poricidal, and contain numerous tiny reniform seeds.

Medicinal uses: Gargle (Japan); cough (Turkey)

*Strong broad-spectrum antibacterial mid-polar extract:* Diethyl ether fraction of aerial parts inhibited the growth of *Staphylococcus aureus* (ATCC 65538), *Staphylococcus epidermidis* (ATCC 12228), and *Escherichia coli* (ATCC 25922) with the MIC values of 39, 156.2, and 1250 µg/mL, respectively (Ünsal et al., 2009).

*Strong broad-spectrum antibacterial alkaloidal fraction:* Alkaloid fraction inhibited the grow of *Staphylococcus aureus*, *Staphylococcus epidermidis*, *Klebsiella pneumoniae*, *Escherichia coli*, and *Pseudomonas aeruginosa* with the MIC values of 1.2, 9.7, 39, 312, and 156 µg/mL, respectively (Coban et al., 2017).

*Moderate antifungal (yeast) mid-polar extract:* Diethyl ether fraction of aerial parts inhibited the growth of *Candida albicans* (ATCC 10231) with an MIC value of 625 µg/mL (Ünsal et al., 2009). Alkaloidal fraction inhibited the growth of *Candida albicans* with an MIC value of 2.4 µg/mL (Coban et al., 2017).

*Strong antibacterial (Gram-positive) amphiphilic aporphine alkaloid:* Roemerine (LogD = 3.5 at pH 7.4; molecular weight = 279.3 g/mol) inhibited the growth of methicillin-resistant *Staphylococcus aureus* (135) and *Staphylococcus aureus* (ATCC25913) with $MIC_{80}$ values of 64 and 64 µg/mL, respectively (Ma et al., 2015).

Roemerine

*Broad-spectrum antifungal aporphine:* Roemerine inhibited the growth of *Candida albicans* (SC 5314), *Candida glabrata* (8535), *Candida krusei* (4996), *Candida tropicalis* (8915), *Candida parapsilosis* (90018), and *Aspergillus fumigatus* (7544) with $MIC_{80}$ values of 256, 64, 32, 64, 32, 32, 64, and 64 µg/mL, respectively (Ma et al., 2015).

Commentary: The plant elaborates isoquinoline alkaloids including protopine and roemerine, the presence and proportions of which vary with location and time of collection; same goes for their antibacterial activity (Coban et al., 2017). The $MIC_{80}$ is the concentration required by an antibacterial or antifungal agent to inhibit the growth of 80% of bacteria or fungi (Rogge & Taft, 2016).

### References

Coban, I., Toplan, G.G., Özbek, B., Gürer, Ç.U. and Sarıyar, G., 2017. Variation of alkaloid contents and antimicrobial activities of *Papaver rhoeas* L. growing in Turkey and northern Cyprus. *Pharmaceutical Biology*, 55(1), pp. 1894–1898.

Ma, C., Du, F., Yan, L., He, G., He, J., Wang, C., Rao, G., Jiang, Y. and Xu, G., 2015. Potent activities of roemerine against *Candida albicans* and the underlying mechanisms. *Molecules*, 20(10), pp. 17913–17928.

Rogge, M. and Taft, D.R. eds., 2016. *Preclinical Drug Development*. CRC Press, USA.

Ünsal, Ç., Özbek, B., Sarıyar, G. and Mat, A., 2009. Antimicrobial activity of four annual *Papaver* species growing in Turkey. *Pharmaceutical Biology*, 47(1), pp. 4–6.

### 4.1.3.17  *Papaver somniferum* L.

Common names: Opium poppy, common poppy; ying su (China); afyum, afing gash, post, chosa, ahiphena, kasakasa (India), kooknar (Iran); posat (Afghanistan); haşhaş, haşkeş (Turkey); keshi-Ikkanshu (Japanese)

Habitat: Grassy open lands

Distribution: South West Asia (Turkey), cultivated in Afghanistan, Pakistan, India, China, Laos, Myanmar, and Thailand

Botanical observation: It is a laticiferous, fleshy, somewhat glaucous herb with a somewhat eery aura that grows to about 1 m tall. The stems are terete and hairy and yield a milky latex upon incision. The leaves are simple, spiral, and exstipulate. Basal leaves are petiolated and cauline leaves are amplexicaul. The blade is ovate to oblong, serrate, cordate at base, and 5–25 cm long. The flowers are showy, 5–12 cm diameter, solitary and terminal on a scape, which is up to 25 cm long. The calyx comprises 2 sepals, which are elliptic, and about 1.5–2.5 cm long. The corolla includes 4 petals, which are white, light or red, 3–7×3–11 cm, obovate, often marked with a dark spot at base, and membranous. The androecium includes numerous linear stamens, which are yellow and up to about 2 cm long. The ovary is globose, 1–2 cm in diameter and with 5–12 stigmas. The capsules are obovoid to subglobose, up to about 9 cm in diameter, glaucous, smooth, poricidal, with 5–12 apical stigma rays, and contain numerous tiny black seeds.

Medicinal use: Bronchitis (India)

*Weak broad-spectrum antibacterial polar extract:* Methanol extract of seeds inhibited the growth of clinical strains of *Enterococcus faecalis*, *Acinetobacter baumannii*, *Klebsiella pneumoniae*, *Proteus mirabilis*, and *Pseudomonas aeruginosa* with MIC/MBC values of 1.5/3.4, 4.2/9.6, 1.5/3.4, 4.2/9.6, and 1.5/3.4 mg/mL, respectively (Rath & Padhy, 2014).

*Strong antiviral (enveloped monopartite linear dimeric single-stranded (+) RNA) polar extract:* Methanol extract of seeds inhibited the replication of Human immunodeficiency virus type-1 (strain NL4.3) in the Human CD4+ T cells by 44.9% with a maximum noncytotoxic concentration of 7.5 μg/mL (Sabde et al., 2011). Methanol extract of roots inhibited the replication of Human immunodeficiency virus type-1 (strain NL4.3) in the Human CD4+ T cells by 40.8% with a maximum noncytotoxic concentration of 10 μg/mL (Sabde et al., 2011).

*Strong antiviral (enveloped monopartite linear dimeric single-stranded (+) RNA) non-polar extract:* Hexane extract of fruits inhibited the replication of Human immunodeficiency virus type-1 (strain NL4.3) in the Human CD4+ T cells by 73.9% with a maximum noncytotoxic concentration of 20 μg/mL (Sabde et al., 2011).

Commentary: The plant contains sanguinarine, protopine, and magnoflorine (Furuya et al., 1972), which are antibacterial and antifungal. Sanguinarine is a benzophenanthridine alkaloid, which intercalates bacterial and fungal DNA (Maiti et al., 1982). The plant also produces meconic acid, which should have some levels of antibacterial or even antifungal or antiviral activities because of its γ-pyrone framework (Rocha et al., 1994). Also, the plant produces the central analgesic alkaloid morphine, which displayed very weak broad-spectrum antibacterial activities (Rota et al., 1997). Morphine is a μ-opioid agonist and could be an antibiotic potentiator because another μ-opioid agonist loperamide, which has no antibacterial activities, potentiates tetracycline effectiveness against Gram-negative bacteria (but not Gram-positive bacteria) (Brown, 2015). Consider that neuroactive (as well as vasoactive) natural products are not uncommonly able to bind to and inhibit efflux pumps in bacteria.

### References

Brown, D., 2015. Antibiotic resistance breakers: can repurposed drugs fill the antibiotic discovery void? *Nature Reviews Drug discovery*, *14*(12), p. 821.

Furuya, T., Ikuta, A. and Syōno, K., 1972. Alkaloids from callus tissue of *Papaver somniferum*. *Phytochemistry*, *11*(10), pp. 3041–3044.

Maiti, M., Nandi, R. and Chaudhuri, K., 1982. Sanguinarine: a monofunctional intercalating alkaloid. *FEBS Letters*, *142*(2), pp. 280–284.

Rath, S. and Padhy, R.N., 2014. Monitoring in vitro antibacterial efficacy of 26 Indian spices against multidrug resistant urinary tract infecting bacteria. *Integrative Medicine Research*, *3*(3), pp. 133–141.

Rocha, L., Marston, A., Auxiliadora, M., Kaplan, C., Stoeckli-Evans, H., Thull, U., Testa, B. and Hostettmann, K., 1994. An antifungal γ-pyrone and xanthones with monoamine oxidase inhibitory activity from *Hypericum brasiliense*. *Phytochemistry*, *36*(6), pp. 1381–1385.

Rota, S., Kaya, K., Timliođlu, O., Karaca, Ö., Yzdeb, S. and Öcal, E., 1997. Do the opioids have an antibacterial effect? *Canadian Journal of Anesthesia/Journal canadien d'anesthésie*, *44*(6), pp. 679–680.

Sabde, S., Bodiwala, H.S., Karmase, A., Deshpande, P.J., Kaur, A., Ahmed, N., Chauthe, S.K., Brahmbhatt, K.G., Phadke, R.U., Mitra, D. and Bhutani, K.K., 2011. Anti-HIV activity of Indian medicinal plants. *Journal of Natural Medicines*, *65*(3–4), pp. 662–669.

## 4.1.4 Family Menispermaceae A.L de Jussieu (1789)

The family Menispermaceae includes about 71 genera and 450 species of tropical or subtropical climbers not uncommonly growing from massive tubers. Transversal section of stems present characteristic radial rays. The leaves are simple, spiral, and exstipulate. The petiole is enlarged at base and apex. The blade is entire and often peltate. The inflorescences are mostly umbelliform cymes of minute flowers. The calyx comprises 3–12 sepals. The corolla includes mostly 6 petals. The androecium includes 6–8 stamens. The gynoecium comprises 1–6, free carpels containing 1 ovule. The fruit is a drupe containing a horseshoe-shaped or moon-shaped seed. Members in this family bring to being enthralling series of antimicrobial benzylisoquinoline alkaloids and/or their bisbenzylisoquinoline, aporphine, and protoberberine derivatives.

### 4.1.4.1 *Cissampelos hirsuta* Buch.-Ham. ex DC.

Synonym: *Cissampelos pareira* L.

 Common names: False pareira root; akanbindi; niltat (Bangladesh); peepra mool (India); tiêt́ dê (Vietnam); katori (Pakistan)

 Habitat: Forests

 Distribution: Tropical Asia

Botanical observation: This woody climber grows to about 5 m long. The leaves are simple, alternate, and exstipulate. The petiole is hairy. The blade is peltate, rotund, 2.5–12 cm × 2.5–11.5 cm, with 5–7 pairs of secondary nerves, base cordate, hairy below, papery, and emarginate at apex. The inflorescence is an axillary or corymbose cyme of minute flowers. The calyx includes 1 or 4, obovate-oblong, hairy sepals. The corolla is tubular and 4-toothed and hairy or presents 1 petal. The androecium includes 4 stamens united into a short column, the anthers connate, encircling the top of the column. The ovary is hairy and develops a trifid style. The drupe is about 5 mm long, red, and shelters a horseshoe-shaped seed.

Medicinal uses: Blennorrhea (Cambodia, Laos, Vietnam); fever, leprosy (India)

*Strong antiviral (enveloped monopartite linear single-stranded (+) RNA) polar extract:* Methanol extract of aerial parts inhibited Dengue virus-1, -2, -3, and -4 with $IC_{50}$ equal to 11.1, 1.9, 3.1, and 1.2 µg/mL, respectively (Sood et al., 2015).

*Strong antiviral (enveloped monopartite linear dimeric single-stranded (+) RNA) extract:* Chloroform fraction of aerial parts inhibited the replication of the Human immunodeficiency virus type-1 (strain NL4.3) in the Human CD4$^+$ T cells by 24.3% with a maximum noncytotoxic concentration of 20 µg/mL (Sabde et al., 2011).

*In vivo antiviral (enveloped monopartite linear single-stranded (+) RNA) polar extract:* Methanol extract of aerial parts given orally twice a day for 5 days at a dose of 125 mg to AG129 mouse infected with Dengue virus-2 increased the survival rate by 50% (Sood et al., 2015).

*Strong antiviral (enveloped monopartite linear dimeric single-stranded (+) RNA) amphiphilic bisbenzylisoquinoline:* Cycleanine (LogD = 4.0 at pH 7.4; molecular mass = 622.7 g/mol) inhibited the replication of the Human immunodeficiency virus type-2 (ROD-strain) with an $IC_{50}$ value of 1.8 µg/mL and a selectivity index of 9 in (Otshudi et al., 2005).

Cycleanine

Commentary: (i) The plant generates an interesting series of isoquinoline alkaloids such as magnoflorine, magnocurarine, cissamine, the bisbenzylisoquinolnes curine, hayatinine, and cycleanine (Kupchan et al., 1965; Bala et al., 2019). Consider that bisbenzyl isoquinolines are not uncommonly antibacterial, antifungal, or antiviral. (ii) Cycleanine has low parenteral toxicity in rodents and preclinical antiretroviral studies are warranted. Being active against the Human immunodeficiency virus, cycleanine is probably of value against COVID-19.

**References**

Bala, M., Kumar, S., Pratap, K., Verma, P.K., Padwad, Y. and Singh, B., 2019. Bioactive isoquinoline alkaloids from *Cissampelos pareira*. *Natural Product Research*, *33*(5), pp. 622–627.

Kupchan, S.M., Patel, A.C. and Fujita, E., 1965. Tumor inhibitors VI. Cissampareine, new cytotoxic alkaloid from *Cissampelos pareira*. Cytotoxicity of bisbenzylisoquinoline alkaloids. *Journal of Pharmaceutical Sciences*, *54*(4), pp. 580–583.

Otshudi, A.L., Apers, S., Pieters, L., Claeys, M., Pannecouque, C., De Clercq, E., Van Zeebroeck, A., Lauwers, S., Frederich, M. and Foriers, A., 2005. Biologically active bisbenzylisoquinoline alkaloids from the root bark of *Epinetrum villosum*. *Journal of Ethnopharmacology*, *102*(1), pp. 89–94.

Sabde, S., Bodiwala, H.S., Karmase, A., Deshpande, P.J., Kaur, A., Ahmed, N., Chauthe, S.K., Brahmbhatt, K.G., Phadke, R.U., Mitra, D. and Bhutani, K.K., 2011. Anti-HIV activity of Indian medicinal plants. *Journal of Natural Medicines*, *65*(3–4), pp. 662–669.

Sood, R., Raut, R., Tyagi, P., Pareek, P.K., Barman, T.K., Singhal, S., Shirumalla, R.K., Kanoje, V., Subbarayan, R., Rajerethinam, R. and Sharma, N., 2015. *Cissampelos pareira* Linn: natural source of potent antiviral activity against all four dengue virus serotypes. *PLoS Neglected Tropical Diseases*, *9*(12), p. e0004255.

### 4.1.4.2 *Cocculus hirsutus* (L.) Diels

Synonyms: *Cebatha hirsuta* Kuntze; *Cocculus hirsutus* (L.) W. Theob.; *Cocculus villosus* DC.; *Menispermum hirsutum* L.

Common names: Broom creeper; huyer, daikhai, jalajmani (Bangladesh); jamti-ki-bel (Pakistan); jal jamini (India)

Habitat: Forests

Distribution: India, Pakistan

Botanical observation: This climber has hairy stems. The leaves are simple, alternate, and exstipulate. The petiole is 0.5–2.5 cm long. The blade is 4–8 cm×2.5–6 cm, somewhat deltoid, cordate at base or truncate, apex obtuse, hairy below, and presents 2–5 pairs of secondary nerves. The inflorescence is axillary and cymose. The calyx comprises 6 sepals, which are minute and hairy. The 6 petals, are triangular, emarginate, and minute. The androecium includes 6 stamens. The gynoecium includes 3 carpels. The drupes are dark purple, 4–8 mm long, and contain a horseshoe-shaped seed.

Medicinal uses: Leucorrhea (Bangladesh); cough, leucorrhea (India)

*Broad-spectrum antibacterial alkaloidal fraction:* Alkaloidal fraction of roots inhibited the growth of *Bacillus subtilis* and *Escherichia coli* with inhibition zone diameters of 18 and 17 mm, respectively (cup plate method; 500 µg) (Mangathayaru & Venkateswarlu, 2005). The effects of this fraction to *Bacillus subtilis* was comparable to that of benzyl penicillin (0.1 µg) (Mangathayaru & Venkateswarlu, 2005).

*Moderate antimycobacterial polar extract:* Ethanol extract inhibited the growth of *Mycobacterium tuberculosis* (H37Rv) with MIC values of 500 µg/mL and hampered the growth of *Mycobacterium tuberculosis* clinical strains with MIC values ranging from 250 to 500 µg/mL (Gupta et al., 2018).

Commentary: The plant produces the isoquinoline alkaloids trilobine, coclaurine, hirsutine, and magnoflorine (Rasheed et al., 1991). Consider that magnoflorine has anticandidal properties (Kim et al., 2018)

**References**

Gupta, V.K., Kaushik, A., Chauhan, D.S., Ahirwar, R.K., Sharma, S. and Bisht, D., 2018. Antimycobacterial activity of some medicinal plants used traditionally by tribes from Madhya

Pradesh, India for treating tuberculosis related symptoms. *Journal of Ethnopharmacology*, *227*, pp. 113–120.

Kim, J., Bao, T.H.Q., Shin, Y.K. and Kim, K.Y., 2018. Antifungal activity of magnoflorine against *Candida strains*. *World Journal of Microbiology and Biotechnology*, *34*(11), p. 167.

Mangathayaru, K. and Venkateswarlu, V., 2005. Antimicrobial activity of *Cocculus hirsutus* (Linn). *Indian Journal of Pharmaceutical Sciences*, *67*(5), p. 619.

Rasheed, T., Khan, M.N.I., Zhadi, S.S.A. and Durrani, S., 1991. Hirsutine: a new alkaloid from *Cocculus hirsutus*. *Journal of Natural Products*, *54*(2), pp. 582–584.

### 4.1.4.3  *Coscinium fenestratum* Colebr.

Common names: Akar kuning (Indonesia), kunyit-kunyit babi, abang asuh, perawan (Malaysia); khruea hen (Thailand), vằng đắng (Vietnam)

Habitat: Forests

Distribution: India, Sri Lanka, Thailand, Cambodia, Vietnam, Peninsular Malaysia, and Indonesia

Botanical observation: This stout woody climber grows to about 10 m long. The cross sections of stems are golden yellow and present radial rays. The leaves are simple, alternate, and exstipulate. The petiole is 3–15 cm long, conspicuously swollen at both ends, and geniculate at base. The blade is broadly lanceolate, glossy above, glaucous light brown below, 10–32×8–22 cm, with 2–5 pairs of secondary nerves sunken above. The inflorescences are 5–11 cm long lax racemes of globose and 6–7 mm across heads on 1–3 cm long peduncles. The flowers are minute, yellowish or whitish. The calyx includes 9 sepals. Corolla none. The androecium comprises 6 stamens. The gynoecium comprises 3 free, subglobose and somewhat hairy carpels. The drupes are globose, brown, orange, or yellow, hairy, up to 3 cm in diameter, and contain subglobose seeds.

Medicinal uses: Wounds (India, Malaysia); fever (India, Indonesia, Vietnam); prevention of tetanus (India); wounds, antiseptic (India)

*Strong antibacterial (Gram-negative) polar extract*: Methanol extract of wood inhibited *Neisseria gonorrhea* with MIC of 47.3 µg/mL (Chomnawang et al., 2009) and inhibited 11 clinical isolates with MIC ranging from 19.5 to 112.1 µg/mL (Chomnawang et al., 2009).

*Strong broad-spectrum antibacterial hydrophilic protoberberine alkaloid:* Berberine (LogD = −1.3 at pH 7.4; molecular weight = 336.4 g/mol) inhibited the growth of *Escherichia coli*, *Pseudomonas aeruginosa*, *Proteus mirabilis*, *Klebsiella pneumonia*, *Acinetobacter baumannii*, *Staphylococcus aureus*, and *Bacillus subtilis* with MIC values of 32, 64, 32, 8, 8, 64, and 128 µg/mL, respectively (Orhan et al., 2007). Berberine inhibited *Neisseria gonorrhea* with an MIC of 13.5 µg/mL and inhibited 11 clinical isolates with MIC ranging from 12.5 to 20.6 µg/mL (Chomnawang et al., 2009).

Commentary: The plant abounds with protoberberine alkaloids, mainly berberine and jatrorrhizine (Pinho et al., 1992), which have both potent antibacterial and antifungal properties.

### References

Chomnawang, M.T., Trinapakul, C. and Gritsanapan, W., 2009. In vitro antigonococcal activity of *Coscinium fenestratum* stem extract. *Journal of Ethnopharmacology*, *122*(3), pp. 445–449.

Nair, G.M., Narasimhan, S., Shiburaj, S. and Abraham, T.K., 2005. Antibacterial effects of *Coscinium fenestratum*. *Fitoterapia*, *76*(6), pp. 585–587.

Orhan, I., Özçelik, B., Karaoğlu, T. and Şener, B., 2007. Antiviral and antimicrobial profiles of selected isoquinoline alkaloids from Fumaria and Corydalis species. *Zeitschrift für Naturforschung C*, *62*(1–2), pp. 19–26.

Pinho, P.M., Pinto, M.M., Kijjoa, A., Pharadai, K., Díaz, J.G. and Herz, W., 1992. Protoberberine alkaloids from *Coscinium fenestratum*. *Phytochemistry*, *31*(4), pp. 1403–1407.

### 4.1.4.4  *Cyclea barbata* Miers

Synonyms: *Cyclea ciliata* Craib; *Cyclea wallichii* Diels

Common names: Green grass jelly; patalpur; wambokhor (Bangladesh); krung kha mao (Thailand); cincau hijau (Indonesia)

Habitat: Forests margins and shrublands

Distribution: China, India, Laos, Myanmar, Thailand, Vietnam, and Indonesia

Botanical observation: This climber grows to about 5 m long. The young stems are hairy. The leaves are simple, spiral, and exstipulate. The petiole is 1–5 cm long and hairy. The blade is peltate, round, 4–10×2.5–8 cm, hairy below, round or emarginated at base, obtuse at apex, and with about 5 pairs of secondary nerves. The inflorescence is axillary or cauliflorous cluster of minute flowers. The calyx is cupular, minute, and 4–5-lobed or includes 2 sepals. The corolla cup-shaped (turbinate), and or includes 2 petals. The ovary is hairy and develops stigmas, which are 3-laciniate. The drupes are red.

Medicinal uses: Tonsilitis (India); fever, typhoid (Indonesia)

*Weak antibacterial (Gram-positive) amphiphilic bisbenzylisoquinoline alkaloid:* Tetrandrine (LogD=4.1 at pH 7.4; molecular weight=622.7 g/mol) inhibited the growth of *Staphylococcus aureus* with an MIC of 250 µg/mL (Lee et al., 2011). Tetrandrine inhibited the growth of *Staphylococcus aureus* (ATCC 25923) and methicillin-resistant *Staphylococcus aureus* (ATCC 33591) (Lee et al., 2011).

*Antibiotic potentiator bisbenzylisoquinoline alkaloid:* Combination of tetrandrine and ethidium bromide (substrate for efflux pumps of which NorA) lowered the MIC of tetrandrine and ethidium bromide by 2–4-folds against clinical strains expressing the mecA gene (Lee et al., 2012).

*Antifungal potentiator isoquinoline alkaloid:* Time-kill curves showed that at 48 hours the viable cell counts of *Candida albicans* treated with ketoconazole and tetrantrine were at least $2 \log_{10}$ CFU/mL lower compared to *Candida albicans* treated with corresponding doses of keto-conazole (Zhang et al., 2010). Synergy was also reported with posaconazole against *Aspergillus fumigatus* (Li et al., 2017).

*Antiviral (enveloped monopartite linear double-stranded DNA) bisbenzylisoquinoline alkaloid:* Tetrandrine given intraperitoneally and daily at a dose of 30 mg/Kg for 7 days evoked some levels of protection in BALB/c mice against cornea Herpes simplex virus type-1 infection (Hu et al., 1997)

Commentaries:

(i) The plant produces bis-benzylisoquinoline alkaloids, such as tetrandrine (Guinaudeau et al., 1993). Tetrandrine, which is a calcium channel antagonist in mammalian cells (Takemura et al., 1996) inhibited efflux pumps in *Staphylococcus aureus* (Takemura et al., 1996). Verapamil is another example of calcium channel antagonist that inhibits efflux pump in bacteria (Gupta et al., 2014). Reserpine is also an inhibitor of bacteria efflux pump (Garvey & Piddock, 2008) and calcium channel antagonist (Greenberg, 1987). The reason why calcium channel antagonists have the tendency to inhibit bacterial efflux pump could be the correlations between bacterial efflux pumps and bacterial calcium transport (Jones et al., 2003). Therefore, natural products known for being calcium channel inhibitors should be screened as antibiotic potentiators. NorA belongs to the major facilitator efflux pumps group.

(ii) Consider that Middle East respiratory syndrome coronavirus translocates in the endoly-sosomal system of host cells via a mechanism involving $Ca^{2+}$-permeable channels (Gunaratne et al., 2018). Hence, natural products known for being calcium channel inhibitors have poten-tial as anti-coronavirus agents. One such natural product is tetrandrine, which impairs Middle East respiratory syndrome coronavirus translocation in host cell (Gunaratne et al., 2018). Examination of tetrandrine against the Severe acute respiratory syndrome-associated coronavi-rus is warranted.

### References

Garvey, M.I. and Piddock, L.J., 2008. The efflux pump inhibitor reserpine selects multidrug-resistant *Streptococcus pneumoniae* strains that overexpress the ABC transporters PatA and PatB. *Antimicrobial Agents and Chemotherapy*, 52(5), pp. 1677–1685.

Greenberg, D.A., 1987. Calcium channels and calcium channel antagonists. *Annals of Neurology: Official Journal of the American Neurological Association and the Child Neurology Society*, 21(4), pp. 317–330.

Guinaudeau, H., Lin, L.Z., Ruangrungsi, N. and Cordell, G.A., 1993. Bisbenzylisoquinoline alkaloids from *Cyclea barbata*. *Journal of Natural Products*, 56(11), pp. 1989–1992.

Gunaratne, G.S., Yang, Y., Li, F., Walseth, T.F. and Marchant, J.S., 2018. NAADP-dependent $Ca^{2+}$ signaling regulates Middle East respiratory syndrome-coronavirus pseudovirus translocation through the endolysosomal system. *Cell Calcium*, 75, pp. 30–41.

Gupta, S., Cohen, K.A., Winglee, K., Maiga, M., Diarra, B. and Bishai, W.R., 2014. Efflux inhibition with verapamil potentiates bedaquiline in *Mycobacterium tuberculosis*. *Antimicrobial Agents and Chemotherapy*, 58(1), pp. 574–576.

Hu, S., Dutt, J., Zhao, T. and Foster, C.S., 1997. Tetrandrine potently inhibits herpes simplex virus type-1-induced keratitis in BALB/c mice. *Ocular Immunology and Inflammation*, 5(3), pp. 173–180.

Jones, H.E., Holland, I.B., Jacq, A., Wall, T. and Campbell, A.K., 2003. *Escherichia coli* lacking the AcrAB multidrug efflux pump also lacks nonproteinaceous, PHB–polyphosphate $Ca^{2+}$ channels in the membrane. *Biochimica et Biophysica Acta (BBA)-Biomembranes*, 1612(1), pp. 90–97.

Lee, Y.S., Han, S.H., Lee, S.H., Kim, Y.G., Park, C.B., Kang, O.H., Keum, J.H., Kim, S.B., Mun, S.H., Shin, D.W. and Kwon, D.Y., 2011. Synergistic effect of tetrandrine and ethidium bromide against methicillin-resistant *Staphylococcus aureus* (MRSA). *The Journal of Toxicological Sciences*, 36(5), pp. 645–651.

Li, S.X., Song, Y.J., Jiang, L., Zhao, Y.J., Guo, H., Li, D.M., Zhu, K.J. and Zhang, H., 2017. Synergistic effects of tetrandrine with posaconazole against aspergillus fumigatus. *Microbial Drug Resistance*, 23(6), pp. 674–681.

Takemura, H., Imoto, K., Ohshika, H. and Kwan, C.Y., 1996. Tetrandrine as a calcium antagonist. *Clinical and Experimental Pharmacology and Physiology*, 23(8), pp. 751–753.

Zhang, H., Wang, K., Zhang, G., Ho, H.I. and Gao, A., 2010. Synergistic anti-candidal activity of tetrandrine on ketoconazole: an experimental study. *Planta Medica*, 76(01), pp. 53–61.

### 4.1.4.5    *Cyclea peltata* Hook. f. & Thomson

Synonyms: *Cocculus burmanni* DC.; *Cyclea burmanni* (DC.) Hook.f. & Thoms.; *Cyclea peltata* (Poiret) Hook.f. & Thoms.; *Menispermum peltatum* Poir.

Common name: Rajpatha, padathali, seenthilkodi (India)

Habitat: Forests

Distribution: India, Andaman and Nicobar Islands

Botanical observation: This slender climber grows from tuberous roots. The young stems are hairy and grooved. The leaves are simple, spiral, and exstipulate. The petiole is 2.5–7 cm long and slender. The blade is peltate, deltoid, glossy, 10–15×4–6 cm, retuse at base, acute at apex, and with about 1–2 pairs of secondary nerves. The inflorescence is axillary racemes of minute flowers. The calyx is campanulate and 5–8-lobate or consists of 1 sepal. The corolla is 5–8-lobate or consists of 1 petal. The 4 stamens are joined in a column. The ovary consists of a single carpel and develops a stigma, which is 3–5-lobate. The drupes are subglobose and dull white, about 4 mm in diameter, and containing a horseshoe-shaped seed.

Medicinal use: Small pox, fever, coryza (India)

*Broad-spectrum antibacterial polar extract*: Ethanol extract of (10 mg/mL/disc) inhibited the growth of *Staphylococcus aureus*, *Bacillus cereus*, *Escherichia coli*, *Klebsiella pneumonia*, *Pseudomonas aeruginosa*, *Salmonella typhi*, *Serratia marsesens*, and *Vibrio cholera* with inhibition zone diameters of 16.3, 10.4, 11.2, 14.5, 9.3, 7.6, 7.1, and 7.9 mm, respectively (Abraham & Thomas, 2012).

Commentary: The plant brings to being the bisbenzyl isoquinoline alkaloids: tetrandrine, disochondrodendrine, and fangchinoline (Kupchan et al., 1961).

### References

Abraham, J. and Thomas, T.D., 2012. Antibacterial activity of medicinal plant *Cyclea peltata* (Lam) Hooks & Thoms. *Asian Pacific Journal of Tropical Disease*, 2, pp. S280–S284.

Kupchan, S.M., Yokoyama, N. and Thyagarajan, B.S., 1961. Menispermaceae alkaloids II. The alkaloids of *Cyclea peltata*. *Journal of Pharmaceutical Sciences*, 50, pp. 164–167.

### 4.1.4.6   *Hypserpa nitida* Miers

Synonyms: *Hypserpa cuspidata* (Hook. f. & Thomson) Miers; *Hypserpa laevifolia* Diels; *Limacia cuspidata* Hook. f. & Thomson

Common names: Serpa (Bangladesh); ye hua teng (China)

Habitat: Forests

Distribution: China, Bangladesh, India, Indonesia, Laos, Malaysia, Myanmar, the Philippines, Sri Lanka, and Thailand

Botanical observation: It is a woody climber. The stems are hairy at apex. The leaves are simple, spiral, and exstipulate. The petiole is 1–2 cm long and somewhat hairy and swollen near the blade. The blade is ovate or broadly elliptic, 4–12×1.5 cm, glossy, rounded at base, acuminate at apex, and with 1 pair of secondary nerves. The inflorescences are axillary and 1–2 cm long clusters of minute flowers. The 7–11 sepals are hairy. The corolla includes 4 or 5 subobovate petals. The 5–10 stamens, are free. The gynoecium includes 2 carpels. The drupes are subglobose, red, and contain horseshoe-shaped seeds.

Medicinal use: Toothache (Bangladesh)

*Strong antiviral (enveloped circular double-stranded DNA) liphophilic proaporphine alkaloids:* Pronuciferine and glaziovine exhibited $IC_{50}$ value of 0.04 and 0.008 mM on Hepatitis B virus (secretion of the Hep G2.2.15 cell line) (Cheng et al., 2007).

Glaziovine

*Weak antiviral (enveloped circular double-stranded DNA) pyrrolidine alkaloid:* Dauricumidine exhibited an IC$_{50}$ value of 0.450 mM (selectivity index = 4.1) on Hepatitis B virus surface antigen secretion by Hep G2.2.15 cell line (Cheng et al., 2007).

### Reference
Cheng, P., Ma, Y.B., Yao, S.Y., Zhang, Q., Wang, E.J., Yan, M.H., Zhang, X.M., Zhang, F.X. and Chen, J.J., 2007. Two new alkaloids and active anti-Hepatitis B virus constituents from *Hypserpa nitida. Bioorganic & Medicinal Chemistry Letters, 17*(19), pp. 5316–5320.

### 4.1.4.7 *Pericampylus glaucus* (Lam.) Merr.

Synonyms: *Cocculus incanus* Colebr.; *Coscinium colaniae* Gagnep.; *Menispermum glaucum* Lam.; *Pericampylus formosanus* Diels; *Pericampylus incanus* (Colebr.) Miers; *Pericampylus omeiensis* W.Y. Lien; *Pericampylus trinervatus* Yamam.

Common names: Barak kant, pipalpati (India); lõi tiê`n (Vietnam)
Habitat: Forests, forest margins, roadsides
Distribution: China, India, Indonesia, Laos, Malaysia, Myanmar, the Philippines, Thailand, and Vietnam.

Botanical observation: This slender and pendulous climber grows to a length of about 10 m long from a massive tuber (difficult to find). The leaves are simple, spiral, pendulous, and exstipulate. The petiole is slender and 3–7 cm long. The blade is triangular to obordate, 3.5–8 cm × 4.5–10 cm, somewhat dull light green, obtuse or rounded at apex, and with 1–3 pairs of secondary nerves. The cymes are corymbose, axillary, pendulous, seldom seen, and 2–10 cm long. The 9 sepals are hairy. The corolla includes 6 petals. The androecium is made of 6 stamens. The ovary develops a 2-lobed stigma. The drupes are red and contain horseshoe-shaped seeds.

Medicinal use: Fever (India)

*Strong antiviral (enveloped monopartite linear dimeric single-stranded (+) RNA) azafluoranthene alkaloid:* Norrufefscine inhibited Human immunodeficiency virus type-1 with the an EC$_{50}$ of 10.9 µM and a selectivity index of 45.7 (Yan et al., 2008)

*Moderate antiviral (enveloped monopartite linear dimeric single-stranded (+) RNA) protoberberine alkaloid* (-)-8-oxotetrahydropalmatine inhibited Human immunodeficiency virus type-1 with an EC$_{50}$ value of 14.1 µM and a selectivity index of 18.8 (Yan et al., 2008).

*Very weak antiviral (enveloped circular double-stranded DNA) azafluoranthene alkaloid:* Norruffescine isolated from the aerial parts inhibited Hepatitis B virus surface antigen expression in Hep G2.2.15 cells with an IC$_{50}$ of 0.9 mM and selectivity index >4.1 (Yan et al., 2008).

*Weak antiviral (enveloped circular double-stranded DNA) protoberberine alkaloid:* (-)-8-oxotetrahydropalmatine isolated from the aerial parts inhibited Hepatitis B virus surface antigen expression in Hep G2.2.15 cells with an IC$_{50}$ of 100 µM and a selectivity index of 22.4 (Yan et al., 2008).

Commentary: *In vivo* antiretroviral examination of Norruffescine is warranted. Consider that members of the family Menispermaceae are not uncommonly capable to inhibit Human immunodeficiency virus reverse transcriptase, this is the case, for instance, of *Sinomenium acutum* (Thunb.) Rehder & E.H. Wilson (Min et al., 2001). Is Norruffescine of value against COVID-19?

### References
Min, B.S., Kim, Y.H., Tomiyama, M., Nakamura, N., Miyashiro, H., Otake, T. and Hattori, M., 2001. Inhibitory effects of Korean plants on HIV -1 activities. *Phytotherapy Research, 15*(6), pp. 481–486.
Yan, M.H., Cheng, P., Jiang, Z.Y., Ma, Y.B., Zhang, X.M., Zhang, F.X., Yang, L.M., Zheng, Y.T. and Chen, J.J., 2008. Periglaucines A-D, anti-HBV and -HIV1 alkaloids from *Pericampylus glaucus. Journal of Natural Products, 71*(5), pp. 760–763.

### 4.1.4.8  *Stephania glabra* (Roxb.) Miers

Synonyms: *Clypea glabra* (Roxb.) Wight & Arn. ex Voigt; *Cissampelos glabra* Roxb.

Common name: Muchi lota (Bangladesh); xi zang di bu rong (China)

Habitat: Shrublands

Distribution: China, Bangladesh, India, Myanmar, Nepal, and Thailand

Botanical observation: This herbaceous climber grows from a massive tuber. The stems are striate and glabrous. The leaves are simple, spiral, and exstipulate. The petiole is slender, 5–15 cm or longer, geniculate, and thickened at base. The blade is peltate, broadly triangular or rotund, 4–14×4–15 cm, membranous, rounded at base, acute at apex, and, palmately veined. The inflorescences are 4–8 cm long axillary clusters of minute umbels. The calyx comprises 6 sepals. The corolla includes 3 petals. The androecium is a peltate synandrium. The gynoecium comprises a single carpel. The drupes are ovoid.

Medicinal use: Mycoses (Bangladesh).

*Strong antibacterial (Gram-positive) hasubanalactam alkaloid*: Glabradine isolated from the tubers inhibited the growth of *Staphylococcus aureus* and *Streptococcus mutans* (Semwal & Rawat, 2009).

*Strong antifungal (filamentous) hasubanalactam alkaloid*: Glabradine isolated from the tubers inhibited the growth of *Microsporium gypseum*, *Microsporium canis*, and *Trichophytum rubrum* with MIC values of 25, 25, and 50 µg/mL (Semwal & Rawat, 2009).

#### Reference

Semwal, D.K. and Rawat, U., 2009. Antimicrobial hasubanalactam alkaloid from *Stephania glabra*. *Planta Medica*, 75(04), pp. 378–380.

### 4.1.4.9  *Stephania japonica* (Thunb.) Miers

Synonyms: *Menispermum japonicum* Thunb.; *Stephania hernandifolia* (Willd.) Walp.

Common names: Akanadi, nimuka, maknadi, thaya nuya (Bangladesh); para (India); cam thảo (Vietnam)

Habitat: Forests and around villages

Distribution: India, Sri Lanka, Nepal, Bangladesh, Laos, Myanmar, Thailand, Vietnam, Malaysia, Indonesia, China, Korea, Japan, Korea, Australia, and Pacific islands

Botanical observation: This climber grows to about 5 m long. The leaves are simple, spiral, and exstipulate. The petiole is 3–12 cm long. The blade is peltate, triangular-rotund, 5–12 cm in diameter, glaucous below, glossy above, rounded at base, acute at apex, and with 8–11 pairs of secondary nerves. The cymes are axillary. The calyx includes 3–4 or 6–8 sepals, which are yellowish green, and minute. The 3 or 4 petals are yellow. The androecium is ovoid, glossy and develops a stigma, which is lacerate. The drupes are red, obovate, 6–8 mm long, and contain a moon-shaped seed.

Medicinal uses: Small-pox, jaundice (India); diarrhea (Vietnam)

*Strong antiviral (enveloped monopartite linear single-stranded (+) RNA) amphiphilic bis-benzylisoquinoline alkaloids:* Tetrandrine (LogD=4.1 at pH 7.4; molecular weight=622.7 g/mol) and fangchinoline (LogD =3.4; molecular weight=608.7 g/mol) inhibited Human Coronavirus OC43 replication in, MRC-5 cells with $IC_{50}$ values of 0.3 and 1 µM, respectively, with selectivity indices of 40.19 and 11.46 (Kim et al., 2019).

Tetrandrine

Fangchinoline

*Strong antiviral (enveloped monopartite linear dimeric single-stranded (+) RNA) amphiphilic bis-benzylisoquinoline alkaloid:* Fangchinoline inhibited the replication of Human immunodeficiency virus type-1 (strain NL4.3), Human immunodeficiency virus type-1 (LAI), and Human immunodeficiency virus type-1 (BaL) with $EC_{50}$ values of 0.8, 1.3, and 1.7 μM by inhibiting gp160 proteolytic processing (Wan et al., 2012).

Commentaries:

(i) The bis-benzisoquinoline alkaloids trilobine isolated from the tubers was as effective as verapamil in decreasing the resistance of adriamicin-resistant human breast adenocarcinoma cells (Hall & Chang, 1997). Extract of the plant inhibited P-glycoprotein (Lee et al., 2018). Natural products decreasing the resistance of cancer cells to chemotherapeutic agents by P-glycoproteins inhibit efflux pumps in bacteria. Note that reserpine, verapamil, and quinidine, which are calcium channel blockers and chemotherapeutic potentiators (Hall & Chang, 1997), are known antibiotic potentiators. Trilobine could be an antibiotic potentiator, and this needs to be ascertained.

(ii) The bis-benzylisoquinoline alkaloids cepharanthine from *Stephania rotunda* Lour. inhibited Human Coronavirus OC43 replication in MRC-5 cells with an $IC_{50}$ value of 0.8 µM and a selectivity index of 13.6 (Kim et al., 2019). (iii) Consider that fangchinoline impairs Middle East respiratory syndrome coronavirus translocation in host cell (Gunaratne et al., 2018). Bisbenzyl isoquinolines represent an interesting reserve of potential antiviral agents or chemical frameworks from which antiviral agents could be developed. Tetrandrine might hold potential for the fight against Severe acute respiratory syndrome-associated coronavirus and fangchinoline need to be assessed *in vivo* for anti-Human immunodeficiency virus effects.

(iv) Consider that tetrandrine inhibited Rv2459(jefA), Rv3728, and Rv3065(mmr) efflux pumps in *Mycobacterium* species (Sharma et al., 2019).

### References

Gunaratne, G.S., Yang, Y., Li, F., Walseth, T.F. and Marchant, J.S., 2018. NAADP-dependent Ca2+ signaling regulates Middle East respiratory syndrome-coronavirus pseudovirus translocation through the endolysosomal system. *Cell Calcium*, 75, pp. 30–41.

Hall, A.M. and Chang, C.J., 1997. Multidrug-resistance modulators from *Stephania japonica*. *Journal of Natural Products*, 60(11), pp. 1193–1195.

Kim, D.E., Min, J.S., Jang, M.S., Lee, J.Y., Shin, Y.S., Park, C.M., Song, J.H., Kim, H.R., Kim, S., Jin, Y.H. and Kwon, S., 2019. Natural bis-benzylisoquinoline alkaloids-tetrandrine, fangchinoline, and cepharanthine, inhibit human coronavirus OC43 infection of MRC-5 human lung cells. *Biomolecules*, 9(11), p. 696.

Lee, G., Joung, J.Y., Cho, J.H., Son, C.G. and Lee, N., 2018. Overcoming P-glycoprotein-mediated multidrug resistance in colorectal cancer: potential reversal agents among herbal medicines. *Evidence-Based Complementary and Alternative Medicine*.

Sharma, A., Gupta, V.K. and Pathania, R., 2019. Efflux pump inhibitors for bacterial pathogens: from bench to bedside. *The Indian Journal of Medical Research*, 149(2), p. 129.

Wan, Z., Lu, Y., Liao, Q., Wu, Y. and Chen, X., 2012. Fangchinoline inhibits human immunodeficiency virus type 1 replication by interfering with gp160 proteolytic processing. *PLoS One*, 7(6), p. e39225.

### 4.1.4.10  *Stephania pierrei* Diels

Synonym: *Stephania erecta* Craib.
  Common name: Poong Mao (Thailand)
  Habitat: Forests
  Distribution: Cambodia, Laos, Thailand, and Vietnam
  Botanical observation: This slender herbaceous climber or erect herb grows to about 3 m long from massive tubers. The leaves are simple, spiral, and exstipulate. The petiole is geniculate and slightly swollen at base and 1.5–4.5 cm long. The blade is peltate, somewhat glaucous below, dull green above, orbicular, 2–6×2–5 cm, membranous, truncate or rounded at base, obtuse at apex, and with 1–2 pairs of secondary nerves. In Thailand, the leaves are eaten to tonify. The cymes are axillary, umbelliform, on 1–4 cm long, peduncles, and bear 4–6 minute flowers. The calyx includes 1–6 sepals, which are yellowish/greenish, fleshy, and orbicular-obovate. Corolla absent. The synandrium is 5–8-locular and sessile. The ovary is ovoid, and develops a 2–10-lobed stigma. The drupes are dull green, obovoid, 6–8×5–7 mm, and shelter horseshoe-shaped seeds.

  Medicinal uses: Apparently none for the treatment of microbial infections

  *Strong antibacterial (Gram-positive) mid-polar extract:* Ethylacetate extract of tubers inhibited the growth of *Streptococcus mutans* (DMST 26095), *Streptococcus mitis* (ATCC 49456T), and *Streptococcus pyogenes* (DNAT 17020) with MIC/MBC values of 1250/>2500, 19.5/19.5, 78.1/78.1 µg/mL, respectively (Meerungrueang & Panichayupakaranant, 2014).

*Antifungal (yeast) polar extract*: Methanol extract of stems at 500 μg/mL inhibited ropoisomerase II in *Saccharomyces cerevisiae* (Sangmalee et al., 2012).

Commentary: Topoisomerase is a target for antibacterial natural products.

### References
Meerungrueang, W. and Panichayupakaranant, P., 2014. Antimicrobial activities of some Thai traditional medical longevity formulations from plants and antibacterial compounds from *Ficus foveolata*. *Pharmaceutical Biology, 52*(9), pp. 1104–1109.

Sangmalee, S., Laorpaksa, A. and Sukrong, S., 2012. A topoisomerase II poison screen of ethnomedicinal Thai plants using a yeast cell-based assay. *Journal of Ethnopharmacology, 142*(2), pp. 432–437.

## 4.1.4.11 *Stephania succifera* H.S. Lo & Y. Tsoong

Common name: Xiao ye di bu rong (China)

Habitats: Stony places in forests

Distribution: China

Botanical observation: This climber grows to 2–5 m long from massive tubers. The stems and leaves contain a red sap. The leaves are simple, spiral, and exstipulate. The petiole is 3–5 cm long. The blade is rotund to triangular-rounded, 5–9 cm×5–9 cm, membranous, truncate or slightly emarginate at base, acuminate at apex, and with 3–4 pairs of secondary nerves. The cymes are axillary, umbelliform, and on a 6–8 cm long peduncle. The calyx includes 6 sepals. The 3 petals are purple. The synandrium includes 6 anthers. The drupe contain an horseshoe-shaped, and about 5 mm seed.

Medicinal use: Detoxification (China)

*Antibacterial (Gram-positive) halo developed by protoberberine alkaloid glycoside*: From the tubers, (-)-1-*O*-β-D-glucoside-8-oxotetrahydropalmatine (500 μg/disc) was active against *Staphylococcus aureus* with 15 mm inhibition diameter zone (Zeng et al., 2017).

*Antibacterial (Gram-positive) phenanthrene*: The phenanthrene alkaloid 1-*N*-monomethylcarbamate-argentinine-3-*O*-β-D-glucoside isolated from the tubers (500 μg/disc) inhibited the growth of methicillin-resistant *Staphylococcus aureus* with an inhibition zone diameter of 8 mm (Zeng et al., 2017).

*Strong antiviral (enveloped monopartite linear single-stranded (–) RNA) hydrophilic protoberberine*: Palmatine (LogD=−0.9 at pH 7.4; molecular weight=352.4 g/mol) inhibited Parainfluenza-3 virus in Vero cells with a maximum nontoxic concentration of and 32 μg/mL, respectively (Orhan et al., 2007).

Palmatine

*Strong antiviral (enveloped monopartite linear double-stranded DNA) hydrophilic protoberberine*: Palmatine inhibited Herpes simplex virus replication in Madin-Darby canine kidney cells with a maximum nontoxic concentration of 32 µg/mL (Orhan et al., 2007).

Commentary: The aporphine crebanine N-oxide, the protoberberine-type alkaloids palmatine, the protoberberine-type alkaloid corydalmine, dehydrocorydalmine, and denitroaristolochic acid and demethylaristofolin C from the tubers exhibited antibacterial activities against *Staphylococcus aureus* and methicillin-resistant *Staphylococcus aureus* (Yang et al., 2013).

### References

Orhan, I., Ozcelik, B., Karaoğlu, T. and Şener, B., 2007. Antiviral and antimicrobial pro-files of selected isoquinoline alkaloids from Fumaria and Corydalis species. *Zeitschrift für Naturforschung C, 62*(1–2), pp. 19–26.

Yang, D.L., Mei, W.L., Zeng, Y.B., Guo, Z.K., Wei, D.J., Liu, S.B., Wang, Q.H. and Dai, H.F., 2013. A new antibacterial denitroaristolochic acid from the tubers of *Stephania succifera*. *Journal of Asian Natural Products Research, 15*(3), pp. 315–318.

Zeng, Y.B., Wei, D.J., Dong, W.H., Cai, C.H., Yang, D.L., Zhong, H.M., Mei, W.L. and Dai, H.F., 2017. Antimicrobial glycoalkaloids from the tubers of *Stephania succifera*. *Archives of Pharmacal Research, 40*(4), pp. 429–434.

### 4.1.4.12   *Stephania venosa* (Blume) Spreng

Synonym: *Clypea venosa* Bl.

Common name: Sa bu leud (Thailand)

Habitat: Forests

Distribution: India, Andaman Islands, Thailand, Vietnam, Indonesia, China, and the Philippines

Botanical observation: It is a climber that grows up to 20 m long from a massive tuber yielding a red sap. The leaves are simple, spiral, and exstipulate. The petiole is 6–15 cm long. The blade is triangular-ovate, membranous, up to 20 cm long, cordate at base, mucronate at apex, and incon-spicuously lobed at margin. The inflorescences are axillary cymes on 5–10 cm long peduncles. The corolla includes 1–6 sepals, which are oblanceolate. The corolla comprises 2–3 petals, which are obovate. Petals and sepals are on 1 side of the perianth. The synandrium is minute. The fruit is an ovoid drupe, which is red, up to about 1 cm long and containing an horseshoe-shaped seed.

Medicinal uses: Apparently none for microbial infections

*Strong antibacterial (Gram-positive) oxoaporphine:* Thailandine was active against *Streptococcus pneumoniae, Staphylococcus aureus*, and *Enterococcus faecalis* with MIC values of 0.03, 0.03 and 0.06 mg/mL, respectively (Makarasen et al., 2011).

*Strong antimycobacterial hydrophilic oxoaporphine:* Thailandine isolated from the tubers inhib-ited *Mycobacterium tuberculosis* ($H_{37}Ra$) with an $IC_{50}$ value of 6.2 µg/mL (Makarasen et al., 2011).

Thailandine

*Strong antiviral (enveloped monopartite linear double-stranded DNA) amphiphilic oxoaporphine alkaloid*: Oxostephanine (Log D=3.2 pH 7.4; molecular mass = 305.3 g/mol) from the tubers inhibited Herpes simplex virus type-1 with an $IC_{50}$ value of 12.6 µg/mL (Makarasen et al., 2011).

*Viral enzyme inhibition by polar extract:* Ethanol extract of tubers inhibited Human immunodeficiency virus type-1 integrase activity with an $IC_{50}$ value of 9.3 µg/mL (Bunluepuech & Tewtrakul, 2011).

Commentary: The plant produces the protoberberine alkaloid tetrahydropalmatine as well as aporphine alkaloids dicentrine, crebanine, and stephanine (Kongkiatpaiboon et al., 2017). Consider that oxostephanine is toxic for human cancer cells, and one could suggest that its antiviral activity might involve some mechanism around DNA.

### References

Bunluepuech, K. and Tewtrakul, S., 2011. Anti-HIV-1 integrase activity of Thai medicinal plants in longevity preparations. *Sonklanakarin Journal of Science and Technology, 33*(6), p. 693.

Kongkiatpaiboon, S., Duangdee, N., Prateeptongkum, S., Tayana, N. and Inthakusol, W., 2017. Simultaneous HPLC analysis of crebanine, dicentrine, stephanine and tetrahydropalmatine in *Stephania venosa*. *Revista Brasileira de Farmacognosia, 27*(6), pp. 691–697.

Makarasen, A., Sirithana, W., Mogkhuntod, S., Khunnawutmanotham, N., Chimnoi, N. and Techasakul, S., 2011. Cytotoxic and antimicrobial activities of aporphine alkaloids isolated from *Stephania venosa* (Blume) Spreng. *Planta Medica, 77*(13), pp. 1519–1524.

### 4.1.4.13 *Tiliacora triandra* Diels

Synonyms: *Cocculus triandrus* Colebr.; *Limacia triandra* (Colebr.) Hook.f. & Thomson

Common names: Bamboo grass; yanang, choi nang (Thailand); akar kusin (Malaysia); xanh tam (Vietnam)

Habitat: Forests

Distribution: Assam, Myanmar, Cambodia, Laos, Thailand, Vietnam, and Malaysia

Botanical observation: It is a climber with striated stems. The leaves are spiral, simple, and exstipulate. The petiole is 0.5–2 cm long. The blade is lanceolate, edible, 6.5–17 cm×2–8.5 cm, dull dark green, rounded at base, acuminate at apex, and marked, with 3–9 pairs of secondary nerves. The inflorescences are axillary slender racemes, which are up to about 15 cm long, and bearing a few minute and yellowish flowers. The calyx comprise 6–12 sepals. The corolla comprises 3 or 6 petals. The androecium consists of 3 stamens. The gynoecium presents 8–9 carpels. The fruits are drupes, which are obovoid, 7–10 mm×6–7 mm, red, and containing, horseshoe-shaped seeds.

Medicinal uses: Fever (Thailand); dysentery (Cambodia)

*Strong antimycobacterial amphiphilic bisbenzylisoquinolines*: Tiliacorinine, 2'-nortiliacorinine, and tiliacorine (LogD=5.7 at pH 7.4; molecular weight=576.6 g/mol) isolated from the roots inhibited the growth of *Mycobacterium tuberculosis* (H37Rv) with MIC values of 6.2, 3.1, and 3.1 µg/mL, respectively (Sureram et al., 2012). These alkaloids were tested against 59 clinical isolates of multidrug-resistant *Mycobacterium tuberculosis* and elicited MIC values ranging from 0.7 to 6.2 µg/mL (Sureram et al., 2012).

Tiliacorine
Commentary: Preclinical trial of plant alkaloidal fraction or tiliacorine for the prevention/treatment
of tuberculosis is warranted.

### Reference
Sureram, S., Senadeera, S.P., Hongmanee, P., Mahidol, C., Ruchirawat, S. and Kittakoop,
   P., 2012. Antimycobacterial activity of bisbenzylisoquinoline alkaloids from *Tiliacora tri-
   andra* against multidrug-resistant isolates of *Mycobacterium tuberculosis*. *Bioorganic &
   Medicinal Chemistry Letters*, 22(8), pp. 2902–2905.

### 4.1.4.14   *Tinospora capillipes* Gagnep.
Synonyms: *Limacia sagittata* Oliver; *Tinospora imbricata* S. Y. Hu; *Tinospora szechuanensis*
S. Y. Hu.
   Habitats: Forests and grasslands
   Distribution: China and Vietnam
   Botanical description: It is a slender climber growing from yellow tuberous roots. The stems are
striated and somewhat hairy at apex. The leaves are simple, spiral, and exstipulate. The petiole is
2.5–6 cm long. The blade is sagittate, 7–20×2–7.5 cm, glossy, dark green, conspicuously reticulate
below, acuminate or caudate at apex, and with 2–3 pairs of secondary nerves. The inflorescences are
axillary fascicles, which are 2–15 cm long. The calyx comprises 6 sepals, which are about 5 mm long,
the inner one elliptic. The corolla presents 6 petals, which are obovate and minute. The gynoecium
includes 3 carpels. The fruits are globose, 6–8 mm across, and wide drupes containing horseshoe-
shaped seeds.
   Medicinal uses: Boils, laryngitis (China)
   *Strong broad-spectrum antibacterial hydrophilic protoberberine*: Palmatine (LogD=−0.9 at
pH 7.4; molecular weight=352.4 g/mol) inhibited the growth of *Escherichia coli*, *Pseudomonas
aeruginosa*, *Proteus mirabilis*, *Klebsiella pneumonia*, *Acinetobacter baumannii*, *Staphylococcus
aureus*, and *Bacillus subtilis* with MIC values of 32, 32, 32, 32, 32, 64, and 64 μg/mL, respectively
(Orhan et al., 2007).
   Commentaries: The plant contains palmatine and jatrorrhizine (Yu et al., 2007).

**Reference**

Orhan, I., Ozcelik, B., Karaoğlu, T. and Şener, B., 2007. Antiviral and antimicrobial profiles of selected isoquinoline alkaloids from Fumaria and Corydalis species. *Zeitschrift für
Naturforschung C, 62*(1–2), pp. 19–26.

Yu, Y., Yi, Z.B. and Liang, Y.Z., 2007. Main antimicrobial components of *Tinospora capil-
lipes*, and their mode of action against *Staphylococcus aureus*. *FEBS Letters, 581*(22),
pp. 4179–4183.

## 4.1.4.15    *Tinospora cordifolia* (Willd.) Miers ex Hook. f. & Thomson

Synonyms: *Menispermum cordifolium* Willd.; *Tinospora malabarica* (Lam.) Hook. f. &
Thomson

  Common names: Heart-leaved moonseed; gulancha (Bangladesh)
  Habitats: Roadsides, forests margins, around villages, cultivated
  Distribution: Pakistan, India, Sri Lanka, Bangladesh, and Myanmar
  Botanical observation: It is a climber that grows to about 8 m long. The stems are terete, smooth,
and glossy. The leaves are simple, spiral, and exstipulate. The petiole is 5–10 cm long. The blade
is cordate, 7.5–13.8 cm×9–17 cm, with about 3–4 pairs of secondary nerves, and hairy below. The
inflorescences are solitary or clusters, which are 7–14 cm long. The calyx includes 6 sepals, which
are membranous and about 5 mm long. The corolla includes 6 petals, which are smaller than the
sepals. The androecium includes 6 stamens. The gynoecium includes 6 carpels. The drupes are
glossy red, ovoid, 6–9 mm long, and contain a horseshoe-shaped seeds.
  Medicinal uses: Urinary tract diseases, fever, leucorrhea (India)
  *Strong broad-spectrum antibacterial polar extract:* Aqueous extract of stems inhibited the
growth of *Bacillus subtilis*, *Klebsiella pneumoniae*, *Proteus mirabilis*, *Staphylococcus aureus*,
*Pseudomonas aeruginosa*, and *Escherichia coli* with MIC values of 27.8, 10.1, 10.9, 7.5, 37.0, and
43.5 µg/mL, respectively (Chakraborty et al., 2014).
  *Strong antiviral (enveloped monopartite linear single-stranded (+) RNA) polar extract:* Methanol
extract of stems inhibited Dengue virus (Sood et al., 2015).
  *Strong antiviral (enveloped monopartite linear dimeric single-stranded (+) RNA) hydrophilic
polar extract:* Methanol extract of stem bark inhibited the replication of Human immunodeficiency
virus type -1 (strain NL4.3) in Human CD4+ T cells by 51% at the maximum noncytotoxic concentration of 15 µg/mL (Sabde et al., 2011).
  Commentary: Siddha healers in South India use the plant to treat COVID-19 (Sasikumar et
al., 2020). Consider that antiretroviral extracts and natural products are not uncommonly able
to inhibit the growth of coronaviruses (see earlier). One could endeavor to examine the anti-
COVID-19 properties of this interesting climber.

**References**

Chakraborty, B., Nath, A., Saikia, H. and Sengupta, M., 2014. Bactericidal activity of selected
    medicinal plants against multidrug resistant bacterial strains from clinical isolates. *Asian
    Pacific Journal of Tropical Medicine, 7*, pp. S435–S441.
Sabde, S., Bodiwala, H.S., Karmase, A., Deshpande, P.J., Kaur, A., Ahmed, N., Chauthe, S.K.,
    Brahmbhatt, K.G., Phadke, R.U., Mitra, D. and Bhutani, K.K., 2011. Anti-HIV activity of
    Indian medicinal plants. *Journal of Natural Medicines, 65*(3–4), pp. 662–669.
Sasikumar, R., Priya, S.D. and Jeganathan, C., 2020. A case study on domestics spread
    of SARS-CoV-2 pandemic in India. *International Journal of Advanced Science and
    Technology, 29*(7), pp. 2570–2574.
Sood, R., Raut, R., Tyagi, P., Pareek, P.K., Barman, T.K., Singhal, S. et al., 2015. Cissampelos
    pareira Linn: Natural source of potent antiviral activity against All four Dengue virus
    serotypes. *PLOS Neglected Tropical Diseases, 9*(12): e0004255.

### 4.1.4.16 *Tinospora crispa* (L.) Hook.f. & Thoms.

Synonyms: *Cocculus crispus* DC; *Menispermum crispum* L.; *Tinospora gibbericaulis* Hand.-Mazz.; *Tinospora mastersii* Diels; *Tinospora rumphii* Boerl.; *Tinospora thorelii* Gagnep.

Common names: Chinese Tinospora; ghol-loai (Bangladesh); bo ye qing niu dan (China); macabuhay, paliavan (the Philippines); akar patawali (Malaysia)

Habitat: Open forests, shrublands

Distribution: China, Cambodia, India, Indonesia, Laos, Malaysia, Myanmar, the Philippines, and Thailand

Botanical observation: It is a woody climber. The stems are tuberculate, somewhat juicy, and glossy, and with a membranous bark. The leaves are simple, spiral, and exstipulate. The petiole is 5–20 cm long and slender. The blade is broadly ovate, 6–13×6–15 cm, cordate at base, acuminate at apex, and with 2–3 pairs of secondary nerves. The inflorescences are slender fascicles, which are cauliflorous and 2–10 cm long. The calyx includes 6 sepals. The corolla comprises 3–6 yellow, and minute petals. The androecium includes 6 stamens. The gynoecium present 3 carpels. The drupes are orange, subglobose, about 2 cm long, and contain horseshoe-shaped seeds.

Medicinal uses: Leprosy, (Bangladesh); cholera, small-pox, wounds (Malaysia); syphilis (Bangladesh; Malaysia)

*Broad-spectrum antibacterial halo developed by mid-polar extract:* Chloroform extract (400 μg/disc) inhibited *Bacillus cereus, Bacillus megaterium, Bacillus subtilis, Staphylococcus aureus, Sarcinia lutea, Escherichia coli, Salmonella paratyphi, Salmonella typhi, Shigella boydii, Shigella dysenteriae, Vibrio mimicus*, and *Vibrio parahemolyticus* with inhibition zone diameters of 8, 8, 7, 7, 9, 7, 7, 9, 7, 8, 8, and 7 mm, respectively (Md et al., 2011).

*Broad-spectrum antifungal (yeast) halo developed by mid-polar extract:* Chloroform extract (400 μg/disc) inhibited the growth of *Candida albicans, Aspergillus niger,* and *Saccharomyces cerevisiae* (Md et al., 2011).

*Commentary:* The plant produces the antimicrobial aporphine alkaloid magnoflorine (Choudhary et al., 2010).

### References

Choudhary, M.I., Ismail, M., Ali, Z., Shaari, K., Lajis, N.H. and Rahman, A.U., 2010. Alkaloidal constituents of *Tinospora crispa*. *Natural Product Communications*, 5, pp. 1747–1750.

Md, H.A., SM, I.A. and Mohammad, S., 2011. Antimicrobial, cytotoxicity and antioxidant activity of *Tinospora crispa*. *Journal of Pharmaceutical and Biomedical Sciences (JPBMS), 13*(13).

### 4.1.4.17 *Tinospora sinensis* (Lour.) Merrill

Synonyms: *Campylus sinensis* Lour., *Cocculus tomentosus* Colebr.; *Menispermum malabaricum* Lam.; *Menispermum tomentosum* (Colebr.) Roxb.; *Tinospora tomentosa* (Colebr.) Hook. f. & Thomson

Common names: Chinese tinospora; guloncho (Bangladesh); zhong hua qing niu dan (China)

Habitat: Forests

Distribution: China, Cambodia, India, Nepal, Sri Lanka, Thailand, and Vietnam

Description: This climber grows to 20 m long. The stems are somewhat fleshy. The leaves are simple, spiral, and exstipulate. The petiole is 6–13 cm long. The blade is broadly ovate, 7–14×5–13 cm, papery, deeply cordate at base, acuminate at apex, and with 2–3 pairs of secondary nerves. The inflorescences are cauliflorous fascicles of minute flowers, which are 1–4 cm long. The calyx includes 6 sepals. The corolla comprises 6 petals. The androecium includes 6 stamens. The drupes are red subglobose and contain horseshoe-shaped seeds.

Medicinal use: Tuberculosis leucorrhea (Bangladesh); sores (Cambodia)

Commentary: The plant produces berberine (Srinivasan et al., 2008).

**Reference**

Srinivasan, G.V., Unnikrishnan, K.P., Shree, A.R. and Balachandran, I., 2008. HPLC esti-
mation of berberine in *Tinospora cordifolia* and *Tinospora sinensis*. *Indian Journal of
Pharmaceutical Sciences*, 70(1), p. 96.

## 4.1.5  FAMILY RANUNCULACEAE A.L. de Jussieu (1789)

The family Ranunculaceae consists of about 71 genera and 2500 species of herbs, shrubs, or climbers, which are mainly found in the hilly areas of Asia Pacific. The leaves are simple or compound,
spiral or opposite, and exstipulate. The inflorescences are solitary or racemes or cymes and often
showy. The calyx comprises 3–6 sepals. The corolla comprises up to 8 petals or is absent. The
androecium includes numerous stamens. The gynoecium includes numerous carpels. The fruits are
follicles, achenes, capsules, or berries. Members of the family Ranunculaceae produce antimicrobial benzylisoquinoline or/and aporphine alkaloids, furanones, quinones, monoterpenes, diterpene
alkaloids, and triterpene saponins.

### 4.1.5.1  *Aconitum carmichaeli* Debeaux

Common names: Azure monkshood; wu tou (China)

Habitat: Forests margins and grassy mountain slopes

Distribution: Vietnam and China

Botanical examination: This poisonous herb grows up to about 1.5 m tall. The leaves are simple,
spiral, and exstipulate. The petiole is up to 2.5 cm long. The blade is incised and 6–11×9–15 cm.
The inflorescence is a terminal and up to about 10 cm long raceme of very characteristically hooded
flowers. The calyx includes 5 sepals which are bluish purple and about 1.5 cm long. The corolla
comprises 2 petals, which are about 1 cm long. The androecium includes numerous stamens. The
gynoecium is made of 2 to 5 free carpels. The follicles are about 1.5 cm long and shelter 3 mm long
seeds.

Medicinal use: COVID-19 (China)

*Very weakly antiviral (non-enveloped linear monopartite single-stranded (+) RNA) anthraqui-
none:* Emodin at a concentration of 500 µg/mL inhibited the replication of the Tobacco mosaic virus
by about 50% (Zhu et al., 2019; Xu et al., 2019).

*Very weakly antiviral (non-enveloped monopartite linear single-stranded (+) RNA) amphiphi-
lic diterpene alkaloids:* Hypaconitine (LogD=2.0 at pH 7.4; molecular weight=615.7 g/mol) and
mesaconitine (LogD=1.2; molecular weight=631.7 g/mol) at a concentration of 500 µg/mL inhibited
the replication of the Tobacco mosaic virus by about 82.4 and 85.6% (Zhu et al., 2019; Xu et al., 2019).

*Very weakly antiviral (enveloped linear monopartite double-stranded DNA) diterpene alka-
loids:* Neochinuline A and mesaconitine at a concentration of 500 µg/mL inhibited the replication
of the Cytomeglovirus by about 50% (Zhu et al., 2019; Xu et al., 2019).

Commentaries: (i) Members of the genus *Aconitum* L. are not uncommonly antiviral. Methanol
extract of root of *Aconitum uchiyamai* Nakai inhibited Human immunodeficiency virus type -1
protease by 52.5% (Park, 2003). (ii) Consider that plants containing emodin are often used to treat
COVID-19, such as *Rheum palmatum* L. (family Polygonaceae). Emodin blocks the Severe acute
respiratory syndrome-associated coronavirus spike protein and angiotensin-converting enzyme 2
interaction (Ho et al., 2007). Once could endeavor to examine the anti-Severe acute respiratory
syndrome -associated coronavirus properties of this anthraquinone despite the fact that anthraquinones have the tendency to be mutagenic (Morooka et al., 1990). (iii) The plant generates a bewildering array of diterpene alkaloids (Xiong et al., 2012; Li et al., 2018). Diterpenes in members of the
genus *Aconitum* L. are antibacterial (Liang et al., 2017).

## References

Ho, T.Y., Wu, S.L., Chen, J.C., Li, C.C. and Hsiang, C.Y., 2007. Emodin blocks the SARS coronavirus spike protein and angiotensin-converting enzyme 2 interaction. *Antiviral Research*, 74(2), pp. 92–101.

Li, Y., Gao, F., Zhang, J.F. and Zhou, X.L., 2018. Four new diterpenoid alkaloids from the roots of aconitum carmichaelii. *Chemistry & Biodiversity*, 15(7), p. e1800147.

Liang, X., Chen, L., Song, L., Fei, W., He, M., He, C. and Yin, Z., 2017. Diterpenoid alkaloids from the root of *Aconitum sinchiangense* WT Wang with their antitumor and antibacterial activities. *Natural Product Research*, 31(17), pp. 2016–2023.

Morooka, N., Nakano, S., Itoi, N. and Ueno, Y., 1990. The chemical structure and the mutagenicity of emodin metabolites. *Agricultural and Biological Chemistry*, 54(5), pp. 1247–1252.

Park, J.C., 2003. Inhibitory effects of Korean plant resources on human immunodeficiency virus type 1 protease activity. *Oriental Pharmacy and Experimental Medicine*, 3(1), pp. 1–7.

Wang, S.X., Wang, Y., Lu, Y.B., Li, J.Y., Song, Y.J., Nyamgerelt, M. and Wang, X.X., 2020. Diagnosis and treatment of novel coronavirus pneumonia based on the theory of traditional Chinese medicine. *Journal of Integrative Medicine*, 18(4), pp. 275–283.

Xiong, L., Peng, C., Xie, X.F., Guo, L., He, C.J., Geng, Z., Wan, F., Dai, O. and Zhou, Q.M., 2012. Alkaloids isolated from the lateral root o f *Aconitum carmichaelii*. *Molecules*, 17(8), pp. 9939–9946.

Xu, W., Zhang, M., Liu, H., Wei, K., He, M., Li, X., Hu, D., Yang, S. and Zheng, Y., 2019. Antiviral activity of aconite alkaloids from *Aconitum carmichaelii* Debx. *Natural Product Research*, 33(10), pp. 1486–1490.

Zhu, Y.Y., Yu, G., Wang, Y.Y., Xu, J.H., Xu, F.Z., Fu, H., Zhao, Y.H. and Wu, J., 2019. Antiviral activity and molecular docking of active constituents from the root of *Aconitum carmichaelii*. *Chemistry of Natural Compounds*, 55(1), pp. 189–193.

### 4.1.5.2 *Anemone biflora* DC.

Synonym: *Anemone tschernjaewii* Regel.

Common name: Batkul (India)

Habitats: Fields in mountains

Distribution: Iran, Tajikistan, Afghanistan, Pakistan, Kashmir, Himalayas, India, and Nepal

Botanical observation: It is a herb that grows up to about 25 cm tall from rhizomes. The leaves are simple and exstipulate. The radical leaves are 1–2. The petiole is up to about 4.5 cm long. The blade is 1.5–2.5 cm × 2.5–4.5 cm, cuneate at base, 3-lobed, and crenate. The inflorescence is a cyme of 1–3 flowers. The 5 tepals are lanceolate, pure white, pinkish, or purplish, and up to about 1.5 cm long. The androecium includes numerous stamens with purple anthers. The gynoecium comprises numerous carpels, which are minute, hairy, and purplish. The fruits are follicles.

Medicinal uses: Skin infections, (Himalayas); boils, cuts, (India), wounds (Himalayas, India)

Commentary: This plant has not been studied for antimicrobial activity. Note that the furanone protoanemonine from *Anemone pulsatilla* L. inhibited the growth of *Staphylococcus aureus* and *Escherichia coli* (Baer et al., 1946).

### Reference

Baer, H., Holden, M. and Seegal, B.C., 1946. The nature of the antibacterial agent from *Anemone pulsatilla*. *Journal of Biological Chemistry*, 162(1), pp. 65–68.

### 4.1.5.3 *Anemone obtusiloba* D. Don

Common names: Dun lie yin lian hua (China); ratanjot, rikabe (Nepal)

Distribution: Afghanistan, Pakistan, Kashmir, Himalayas, India, Bhutan, Nepal, and Mongolia

Habitats: Fields in mountains

Botanical observation: This hairy herb grows up to about 25 cm tall from rhizomes. The leaves are simple and exstipulate. The radical leaves are 5–10. The petiole is up to about 15 cm long. The blade is 2–6 cm×2–8 cm, broadly ovate, 3-lobed, cordate at base, and lobate. The inflorescence is a cyme with 1–3 flowers. The perianth comprises 5–6 tepals, which are yellowish, or bluish, obovate, and up to about 1.5 cm long. The androecium includes numerous stamens, which are yellowish brown. The gynoecium comprises numerous carpels, which are minute, hairy, and brown. The fruits are follicles which are about 5 mm long.

Medicinal uses: Cough cold (Nepal)

*Broad-spectrum antibacterial halo developed by polar extract:* Methanol extract (40 µg/disc) inhibited the growth of *Bacillus subtilis, Staphylococcus aureus, Escherichia coli,* and *Mycobacter phlei* (Taylor et al., 1995).

*Antifungal (filamentous) hydrophilic halo developed by polar extract:* Methanol extract (40 µg/disc) inhibited the growth *Microsporium gypseum* (Taylor et al., 1995).

### Reference

Taylor, R.S., Manandhar, N.P. and Towers, G.H.N., 1995. Screening of selected medicinal plants of Nepal for antimicrobial activities. *Journal of Ethnopharmacology, 46*(3), pp. 153–159.

### 4.1.5.4   *Aquilegia vulgaris* L.

Synonyms: *Aquilegia longisepala* Zimm.; *Aquilina vulgaris* (L.) Bubani

Common name: Common columbine

Habitat: Cultivated

Distribution: Temperate Asia

Botanical observation: This beautiful erect herb grows to about 50 cm tall. The leaves are compound and exstipulate. The basal leaves are about up to 30 cm long, with a slender petiole with 3 folioles, which are somewhat obovate and incised. Stem leaves are shorter than basal leaves. The inflorescences are terminal cymes of a few magnificent and nodding flowers. The calyx includes 5 bluish sepals. The corolla includes 5 bluish petals, which are prolonged into a spur and 1–1.5 cm×0.5–1 cm. The numerous stamens are linear-filiform and up to about 1.3 cm long. The gynoecium comprises 5 carpels. The follicles are narrowly cylindrical, beaked, and contain numerous minute seeds.

Medicinal uses: Apparently none for microbial infections.

*Strong antibacterial (Gram-positive) hydrophilic C-glycosyl flavone:* Isocytisoside (LogD=−0.3 at pH 7.4; molecular mass=446.4 g/mol) isolated from the leaves inhibited the growth of *Staphylococcus aureus* (ATCC 25923), *Staphylococcus epidermidis* (NCTC 11047), *Micrococcus luteus* (ATCC 9341), *Bacillus subtilis* (ATCC 6633), *Bacillus pumilus* (NCTC 8241), *Enterococcus faecalis* (ATCC 14428), *Escherichia coli* (ATCC 25922), *Proteus mirabilis* (NCTC 6635), *Klebsiella pneumonia* (ATCC 27736), and *Pseudomonas aeruginosa* (ATCC 27853) (Bylka et al., 2004).

*Moderate broad-spectrum antifungal hydrophilic C-glycosyl flavone:* Isocytisoside inhibited the growth of *Candida albicans* (ATCC 10231) and *Aspergillus niger* (ATCC 16404) with MIC values of 125 and 62.5 µg/mL, respectively (Bylka et al., 2004).

Commentary: *C*-glycosyl flavonoids are not uncommon in plants and can be found notably in members of the family Poaceae. These have in general strong antibacterial activities *in vitro*.

### Reference

Bylka, W., Szaufer-Hajdrych, M., Matławska, I. and Goślińska, O., 2004. Antimicrobial activity of isocytisoside and extracts of *Aquilegia vulgaris* L. *Letters in Applied Microbiology, 39*(1), pp. 93–97.

#### 4.1.5.5   *Cimicifuga foetida*

Synonyms: *Actaea cimicifuga* L.; *Actaea frigida* (Wall. ex Royle) Prantl; *Actinospora frigida* Fisch. & C.A. Mey.; *Cimicifuga europaea* Schipcz.; *Cimicifuga frigida* Wall. ex Royle

   Common name: Sheng ma (China)

   Distribution: China, Bhutan, India, Kazakhstan, and Mongolia

   Habitat: Mountains, forests, and open grassy fields

   Botanical observation: This herb grows up to about 2 m tall from a rhizome. The basal leaves are compound and exstipulate. The petiole is up to about 15 cm long. The blade is bi or triternately pinnate. The folioles are lanceolate to rhombic, incised, dark green, glossy, 2.5–10 cm × 1–7 cm, and serrate. The inflorescences are about 45 cm long racemes of whitish spikes of minute flowers. The perianth comprises 5 tepals, which are up to about 4 mm long, obovate, and whitish. The androecium comprises up to 20 stamens, which are conspicuous and yellowish white. The gynoecium includes 2–5 carpels, which are hairy. The follicles are up to about 1.5 cm long, appressed, and containing numerous minute seeds.

   Medicinal uses: Apparently none for microbial infections.

   *Strong antiviral (enveloped monopartite linear dimeric single-stranded (+) RNA) amphiphilic triterpene saponin*: The plant contains the cycloartane saponin actein (LogD = 4.4 at pH 7.4; 676.8 g/mol), which inhibited the replication of Human immunodeficiency virus in H9 lymphocytes with an $EC_{50}$ of 0.3 µg/mL and a selectivity index of 144 (Sakurai et al., 2004).

   *Antiviral (enveloped monopartite linear single-stranded (-) RNA) lipophilic triterpene*: The cycloartane cimicifugin (LogD = 5.1 at pH 7.4; molecular mass = 472.7 g/mol) inhibited Human respiratory syncytial virus-induced plaque formation in both HEp-2 and A549 cells by inhibiting viral attachment to hosts cells (Wang et al., 2012).

   Commentary: Aqueous extract of rhizomes of *Cimicifuga heracleifolia* Komarov at a concentration of 250 µg/mL inhibited Human immunodeficiency virus type-1 protease by 50.2% (Xu et al., 1996).

#### References

Naika, R H. and Krishna V., 2007. Antimicrobial activity of extracts from the leaves of *Clematis gouriana* Roxb. *International Journal of Biomedical and Pharmaceutical Sciences*. 69–72.

Sakurai, N., Wu, J.H., Sashida, Y., Mimaki, Y., Nikaido, T., Koike, K., Itokawa, H. and Lee, K.H., 2004. Anti-AIDS agents. Part 57: actein, an anti-HIV principle from the rhizome of *Cimicifuga racemosa* (black cohosh), and the anti-HIV activity of related saponins. *Bioorganic & Medicinal Chemistry Letters*, *14*(5), pp. 1329–1332.

Wang, K.C., Chang, J.S., Lin, L.T., Chiang, L.C. and Lin, C.C., 2012. Antiviral effect of cimicifugin from *Cimicifuga foetida* against human respiratory syncytial virus. *The American Journal of Chinese Medicine*, *40*(05), pp. 1033–1045.

Xu, H.X., Wan, M., Loh, B.N., Kon, O.L., Chow, P.W. and Sim, K.Y., 1996. Screening of traditional medicines for their inhibitory activity against HIV-1 protease. *Phytotherapy Research*, *10*(3), pp. 207–210.

#### 4.1.5.6   *Clematis gouriana* Roxb. ex DC.

Synonyms: *Clematis martini* H. Lév.; *Clematis simplicifolia* Qureshi & Chaudhri

   Common names: Iao suo yi teng (China); belkangu (India); kureni (Nepal)

   Habitat: Road sides, along streams, and open lands

   Distribution: Pakistan, Himalaya, India, Nepal, Sri Lanka, Myanmar, China, the Philippines, Papua New Guinea

   Botanical observation: This woody and poisonous climber grows up to 5 m long. The stems are somewhat hairy when young. The leaves are pinnate or bipinnate, opposite, and exstipulate. The petiole is 1–7 cm long. The folioles are lanceolate, cordate at base, 2.5–10.5 cm × 1.5–5.5 cm, and attenuate at apex. The cymes are axillary, terminal, and many flowered. The 4 tepals are white, about 5 mm long, and obovate. The androecium includes numerous minute stamens. The gynoecium comprises numerous carpels, which are minute and hairy. The follicles are hairy and linear and are up to 3 cm long.

Medicinal uses: Leprosy, burns (India); fever (India, Nepal); wounds (India; Taiwan)

*Strong antibacterial (Gram-negative) polar extract:* Methanol extract of leaves elicited antibacterial effects against 9 clinical strains of *Pseudomonas aeruginosa* and *Klebsiella pneumoniae.*

Commentaries: (i) Members of the genus *Clematis* L. produce protoanemonin, which is antibacterial and antifungal, as well as magnoflorine (Chawla et al., 2012). (ii) Aqueous extract of roots of *Clematis chinensis* Retz. at a concentration of 250 µg/mL inhibited Human immunodeficiency virus type-1 protease by 38.2% (Xu et al., 1996) and methanol extract of *Clematis heracleifolia* DC at a concentration of 100 µg/mL inhibited RNA-dependent DNA polymerase and ribonuclease H activities of Human immunodeficiency virus type-1 reverse transcriptase and Human immunodeficiency virus type-1 protease by 41.4, 16.8, and 45.3%, respectively (Min et al., 2001). (iii) *Clematis javana* DC is used in the Philippines (local name: *Kalupa*t) to heal wounds. (iv) Consider that natural products from the plants in the clades in this book with strong antibacterial activity against Gram-negative bacteria are mainly medium molecular weight and amphiphilic alkaloids.

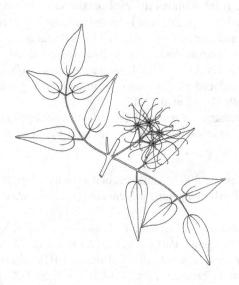

*Clematis gouriana* Roxb. ex DC.

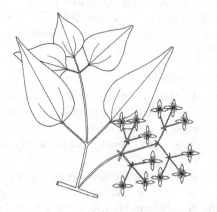

*Clematis chinensis* Retz.

### References

Chawla, R., Kumar, S. and Sharma, A., 2012. The genus Clematis (Ranunculaceae): chemical and pharmacological perspectives. *Journal of Ethnopharmacology*, *143*(1), pp. 116–150.

Min, B.S., Kim, Y.H., Tomiyama, M., Nakamura, N., Miyashiro, H., Otake, T. and Hattori, M., 2001. Inhibitory effects of Korean plants on HIV -1 activities. *Phytotherapy Research*, *15*(6), pp. 481–486.

Shalika, R., Kiran, R., Madhavi, M., et al., 2015. Screening of bioconstituents and in vitro cytotoxicity of Clematis gouriana leaves. *Natural product research*, *29*(23), pp. 2242–2246.

Xu, H.X., Wan, M., Loh, B.N., Kon, O.L., Chow, P.W. and Sim, K.Y., 1996. Screening of traditional medicines for their inhibitory activity against HIV -1 protease. *Phytotherapy Research*, *10*(3), pp. 207–210.

### 4.1.5.7 *Coptis chinensis* Franch.

Synonym: *Coptis teeta* var. *chinensis* (Franch.) Finet & Gagnep.

Common names: Chinese goldthread; huang lian (China)

Distribution: China

Habitats: Forests

Botanical observation: It is a rhizomatous herb that grows up to about 50 cm tall. The rhizome is bright orange within. The leaves are spiral and exstipulate. The petiole is about 10 cm long and originates from the rhizome. The blade is ovate-triangular, 3–7.5 × 2.5–8 cm, somewhat ternatisect, cordate at base, incised, serrate, and acute at apex. The inflorescences are terminal cymes of about 8 flowers on 25 cm tall peduncles. The calyx comprises 5 sepals, which are greenish, lanceolate, about, and 1.2 cm long. The 5 petals are linear-lanceolate, about 5 mm long, and green. The numerous stamens are about 5 mm long. The gynoecium comprises 8–12 carpels, which grow up to about 8 mm long. The follicles are long-peduncled, verticilled, recurved at apex, and contain numerous minute seeds.

Medicinal use: Antiseptic (China)

*Strong broad-spectrum antibacterial polar extract:* Aqueous extract of rhizome inhibited *Klebsiella pneumoniae, Proteus vulgaris, Staphylococcus aureus,* and *Mycobacterium smegmatis* (Franzblau & Cross, 1986).

*Moderate antifungal (filamentous) polar extract:* Aqueous extract of the plant inhibited the growth of *Trichophyton rubrum* (11788), *Trichophyton mentagrophytes* (11115), *Microsporium canis* (11883), *Epidermophyton floccosum* (10342), *Trichophyton schoenleinii* (0300), *Microsporium gypseum* (10079), *Trichophyton tonsurans* (11542), and *Trichophyton violaceum* (8001) with MIC values of 100, 100, 50, 100, 100, 100, 100, and 100 µg/mL, respectively (Yang et al., 2015).

*Strong anticandidal polar extract:* Aqueous extract of rhizome inhibited *Candida albicans* with an MIC value of 2 µg/mL (Franzblau & Cross, 1986).

*Viral enzyme inhibition:* Aqueous extract of rhizomes at a concentration of 250 µg/mL inhibited Human immunodeficiency virus type-1 protease by 100% (Xu et al., 1996).

Commentaries: (i) The plant produces berberine and jatrorrhizine, which are broad-spectrum antibacterial and antifungal protoberberine alkaloids (Wang et al., 2008). The plant is used in China for the treatment of COVID-19 (Luo, 2020; Wang et al., 2020). Are berberine and jatrorrhizine of clinical value for the treatment of COVID-19.

### References

Franzblau, S.G. and Cross, C., 1986. Comparative in vitro antimicrobial activity of Chinese medicinal herbs. *Journal of Ethnopharmacology*, *15*(3), pp. 279–288.

Luo, A., 2020. Positive SARS-Cov-2 test in a woman with COVID-19 at 22 days after hospital discharge: a case report. *Journal of Traditional Chinese Medical Sciences*, *7*(4), pp. 413–417.

Wang, S.X., Wang, Y., Lu, Y.B., Li, J.Y., Song, Y.J., Nyamgerelt, M. and Wang, X.X., 2020. Diagnosis and treatment of novel coronavirus pneumonia based on the theory of traditional Chinese medicine. *Journal of Integrative Medicine, 18*(4), pp. 275–283.

Wang, L.J., Ye, X.L., Li, X.G., Sun, Q.L., Yu, G., Cao, X.G., Liang, Y.T. and Zhou, J.Z., 2008. Synthesis and antimicrobial activity of 3-alkoxyjatrorrhizine derivatives. *Planta Medica, 74*(03), pp. 290–292.

Xu, H.X., Wan, M., Loh, B.N., Kon, O.L., Chow, P.W. and Sim, K.Y., 1996. Screening of traditional medicines for their inhibitory activity against HIV -1 protease. *Phytotherapy Research, 10*(3), pp. 207–210.

Yang, F., Ding, S., Liu, W., Liu, J., Zhang, W., Zhao, Q. and Ma, X., 2015. Antifungal activity of 40 TCMs used individually and in combination for treatment of superficial fungal infections. *Journal of Ethnopharmacology, 163*, pp. 88–93.

### 4.1.5.8 *Coptis teeta* Wall.

Synonym: *Coptis teetoides* C.Y. Cheng

Common names: Golden thread; tita (Bangladesh); mamira, mahatita (India); yun nan huang lian, hung ling (China)

Distribution: India, Tibet, Bangladesh, and China

Botanical observation: This rhizomatous herb grows up to about 50 cm tall. The rhizome is bright orange within. The leaves are spiral and exstipulate. The petiole is about 15 cm long and originates from the rhizome. The blade is ovate-triangular, 6–12×5–9 cm, somewhat ternatisect, cordate at base, incised, serrate, and attenuate at apex. The inflorescences are terminal cymes of about 3–5 flowers on a 25 cm tall peduncle. The 5 sepals are linear-lanceolate, about 5 mm long, and elliptic. The 5 petals are pure white, spathulate, about 5 mm long, and round at apex. The androecium includes numerous minute stamens. The gynoecium comprises 11–14 carpels. The follicles are long-peduncled, verticilled, and are up to about 1 cm long, recurved at apex, and contain numerous minute seeds.

Medicinal uses: Fever and cough (India)

*Strong antiviral (enveloped monopartite linear single-stranded (+) RNA) polar extract*: Methanol extract of roots inhibited the replication of Dengue virus-1, -2, -3, and -4 with $IC_{50}$ values equal to 11.1, 5, 7.8, and 33.3 µg/mL, respectively (Sood et al., 2015).

*Strong antiviral (enveloped, segmented, single-stranded (-) RNA) amphiphilic protoberberine*: Berberine (LogD = 1.3 at pH 7.4; molecular mass = 336 g/mol) (Goel et al., 2012) inhibited the growth of Influenza A (PR/8/34) virus and Influenza A (WS/33) in RAW 264.7 cells with $IC_{50}$ values of 0.01 and 0.4 µM (amantadine $IC_{50}$: 27 µM) (Cecil et al., 2011). Berberine act post-translationally to inhibit virus protein trafficking/maturation, which in turn inhibits virus growth (Cecil et al., 2011)

**References**

Cecil, C.E., Davis, J.M., Cech, N.B. and Laster, S.M., 2011. Inhibition of H1N1 influenza A virus growth and induction of inflammatory mediators by the isoquinoline alkaloid berberine and extracts of goldenseal (*Hydrastis canadensis*). *International Immunopharmacology, 11*(11), pp. 1706–1714.

Goel, A., Baboota, S., Sahni, J.K., Srinivas, K.S., Gupta, R.S., Gupta, A., Semwal, V.P. and Ali, J., 2012. Development and validation of stability-indicating assay method by UPLC for a fixed dose combination of atorvastatin and ezetimibe. *Journal of Chromatographic Science, 51*(3), pp. 222–228.

Sood, R., Raut, R., Tyagi, P., Pareek, P.K., Barman, T.K., Singhal, S., Shirumalla, R.K., Kanoje, V., Subbarayan, R., Rajerethinam, R. and Sharma, N., 2015. *Cissampelos pareira* Linn: natural source of potent antiviral activity against all four dengue virus serotypes. *PLoS Neglected Tropical Diseases, 9*(12), p. e0004255.

### 4.1.5.9 *Delphinium denudatum* Wall. ex Hook. f. & Thomson

Synonym: *Delphinium pauciflorum* Nutt.

Common names: Nirbisis, nirbasi (India)

Habitat: Open fields in mountains

Distribution: Himalayas, Pakistan, Kashmir, and India

Botanical observation: This pleasant-looking yet poisonous erect herb grows to 40–80 cm high The petides of lower leaves are about 15 cm long, and the upper ones much shorter. The blade of basal leaves are 5–15 mm wide, rounded, 3–5 parted into broadly obovate segments, segments pinnately and divaricately laciniate into oblong lobes or teeth. The cauline leaves are similar but smaller. The racemes are paniculate. The calyx presents 5 sepals which are blue to violet, about 1.5 cm long including a spur. The corolla present 4 blue to violet petals. The numerous stamens are about 5 mm long. The follicles are about 1 cm long, and shelter numerous tiny and black seeds.

Medicinal uses: Fever, cough, ulcers (India)

*Weak broad-spectrum antibacterial polar extract:* Ethanol extract of stems inhibited the growth of *Bacillus subtilis*, *Bacillus megaterium*, *Serratia marscesnens* (MTCC4822), *Pseudomonas chlororaphis* (MCC2693), and *Escherichia coli* with MIC values of 400, 600, 300, 200, and 500 µg/mL, respectively (Kumari et al., 2019).

*Weak antifungal (filamentous) diterpene alkaloids:* 8-acetylheterophyllisine, vilmorrianone, and panicutine inhibited the growth of *Allescheria boydii* with MIC values of 100, 150, and 75 µg/mL, respectively (Atta-ur-Rahman et al., 1997). 8-Acetylheterophyllisine, vilmorrianone, and panicutine inhibited the growth of *Aspergillus niger* with MIC values of 200, 100, and 125 µg/mL, respectively (Atta-ur-Rahman et al., 1997). 8-Acetylheterophyllisine, vilmorrianone, and panicutine inhibited the growth of *Epidermophyton floccosum* with MIC values of 250, 225, and 200 µg/mL, respectively (Atta-ur-Rahman et al., 1997). 8-Acetylheterophyllisine, vilmorrianone, and panicutine inhibited the growth of *Pleurotus ostreatus* with MIC values of 150, 175, and 125 µg/mL, respectively (Atta-ur-Rahman et al., 1997).

Commentaries: (i) Members of the genus *Delphinium* L. produce rare and beautiful, yet poisonous, series of diterpene alkaloids such as ajacisine E from *Delphinium ajacis* L., which inhibited the replication of the Respiratory syncytial virus (A2 strain) with an $IC_{50}$ value of 10.1 µM and a selectivity index superior to 9.9 (ribavirin: 3.1 µM; selectivity index: >32.3) (Yang et al., 2017). In Nepal, *Delphimium altissimum* Wall. (local name: *Junge lahara*) is used to heal putrefied wounds in veterinary medicine.

### References

Atta-ur-Rahman, Nasreen, A., Akhtar, F., Shekhani, M.S., Clardy, J., Parvez, M. and Choudhary, M.I., 1997. Antifungal diterpenoid alkaloids from *Delphinium denudatum*. *Journal of Natural Products*, 60(5), pp. 472–474.

Kumari, K., Adhikari, P., Pandey, A., Samant, S.S. and Pande, V., 2019. Antimicrobial potential of *Delphinium denudatum* (Wall Ex Hook & Thom). *Bulletin of Environment, Pharmacology and Life Sciences*, 8, pp. 152–158.

Yang, L., Zhang, Y.B., Zhuang, L., Li, T., Chen, N.H., Wu, Z.N., Li, P., Li, Y.L. and Wang, G.C., 2017. Diterpenoid alkaloids from *Delphinium ajacis* and their anti-RSV activities. *Planta Medica*, 83(01/02), pp. 111–116.

### 4.1.5.10 *Nigella sativa* L.

Common names: Black cumin, black seeds; kalonji (Pakistan); jintan hitam (Malaysia); hei xian hao (China); kuro tanetsu (Japan); thian dam (Thailand); ekilen (Turkey); thi la den (Vietnam)

Habitat: Waste places, roadsides, or cultivated

Distribution: Southwest Asia

Botanical observation: This heavenly-looking herb grows to about 50 cm tall. The leaves are simple, spiral, and exstipulate. The petiole of basal leaves is up to 6 cm long. The blade is about

$7 \times 5$ cm, deeply dissected with linear lobes. The inflorescences are solitary and terminal on 4–8 mm long peduncles. The calyx comprises 5 whitish sepals. The corolla comprises 5–10 petals, which are white or bluish. The androecium includes numerous stamens, which are up to about 1 cm long. The gynoecium comprises 8 carpels fused in a compound ovary, which is 4–9 mm long and with free stigmas. The capsules are green, glossy, $6–16 \times 5–12$ mm, with a long apical appendage and contain numerous minute black, trigonal, aromatic, and about 3 mm long seeds.

Medicinal uses: Cold and fever (India)

*Antibacterial (Gram-positive) polar extract:* Aqueous decoction of seeds (10 g/100 mL to 6 mm disc) inhibited *Staphylococcus aureus* with the inhibition zone diameters of 19.6 mm (Chaudhry & Tariq, 2008).

*Antibacterial (Gram-positive) mid-polar extract:* Ether extract of seeds (300 mg/mL in agar wells) inhibited the growth of *Staphylococcus aureus* with an inhibition zone diameter of 15 mm, and *Micrococcus luteus* with 12 mm (Özmen et al., 2007).

*Antibacterial non-polar extract (Gram-positive):* Petroleum ether extract of seeds (5 mg/disc) inhibited the growth of *Bacillus subtilis* and *Staphylococcus aureus* with inhibition zone diameters of 15 and 12 mm, respectively (Kökdil et al., 2005).

*Strong broad-spectrum antibacterial amphiphilic monoterpene:* Thymoquinone (LogD = 1.9 at pH 7.4; molecular weight = 164.2 g/mol) (a major constituent of seed oil) inhibited the growth of *Vibrio parahaemolyticus, Bacillus cereus, Listeria monicytogenes, Enterococcus faecalis,* and *Staphylococcus aureus* with the MIC/MBC values of 32/64, 8/8, 16/32, 32/64, and 8/16 µg/mL, respectively (Chaieb et al., 2011). Thymoquinone at $2 \times$ MIC inhibited biofilm formation by *Staphylococcus aureus* (Chaieb et al., 2011). Thymoquinone inhibited the growth of *Staphylococcus aureus, Streptococcus pneumonia, Streptococcus pyogenes, Enterococcus faecalis, Streptococcus pneumonia, Streptococcus agalactiae, Bacillus subtilis, Bacillus cereus, Enterococcus faecalis, Proteus vulgaris, Salmonella typhii, Salmonella paratyphi* A, *Morexella cataeehalis, Shigella flexneri, Burkholderia cepacia,* and *Stenotrophomonas maltophilia* (Dey et al., 2014).

Thymoquinone

*Antibiotic potentiator monoterpene:* Combination of thymoquinone with ampicillin, cephalexin, chloramphenicol, tetracycline, gentamicin, and ciprofloxacin exerted synergism *in Staphylococcus aureus* (Halawani, 2009).

*Moderate antifungal (filamentous) monoterpene:* Thymoquinone inhibited the growth of clinical strains of *Trichophyton rubrum, Trichophyton mentagrophytes, Trichophyton interdigitale, Epidermophyton floccosum,* and *Microsporium canis* with the MIC values of 0.2, 0.2, 0.2, 0.1, and 0.1 mg/mL, respectively (Aljabre et al., 2005).

Commentary: (i) Methoxynigeglanine and 6-methoxythymol-3-*O*-β-D-glucopyranoside from the seeds of *Nigella glandulifera* Freyn & Sint. inhibited the growth of *Mycobacterium tuberculosis* (H37Rv) with MIC values of 32 and 62.5 µg/mL (isoniazide: 0.06 µg/mL) (Sun et al., 2013). Note that Onifade et al. (2013) suggest that *Nigella sativa* L. may effectively control Human immunodeficiency virus infection. (ii) Consider that thymoquinone inhibited *Escherichia coli* ATP synthase with an $IC_{50}$ value of about $50\,\mu M$ (Ahmad et al., 2015). (iii) According to the cold, hot, humid, and dry classification of diseases of Hippocrates (Lloyd, 1964), flu is wet and cold and is treated by hot and dry medicinal plants. One of these plants was *Nigella sativa* L. known to the Ancient Greeks as "*Melanthion*". Is *Nigella sativa* L. of any use to fight COVID-19?

### References

Ahmad, Z., Laughlin, T.F. and Kady, I.O., 2015. Thymoquinone inhibits *Escherichia coli* ATP synthase and cell growth. *PLoS One, 10*(5), p. e0127802.

Aljabre, S.H.M., Randhawa, M.A., Akhtar, N., Alakloby, O.M., Alqurashi, A.M. and Aldossary, A., 2005. Antidermatophyte activity of ether extract of *Nigella sativa* and its active principle, thymoquinone. *Journal of Ethnopharmacology, 101*(1–3), pp. 116–119.

Chaieb, K., Kouidhi, B., Jrah, H., Mahdouani, K. and Bakhrouf, A., 2011. Antibacterial activity of Thymoquinone, an active principle of *Nigella sativa* and its potency to prevent bacterial biofilm formation. *BMC Complementary and Alternative Medicine, 11*(1), p. 29.

Chaudhry, N.M.A. and Tariq, P., 2008. In vitro antibacterial activities of kalonji, cumin and poppy seed. *Pakistan Journal of Botany, 40*(1), p. 461.

Dey, D., Ray, R. and Hazra, B., 2014. Antitubercular and antibacterial activity of quinonoid natural products against multi-drug resistant clinical isolates. *Phytotherapy Research, 28*(7), pp. 1014–1021.

Halawani, E., 2009. Antibacterial activity of thymoquinone and thymohydroquinone of *Nigella sativa* L. and their interaction with some antibiotics. *Advances in Biological Research, 3*(5–6), pp. 148–152.

Kökdil, G., Delialioğlu, N., Özbilgin, B. and Emekdaş, G., 2005. Antilisterial activity of ballotaspecies growing in turkey-antibacterial activity screening of *Nigella* L. species growing in Turkey. *Journal of Faculty of Pharmacy, 34*, pp. 183–190.

Lloyd, G.E.R., 1964. The hot and the cold, the dry and the wet in Greek philosophy. *Journal of Hellenic Studies*, p. 92–106.

Onifade, A.A., Jewell, A.P. and Adedeji, W.A., 2013. *Nigella sativa* concoction induced sustained seroreversion in HIV patient. *African Journal of Traditional, Complementary and Alternative Medicines, 10*(5), pp. 332–335.

Özmen, A., Basbülbül, G. and Aydin, T., 2007. Antimitotic and antibacterial effects of the *Nigella sativa* L. Seed. *Caryologia, 60*(3), pp. 270–272.

Sun, L.L., Luan, M., Zhu, W., Gao, S., Zhang, Q.L., Xu, C.J., Lu, X.H., Xu, X.D., Tian, J.K. and Zhang, L., 2013. Study on antitubercular constituents from the seeds of *Nigella glandulifera*. *Chemical and Pharmaceutical Bulletin, 61*(8), pp. 873–876.

### 4.1.5.11 *Naravelia zeylanica* (L.) DC.

Synonym: *Atragene zeylanica* L.

Common names: Ceylon Naravelia; toilakti (Bangladesh); dhanavali (India)

Habitat: Thickets

Distribution: India and Bangladesh

Botanical observation: It is a woody climber that grows to about 4 m long. The stem is terete, striated, and subglabrous. The leaves comprise basally 2 folioles with 1.5–2.5 cm long petiolules, and apically 3 folioles developed into tendrils with 3–7 cm long petiolules. The folioles are ovate, 6–11×6–10 cm, rounded to cordate at base; and present 2 pairs of secondary nerves originating from the base. The racemes are terminal or axillary, up to 40 cm long, and bear a few flowers, which are about 1 cm in diameter. The flower peduncles are 1–1.5 cm long. The 4 sepals are light yellowish, elliptic, about 1 cm long, and hairy. The 8–10 petals, are about 1 cm long, somewhat greenish, and clavate. The numerous stamens are linear and minute. The gynoecium consists of numerous carpels, which are hairy and showy. The achenes are fusiform about 5 mm long and pilose.

Medicinal uses: Leprosy wounds (India)

Commentary: Not much is apparently on the antimicrobial properties of this plant.

### 4.1.5.12   *Pulsatilla koreana* (Yabe ex Nakai) Nakai ex T. Mori

Synonyms: *Anemone cernua* var. *koreana* Yabe ex Nakai; *Pulsatilla cernua* (Thunb.) Bercht. ex J. Presl; *Pulsatilla cernua* var. *koreana* (Yabe ex Nakai) Y.N. Lee

Common name: Korean Pulsatilla, Korean pasque flower; hal-me-kot (Korea); chao xian bai tou weng (China)

Habitats: Grassy slopes

Distribution: Japan, Korea, and China

Botanical observation: This graceful rhizomatous herb grows to about 30 cm tall. The leaves are simple, in rosettes, and exstipulate. The petiole is slender, 4.5–14 cm long and hairy. The blade is 3-lobed, the lobes deeply incised, 3–7.5×4.5–6.5 cm, and trilobed. The inflorescence is solitary and terminal, with involucral bracts, which are incised and up to about 4.5 cm long. The 5 tepals are purplish red to dark purple, hairy below, oblong to ovate-oblong, and 1.8–3×0.5–1.5 cm androecium is conspicuous, yellow, and includes numerous stamens, which are about 1 cm long. The gynoecium includes numerous purplish carpels. The achenes are obovate-oblong, 3 mm long, and develop a 4 cm long, slender, and curved hairy tail.

*Strong broad-spectrum amphiphilic antibacterial benzoquinone:* Pulsaquinone isolated from the roots inhibited the growth of *Propionibacterium acnes*, *Bacillus subtilis*, *Staphylococcus aureus*, *Streptococcus mutans*, *Pseudomonas aeruginosa*, and *Shigella sonnei* with MIC values of 2, 2.7, 2, 2, 3.3, and 2 µg/mL, respectively (Cho et al., 2009).

Pulsaquinone

*Strong broad-spectrum antifungal quinone*: Pulsaquinone isolated from the roots inhibited the growth of *Trichophyton mentagrophytes*, *Candida albicans*, *Candida glabrata*, and *Candida tropicalis* with MIC values of 21.3, 74.7, 16.0, and 26.7 µg/mL, respectively (Cho et al., 2009).

Commentary: Aqueous extract of root of *Pulsatilla chinensis* (Bunge) Regel at a concentration of 250 µg/mL inhibited Human immunodeficiency virus type-1 protease by 58.9% (Xu et al., 1996).

### References

Cho, S.C., Sultan, M.Z. and Moon, S.S., 2009. Anti-acne activities of pulsaquinone, hydro-pulsaquinone, and structurally related 1, 4-quinone derivatives. *Archives of Pharmacal Research*, 32(4), pp. 489–494.

Xu, H.X., Wan, M., Loh, B.N., Kon, O.L., Chow, P.W. and Sim, K.Y., 1996. Screening of traditional medicines for their inhibitory activity against HIV-1 protease. *Phytotherapy Research*, 10(3), pp. 207–210.

### 4.1.5.13  *Ranunculus sceleratus* L.

Synonyms: *Hecatonia palustris* Lour.; *Hecatonia scelerata* (L.) Fourr.; *Ranunculus holophyllus* Hance; *Ranunculus oryzetorum* Bunge

   Common names: Cursed buttercup; shi long rui (China); nakkore (Nepal); kandakatuka (India)
   Habitat: Watery grasslands, riverbanks, and lakes
   Distribution: Turkey, Afghanistan, Iran, Kazakhstan, Pakistan, India, Bhutan, Nepal, Thailand, China, Korea, Taiwan, and Japan
   Botanical observation: This herb grows to about 75 cm tall. The roots are fibrous. The leaves are simple, spiral, and exstipulate. The petiole is up to 15 cm long and sheathing at base. The blade is incised, 3-lobed, reniform, broadly ovate, or elliptic, and 1–4.5×1.5–5.5 cm. The cymes are lax and terminal. The calyx includes 5 minute sepals. The 5 petals are obovate, up to about 4.5 cm long, and dull yellow. The androecium includes 10–19 minute stamens. The numerous carpels are minute, hooked at apex, and free. The achenes are elongated, ovoid, and up to about 2 cm long receptacle.
   Medicinal use: Fever (India)
   *Moderate antiviral (enveloped circular double-stranded DNA virus) coumarin:* isoscopoletin at 100 µg/mL inhibited the secretion of Hepatitis B surface antigen and Hepatitis B virus e-antigen in HepG2.2.15 cells by 11.33 and 21.03% (Li et al., 2005).
   *Strong antiviral (enveloped monopartite linear double-stranded DNA) simple phenol:* Protocatechuic aldehyde inhibited Herpes simplex virus type-1 in Vero cells with an $IC_{50}$ value of 17.3 µg/mL (selectivity index > 11.53) (Li et al., 2005).

Protocatechuic aldehyde

Commentaries: (i) The plant contains the bactericidal lactone anemonin (Neag et al., 2018; Osborn, 1943). Misra and Dixit (1978) demonstrated the fungitoxic effects of leaf extract on account of anemonin (Misra & Dixit, 1980). (ii) In Nepal, *Ranunculus laetus* Wall. ex D.Don (local name: "*Bokua*") is used to treat skin infection and sinusitis.

### References

Li, H., Zhou, C.X., Pan, Y., Gao, X., Wu, X., Bai, H., Zhou, L., Chen, Z., Zhang, S., Shi, S. and Luo, J., 2005. Evaluation of antiviral activity of compounds isolated from *Ranunculus sieboldii* and *Ranunculus sceleratus*. *Planta Medica, 71*(12), pp. 1128–1133.
Misra, S.B. and Dixit, S.N., 1978. Antifungal properties of leaf extract of *Ranunculus sceleratus* L. *Experientia, 34*(11), pp. 1442–1443.
Misra, S.B. and Dixit, S.N., 1980. Antifungal principle of *Ranunculus sceleratus*. *Economic Botany, 34*(4), pp. 362–367.
Neag, T., Olah, N.K., Hanganu, D., Benedec, D., Pripon, F.F., Ardelean, A. and Toma, C.C., 2018. The anemonin content of four different *Ranunculus* species. *Pakistan Journal of Pharmaceutical Sciences, 31*(5 (Supplementary)), pp. 2027–2032.

Osborn, E.M., 1943. On the occurrence of antibacterial substances in green plants. *British Journal of Experimental Pathology*, 24(6), p. 227.

### 4.1.5.14 *Ranunculus ternatus* Thunb.

Common names: Ternate-leaved crowfoot; mao zhua cao (China)
  Habitat: Open grassy lands
  Distribution: China and Japan
  Botanical observation: It is a herb that grows to about 30 cm tall from tubers. The leaves are simple, spiral, and exstipulate. The petiole is up to 6 cm long, somewhat hairy, and sheathing at base. The blade is ternate, broadly ovate, and 0.5–1.5 × 1.5–2.5 cm. The inflorescences are lax terminal cymes with leaflike bracts. The 5 sepals are about 5 mm long. The 5 petals are oblong, up to about 7 mm long, and dull yellow. The androecium includes numerous minute stamens. The gynoecium includes numerous minute carpels, hooked at apex, and free. The fruits are ovoid achenes on elongated receptacle.
  Medicinal use: Tuberculosis (China)
  *Strong antimycobacterial benzophenone:* Methyl (*R*)-3-[2-(3,4-dihydroxybenzoyl)-4,5-dihydroxyphenyl]-2-hydroxypropanoate from the roots inhibited *Mycobacterium tuberculosis* (H37Rv) with an MIC value of 41.6 µg/mL (Deng et al., 2013).
  *Moderate antimycobacterial simple phenolic acids:* Vanillic acid and gallic acid from the roots inhibited *Mycobacterium tuberculosis* (H37Rv) with MIC values of 83.3 and 66.6 µg/mL, respectively (Deng et al., 2013).

### Reference

Deng, K.Z., Xiong, Y., Zhou, B., Guan, Y.M. and Luo, Y.M., 2013. Chemical constituents from the roots of *Ranunculus ternatus* and their inhibitory effects on *Mycobacterium tuberculosis*. *Molecules*, 18(10), pp. 11859–11865.

### 4.1.5.15 *Thalictrum simplex* L.

Common name: Jian tou tang song cao (China)
  Habitat: Grassy slopes, meadows
  Distribution: Turkey, Iran, Afghanistan, Turkmenistan, Tajikistan, Pakistan, China, Korea, and Japan
  Botanical observation: It is a herb that grows to about 1 m tall. The leaves are simple, basal and cauline, spiral, and exstipulate. The blade of basal leaves is bipinnate, with rhombic, ovate, or obovate, folioles, which are 2–4 cm × 1.5–4.5 cm. The inflorescence is a terminal and lax cyme, which bears numerous slender and yellowish flowers. The calyx comprises 4 sepals, which are elliptic, greenish yellow, and minute. The gynoecium is conspicuous and comprises 15 elongated stamens. The gynoecium includes 3–6 carpels. The achenes minute and ellipsoid.
  Medicinal uses: Wounds, infections (China)
  *Strong antiviral (enveloped segmented single-stranded (-) RNA) pavine alkaloids:* (-)-Thalimonine from aerial parts inhibited the replication of Influenza virus A/Germany/27, strain Weybridge (H7N7) and Influenza virus A/Germany/34, str. Rostock (H7N1) in chicken embryo fibroblast with $EC_{50}$ values of 0.1 and 0.6 µM and selectivity indices of 640 and 106.6, respectively (Serkedjieva et al., 2003a,b). The alkaloid was most active 4–5 hours post infection, suggesting the inhibition of intracellular-specific steps (Serkedjieva et al., 2003a,b). (-)-Thalimonine *N*-oxide from aerial part inhibited Influenza virus A/Germany/27, strain Weybridge (H7N7) and Influenza virus A/Germany/34, str. Rostock (H7N1) in cell cultures of chicken embryo fibroblast with $EC_{50}$ values of 4 and 1 µM and selectivity indices of 15 and 60, respectively (Serkedjieva et al., 2003a,b).

(-)-Thalimonine

Commentaries: (i) The plant brings being aporphines, including (+)-ocoteine as well as phenan-there alkaloids (Velcheva et al., 1996). Aporphines may at least partially account for the traditional use. In Nepal, *Thalictrum reniforme* Wall. (local name: *Pajeni*) is used to treat leucorrhea.

### References

Serkedjieva, J. and Velcheva, M., 2003a. In vitro anti-influenza virus activity of the pavine alkaloid (-)-thalimonine isolated from *Thalictrum simplex* L. *Antiviral Chemistry and Chemotherapy*, *14*(2), pp. 75–80.

Serkedjieva, J. and Velcheva, M., 2003b. In vitro anti-influenza virus activity of isoquinoline alkaloids from *Thalictrum* species. *Planta Medica*, *69*(02), pp. 153–154.

Velcheva, M.P., Petrova, R.R., Samdanghiin, Z., Danghaaghiin, S., Yansanghiin, Z., Budzikiewicz, H. and Hesse, M., 1996. Isoquinoline alkaloid N-oxides from *Thalictrum simplex*. *Phytochemistry*, *42*(2), pp. 535–537.

### 4.1.5.16   *Trollius chinensis* Bunge

Synonyms: *Trollius asiaticus* var. *chinensis* (Bunge) Maxim.; *Trollius macropetalus* (Regel) F. Schmidt

Common name: Jin lian hua (China)

Habitat: Grassy slopes

Distribution: China

Botanical observation: This showy herb grows to 80 cm tall. The leaves are simple, spiral, and exstipulate. The petiole is sheathing at base and 12–30 cm long. The blade is deeply incised 3.8–6.5×6.5–12.5 cm, trilobed, and dentate. The inflorescence is a terminal cyme of a few bright orange flowers, which are 5.5 cm across. The calyx includes 10–15 yellow, elliptic, and 1.5–2.5×0.7–1.5 cm sepals. The corolla includes about 20, 1.5–2.5 cm×1.2–1.5 mm petals. The stamens are numerous, about 1 cm long and filiform. The gynoecium includes 20–30 carpels. The follicles are up to about 1 cm long.

Medicinal use: Respiratory infections, tonsillitis, pharyngitis (China)

*Strong antiviral (enveloped monopartite linear single-stranded (−) RNA) hydrophilic C-glycosyl flavones:* Orientin (LogD=−0.8 at pH 7.4; molecular weight=448.3 g/mol) and vitexin (LogD=−0.6 pH at 7.4; molecular weight=432.3 g/mol) isolated from the flowers inhibited the replication of Parainfluenza type 3 virus in Hp2-cells with $IC_{50}$ values of 11.7 and 20.8 μg/mL and selectivity indices of 32.1 and 16, respectively (Li et al., 2002).

Orientin

Vitexin

*Moderate antiviral (enveloped, segmented, single-stranded (-) RNA) C-glycosyl flavones: 2″-O-(2‴-methylbutyryl)isoswertisin from the flowers inhibited Influenza virus A with an $IC_{50}$ value of 74.3 μg/mL and a selectivity index of 7.1 (Cai et al., 2006).*

Commentary: The flowers shelter phenolic compounds with antibacterial and antifungal-activities (Li et al., 2014).

### References

Cai, S.Q., Wang, R., Yang, X., Shang, M., Ma, C. and Shoyama, Y., 2006. Antiviral flavonoid-type C-glycosides from the flowers of *Trollius chinensis*. *Chemistry & Biodiversity, 3*(3), pp. 343–348.

Li, Y.L., Ma, S.C., Yang, Y.T., Ye, S.M. and But, P.P.H., 2002. Antiviral activities of flavonoids and organic acid from *Trollius chinensis* Bunge. *Journal of Ethnopharmacology, 79*(3), pp. 365–368.

Li, D.Y., Wei, J.X., Hua, H.M. and Li, Z.L., 2014. Antimicrobial constituents from the flowers of *Trollius chinensis*. *Journal of Asian Natural Products Research, 16*(10), pp. 1018–1023.

## 4.2 ORDER PROTEALES JUSS. EX BERCHT. ET J. PRESL (1820)

### 4.2.1 FAMILY NELUMBONACEAE BERCHT. ET J. PRESL (1820)

The family Nelumbonaceae consists of the single Asian genus *Nelumbo* Adans.

#### 4.2.1.1 *Nelumbo nucifera* Gaertn.

Synonyms: *Nelumbium nuciferum* Gaertn.; *Nelumbo caspica* Fisch. ex DC.; *Nelumbo komarovii* Grossh.; *Nelumbo speciosa* Willd.; *Nymphaea nelumbo* L.

Common names: Sacred lotus; lan (China); kamala, pushkara (India); kesorn bua luang (Thailand); chuk (Cambodia); sukaw (the Philippines); hoa sen (Vietnam)

Habitat: Lakes, ponds, and slow rivers

Distribution: Tropical Asia and Pacific

Botanical observation: This heavenly nuphar grows from a tuberous and edible rhizome. The flower is a powerful positive and holly symbol in the teachings of Buddha and Hinduism. The leaves are simple, alternate, exstipulate, and arise from the rhizome. The petiole 1–2 m long, terete, and fleshy. The blade is peltate, glaucous below, somewhat dull light green above, orbicular, 25–90 cm in diameter, fleshy, glabrous, and entire. The inflorescence is solitary, axillary with a flower peduncle, which reaches about 1 m long. The perianth is showy, 10–25 cm in diameter, and consists of numerous pinkish, oblong-elliptic to obovate, 5–10×3–5 cm tepals. The androecium includes numerous anthers, which are linear and about 1 cm long. The gynoecium includes numerous carpels merged at the apex of an obconic receptacle. The achenes are merged in a somewhat spongy green-glaucous showerhead-like receptacle, which is up to 15 cm in diameter. The seeds are about 1 cm long, elliptic, and edible.

Medicinal uses: Diarrhea, pneumonia, cough, gonorrhea, leucorrhea (China); cholera (Cambodia, Laos, Vietnam)

*Broad-spectrum antibacterial polar extract:* Ethanol extract of flowers inhibited the growth of *Staphylococcus aureus* (IOA-106) (resistant to amoxicillin, cefuroxime, cloxacillin, nalidixic acid), *Bacillus subtilis* (MTCC-121), *Escherichia coli* (UP 2566) resistant to methicillin, cloxacillin, doxycycline, novobiocin), and *Shigella dysenteriae* (IOA-108) (resistant to methicillin, tetracycline, doxycycline), and with inhibition zone diameters between 21 and 30 mm, 10 and 20 mm, below 10 mm, and between 10 and 20 mm, respectively (8 mm diameter well, 100 µL of 150 mg/mL solution) (Ahmad & Beg, 2001).

*Weak antibacterial (Gram-positive) polar extract:* Extract of rhizome inhibited the growth of *Streptococcus mutans* (MT 5091) and *Streptococcus mutans* (OMZ 176) with MIC values of 2750 and 2750 µg/mL, respectively (Chen et al., 1989).

*Anticandidal halo developed by polar extract:* Ethanol extract of flowers (8 mm diameter well, 100 µL of 150 mg/mL solution) inhibited the growth of *Candida albicans* (IOA-109) with inhibition zone diameters between 10 and 20 mm (Ahmad & Beg, 2001).

*Strong anticandidal amphiphilic aporphine:* Roemerine (also known as aporheine; LogD = 3.5 at pH 7.4; molecular weight = 279.3 g/mol) isolated from the leaves inhibited *Candida albicans* (ATCC 90028) with an MIC value of 10 µg/mL (Agnihotri et al., 2008).

*Antiviral (non-enveloped segmented linear double-stranded RNA) polar extract:* Aqueous extract inhibited Rhesus Rotavirus replication in MA-104 cells with an $IC_{50}$ value below 300 µg/mL (Knipping et al., 2012).

*Strong antiviral (enveloped monopartite linear dimeric single-stranded (+) RNA) hydrophilic benzylisoquinoline alkaloids:* (*R*)-Coclaurine (LogD = 0.2 at pH 7.4; molecular weight = 285.3 g/mol) and (*S*)-norcoclaurine (also known as higenamine; LogD = 0.2 at pH 7.4; molecular weight = 285.3 g/mol) inhibited the replication of Human immunodeficiency virus type-1 (strain IIIB) replication in H9 cells with $EC_{50}$ values of 0.8 µg/mL and below 0.8 µg/mL with selectivity indices above 20 (Kashiwada et al., 2005).

(S)-norcoclaurine

*Strong antiviral (enveloped monopartite linear dimeric single-stranded (+) RNA) amphiphilic bis-benzylisoquinoline alkaloids:* Liensinine (LogD=3.8 at pH 7.4; molecular weight=610.7 g/mol), negferine, and isoliensinin isolated from the leaves inhibited the replication of Human immunodeficiency virus type-1 (strain IIIB) with IC$_{50}$ values below 0.8 µg/mL and selectivity indices >9.9, >8.6, and >6.5, respectively (Kashiwada et al., 2005).

Liensinine

*Strong antiviral (enveloped monopartite linear dimeric single-stranded (+) RNA) amphiphilic aporphine alkaloid:* Nornuciferine (LogD=2.7 at pH 7.4; molecular weight=281.3 g/mol) inhibited Human immunodeficiency virus (strain IIIB) replication in H9 cells with an EC$_{50}$ value below 0.8 µg/mL and a therapeutic index above 20 (Kashiwada et al., 2005).

Nornuciferine

*Strong antiviral (enveloped monopartite linear dimeric single-stranded (+) RNA) flavonol gly-cosides:* Quercetin 3-O-$\beta$-D-glucuronide and quercetin 3-O-$\beta$-D-xylopyranosyl-(1→2)-$\beta$-D-galactopyranoside inhibited the replication the Human immunodeficiency virus type-1 in H9 cells with $EC_{50}$ values of 2 and 4 µg/mL and selectivity indices above 20 (Kashiwada et al., 2005).

Commentary: The plant is used for the treatment of COVID-19 in China (Wang et al., 2020). It contains isoquercitrin, which is antiviral (Gaudry et al., 2018). Consider that plants with anti-Human immunodeficiency virus properties are not uncommonly used for the treatment of COVID-19. The Severe acute respiratory syndrome-associated coronavirus type 2 and Human immunodeficiency virus type-1 are both enveloped linear single-stranded (+) RNA viruses.

## References

Agnihotri, V.K., ElSohly, H.N., Khan, S.I., Jacob, M.R., Joshi, V.C., Smillie, T., Khan, I.A. and Walker, L.A., 2008. Constituents of *Nelumbo nucifera* leaves and their antimalarial and antifungal activity. *Phytochemistry Letters*, *1*(2), pp. 89–93.

Ahmad, I. and Beg, A.Z., 2001. Antimicrobial and phytochemical studies on 45 Indian medicinal plants against multi-drug resistant human pathogens. *Journal of Ethnopharmacology*, *74*(2), pp. 113–123.

Chen, C.P., Lin, C.C. and Tsuneo, N., 1989. Screening of Taiwanese crude drugs for antibacterial activity against *Streptococcus* mutans. *Journal of Ethnopharmacology*, *27*(3), pp. 285–295.

Gaudry, A., Bos, S., Viranaicken, W., Roche, M., Krejbich-Trotot, P., Gadea, G., Desprès, P. and El-Kalamouni, C., 2018. The flavonoid isoquercitrin precludes initiation of Zika virus infection in human cells. *International Journal of Molecular Sciences*, *19*(4), p. 1093.

Kashiwada, Y., Aoshima, A., Ikeshiro, Y., Chen, Y.P., Furukawa, H., Itoigawa, M., Fujioka, T., Mihashi, K., Cosentino, L.M., Morris-Natschke, S.L. and Lee, K.H., 2005. Anti-HIV-1 benzylisoquinoline alkaloids and flavonoids from the leaves of *Nelumbo nucifera*, and structure–activity correlations with related alkaloids. *Bioorganic & Medicinal Chemistry*, *13*(2), pp. 443–448.

Knipping, K., Garssen, J. and van't Land, B., 2012. An evaluation of the inhibitory effects against rotavirus infection of edible plant extracts. *Virology Journal*, *9*(1), p. 1.

Yuan, L.C., Liu, B. and Shi, R.B., 2010. HPLC determination of hyperin and isoquercitrin in *Nelumbo nucifera* Gaertn. of different sources [J]. *Chinese Journal of Pharmaceutical Analysis*, *1*.

# 5 The Clade Core Eudicots

## 5.1 ORDER DILLENIALES DC. EX BERCHT. & J. PRESL (1820)

The order Dilleniales consists of the single family Dilleniaceae.

### 5.1.1 FAMILY DILLENIACEAE R.A. SALISBURY (1807)

The family Dillenidae consists of 12 genera and 400 species of trees, shrubs, and woody climbers. The leaves are simple, exstipulate, and spiral. The blades are not uncommonly coriaceous, serrate, and often very characteristically marked with broad, conspicuous, and straight secondary nerves. The inflorescence is a cyme, raceme, or solitary. The calyx consists of 4–5 sepals, which are fleshy and imbricate. The corolla includes 3–5 membranous and ephemeral petals. The androecium comprises numerous stamens. The gynoecium consists of 2–7 carpels, which are free and sheltering each 1 to many ovules on submarginal or basal placentas. The fruits are follicles, berries, or capsules, and contain 1 to many seeds. Not much is apparently known about the antimicrobial principles in this family.

#### 5.1.1.1 *Acrotrema costatum* Jack

Common names: Yellow jungle star; meroyan punai tanah (Malaysia); pot khon (Thailand)

Habitat: Forests on wet soil or moist shady rocks

Distribution: Myanmar, Thailand, Malaysia, and Indonesia

Botanical observation: This herb grows from horizontal woody rhizomes. The leaves are simple, spiral, in rosette, and exstipulate. The petiole is stout, with amplexicaul wings, and about 1 cm long. The blade is obovate, auriculate at base, coriaceous, dark green glossy, with a whitish line along the midrib, 7–30×3–12 cm, hairy below, with about 12 pairs of conspicuous secondary nerves, tapering at base, acute or rounded, serrate, and obtuse at apex. The racemes are terminal and up to about 13 cm tall, reddish, and hairy. The 5 sepals are lanceolate, hairy, and light green. The corolla comprises 5 spathulate petals, which are yellow, about 1.5 cm long, and membranous. The androecium comprises numerous stamens arranged in 3 bundles. The gynoecium includes 3 carpels. The follicles are dehiscent and containing numerous tiny seeds with white membranous arils.

Medicinal uses: Post-partum remedy (Malaysia)

Commentary: (i) This plant has apparently not been studied for its potential antimicrobial constituents. (ii) Condensed tannins and/or triterpenes are probably, at least partially, responsible for the medicinal use. (iii) The plant produces apigenin and luteolin (Hegnauer, 1989). (iv) The traditional Malay system of medicine uses numerous plants for post-partum compared to other Southeast Asian medicinal systems.

**Reference**

Hegnauer, R., 1989. Dilleniaceae. In *Chemotaxonomie der Pflanzen*. Verlag, Germany.

#### 5.1.1.2 *Dillenia indica* L.

Synonyms: *Dillenia speciosa* Thunb., *Dillenia speciosa* Thunb.

Common names: Elephant apple; chalta, chalita, thigi (Bangladesh); san (Cambodia); wu ya guo (China); avartaki, chalota, chalta, ruvya, kawrthindeng, peddakalinga (India); sempu (Indonesia); simpoh (Malaysia); thabyu (Myanmar); panch phal (Nepal); handapara (the Philippines); wampara (Sri Lanka); sompru (Thailand); so ba (Vietnam)

Habitat: Forest near streams and rivers, or cultivated

Distribution: India, Sri Lanka, Bangladesh, Myanmar, Cambodia, Laos, Vietnam, Thailand, Malaysia, Indonesia, and the Philippines

Botanical observation: It is a magnificent timber that grows to 30 m tall and used to ornament parks. The bole is straight. The bark is reddish brown and exfoliating. The leaves are simple, spiral, and exstipulate. The petiole is narrowly winged, stout, and 2–4 cm long. The blade obovate-oblong, coriaceous, 15–40×7–14 cm, with 30–40 pairs of conspicuous secondary nerves, tapering at base, acute or rounded, serrate, and acuminate at apex. The flowers are showy, 12–20 cm in diameter, terminal, and solitary. The calyx comprises 5 sepals, which are rounded, imbricate, and 4–6 cm in diameter, fleshy, dull green and coriaceous. The corolla comprises 5 petals, which are pure white, obovate, to oblong membranous and 7–9 cm long. The androecium comprises numerous stamens. The gynoecium includes up to 20 linear, lanceolate, and recurved carpels. The fruit is globose, heavy, 10–25 cm in diameter, with persistent sepals and somewhat glossy to dull green.

Medicinal uses: Jaundice, cough, diarrhea, dysentery, abscess (India); fever (Bangladesh; India; Myanmar)

*Weak broad-spectrum antibacterial non-polar extract:* Hexane extract of bark inhibited the growth of *Bacillus cereus* (ATCC 14579), *Bacillus subtilis* (ATCC 6059), *Staphylococcus aureus* (ATCC 6538), *Escherichia coli* (ATCC 25922), *Pseudomonas aeruginosa* (ATCC 27853), *Salmonella paratyphi* (ATCC 9150), and *Shigella dysenteriae* (ATCC 9361) with MIC values of 1.2, 1.2, 0.3, 5, 0.6, 5, and 0.3 mg/mL, respectively (Alam et al., 2011).

*Weak broad-spectrum antifungal non-polar extract:* Hexane extract of bark inhibited the growth of *Candida albicans* (ATCC 90028), *Aspergillus niger* (ATCC 1004), and *Saccharomyces cerevisiae* (ATCC 60782) with MIC values of 2.5, 5, and 2.5 mg/mL, respectively (Alam et al., 2011).

*Moderate antiviral (enveloped monopartite linear double-stranded DNA) polar extract:* Methanol extract inhibited the growth of Herpes simplex virus (ATCC 733) and Herpes simplex virus type-1 (clinical strain) with $IC_{50}$ values of 56.1 and 61.2 µg/mL, respectively (Goswami et al., 2016).

*Viral enzyme inhibition by triterpene:* Betulinic acid inhibited aminopeptidase N with an $IC_{50}$ value of 7.3 µM (Bauvois & Dauzonne, 2006).

Commentary: (i) The plant produces proanthocyanidins and condensed tannins (Fu et al., 2015) as well as lupane-type pentacyclic triterpenes, including betulinic acid (Ghosh et al., 2014) which probably account for the medicinal uses (Ghosh et al., 2014).

### References

Alam, M.B., Chowdhury, N.S., Mazumder, M.E.H. and Haque, M.E., 2011. Antimicrobial and toxicity study of different fractions of *Dillenia indica* Linn. bark extract. *International Journal of Pharmaceutical Sciences and Research*, 2(4), p. 860.

Bauvois, B. and Dauzonne, D., 2006. Aminopeptidase-N/CD13 (EC 3.4. 11.2) inhibitors: chemistry, biological evaluations, and therapeutic prospects. *Medicinal Research Reviews*, 26(1), pp. 88–130.

Fu, C., Yang, D., Peh, W.Y.E., Lai, S., Feng, X. and Yang, H., 2015. Structure and antioxidant activities of proanthocyanidins from elephant apple (*Dillenia indica* Linn.). *Journal of Food Science*, 80(10), pp. C2191–C2199.

Ghosh, P.S., Sarma, I.S., Sato, N., Harigaya, Y. and Dinda, B., 2014. Lupane-triterpenoids from stem bark of *Dillenia indica*. *Indian Journal of Chemistry*, 53B, pp. 1284–1287.

Goswami, D., Mukherjee, P.K., Kar, A., Ojha, D., Roy, S. and Chattopadhyay, D., 2016. Screening of ethnomedicinal plants of diverse culture for antiviral potentials. *Indian Journal of Traditional Medicine*, 5(3), pp. 474–481.

### 5.1.1.3 *Dillenia papuana* Martelli

Common names: Kaigabar, simpoh (Indonesia); mayonga (Papua)

 Habitat: Forests

 Distribution: East Indonesia and Papua New Guinea

 Botanical observation: This handsome timber grows to 40 m tall. The bole is straight and presents massive buttresses. The bark is brown, somewhat mucilaginous, and scaly. The leaves are simple, spiral, and exstipulate. The petiole is stout, and up to about 5 cm long. The blade is broadly elliptic, coriaceous, dull dark green above, 15–40×10–30 cm, tapering or cordate at base, wavy, with up to about 25 pairs of conspicuous secondary nerves, and acuminate or rounded at apex. The flowers are showy, 12–20 cm in diameter, terminal and axillary. The calyx comprises 5 sepals, which are rounded, imbricate, and about 2 cm long, fleshy, red, and coriaceous. The 5 petals are white, oblong membranous and about 3 long. The androecium comprises numerous stamens, which are pinkish. The gynoecium includes up to 15 linear, lanceolate, and recurved carpels. The fruit is an aggregate of follicles, which is up to 3.5 cm long and contains up to 15 seeds, which are about 5 mm long.

 Medicinal uses: Child delivery (East Indonesia, Papua New Guinea)

 *Broad-spectrum antibacterial halo elicited by triterpenes:* The lupane-type triterpenes dillenic acid D and the oleanane-type triterpene 3-oxoolean-12-en-oic acid inhibited the growth of *Bacillus subtilis* with minimum growth inhibition amount on thin layer chromatography of 0.5 and 0.8 µg (Nick et al., 1995). Dillenic acid D and 3-oxoolean-12-en-oic acid inhibited the growth of *Escherichia coli* with minimum growth inhibition amount of 0.5 and 1.8 µg, respectively, on thin layer chromatography (Nick et al., 1995). Dillenic acid D and 3-oxoolean-12-en-oic acid inhibited the growth of *Micrococcus luteus* with minimum growth inhibition amount of 1 and 0.8 µg, respectively, on thin layer chromatography (Nick et al., 1995).

 Commentary: The precise antibacterial mode of action of pentacyclic triterpenes is apparently not fully understood. Being mostly lipophilic, these principles may accumulate in the bacterial membranes and destabilize them (Broniatowski et al., 2015).

**References**

Broniatowski, M., Flasiński, M., Wydro, P. and Fontaine, P., 2015. Grazing incidence diffraction studies of the interactions between ursane-type antimicrobial triterpenes and bacterial anionic phospholipids. *Colloids and Surfaces B: Biointerfaces*, 128, pp. 561–567.

Nick, A., Wright, A.D., Rali, T. and Sticher, O., 1995. Antibacterial triterpenoids from *Dillenia papuana* and their structure-activity relationships. *Phytochemistry*, 40(6), pp. 1691–1695.

### 5.1.1.4 *Tetracera scandens* (L.) Merr.

Synonym: *Tragia scandens* L.

 Common names: Stone leaf; mao guo xi ye teng (China); kiasahan (Indonesia); akar mempelas (Malaysia); malakatmon (the Philippines); chac chiu, cây y hít (Vietnam)

 Habitat: Roadsides and disturbed forests

 Distribution: India, Myanmar, Indonesia, Malaysia, Thailand, Vietnam, and the Philippines

 Botanical observation: It is a woody climber that grows to 30 m long. The stem is terete, and hairy at apex. Reddish and smooth when young. The leaves are simple, spiral, and exstipulate. The petiole is up to about 1.5 cm long. The blade is ovate to obovate, coriaceous, 4–10×2.5–5 cm, dark green, somewhat glossy, somewhat hairy below, with 9–12 pairs of conspicuous secondary nerves, rounded to tapering at base, serrate, and obtuse to round at apex. The inflorescence is a lax terminal panicle, which is up to 20 cm long. The 4–5 sepals are ovate, about 4 mm long, and persistent in fruits. The 3 petals are white, and about 5 mm long. The androecium comprises numerous stamens. The gynoecium includes 1–2 carpels. The fruits are follicles, which are ovate and up to about 1 cm long and containing 1–2 arillate seeds.

Medicinal use: Sore throat (the Philippines)

*Strong antiviral (enveloped monopartite linear dimeric single stranded (+)RNA) polar extract:* Ethanol extract inhibited the replication of Human immunodeficiency virus at 4 µg/mL by 87% and inhibited Human immunodeficiency virus reverse transcriptase with $IC_{50}$ of 0.7 µg/mL (Kwon et al., 2012).

Commentaries: (i) The plant produces series of lupane-type triterpenes (Nguyen & Nguyen, 2013) and may contain condensed tannins. (ii) Note that betulinic acid, 3-cis-p-coumaroyl maslinic acid, and 3-trans-p-coumaroyl maslinic acid isolated from a member of the genus *Tetracera* L. inhibited recombinant rat liver DNA polymerase β with $IC_{50}$ values of 14, 15, and 4.2 µM, respectively (Ma et al., 1999). DNA polymerase is bacteria is responsible for DNA replication (Lemon & Grossman, 1998), and one could have some interest to look for antibacterial natural products inhibiting this enzyme, like lupane-triterpenes, in *Tetracera scandens* (L.) Merr., and other members of the family Dilleniaceae.

### References

Kwon, H.S., Park, J., Kim, J.H. and You, J.C., 2012. Identification of anti-HIV and anti-reverse transcriptase activity from *Tetracera scandens*. *BMB Reports, 45*(3), pp. 165–170.

Lemon, K.P. and Grossman, A.D., 1998. Localization of bacterial DNA polymerase: evidence for a factory model of replication. *Science, 282*(5393), pp. 1516–1519.

Ma, J., Starck, S.R. and Hecht, S.M., 1999. DNA polymerase β inhibitors from *Tetracera boiviniana*. *Journal of Natural Products, 62*(12), pp. 1660–1663.

Nguyen, M.T.T. and Nguyen, N.T., 2013. A new lupane triterpene from *Tetracera scandens* L., xanthine oxidase inhibitor. *Natural Product Research, 27*(1), pp. 61–67.

## 5.2   ORDER SAXIFRAGALES BERCHT. & PRESL (1820)

### 5.2.1   FAMILY ALTINGIACEAE HORANINOW (1841)

The family Altingiaceae comprises 3 genera and 17 species of resinous trees. The stems are terete and often covered with stellate hairs. The leaves are alternate, simple, and stipulate. The flowers are minute and arranged in terminal racemes or heads. The perianth comprises 12–50 minute lobes or is absent or vestigial. The androecium includes 12–50 stamens. The gynoecium consists of 2 carpels, each containing 20–50 ovules. The fruits are dehiscent capsules. Members of the family Altingiaceae produce antibacterial gallotannins, iridoids, flavonols, ellagic acid, and essential oils.

### 5.2.1.1   *Altingia excelsa* Noronha

Synonym: *Liquidambar altingiana* Bl.

   *Common names:* Rasamala; xi qing pi (China); nantayok (Myanmar); mandung (Indonesia); satu (Thailand); Silaras (India); patharkumkum (Nepal)

   Distribution: China, India, Myanmar, Nepal, Bhutan; India, Malaysia, and Indonesia

   Habitat: Forests

   Botanical observation: It is a resinous tree that grows up to about 20 m tall. The stems are hairy. The leaves are simple, spiral, and stipulate. The stipules are up to about 5 mm long. The petiole is slender, terete, hairy, and 2–4 cm long. The blade is ovate, 8–13.5×4–7.5 cm, coriaceous, glossy, rounded at base, crenate, acuminate at apex, and with 6–8 pairs of secondary nerves. The male inflorescence is an axillary raceme of up to about 20 flowers. The female flowers are a terminal globose head. The androecium includes numerous stamens. The gynoecium is about 5 mm long and bifid. The gynoecium is minute and glabrous. The fruits are dehiscent tetralocular capsules, which are subglobose, up to about 2 cm in diameter, and containing numerous winged seeds.

Medicinal uses: Bronchitis, fever (India); cough (Indonesia)

*Antibacterial (Gram-negative) halo developed by cyclic monoterpenes:* α-Pinene inhibited *Staphylococcus aureus* (ATCC 9144), *Proteus vulgaris* (ATCC 13315), *Bacillus cereus* (NCIMB 6349), and *Staphylococcus epidermidis* (ATCC 12228) with inhibition zone diameters of 11, 13, 17, and 10 mm, respectively (10 μL, 4 mm well) (Lis-Balchin et al., 1998). Limonene inhibited *Bacillus cereus* (NCIMB 6349) with an inhibition zone diameter of 9 mm (10 μL per well 4 mm well) (Lis-Balchin et al., 1998).

*Weak antiviral (enveloped monopartite linear single-stranded (+) RNA) cyclic monoterpene:* β-Pinene inhibited the replication of Avian infectious bronchitis virus with an $IC_{50}$ of value of 250 μM, and the activity was much stronger after penetration of the virus into the cell (Yang et al., 2011).

Commentaries: (i) Essential oil in the leaves is made of mainly α-pinene, β-pinene, α-phellandrene, limonene, and β-phellandrene (Kanjilal et al., 2003). (ii) Avian infectious bronchitis virus is a positive-sense, single-stranded RNA virus that belongs to the family Coronaviridae, infecting poultry (Kasmi et al., 2020)

### References

Kanjilal, P.B., Kotoky, R. and Singh, R.S., 2003. Chemical composition of the leaf oil of Altingia excelsa Nornha. *Flavour and Fragrance Journal*, 18(5), pp. 449–450.

Kasmi, Y., Khataby, K., Souiri, A. and Ennaji, M.M., 2020. Coronaviridae: 100,000 years of emergence and reemergence. In *Emerging and Reemerging Viral Pathogens* (pp. 127–149). Academic Press.

Lis-Balchin, M., Buchbauer, G., Ribisch, K. and Wenger, M.T., 1998. Comparative antibacterial effects of novel Pelargonium essential oils and solvent extracts. *Letters in Applied Microbiology*, 27(3), pp. 135–141.

Yang, Z., Wu, N., Zu, Y. and Fu, Y., 2011. Comparative anti-infectious bronchitis virus (IBV) activity of (-)-pinene: effect on nucleocapsid (N) protein. *Molecules*, 16(2), pp. 1044–1054.

### 5.2.1.2 *Liquidambar formosana* Hance

Synonyms: *Liquidambar acerifolia* Maxim.; *Liquidambar maximowiczii* Miq.; *Liquidambar tonkinensis* A. Chev.

Common names: Formosa sweet gum; feng xiang shu (China); sau sau, chao, cha phai (Vietnam)

Habitat: Forests

Distribution: Cambodia, Laos, Vietnam, China, Korea, Taiwan, and Japan

Botanical observation: It is a resinous tree that grows up to about 40 m tall. The stems are slender, hairy or not, and lenticelled. The leaves are simple, spiral, and stipulate. The stipules are up to about 2 long, caducous, reddish, and linear. The petiole is 8–12 cm long and somewhat hairy. The blade is mostly 3-lobed, about 10–13 cm × 10–16 cm, glossy, dark green above, turning red, cordate at base, serrulate, the lobes somewhat deltoid and acute, acuminate, or caudate at apex. The male flowers are organized in terminal racemes. The female inflorescence is a solitary peduncled globose head. The calyx consists of minute scales. The androecium includes numerous stamens. The gynoecium is hairy and comprises 2 carpels, bifid and up to about 1 cm long. The fruits are merged in a globose head, which is up to about 3.5 cm in diameter, spiny, and comprising numerous dehiscent capsules containing minute and brown winged seeds are minute and brown.

Medicinal uses: Boils, wounds (China); tuberculosis (Japan)

*Weak broad-spectrum antibacterial polar extract:* Aqueous extract inhibited the growth of *Bacillus cereus*, *Bacillus subtilis*, *Clostridium perfringens*, *Staphylococcus aureus*, *Vibrio parahaemolyticus*, *Klebsiella pneumoniae*, *Shigella sonnei*, and *Pseudomonas aeruginosa* with MIC values of 1600, 2667, 133, 800, 133, 600, 2133, and 1067 μg/mL, respectively (Taguri et al., 2006).

*Weak antibacterial (Gram-positive) hydrolysable tannin:* The ellagitannin oligomer isorugosin A isolated from the fresh leaves inhibited the growth of *Staphylococcus aureus* (209P) and methicillin-resistant *Staphylococcus aureus* (OM481) with MIC values of 58 and 29 µM, respectively (Shimozu et al., 2017).

*Antifungal triterpene:* 3α, 25-dihydroxyolean-12-en-28-oic acid from the resin inhibited the growth of *Lenzites betulina* and *Laetiporus sulphureus* (Chien et al., 2013).

*Antifungal phenylpropanoid:* Bornyl cinnamate from the resin inhibited the growth of *Lenzites betulina* and *Laetiporus sulphureus* (Chien et al., 2013).

*Weak antiviral (enveloped segmented linear single-stranded (-) RNA virus) polar extract:* Aqueous extract of infructescences inhibited the Influenza A (California/7/2009 (H1N1) NYMC X-179A) of virus with an $IC_{50}$ value of 102.2 µg/mL (Tian et al., 2011).

*Strong antiviral (enveloped monopartite linear single-stranded (+) RNA virus) hydrophilic flavonol:* Quercetin inhibited the Influenza A (California/7/2009 (H1N1) NYMC X-179A) of virus with an $IC_{50}$ value of 5.4 µg/mL (Tian et al., 2011).

*Viral enzyme inhibition by aqueous extract:* Aqueous extract of infructescences inhibited neuraminidase activity with an $IC_{50}$ value of 1 µg/mL (Tian et al., 2011).

*Commentary:* Ellagitannins, gallic acids, and other phenolics probably account for the antibacterial property. According to Taguri et al. (2006) the pyrogallol group is important for the antibacterial activity of the polyphenols, and the catechol and resorcinol groups are less important. Consider that pyrogallol (LogD = 0.4 at pH 7.4; molecular weight = 126.1 g/mol) inhibited the replications of Herpes simplex virus type-2 with an $IC_{50}$ value of 0.1 µg/mL (Kane et al., 1988).

Pyrogallol

### References

Chien, S.C., Xiao, J.H., Tseng, Y.H., Kuo, Y.H. and Wang, S.Y., 2013. Composition and antifungal activity of balsam from *Liquidambar formosana* Hance. *Holzforschung, 67*(3), pp. 345–351.

Kane, C.J., Menna, J.H., Sung, C.C. and Yeh, Y.C., 1988. Methyl gallate, methyl-3, 4, 5-trihydroxybenzoate, is a potent and highly specific inhibitor of herpes simplex virusin vitro. II. Antiviral activity of methyl gallate and its derivatives. *Bioscience Reports, 8*(1), pp. 95–102.

Shimozu, Y., Kuroda, T., Tsuchiya, T. and Hatano, T., 2017. Structures and antibacterial properties of Isorugosins H–J, oligomeric ellagitannins from Liquidambar formosana with characteristic bridging groups between sugar moieties. *Journal of Natural Products, 80*(10), pp. 2723–2733.

Taguri, T., Tanaka, T. and Kouno, I., 2006. Antibacterial spectrum of plant polyphenols and extracts depending upon hydroxyphenyl structure. *Biological and Pharmaceutical Bulletin, 29*(11), pp. 2226–2235.

Tian, L., Wang, Z., Wu, H., Wang, S., Wang, Y., Wang, Y., Xu, J., Wang, L., Qi, F., Fang, M. and Yu, D., 2011. Evaluation of the anti-neuraminidase activity of the traditional Chinese medicines and determination of the anti-influenza A virus effects of the neuraminidase inhibitory TCMs in vitro and in vivo. *Journal of Ethnopharmacology, 137*(1), pp. 534–542.

### 5.2.1.3 *Liquidambar orientalis* Mill.

Common names: Oriental sweetgum, Turkish sweet gum; sığala ağacı (Turkey); silaras (Bangladesh); sihlahka (India)

Distribution: Turkey

Habitat: Riverbanks, stream valleys, and marshes

Botanical observation: This resinous tree grows up to about 30m tall. The stems are hairy. The leaves are simple, spiral, and stipulate. The stipules are up to about 5mm long. The petiole is slender, channeled, and 3.4–6.5cm long. The blade is palmatifid, 5–10×6–12.5cm, glossy, base subcordate at base, serrulate, the lobes somewhat deltoid and acute at apex. The male flowers are in terminal racemes, which are up to about 4.5cm long. The female inflorescence is a solitary peduncled globose head. The androecium includes numerous stamens. The gynoecium comprises 2 carpels and is bifid and reddish. A disc is present. The infructescences are globose heads, which are up to about 2.5cm in diameter, spiny, and comprising about 40 dehiscent capsules containing about 8 winged and glossy seeds, which are up to about 5mm long.

Medicinal uses: Bronchitis, fever, leprosy, fever (India); cuts (Turkey); wounds (India; Turkey)

*Broad-spectrum antibacterial halo developed by resin:* The resin (50μL of a solution containing 10.0%, 4mm diameter agar well) inhibited the growth of *Bacillus brevis*, *Bacillus cereus*, *Bacillus subtilis*, *Corynebacterium xerosis* (UC 9165), *Enterobacter aerogenes* (CCM 2531), *Enterococcus faecalis* (ATCC 15753), *Klebsiella pneumoniae* (FMC 5), *Micrococcus luteus* (LA 2971), *Mycobacterium smegmatis* (RUT), *Proteus vulgaris* (FMC 1), *Pseudomonas aeruginosa* (ATCC 27853), *Pseudomonas fluorescens* (EU), and *Staphylococcus aureus* (Cowan 1) with inhibition zone diameters of 12, 16, 10, 8, 12, 13, 6, 14, 11, 12, 12, 11, and 14mm, respectively (Sağdıç et al., 2005).

*Weak broad-spectrum antibacterial polar extract:* Ethanol extract of leaves inhibited the growth of *Bacillus subtilis* (RSKK245), *Staphylococcus aureus* (RSKK2392), *Listeria monocytogenes* (ATCC7644), *Enterococcus faecalis* (ATCC8093), *Escherichia coli* (ATCC11229), *Yersinia enterocolitica* (NCTC11174), and *Salmonella typhimurium* (RSKK19) with MIC values ranging from 1625 to 3950 μg/mL, respectively (Okmen et al., 2014).

*Weak anticandidal polar extract:* Ethanol extract of leaves inhibited the growth of *Candida albicans* (RSKK02029 with an MIC value of 3250 μg/mL (Okmen et al., 2014).

Commentary: (i) The essential oil of the plant consists mainly of styrene, α-pinene, β-pinene, and limonene (Fernandez et al., 2005). α-Pinene and limonene are antibacterial as discussed previously. (ii) The plant is used for the treatment of COVID-19 in China (Wang et al., 2020).

### References

Fernandez, X., Lizzani-Cuvelier, L., Loiseau, A.M., Perichet, C., Delbecque, C. and Arnaudo, J.F., 2005. Chemical composition of the essential oils from Turkish and Honduras Styrax. *Flavour and Fragrance Journal*, *20*(1), pp. 70–73.

Okmen, G., Turkcan, O., Ceylan, O. and Gork, G., 2014. The antimicrobial activity of *Aliquidambar orientalis* mill. against food pathogens and antioxidant capacity of leaf extracts. *African Journal of Traditional, Complementary and Alternative Medicines*, *11*(5), pp. 28–32.

Sağdıç, O., Özkan, G., Özcan, M. and Özçelik, S., 2005. A study on inhibitory effects of sığla tree (*Liquidambar orientalis* Mill. var. orientalis) storax against several bacteria. *Phytotherapy Research*, *19*(6), pp. 549–551.

Wang, S.X., Wang, Y., Lu, Y.B., Li, J.Y., Song, Y.J., Nyamgerelt, M. and Wang, X.X., 2020. Diagnosis and treatment of novel coronavirus pneumonia based on the theory of traditional Chinese medicine. *Journal of Integrative Medicine*, *18*(4), pp. 275–283.

## 5.2.2 FAMILY CRASSULACEAE J. ST.-HIL. (1805)

The family Crassulaceae consists of 35 genera and about 1500 of herbs or shrubs, which are fleshy or succulent and growing among sun exposed stones or walls. The leaves are simple or pinnate, alternate, opposite, or verticillate, and exstipulate. The inflorescences are terminal or axillary cymes, spikes, or solitary. The calyx comprises 4–6 sepals free or united, and the corolla includes 4–6 petals free or partially connate in a tube. The androecium is made of 4–30 stamens. The gynoecium consists of free carpels. Plants in this family generate interesting series of antimicrobial phenolics.

### 5.2.2.1 *Kalanchoe pinnata* (Lam.) Pers.

Synonyms: *Bryophyllum calycinum* Salisb.; *Bryophyllum germinans* Blanco; *Bryophyllum pinnatum* (Lam.) Asch. & Schweinf.; *Bryophyllum pinnatum* (Lam.) Kurz; *Bryophyllum pinnatum* (Lam.) Oken; *Calanchoe pinnata* Pers.; *Cotyledon calycina* Roth; *Cotyledon calyculata* Solander; *Cotyledon pinnata* Lam.; *Cotyledon rhizophilla* Roxb.; *Crassula pinnata* L. f.; *Crassuvia floripendia* Comm. ex Lam.; *Crassuvia floripenula* Comm.; *Kalanchoe calycina* Salisb.; *Sedum madagascariense* Clus.; *Verea pinnata* (Lam.) Spreng.

Common names: Air plant, life plant; gios, patharkuchi (Bangladesh); lao di sheng gen (China); amar poi (India); daun sejuk (Indonesia); setawar padang (Malaysia); karitana (the Philippines); ton tai bai pen (Thailand); lac dia sinh căn (Vietnam)

Habitat: Cultivated

Distribution: Tropical Asia

Botanical observation: It is fleshy and erect herb that grows up to 1.5 m tall. The stems are terete, smooth, and purplish. The leaves are imparipinnate, decussate, and exstipulate. The blade is 10–30 cm long, comprises 3–5 folioles, the petiolules 2–4 cm, the folioles oblong to elliptic, 5–20×2.5–12 cm, crenate, rounded at base, dull green, and obtuse at apex. The inflorescences are axillary or terminal, panicles, which are 10–40 cm long and many flowered. The flowers are pendulous and showy. The calyx is tubular, cylindrical, purplish or light green, 2–4 cm long, and produces 4 lobes, which are triangular. The corolla is reddish to purple, up to 5 cm long and presents 4 lobes which are ovate-lanceolate. The 8 stamens are inserted basally on corolla. Nectar scales are present. The ovary is ovoid, minute, and develops a slender style. The 4 follicles are about 1.5 cm long, included in calyx and corolla tube, and contain numerous minute seeds.

Medicinal uses: Cold, cough, wounds (India), boil, sores, cholera (Myanmar); otitis (Indonesia)

*Moderate antibacterial (Gram-negative) polar extract:* Ethanol extract of leaves inhibited *Pseudomonas aeruginosa* (ATCC 27853) with an MIC value of 512 µg/mL, (Romulo et al., 2018).

*Strong broad-spectrum antibacterial hydrophilic flavonol glycosides:* Kaempferitrin (also known as Kaempferol 3,7-dirhamnoside; LogD=−1.4 at pH 7.4; molecular mass=578.5 g/mol) inhibited *Staphylococcus aureus*, *Pseudomonas aeruginosa*, and *Salmonella typhi* with MIC values of 32, 32, and 32 µg/mL, respectively (Tatsimo et al., 2012). Kaempferol 3-*O*-α-L-(4-acetyl) rhamnopyranoside-7- *O* - α -L-rhamnopyranoside inhibited *Staphylococcus aureus*, *Pseudomonas aeruginosa*, and *Salmonella typhi* with an MIC value of 64, 128, and 32 µg/mL, respectively (Tatsimo et al., 2012). Afzelin inhibited *Staphylococcus aureus*, *Pseudomonas aeruginosa*, and *Salmonella typhi* with MIC values of 8, 16, and 2 µg/mL, respectively (Tatsimo et al., 2012). α-Rhamnoisorobin inhibited *Staphylococcus aureus*, *Pseudomonas aeruginosa*, and *Salmonella typhi* with MIC values of 2, 1, and 1 µg/mL, respectively (Tatsimo et al., 2012).

Kaempferitrin

*Strong broad-spectrum antifungal flavonol glycosides:* Kaempferitrin inhibited *Candida parap-silosis, Candida parapsilosis,* and *Cryptococcus neoformans* with MIC values of 32, 16, and 16 µg/mL, respectively (Nyaa et al., 2009). Kaempferol 3-*O*-α-L-(4-acetyl) rhamnopyranoside-7-*O*-α-L-rhamnopyranoside inhibited the growth of *Candida parapsilosis, Candida parapsilosis,* and *Cryptococcus neoformans* with MIC values of 64, 32, and 32 µg/mL, respectively (Tatsimo et al., 2012). Afzelin (LogD=−0,2; molecular weight=432.3 g/mol) inhibited the growth of *Candida parapsilosis, Candida parapsilosis,* and *Cryptococcus neoformans* with MIC values of 16, 4, and 4 µg/mL, respectively (Tatsimo et al., 2012). α-Rhamnoisorobin inhibited the growth of *Candida parapsilosis, Candida* parapsilosis, and *Cryptococcus neoformans* with MIC values of 2, 1, and 1 µg/mL, respectively (Tatsimo et al., 2012).

Afzelin

*Strong antiviral (enveloped monopartite linear double-stranded DNA) aryltetraline lignan:* KPB-100 isolated from the roots inhibited Vaccinia virus in Vero cells with $IC_{50}$ values of 3.1 µg/mL (cidofovir: 9.7 µg/mL) (Cryer et al., 2017). KPB-100 inhibited Herpes simplex virus type-2 (HHV-2) (E194) in Vero cells with $IC_{50}$ values of 2.5 µg/mL (Cryer et al., 2017).

KPB-100

*Strong antiviral (enveloped monopartite linear double-stranded DNA) simple phenol glycoside:* KPB-200 isolated from the roots inhibited Vaccinia virus in Vero cells with $IC_{50}$ value of 7.4 µg/mL (Cryer et al., 2017). KPB-200 inhibited Herpes simplex virus type -2 (HHV-2 (E194) in Vero cells with an $IC_{50}$ value of 2.9 µg/mL (Cryer et al., 2017).

KPB-200

Commentary: Consider that afzelin at a concentration of 20μM inhibited coronavirus ion channel 3a-mediated current by almost 100% (Schwarz et al., 2014).

### References

Cryer, M., Lane, K., Greer, M., Cates, R., Burt, S., Andrus, M., Zou, J., Rogers, P., Hansen, M.D., Burgado, J. and Satheshkumar, P.S., 2017. Isolation and identification of compounds from *Kalanchoe pinnata* having human alphaherpesvirus and vaccinia virus antiviral activity. *Pharmaceutical Biology*, 55(1), pp. 1586–1591.

Romulo, A., Zuhud, E.A., Rondevaldova, J. and Kokoska, L., 2018. Screening of in vitro antimicrobial activity of plants used in traditional Indonesian medicine. *Pharmaceutical Biology*, 56(1), pp. 287–293.

Schwarz, S., Sauter, D., Wang, K., Zhang, R., Sun, B., Karioti, A., Bilia, A.R., Efferth, T. and Schwarz, W., 2014. Kaempferol derivatives as antiviral drugs against the 3a channel protein of coronavirus. *Planta Medica*, 80(02/03), pp. 177–182.

Tatsimo, S.J.N., de Dieu Tamokou, J., Havyarimana, L., Csupor, D., Forgo, P., Hohmann, J., Kuiate, J.R. and Tane, P., 2012. Antimicrobial and antioxidant activity of kaempferol rhamnoside derivatives from *Bryophyllum pinnatum*. *BMC Research Notes*, 5(1), p. 158.

### 5.2.2.2 *Rhodiola rosea* L.

Synonym: *Sedum rosea* (L.) Scop.

Common name: Hong jing tian (China)

Habitat: Mountain slopes

Distribution: Kazakhstan; China, Korea, Japan, and Korea

Botanical observation: It is a succulent herb that grows from a rhizome up to a height of 30 cm. The leaves are simple, opposite, sessile, and exstipulate. The blade is broadly lanceolate to obovate, 0.7–3.5×0.5–1.8 cm, somewhat amplexicaul, laxly serrate, and acute to acuminate at apex. The inflorescence is a terminal and showy cyme of numerous yellow flowers. The calyx presents 4 minute sepals. The 4 petals are oblong to oblanceolate and about 3 mm long. The androecium includes 8 stamens. The gynoecium presents free carpels. The follicles are narrowly lanceolate, about 8 mm long, and beaked at apex.

Medicinal use: Apparently none for microbial infections

*Moderate antibacterial (Gram-positive) flavonol glycoside:* Gossypetin-7-O-L-rhamnopyranoside (also known as rhodiolgin) and rhodioflavonoside inhibited the growth of *Staphylococcus aureus* with the MIC of 50 and 100 μg/mL, respectively (Ming et al., 2005).

*Antiviral (enveloped monopartite linear single-stranded (+) RNA) hydrophilic phenolic glycoside:* Salidroside (LogD=−0.7 at pH 7.4; molecular mass=300.3 g/mol) at 166 μM inhibited Dengue virus replication in THP-1 cells (Sharma et al., 2016).

Salidroside

*Antiviral (non-enveloped monopartite linear single-stranded (+)RNA virus) phenolic glycoside:* Salidroside at 60 μg/mL decreased virus titer in primarily cultured myocardial cells infected with Coxsackievirus B3 (Nancy strain).

*In vivo (non-enveloped monopartite linear single-stranded (+)RNA virus) antiviral phenolic glycoside:* Salidroside given daily at 80 mg/Kg/day for 7 days orally to BALB/c mice infected with Coxsackievirus B3 prevented cardiac insults (Wang et al., 2009).

*Virus enzyme inhibition by polar extract:* Methanol extract of roots inhibited Human immunodeficiency virus type-1 protease (Park, 2003).

Commentary: Members of the family Crassulaceae often inhibit the Human immunodeficiency virus protease *in vitro* as observed for instance with *Sedum sarmentosum* Bunge, *Sedum polytrichoid es*Hemsl., and *Orostachys japonica* (Maxim.) A. Berger (Park, 2003).

### References

Ming, D.S., Hillhouse, B.J., Guns, E.S., Eberding, A., Xie, S., Vimalanathan, S. and Towers, G.H., 2005. Bioactive compounds from *Rhodiola rosea* (Crassulaceae). *Phytotherapy Research, 19*(9), pp. 740–743.

Park, J.C., 2003. Inhibitory effects of Korean plant resources on human immunodeficiency virus type 1 protease activity. *Oriental Pharmacy and Experimental Medicine, 3*(1), pp. 1–7.

Sharma, N., Mishra, K.P. and Ganju, L., 2016. Salidroside exhibits anti-dengue virus activity by upregulating host innate immune factors. *Archives of Virology, 161*(12), pp. 3331–3344.

Wang, H., Ding, Y., Zhou, J., Sun, X. and Wang, S., 2009. The in vitro and in vivo antiviral effects of salidroside from *Rhodiola rosea* L. against coxsackievirus B3. *Phytomedicine, 16*(2–3), pp. 146–155.

### 5.2.2.3  *Sedum aizoon* L.

Synonyms: *Phedimus aizoon* (L.) 't Hart; *Sedum hyperaizoon* Kom.

Common names: Fei cai, tusanqi (China)

Habitat: Mountain slopes

Distribution: China, Mongolia, Korea, and Japan

Botanical examination: It is a herb that grows from a taperoot to a height of 50 cm. The leaves are simple and alternate. The blade is flat, narrowly lanceolate, 3.5–8×0.5–3 cm, cuneate at base, laxly serrate, and obtuse to acuminate at apex. The inflorescences are terminal cymes. The calyx presents 5 linear sepals, which are about 5 mm long. The corolla includes 5 petals, which are yellow, oblong to elliptic-lanceolate, and up to 1 cm long and mucronate at apex. The androecium includes 10 stamens. The gynoecium presents 5 carpels. The follicles are 7 mm long and contain numerous tiny seeds.

Medicinal use: Cough (China)

*Strong broad-spectrum amphiphilic simple phenolic:* Methyl gallate (LogD = 1.1 at pH = 7.4; molecular weight = 184.1 g/mol) inhibited *Pseudomonas aeruginosa* (Z61), *Escherichia coli* (DC2), and *Staphylococcus aureus* (RN4220) with MIC values of 12.5, 25, and 200 μg/mL, respectively (Saxena et al., 1994). Methyl gallate from the aerial parts inhibited the growth of *Escherichia coli* [CMCC (B) 441020], with the MIC of 93 μg/mL (Xu et al., 2015). Gallic acid and methyl gallate inhibited the growth of *Staphylococcus aureus* [CMCC (B) 26003) with MIC values of 232 and 7.8 μg/mL, respectively (Xu et al., 2015). Gallic acid inhibited the growth of *Bacillus subtilis* (CMCC (B) 63501) with an MIC value of 464 μg/mL (Xu et al., 2015).

*Moderate antibacterial (Gram-positive) flavonol glycoside:* Herbacetin-3-O-α-L-rhamnopyranosy l-8-O-α-D-lyxopyranoside from the aerial parts inhibited the growth of *Staphylococcus aureus* (CMCC (B) 26003) with MIC of 260 μg/mL and *Bacillus subtilis* (CMCC (B) 63501) with MIC of 260 μg/mL (Xu et al., 2015).

*Viral enzyme inhibition by flavonol:* Herbacetin (8-hydroxykaempferol) inhibited Severe acute respiratory syndrome (Severe acute respiratory syndrome)-associated coronavirus chymotrypsin-like protease with an $IC_{50}$ value of 33.1 µM (Jo et al., 2020). Herbacetin inhibited Middle East respiratory syndrome coronavirus chymotrypsin-like protease with $IC_{50}$ values of 40.5 µM (Jo et al., 2019)

*Viral enzyme inhibition by flavonol glycosides:* Rhoifolin (apigenin-7-O-rhamnoglucoside) and pectolinarin inhibited the Severe acute respiratory syndrome-associated coronavirus chymotrypsin-like protease with $IC_{50}$ values of 27.4 and 37.7 µM, respectively (Jo et al., 2020). Quercetin 3-β-d-glucoside inhibited Middle East respiratory syndrome coronavirus chymotrypsin-like protease with an $IC_{50}$ value of 37 µM (Jo et al., 2019)

*Viral enzyme inhibition by chalcones:* Isobavachalcone and helichrysetin inhibited Middle East respiratory syndrome coronavirus chymotrypsin-like protease with $IC_{50}$ values of 35.8 and 67 µM, respectively (Jo et al., 2019).

Commentary: *In vitro* and *in vivo* anti-Severe acute respiratory syndrome-associated coronavirus of polar extracts from this plant are warranted. Consider that phenolic and flavonoid glycosides are often antiviral *in vitro.*

### References

Jo, S., Kim, H., Kim, S., Shin, D.H. and Kim, M.S., 2019. Characteristics of flavonoids as potent MERS-CoV 3C-like protease inhibitors. *Chemical Biology & Drug Design, 94*(6), pp. 2023–2030.

Jo, S., Kim, S., Shin, D.H. and Kim, M.S., 2020. Inhibition of SARS-CoV3CL protease by flavonoids. *Journal of Enzyme Inhibition and Medicinal Chemistry, 35*(1), pp. 145–151.

Saxena, G., McCutcheon, A.R., Farmer, S., Towers, G.H.N. and Hancock, R.E.W., 1994. Antimicrobial constituents of *Rhus glabra. Journal of Ethnopharmacology, 42*(2), pp. 95–99.

Xu, T., Wang, Z., Lei, T., Lv, C., Wang, J. and Lu, J., 2015. New flavonoid glycosides from *Sedum aizoon* L. *Fitoterapia, 101*, pp. 125–132.

## 5.2.3   FAMILY GROSSULARIACEAE A.P. DE CANDOLLE (1805)

The family Grossulariaceae consists of the single genus *Ribes* L.

### 5.2.3.1   *Ribes nigrum* L.

Common name: Black current; nabar (India); karan (Pakistan); kurokarin (Japan)

Habitat: Mountains or cultivated

Distribution: Turkey, Iran, Afghanistan, Pakistan, and Himalayas

Botanical observation: It is a shrub that grows up to 2 m tall. The stems are terete. The leaves are simple, spiral, and exstipulate. The petiole grows up to about 7 cm long. The blade is 2.5–10.5 cm × 2.5–12 cm broad, dull green, cordate at base, 2–5-lobed, and serrate. The inflorescence is a raceme. The calyx is tubular, 2 mm long, and develops 5 lobes, which are oblong, somewhat burgundy to purple or greenish, recurved, and obtuse to round at apex. The corolla presents 5 petals which are spathulate, membranous, ephemeral, and 1 mm long. The androecium includes 5 stamens, which are about 1.5 mm long. The ovary develops 2 styles. The berries are black, glossy, globose, crowned by the persistent calyx, edible, aromatic, and about 1 cm in diameter.

Medicinal uses: Cough, flu, and sore throat (India)

*Strong antibacterial (Gram-positive) polar extract:* Methanol extract of leaves inhibited the growth of *Staphylococcus aureus, Enterococcus faecalis, Escherichia coli,* and *Klebsiella pneumoniae* with MIC values of 7.8, 15.6, 500, and 500 µg/mL, respectively (Kendir et al., 2016).

*Very weakly anticandidal polar extract:* Methanol extract of leaves inhibited the growth of *Candida albicans* and *Candida parapsilopsis* with MIC values of 500 and 500 µg/mL, respectively (Kendir et al., 2016).

*Strong antiviral (enveloped segmented linear single-stranded (-) RNA virus) extract:* Extract inhibited Influenza type A virus (and Influenza type B virus with IC$_{50}$ of 3.2 µg/mL, respectively (Knox et al., 2003).

*Antiviral (enveloped segmented linear single-stranded (-) RNA virus) juice:* Juice of fruits inhibited the replication of Influenza virus type-A (A/Yamagata/120/86) and Influenza virus type-B (B/Singapore) with IC$_{50}$ values of 0.1 and 0.1 µg/mL, respectively (Ikuta et al., 2012). Juice of fruits at 3% inhibited the absorption of the Influenza virus (AH1), Influenza virus (AH3) (Hong Kong flu), and Influenza virus (AH1tam$^r$) (Russian flu) to Madin-Darby canine kidney cells by 100% (Ikuta et al., 2013). Juice of fruits inhibited the replication of Influenza virus type-A (A/PR/8/34) and Influenza virus type-B (B/Gifu/2/73), with the IC$_{50}$ values of 0.1 and 0.1%, respectively (Ikuta et al., 2012).

*Antiviral (enveloped monopartite linear double-stranded DNA) juice:* Juice of the fruits inhibited the replication of Herpes simplex virus type-1 (VR-3 strain) with IC$_{50}$ values of 0.4% (Ikuta et al., 2012).

*Antiviral (non-enveloped monopartite double-stranded DNA virus):* Juice of the fruits inhibited the replication of the Adenovirus (type 5 prototype strain) with an IC$_{50}$ value of 2.5% (Ikuta et al., 2012).

*Antiviral (enveloped linear monopartite single-stranded (-)RNA virus) polar extract:* Juice of the fruits inhibited the replication of the Respiratory syncytial virus with an IC$_{50}$ value of 0.08% (Ikuta et al., 2012).

Commentary: The fruits abound with flavan glycosides, principally glucosides of cyanidine and delphinidine (Slimestad & Solheim, 2002). Black current could be developed at a prophylactic phytomedication for Influenza. Could it be of value in the prophylaxis of COVID-19?

### References

Ikuta, K., Hashimoto, K., Kaneko, H., Mori, S., Ohashi, K. and Suzutani, T., 2012. Anti-viral and anti-bacterial activities of an extract of blackcurrants (*Ribes nigrum* L.). *Microbiology and Immunology*, 56(12), pp. 805–809.

Ikuta, K., Mizuta, K. and Suzutani, T., 2013. Anti-influenza virus activity of two extracts of the blackcurrant (*Ribes nigrum* L.) from New Zealand and Poland. *Fukushima Journal of Medical Science*, 59(1), pp. 35–38.

Kendir, G., Köroglu, A., Özkan, S., Özgen Özgacar, S., Karaoglu, T. and Gargari, S., 2016. Evaluation of antiviral and antimicrobial activities of ribes species growing in turkey. *Journal of Biologically Active Products from Nature*, 6(2), pp. 136–149.

Knox, Y.M., Suzutani, T., Yosida, I. and Azuma, M., 2003. Anti-influenza virus activity of crude extract of *Ribes nigrum* L. *Phytotherapy Research*, 17(2), pp. 120–122.

Slimestad, R. and Solheim, H., 2002. Anthocyanins from black currants (*Ribes nigrum* L.). *Journal of Agricultural and Food Chemistry*, 50(11), pp. 3228–3231.

## 5.2.4 FAMILY HALORAGACEAE R. BROWN (1814)

The family Halogaraceae consists of about 8 genera and 100 species of herbs. The leaves are spiral, whorled or opposite, and exstipulate. The inflorescences are cymes of minute flowers. The calyx comprises 4 sepals. The corolla comprises up to 4 petals. The androecium includes up to 8 stamens. The gynoecium includes up to 4 carpels. The fruits are nuts or drupes. Members in this family Halogaceae produce hydrolyzable tannins

### 5.2.4.1 *Myriophyllum spicatum* L.

Common names: Eurasian water–milfoil, spiked water–milfoil, sui zhuang hu wei zao (China), mul su se mi (Korea)

Habitat: Ponds, lakes, and slow rivers

Distribution: Tropical Asia

Botanical observation: It is a graceful filamentous aquatic herb. The stems are much branched, leafy and grow to 2.5 m long. The submerged leaves are 4–or 5–whorled, pectinate, and 3–5 × 1–3.5 cm. The segments are arranged in 15 pairs, filiform and 1–1.5 cm long. The inflorescence is a 6–10 cm long terminal spike of whorled flowers. The male flowers are minute and present a broadly campanulate minute calyx. The corolla presents 4 petals, which are pale pink, elliptic, and 0.2 cm long. The androecium comprises 8 stamens. The female flowers are minute and present a calyx, which is tubiform. The fruits are cylindrical, 0.2 cm×0.1 cm and made of 4 locules.

Medicinal use: Dysentery (China)

Commentary: The plant has apparently not been examined for possible antibacterial properties. Note the presence of hydrolyzable tannins β–1,2,3–tri–$O$–galloyl–4,6–(S)–hexahydroxydiphenyl–D–glucose (also known as tellimagrandin II) used by the plant as algicidal and antibacterial weapons (Leu et al., 2002). Is tellimagrandin II antibacterial? Probably yes.

### Reference

Leu, E., Krieger-Liszkay, A., Goussias, C. and Gross, E.M., 2002. Polyphenolic allelochemicals from the aquatic angiosperm *Myriophyllum spicatuminhibit* photosystem II. *Plant Physiology*, *130*(4), pp. 2011–2018.

### 5.2.5  FAMILY HAMAMELIDACEAE R. BROWN (1818)

The family Hamamelidaceae comprises 28 genera and about 110 species of trees or shrubs. The stems are terete and often covered with stellate hairs. The leaves are alternate, simple, and stipulate. The flowers are minute, and arranged in various types of inflorescence. The calyx comprises 4–5 free sepals. The corolla comprises up to 4–5 free petals or none. The androecium includes 4–32 stamens. The gynoecium consists of 2–3 free carpels, each containing mostly a single anatropous ovule. The fruits are minute achenes or capsules. Members of the family produce gallotannins, ellagic acid, flavonoids (quercetin, kaempferol, and frequently also myricetin), and C-glycosyl flavones.

### 5.2.5.1  *Corylopsis coreana* Uyeki

Synonym: *Corylopsis gotoana* var. *coreana* (Uyeki) T. Yamaz.

Common name: Korean winter hazel

Distribution: Korea

Habitats: Forest edges

Botanical observation: It is a tree that grows up to about 10 m tall. The stems are glabrous, terete, zigzag shaped, and lenticelled. The leaves are simple, spiral, and stipulate. The stipules are showy and caducous, leaving a conspicuous scar. The petiole is 1.5–2.8 cm long. The blade is ovate to obovate, 6–7.5×7.5–8.5 cm, somewhat glaucous below, membranous, asymmetrical and cordate at base, serrate, acute at apex, and with 8–9 pairs of secondary nerves. The inflorescence is a terminal and 4–5 cm long pendulous raceme. The calyx includes 5 cream-colored sepals, which are minute. The 5 cream-colored petals are oblong to spathulate and about 4 mm long. The androecium includes 5 stamens. Disc scales are present. The gynoecium is minute and glabrous. The capsules are dehiscent, globose, and up to about 7 mm long.

Medicinal use: Apparently none for microbial infections.

*Moderate broad-spectrum antibacterial polar extract:* Ethanol extract of flowers inhibited the growth of *Staphylococcus aureus*, *Micrococcus luteus*, *Mycobacterium smegmatis*, and methicillin-resistant *Staphylococcus aureus* (Park et al., 2017). Extract inhibited the growth of vancomycin-resistant *Staphylococcus aureus*, with MIC of 500 µg/mL (Park et al., 2017).

Commentary: (i) The plant produces phenolics including chalcones, such as isosalipur-poside (Han et al., 2015), bergenin, 6'-O-galloylbergenin, 3'-O-galloylbergenin, (-)-catechin, (-)-epicatechin, (-)-epicatechin-3-O-galloyl ester, 4-methoxy-3,-5-dihydroxybenzoic acid, gallic acid, 2,4,6-trimethoxyphenol-1-O-β-D-glucopyranoside, and 2,4,6-trimethoxyphenol-1-O-β-D-(6-O-galloyl)-glucopyranoside (Kwon et al., 2016), which contribute to the antibacterial activity of the plant. (ii) Note that plant can sense the presence of pathogenic fungi and bacteria, which trigger chalcone synthesis as a precursor for flavonoids and stilbene (Dao et al., 2011). (iii) Hamamelidaceae are rich in phenolic compounds, explaining their ability to inhibit viral enzyme *in vitro* as with for example aqueous extract of stems of *Distylium racemosum* Siebold & Zucc. which inhibited Human immunodefiency virus type-1 protease (Park, 2003).

### References

Dao, T.T.H., Linthorst, H.J.M. and Verpoorte, R., 2011. Chalcone synthase and its functions in plant resistance. *Phytochemistry Reviews*, 10(3), p. 397.

Han, J.Y., Cho, S.S., Yang, J.H., Kim, K.M., Jang, C.H., Park, D.E., Bang, J.S., Jung, Y.S. and Ki, S.H., 2015. The chalcone compound isosalipurposide (ISPP) exerts a cytoprotective effect against oxidative injury via Nrf2 activation. *Toxicology and Applied Pharmacology*, 287(1), pp. 77–85.

Kwon, O.K., Kim, C.S., Suh, W.S., Park, K.J., Cha, J.M., Choi, S.U., Kwon, H.C. and Lee, K.R., 2016. Phenolic compounds from the twigs of *Corylopsis coreana* Uyeki and their cytotoxic activity. *Korean Journal of Pharmacognosy*, 47(1), pp. 1–6.

Park, J.C., 2003. Inhibitory effects of Korean plant resources on human immunodeficiency virus type 1 protease activity. *Oriental Pharmacy and Experimental Medicine*, 3(1), pp. 1–7.

Park, D.E., Yoon, I.S., Kim, J.E., Seo, J.H., Yoo, J.C., Bae, C.S., Lee, C.D., Park, D.H. and Cho, S.S., 2017. Antimicrobial and anti-inflammatory effects of ethanol extract of *Corylopsis coreana* uyeki flos. *Pharmacognosy Magazine*, 13(50), p. 286.

### 5.2.6  FAMILY PAEONIACEAE RAFINESQUE (1815)

The family Paeoniaceae comprises a single genera *Paeonia* L.

#### 5.2.6.1  *Paeonia emodi* Wall. ex Royle

Common names: Himalayan paeony; duo hua shao yao (China); mamekh (Pakistan); ood salib (India); Dhandra (Western Himalayas)

Habitats: Thickets

Distribution: Afghanistan, Pakistan, India, Kashmir, Himalaya, and Nepal

Botanical observation: This magnificent herb grows to about 50 cm tall. The petiole is 2–8 cm long. The leaves are compound, spiral, and exstipulate. The blade is 2-ternate. The folioles are lanceolate to elliptic, 9–14 cm×2–4.5 cm, cuneate at base and acuminate at apex. The flowers are 8–12 cm in diameter. The calyx includes 3 sepals, which are lanceolate, up to 2.5 cm long, and caudate at apex. The corolla comprises 8 pure white petals, which are obovate, wavy, and up to about 5 cm long. The androecium includes numerous stamens, which are 1.5–2 cm long. An annular disc is present. The gynoecium includes 2 carpel, which are yellowish and hairy. The follicles are dehiscent, ovoid, beaked at apex, up to 3.5 cm long, and contain numerous black and globose seeds.

Medicinal uses: Diarrhea (India); wounds (Pakistan)

*Broad-spectrum antibacterial halo developed by polar extract:* Methanol extract (10 mg/mL, 200 μL well) inhibited the growth of *Escherichia coli, Pseudomonas aeruginosa, Klebsiella pneumonia*, methicillin-resistant *Staphylococcus aureus, Staphylococcus aureus, Staphylococcus epidermidis*, and *Salmonella typhi* (Mufti et al., 2012).

*Weak anticandidal polar extract:* Methanol extract of seeds inhibited *Candida glabrata* (MTCC3019/ATCC-90030) with an MIC value of 3100 μg/mL via permeabilization of the cell wall (Sharma et al., 2018). This extract evoked a synergistic effect with amphotericin B, nystatin, and fluconazole towards *Candida glabrata* (MTCC3019/ATCC-90030) with FICI values of 0.4, 0.3, and 0.5 mg/mL, respectively (Sharma et al., 2018).

*Antiviral (enveloped circular double-stranded DNA virus) monoterpene phenyl glycoside:* Paeonins B inhibited Hepatitis B virus surface antigen and Hepatitis B virus e-antigen production by HepG2215 by about 30 and 20%, respectively, at a concentration of 25 µM (Bi et al., 2013).

Commentaries: (i) The antibacterial and antifungal constituents of this plant are apparently unknown. (ii) The ability of an extract or compound to increase the sensitivity of bacteria or fungi towards an antibiotic or antifungal agent can be measured by fractional inhibitory concentration (FIC) indices (FICI) (Giacometti et al., 1999; Schelz et al., 2006). FICI is the sum of the FIC of a compound and the FIC of an antibiotic or antifungal drug calculated according to the following formula (Berenbaum, 1978):

$$FIC\left(\text{extract or compound}\right)=\frac{MIC\left(\text{extract or compound in the presence of antibiotic or antifungal agent}\right)}{MIC\left(\text{extract or compound alone}\right)}$$

$$FIC\left(\text{antibiotic or antifungal drug}\right)=\frac{MIC\left(\text{antibiotic in the presence of extract or compound}\right)}{MIC\left(\text{antibiotic or antifungal agent alone}\right)}$$

$$FICI = FIC\left(\text{extract or compound}\right) + FIC\left(\text{antibiotic or antifungal drug}\right)$$

FICI ≤ 0.5 = synergistic
  FICI from 0.5 to 1 = additive
  1 < FICI ≤ 4 = indifferent
  FICI ≥ 4 antagonist
(iii) Consider that much attention has been given to resistant bacteria but one must note that multidrug-resistant fungi are emerging, including the infamous *Candida auris*, which is currently almost untreatable in hospital settings (ElBaradei, 2020).

### References

Berenbaum, M.C., 1978. A method for testing for synergy with any number of agents. *Journal of Infectious Diseases*, 137(2), pp. 122–130.

Bi, M., Tang, C., Yu, H., Chen, W. and Wang, J., 2013. Monoterpenes from *Paeonia sinjiangensis* inhibit the replication of Hepatitis B virus. *Records of Natural Products*, 7(4), p. 246.

ElBaradei, A., 2020. A decade after the emergence of *Candida auris*: what do we know? *European Journal of Clinical Microbiology & Infectious Diseases*, pp. 1–11.

Giacometti, A., Cirioni, O., Barchiesi, F., Fortuna, M. and Scalise, G., 1999. In vitro anti-cryptosporidial activity of ranalexin alone and in combination with other peptides and with hydrophobic antibiotics. *European Journal of Clinical Microbiology & Infectious Diseases*, 18, pp. 827–829.

Mufti, F.U.D., Ullah, H., Bangash, A., Khan, N., Hussain, S., Ullah, F., Jamil, M. and Jabeen, M., 2012. Antimicrobial activities of *Aerva javanica* and *Paeonia emodi* plants. *Pakistan Journal of Pharmaceutical Sciences*, 25(3), pp. 565–569.

Schelz, Z., Molnar, J. and Hohmann, J., 2006. Antimicrobial and antiplasmid activities of essential oils. *Fitoterapia*, 77, pp. 279–285.

Sharma, A., Singh, S., Tewari, R., Bhatt, V.P., Sharma, J. and Maurya, I.K., 2018. Phytochemical analysis and mode of action against *Candida glabrata* of *Paeonia emodi* extracts. *Journal de Mycologie Medicale*, 28(3), pp. 443–451.

### 5.2.6.2 *Paeonia lactiflora* Pall.

Synonyms: *Paeonia albiflora* Pall.; *Paeonia chinensis* L. Vilmorin; *Paeonia sinensis* Steud.; *Paeonia yui* W.P. Fang

Common names: Chinese paeony; shao yao (China)
Habitats: Forests and open grassy lands
Distribution: China, Russia, Mongolia, Korea, and Japan
Botanical observation: It is a graceful herb that grows to about 70 cm tall from somewhat large and fleshy tuberous roots. The leaves are spiral, compound, and exstipulate. The petiole is up to about 10 cm long. The blade is 2-ternate, the folioles are tapering and asymmetrical at base, lanceolate-elliptic, 4.5–16 cm × 1.5–5 cm, minutely dentate, and acute to acuminate at apex. The flowers are 8–13 cm in diameter. The calyx includes 3–4 sepals, which are ovate, up to 1.8 cm long, and caudate at apex. The 9–13 petals are pure white to light pink, obovate, notched at apex, and up to about 4.5 cm long. The numerous stamens are yellow and 0.7–1.2 cm long. An annular disc is present. The gynoecium includes 2–4 carpels, which are bottle shaped, purplish green, and curved at apex. The follicles are dehiscent, oblong, beaked at apex, up to 3 cm long and contain numerous black and globose seeds.

Medicinal use: Hepatitis (China)

*Moderate broad-spectrum antibacterial polar extract:* Aqueous extract inhibited the growth of *Bacillus cereus*, *Bacillus subtilis*, *Clostridium perfringens*, *Staphylococcus aureus*, *Vibrio para-haemolyticus*, *Klebsiella pneumoniae*, *Shigella flexneri*, and *Pseudomonas aeruginosa* with MIC values of 167, 167, 67, 133, 67, 500, 333, and 267 µg/mL, respectively (Taguri et al., 2006).

*Strong antiviral (enveloped monopartite linear single-stranded (-)RNA) polar extract:* Aqueous extract at 10 µg/mL inhibited the replication of the Human respiratory syncitial virus larynx epidermoid carcinoma cell line (HEp-2) as efficiently as 10 µg/mL of ribavirin (Lin et al., 2013). This treatment inhibited viral attachment and internalization (Lin et al., 2013).

*Weak antiviral (enveloped circular double-stranded DNA virus) monoterpene phenyl glycosides:* Paeoniflorin and albiflorin inhibited Hepatitis B virus surface antigen secretion by HepG2215 by about 20 and 50% at a concentration of 25 µM (Bi et al., 2013). Albiflorin also inhibited Hepatitis B virus e-antigen production by HepG2215 by about 30% (Bi et al., 2013).

Commentary: Plant phenolic or terpenic glycosides are often antiviral by inhibiting the entry of virus to host cells. It seems that the glycosidic moiety is an important pharmacophore. The plant produces monoterpene phenolic glycoside (Yang et al., 2004). These rare types of compounds are mainly produced by members of the genus *Paeonia* and should be thoroughly studied for their antiviral properties. It would be of interest to assess paeoniflorin and albiflorin against the Severe acute respiratory syndrome-associated coronavirus. The plant is used for the treatment of COVID-19 in China (Wang et al., 2020). Aqueous extract of root bark of *Paeonia suffruticosa* Andrews at a concentration of 250 µg/mL inhibited human immunodeficiency virus type-1 protease by 62.6% (Xu et al., 1996).

### References

Bi, M., Tang, C., Yu, H., Chen, W. and Wang, J., 2013. Monoterpenes from *Paeonia sinjiangensis* inhibit the replication of Hepatitis B virus. *Records of Natural Products*, 7(4), p. 246.

Lin, T.J., Wang, K.C., Lin, C.C., Chiang, L.C. and Chang, J.S., 2013. Anti-viral activity of water extract of *Paeonia lactiflora* pallas against human respiratory syncytial virus in human respiratory tract cell lines. *The American Journal of Chinese Medicine*, 41(03), pp. 585–599.

Taguri, T., Tanaka, T. and Kouno, I., 2006. Antibacterial spectrum of plant polyphenols and extracts depending upon hydroxyphenyl structure. *Biological and Pharmaceutical Bulletin*, 29(11), pp. 2226–2235.

Wang, S.X., Wang, Y., Lu, Y.B., Li, J.Y., Song, Y.J., Nyamgerelt, M. and Wang, X.X., 2020. Diagnosis and treatment of novel coronavirus pneumonia based on the theory of traditional Chinese medicine. *Journal of Integrative Medicine*.

Xu, H.X., Wan, M., Loh, B.N., Kon, O.L., Chow, P.W. and Sim, K.Y., 1996. Screening of traditional medicines for their inhibitory activity against Human Immunodeficiency Virus type-1 -1 protease. *Phytotherapy Research*, *10*(3), pp. 207–210.

Yang, H.O., Ko, W.K., Kim, J.Y. and Ro, H.S., 2004. Paeoniflorin: an antihyperlipidemic agent from *Paeonia lactiflora. Fitoterapia*, *75*(1), pp. 45–49.

## 5.2.7 FAMILY SAXIFRAGACEAE A.L DE JUSSIEU (1789)

The family Saxifragaceae consists of about 80 genera and 1200 species of herbs or shrubs, which often grow in rocky soils. The leaves are simple or compound, spiral or opposite, and exstipulate. The inflorescences are cymes, panicles, or racemes. The calyx includes 4–5 sepals and the corolla includes 4–5 petals. The androecium presents 4 to numerous stamens. The gynoecium presents a pair of carpels fused in a 2- to many-locular ovary, each locule sheltering 1 to many ovules growing on parietal or axil placentas. The fruits are mainly capsules or berries. Consider that plants growing among rocks and in rock fissures often produce phenolics. Phenolics in this family have antimicrobial effects.

### 5.2.7.1 *Bergenia ciliata* Sternb.

Synonym: *Bergenia ciliata* A. Br. ex Engl; *Megasea ciliata* Haw.

Common names: Hairy bergenia, stone breaker; patharkuchi (Bangladesh, India); paashaanabheda (India); pakhanbed, silpauri (Nepal); butpeewa, zakhm-e-hayat (Pakistan)

Habitat: Mountain watery rocky lands

Distribution: Pakistan, Himalayas, India, and Nepal

Botanical examination: It is a magnificent herb that grows up to about 35 cm tall from a woody rhizome. The leaves are simple, basal in, rosette, and exstipulate. The petiole is up to 5 cm long. The blade is somewhat fleshy, glossy, cordate or round at base, rounded or acute at apex, hairy or glabrous, ciliate at margin, dark green, broadly elliptic to spathulate, and 4–17 cm×3–13 cm. The inflorescence is a cyme on top of a scape. The calyx includes 5 sepals, which are 7 mm long and, broadly lanceolate. The 5 petals are spathulate, 1 cm long, and heavenly light pink. The 10 stamens are about 1 cm long. The gynoecium includes a pair of carpels. The capsule is about 1 cm long and shelters numerous tiny and tuberculate seeds.

Medicinal use: Cough, colds, post-partum (Nepal); wounds (India, Pakistan); urinary tract infection (India)

*Weak broad-spectrum antibacterial polar extract:* Methanol extract of rhizomes inhibited the growth of *Bordetella bronchiseptica* (ATCC 4617), *Escherichia coli* (ATCC 14224), *Micrococcus luteus* (ATCC 10240), and *Staphylococcus aureus* (ATCC 6538) with MIC values of 800, 5000, 600, and 2000 µg/mL, respectively (Ahmed et al., 2016).

*Strong antiviral (enveloped monopartite linear dimeric single-stranded (+)RNA) hydrophilic phenolic:* Bergenin (LogD=0.03 at pH=7.3; molecular weight=328.2 g/mol) inhibited the replication of the Human immunodeficiency virus type-1 (MN strain) with an $EC_{50}$ value of 40 µg/mL and a selectivity index above 25 in C8166 cells (Piacente et al., 1996).

Bergenin

*Weak antifungal (filamentous) simple phenolic:* Bergenin inhibited the growth of *Trychophyton mentagrophytes*, *Epidermophyton floccosum*, *Trychophyton rubrum* (MTCC 296), *Aspergillus niger* (MTCC 1344), and *Botrytis cinerea* with MIC values of 250, 500, 500, 500, and 250 µg/mL, respectively (Raj et al., 2012).

  Commentary: (i) The plant abounds with simple phenolics: bergenin, gallic acid, gallicin, and arbutin, as well as catechin, (–)-3-*O*-galloylcatechin, and (–)-3-*O*-galloylepicatechin (Ahmad et al., 2018), which most probably account for the medicinal uses. Consider that natural products inhibiting the replication of Human immunodeficiency virus are not uncommonly inhibiting Severe acute respiratory syndrome-associated coronavirus, one could have the curiosity to examine the anti-COVID-19 properties of bergerin. (ii) Catechin inhibits the replication of the Transmissible gastroenteritis virus (member of the family Coronaviridae) *in vitro* and as such may very well be active against the Severe acute respiratory syndrome-associated coronavirus. (+)-Catechin gallate (LogD=2.1 at pH 7.4; molecular weight=442.3 g/mol) inhibited H3N2 (A/Perth/16/09) with an $EC_{50}$ of 2.2 µM and a selectivity index above 31.8 (Chang et al., 2016).

(+)-Catechin gallate

Methanol extract of *Bergenia ligulata* Engl. (Local name: *Pashanved* (Nepal)) inhibited the replication of Human Influenza virus (A/WSN/33 (H1N1) London) and Herpes simplex virus type-1 with $IC_{50}$ values of 10 and 70 µg/mL, respectively (Rajbhandari et al., 2001). Methanol extract of rhizomes of *Bergenia ligulata* at a concentration of 25 µg/mL protected Madin-Darby canine kidney cells by 60% against Human Influenza virus (A/WSN/33 (H1N1)). At 100 µg/mL, the extract abrogated viral RNA synthesis as well as viral protein synthesis (Rajbhandari et al., 2003).

(iii) Hydrolyzable tannins, flavonoids, and simple phenolic are often able to inhibit the growth of viral enzymes. 1,2,3,4,6-Penta-*O*-galloyl-β-D-glucoside, 1,3,4,6-tetra-*O*-galloyl-β-D-glucoside, punicafolin, quercetin, rutin, kaempferol-3-β-D-glucoside, methylgallate, kaempferol, bergenin, gallic acid, and ethyl gallate isolated from the aerial parts of *Saxifraga melanocentra* Franch. (locally known in China *as hei rui hu er cao*) inhibited Hepatitis C virus NS3 serine protease with $IC_{50}$ values of 0.6, 0.8, 0.8, 33.1, 77.0, 1.0, 1.3, 1.6, 1.7, 1.7, and 3.9 µM, respectively (Zuo et al., 2005). The Hepatitis C virus, the Human immunodeficiency virus as well as the Severe acute respiratory syndrome-associated coronavirus are single-stranded (+)RNA viruses, and hence natural products active against Hepatitis C virus and Human immunodeficiency virus, can be seen as interesting anti-COVID-19 candidates.

## References

Ahmad, M., Butt, M.A., Zhang, G., Sultana, S., Tariq, A. and Zafar, M., 2018. *Bergenia ciliata*: a comprehensive review of its traditional uses, phytochemistry, pharmacology and safety. *Biomedicine & Pharmacotherapy*, 97, pp. 708–721.

Ahmed, M., Phul, A.R., Bibi, G., Mazhar, K., Ur-Rehman, T., Zia, M. and Mirza, B., 2016. Antioxidant, anticancer and antibacterial potential of Zakhm-e-hayat rhizomes crude extract and fractions. *Pakistan Journal of Pharmaceutical Sciences*, 29(3).

Chang, S.Y., Park, J.H., Kim, Y.H., Kang, J.S. and Min, J.Y., 2016. A natural component from *Euphorbia humifusa* Willd displays novel, broad-spectrum anti-influenza activity by blocking nuclear export of viral ribonucleoprotein. *Biochemical and Biophysical Research Communications*, 471(2), pp. 282–289.

Piacente, S., Pizza, C., De Tommasi, N. and Mahmood, N., 1996. Constituents of *Ardisia japonica* and their in vitro anti-HIV activity. *Journal of Natural Products*, 59(6), pp. 565–569.

Raj, M.K., Duraipandiyan, V., Agustin, P. and Ignacimuthu, S., 2012. Antimicrobial activity of bergenin isolated from Peltophorum pterocarpum DC. flowers. *Asian Pacific Journal of Tropical Biomedicine*, 2(2), pp. S901-S904.

Rajbhandari, M., Wegner, U., Jülich, M., Schoepke, T. and Mentel, R., 2001. Screening of Nepalese medicinal plants for antiviral activity. *Journal of Ethnopharmacology*, 74(3), pp. 251–255.

Rajbhandari, M., Wegner, U., Schoepke, T., Lindequist, U. and Mentel, R., 2003. Inhibitory effect of *Bergenia ligulata* on influenza virus A. *Die Pharmazie-An International Journal of Pharmaceutical Sciences*, 58(4), pp. 268–271.

Zuo, G.Y., Li, Z.Q., Chen, L.R. and Xu, X.J., 2005. In vitro anti-HCV activities of *Saxifraga melanocentra* and its related polyphenolic compounds. *Antiviral Chemistry and Chemotherapy*, 16(6), pp. 393–398.

### 5.2.7.2 *Aceriphyllum rossii* (Oliv.) Engl.

Synonyms: *Mukdenia rossii* (Oliv.) Koidz.; *Saxifraga rossii* Oliv.

Common name: Qi ye cao (China)

Habitat: Mountain rocky slopes, cultivated

Distribution: China and Korea

Botanical observation: It is a herb that grows up to about 35 cm tall from a rhizome. The leaves are simple, in rosette, and exstipulate. The petiole grows up to about 15 cm long. The blade is palmately lobed, cordate at base, somewhat dark green, 10–14.3 × 12–14.5 cm, the 5–7 lobes elliptic, incised and acute at apex. The inflorescence is a cyme at apex of a scape. The calyx presents 6 sepals, which are white, broadly lanceolate, and up to 5 mm long. The corolla presents 6 white petals, which are lanceolate and about 2.5 mm long. The androecium includes 6 stamens. The gynoecium includes a pair of carpels, which are about 4 mm long. The capsule is about 7 mm long.

Medicinal uses: Apparently none for microbial infections.

*Strong broad-spectrum antibacterial amphiphilic oleanane-type triterpenes:* Aceriphyllic acid A (also known as (3α)-3,23-Dihydroxyolean-12-en-27-oic acid; Log D=4.3 at pH 7.4; molecular weight 472.7 g/mol) isolated from the rhizome inhibited the growth of *Staphylococcus aureus* (503), Methicillin-resistant *Staphylococcus aureus* (CCARM 3167), quinolone-resistant *Staphylococcus aureus* (CCARM 3505), *Bacillus subtilis* (KCTC 1021), *A. calcoaceticus* (KCTC 2357), and *Micrococcus luteus* (KCTC 1056) with MIC values of 4, 8, 8, 8, 4, and 8 µg/mL, respectively (Zheng et al., 2008). 3-Oxoolean-12-en-27-oic acid (LogD=5.3 at pH 7.4; molecular weight 454.6 g/mol) isolated from the rhizome inhibited the growth of *Staphylococcus aureus* (503), Methicillin-resistant *Staphylococcus aureus* (CCARM 3167), quinolone-resistant *Staphylococcus aureus* (CCARM 3505), *Bacillus subtilis* (KCTC 1021), *A. calcoaceticus* (KCTC 2357), and *Micrococcus luteus* (KCTC 1056) with MIC values of 42, 4, 16, 8, 4, and 8 µg/mL, respectively (Zheng et al., 2008).

Aceriphyllic acid A

Commentary: Triterpenes generated from members of the family Saxifragaceae are of antiviral interest. This is the case with 16β-hydroxybryodulcosigenin, oleanic acid, 3 β-hydroxyolean-12-en-27-oic acid, and 3α-O-feruloylolean-12-en-27-oic acid isolated from *Saniculiphyllum guangxiense* C.Y. Wu & T.C. Ku (Local name: Bian-dou-ye-cao (China), which inhibited Hepatitis B virus surface antigen with IC$_{50}$ values of 0.1, 0.03, 0.01, and 0.04 mM and selectivity indices of <1, <1, 1.6, and <1, respectively, in Hep G 2.2.15 cells (Geng et al., 2012). 3 β-hydroxyolean-12-en-27-oic acid inhibited Hepatitis B virus e-antigen with an IC$_{50}$ value of 0.032 mM and selectivity index <1 in Hep G 2.2.15 cells (Geng et al., 2012). Oleanic acid, 3 β-hydroxyolean-12-en-27-oic acid (LogD=5.7 at pH 7.4; molecular mass=456.7 g/mol), and 3α-O-feruloylolean-12-en-27-oic acid isolated from *Saniculiphyllum guangxiense* inhibited Hepatitis B Virus DNA with IC$_{50}$ values of 0.06, 0.01, and 0.1, mM, respectively (Geng et al., 2012).

3 β-hydroxyolean-12-en-27-oic acid

**References**

Geng, C.A., Huang, X.Y., Lei, L.G., Zhang, X.M. and Chen, J.J., 2012. Chemical constituents of *Saniculiphyllum guangxiense. Chemistry & Biodiversity*, 9(8), pp. 1508–1516.

Zheng, C.J., Sohn, M.J., Kim, K.Y., Yu, H.E. and Kim, W.G., 2008. Olean-27-carboxylic acid-type triterpenes with potent antibacterial activity from *Aceriphyllum rossii. Journal of Agricultural and Food Chemistry*, 56(24), pp. 11752–11756.

# 6 The Clade Rosids

## 6.1 VITALES JUSS. EX BERCHT. & J.PRESL (1820)

### 6.1.1 FAMILY LEEACEAE DUMORTIER (1829)

The family Leeaceae consists of the single genus *Leea* D. Royen ex L.

#### 6.1.1.1 *Leea indica* (Burm. f.) Merr.

Synonyms: *Aquilicia otillis* Gaertn.; *Aquilicia sambucina* L.; *Leea biserrata* Miq.; *Leea celebica* C.B. Clarke; *Leea divaricata* T. & B.; *Leea expansa* Craib; *Leea fuliginosa* Miq.; *Leea gigantea* Griff.; *Leea gracilis* Lauterb.; *Leea longifolia* Merr.; *Leea naumannii* Engl.; *Leea novoguineensis* Val.; *Leea ottilis* (Gaertn.) DC.; *Leea palambanica* Miq.; *Leea pubescens* Zipp. ex Miq.; *Leea ramosii* Merr.; *Leea robusta* Blume; *Leea roehrsiana* Sanders ex Masters; *Leea sambucifolia* Salisb.; *Leea sambucina* (L.) Willd.; *Leea staphylea* Roxb.; *Leea sumatrana* Miq.; *Leea sundaica* Miq.; *Leea umbraculifera* C.B. Clarke; *Leea viridiflora* Planch.; *Staphylea indica* Burm. f.

Common names: Bandicoot berry; chila-khor, kukur-jhiwa, hoti gach (Bangladesh); ken tun, huo tong shu (China); kukura-thengia (India); ki tuwa (Indonesia); jolok-jolok (Malaysia), paikoro (Papua New Guinea); amamali (the Philippines); bangbaa (Thailand); cu roi den (Vietnam)

Habitat: Forests

Distribution: India, Nepal, Sri Lanka, Bangladesh, Myanmar, Cambodia, Laos, Thailand, Vietnam, China, Malaysia, Indonesia, the Philippines, Papua New Guinea, Australia, and Pacific Islands

Botanical observation: This shrub grows to 3 m tall. The stems are terete, glossy, and glabrous. The leaves are compound, alternate, and stipulate. The stipules are broadly obovate, 2.5–4.5×2–3.5 cm, rounded at apex, and, glabrous. The petiole is 13–23 cm long. The blade is 2- or 3-pinnate, the rachis is 14–30 cm long. The folioles are elongate elliptic, 6–32×2.5–8 cm, rounded at base, serrate, acuminate at apex, and with 6–11 pairs of secondary nerves. The inflorescences are opposite to the leaves and corymbose. The calyx is tubular and develops 5 lobes, which are minute and triangular. The 5 petals are basally united, elliptic, minute, and greenish white. The androecium includes 5 minute stamens. The ovary is globose and develops a short style. The berries are 1 cm across and contain 4–6 seeds.

Medicinal uses: Dysentery, boils, abscesses (Bangladesh); external injuries (China)

*Antifungal (filamentous) polar extract:* Ethanol extract of leaves (30 µL from a extract prepared from 0.5 g of leaves in 1 mL of ethanol) inhibited the growth of *Pestalotiopsis theae*, *Colletotrichum camelliae*, *Curvularia eragrostridis*, and *Botryodiplodia theobrome* by 91.7, 26.6, 97.6, and 74.7%, respectively (Saha et al., 2005).

*Strong antiviral (enveloped monopartite linear double-stranded DNA) polar extract:* Ethanol extract inhibited the replication of the Herpes simplex virus type-1 in Vero cells with an MIC value of 50 µg/mL (Hamidi et al., 1996).

Commentaries: (i) No study is apparently available to describe the chemical constituents of this plant. Lakornwong and coworkers reported the presence of the simple phenolics bergenin, 11-*O*-acetyl bergenin, 11-*O*-(4′-*O*-methylgalloyl) bergenin, and 3,5-dihydroxy-4-methoxybenzoic acid, as well as the flavanols (−)-epicatechin, 4″-*O*-methyl-(−)-epicatechin gallate, (−)-epicatechin gallate, and the coumarin microminutinin from *Leea thorelii* Gagnep. Bergenin is antiviral. Note that butanol fraction of root bark of *Leea tetramera* Burtt from Papua New Guinea inhibited the growth of *Bacillus cereus*, *Staphylococcus aureus*, *Escherichia coli*, *Klebsiella pneumoniae*, *Neisseria gonorrhoeae*, *Pseudomonas aeruginosa*, and *Salmonella typhimurium* (Khan et al., 2003). (ii) The medicinal use is most probably owed to synergistic antibacterial effects of phenolics and flavonoids.

(iii) Consider that epicatechin gallate inhibited tetK efflux pump in *Staphylococcus aureus* (Sharma et al., 2019).

### References

Hamidi, J.A., Ismaili, N.H., Ahmadi, F.B. and Lajisi, N.H., 1996. Antiviral and cytotoxic activities of some plants used in Malaysian indigenous medicine. *Pertanika Journal of Tropical Agricultural Science, 19*(2/3), pp. 129–136.

Khan, M.R., Omoloso, A.D. and Kihara, M., 2003. Antibacterial activity of *Alstonia scholaris* and *Leea tetramera. Fitoterapia, 74*(7–8), pp. 736–740.

Saha, D., Dasgupta, S. and Saha, A., 2005. Antifungal activity of some plant extracts against fungal pathogens of tea (*camellia sinensis.*). *Pharmaceutical Biology, 43*(1), pp. 87–91.

Sharma, A., Gupta, V.K. and Pathania, R., 2019. Efflux pump inhibitors for bacterial pathogens: from bench to bedside. *The Indian Journal of Medical Research, 149*(2), p. 129.

### 6.1.1.2 *Leea macrophylla* Roxb.

Synonyms: *Leea angustifolia* P. Lawson; *Leea aspera* Wall. ex G. Don; *Leea cinarea* P. Lawson; *Leea coriacea* P. Lawson; *Leea diffusa* P. Lawson; *Leea integrifolia* Roxb.; *Leea latifolia* Wall. ex Kurz; *Leea pallida* Craib.; *Leea parallela* Wallich ex Lawson; *Leea robusta* Roxb.; *Leea simplicifolia* Griff.; *Leea talbotii* King ex Talbot; *Leea venkobarowii* Gamble

Common names: Ash gach, harmadare (Bangladesh); bulyettra, kath tenga, motalidanhi (India); mali mali (Malaysia); kya hpetgyi (Myanmar)

Habitat: Forests

Distribution: India, Bangladesh, Myanmar, Cambodia, Laos, Vietnam, Malaysia, and Indonesia

Botanical observation: This shrub grows to 5 m tall. The stems are terete, with longitudinal ridge, and hairy. The leaves are simple or compound, alternate, and stipulate. The stipules are obovate, 4–6×2–6 cm, and caducous. The petiole is 15–20 cm long and hairy. The blade is broadly ovate, 3-foliolate, or 1–3-pinnate, 40–65×35–60 cm, rounded at base, serrate, acuminate at apex, hairy, and with 12–15 pairs of secondary nerves. The inflorescences are opposite to leaves, and corymbose on 20–25 cm hairy peduncles. The calyx is tubular, hairy, and develops 5 minute lobes. The 5 petals are basally united, elliptic, about 4 mm long, and greenish white. The androecium includes 5 minute stamens. The ovary is globose and develops a short style and a capitate stigma. The berries are ovoid 1.3 cm long and contain 6 seeds.

Medicinal uses: Cuts (Bangladesh); dog bites, viral fever, ringworm, wounds (India); tonsillitis (Bangladesh, India); tetanus (Bangladesh; India)

Commentaries: (i) The wound-healing properties of this plant have been substantiated (Joshi et al., 2016a,b). Consider that plants generating mainly phenolics, tannins, and flavonoids are astringent and thus styptic because they precipitate proteins at high concentration. Apparently, constituents in this plant are as of yet unknown. (ii) In the Philippines, *Leea philippinensis* Merr. (local name: *Aniblawun*) is used for the treatment of skin eruptions.

### References

Joshi, A., Joshi, V.K., Pandey, D. and Hemalatha, S., 2016a. Systematic investigation of ethanolic extract from *Leea macrophylla*: implications in wound healing. *Journal of Ethnopharmacology, 191*, pp. 95–106.

Joshi, A., Prasad, S.K., Joshi, V.K. and Hemalatha, S., 2016b. Phytochemical standardization, antioxidant, and antibacterial evaluations of *Leea macrophylla*: a wild edible plant. *Journal of Food and Drug Analysis, 24*(2), pp. 324–331.

### 6.1.2 Family Vitaceae A.L. de Jussieu (1789)

The family Vitaceae consists of 15 genera and 900 species of woody climbers. The leaves are simple or compound, spiral, and exstipulate. The inflorescences are panicles, corymbs, or spikes, which are often leaf-opposite, terminal, or axillary. The flowers are minute. The calyx develops 4 or 5 lobes. The corolla includes 4 or 5 petals. The androecium presents 4–5 stamens. A disk is present. The gynoecium presents a 2-loculed ovary. Each locule sheltering a pair of ovules growing on axile placentas. The fruits are drupes or berries containing up to 4 seeds. Members in this family generate an interesting series of antimicrobial hydrolysable tannins and stilbenes (phytoalexins).

#### 6.1.2.1 *Cayratia trifolia* (L.) Domin

Common names: Fox grape; amla lata (Bangladesh); san ye wu lian mea (China); amlavetash, gandira, vathakkodi (India); galing galing (Indonesia); kalit kalit (the Philippines); thao khan khao (Thailand)

Habitat: Forests

Distribution: Pakistan, India, Nepal, Bangladesh, Myanmar, Laos, Cambodia, Vietnam, China, Thailand, Malaysia, the Philippines, Papua New Guinea, and Pacific Islands

Botanical examination: This discrete woody climber can grow up to a length of 20 m. The stems are terete and develop tendrils, which are 3–5 branched. The leaves are trifoliolate, spiral, and exstipulate. At very first glance the leaves are Fabaceous. The petiole is 2.5–6 cm long. The middle petiolules are 0.4–2.5 cm long. The folioles are elliptic to somewhat asymmetrical (lateral ones), 3–6×1.5–4 cm, hairy below, marked with 7–8 pairs of secondary nerves, serrate, rounded at base, and acute or obtuse at apex. The cymes are axillary on 7.5 cm peduncles. The calyx is minute and cupular. The 5 petals are minute and somewhat whitish. A disk is present. The androecium comprises 5 stamens. The berries are globose, 7–8 mm across, glossy, black, and shelter 2–3 seeds

Medicinal uses: Abscess (Thailand); boils, fever, leucorrhea, ulcer (India)

*Antibacterial (Gram-positive) polar extract:* Ethanol extract of leaves inhibited the growth of *Staphylococcus aureus* with an MIC value of 125 ppm and an MBC value of 250 ppm (Sari et al., 2018).

*Broad-spectrum antibacterial halo developed by amphiphilic stilbene:* ε-Viniferin (Log D=4.1 at pH 7.4, molecular mass=454.4 g/mol) (200 μL per well from 1 mg/mL solution) inhibited the growth of *Escherichia coli* and *Staphylococcus aureus* with inhibition zone diameters of 11 and 7 mm, respectively (Sahidin et al., 2017).

ε-Viniferin

*Strong antibacterial (Gram-positive) amphiphilic stilbenes:* ε-Viniferin inhibited the growth of methicillin-resistant *Staphylococcus aureus* (ATCC 33591) with an MIC value of 25 µg/mL (Basri et al., 2014). ε-Viniferin inhibited the growth of *Streptococcus mutans* and *Streptococcus sanguis* with MIC/MBC values of 25/50 and 50/50 µg/mL, respectively (Yim et al., 2010). Resveratrol (LogD=2.8 at pH 7.4; molecular weight=228.2 g/mol) inhibited the growth of *Streptococcus mutans* and *Streptococcus sanguis* with MIC/MBC values of 50/50 and 50/100 µg/mL, respectively (Yim et al., 2010). In a subsequent study, resveratrol inhibited the growth of *Staphylococcus aureus* (ATCC) with an MIC value of 512 µg/mL and ε-viniferin inhibited the growth of *Staphylococcus aureus* (ATCC) and *Pseudomonas aeruginosa* (ATCC 27853) with MIC values of 512 and 256 µg/mL, respectively (Mattio et al., 2019).

*Antibiotic potentiator (Gram-positive) stilbene:* ε-Viniferin increased the sensitivity of methicillin-resistant *Staphylococcus aureus* (ATCC 33591) to vancomycin with an FICI value of 0.2 (Basri et al., 2014).

*Strong antifungal (filamentous) amphiphilic stilbene:* Resveratrol (LogD=2.8 at pH 7.4; molecular weight=228.2 g/mol) inhibited the germination of *Botrytis cinerea* with an $IC_{50}$ value of about 90 µg/mL with formation of curved germ tubes and cessation of growth of some germ tubes (Adrian et al., 1997).

*Strong antiviral (enveloped monopartite linear single-stranded (+) RNA)) amphiphilic stilbene:* ε-Viniferin inhibited the replication of the Hepatitis C virus with an $EC_{50}$ value of 0.1 µM (Lee et al., 2019).

*Moderate antiviral (enveloped segmented linear single-stranded (-) RNA) stilbene:* ε-Viniferin inhibited the replication of Influenza A/PR/8/34 (H1N1), Novel H1N1 (WT), and oseltamivir-resistant novel H1N1 (H274Y mutant) with $IC_{50}$ values of 88.5, 129.3, and 173.7 µg/mL, respectively (Nguyen et al., 2011).

*Moderate antiviral (enveloped segmented linear single-stranded (-) RNA virus)stilbene glycoside:* Piceid inhibited the replication of Influenza A/PR/8/34 (H1N1), Novel H1N1 (WT), and oseltamivir-resistant novel H1N1 (H274Y mutant) with $IC_{50}$ values of 110.7, 110.2, and 150.8 µg/mL, respectively (Nguyen et al., 2011).

Commentary: (i) Roots cultured *in vitro* produce a series of stilbenes, including resveratrol, piceid (resveratrol-3-O-glucoside), ε-viniferin, and ampelopsins (Arora et al., 2009). Consider that resveratrol is antiviral and antibacterial. Vitaceae are known to produce stilbene phytoalexins. In response to *Botrydis cinerea* infestation, *Vitis vinifera* L. (Grapes; *Angur, draksha* (India); *Inab* (Iraq); *Kishmish* (Iran)) produces the phyoalexin stilbenes resveratrol, piceid, pterostilbene, and ε-viniferin (dehydrodimer of resveratrol), which are fungitoxic (Breuil et al., 1999). Ethanol extract of seeds of *Vitis vinifera* L. inhibited the growth of *Candida albicans* (CA-1 strain) and *Candida albicans* (A9 strain) (Han, 2007).

(ii) Members of the family Vitaceae are not uncommonly active against linear single-stranded (+) RNA virus. Acetone extract of the stem and roots of a member of the genus *Ampelopsis* Michx. inhibited the replication of Enterovirus 71 (clinical strain CMUH-527-1), Enterovirus 71 (clinical strain CMUH-V4079), and Enterovirus71 (ATCC-VR784) with $EC_{50}$ values of 18.5, 39.9, and 19.8 µg/mL, respectively, in Human embryonal rhabdomyosarcoma cells ($CC_{50}$ value of 128.8 µg/mL) (Hsu et al., 2011). The amphiphilic flavanol ampelopsin (also known as dihydromyricetin, LogD=0.7 at pH 7.4, molecular weight=320.2 g/mol) protected MT-4 cells from Human immunodeficiency virus type-1 with an $EC_{50}$ of 100 µg/mL (Liu et al., 2004).

(iii) Severe acute respiratory syndrome-associated coronavirus and the Hepatitis C virus, are enveloped linear single-stranded (+)RNA virus, thus, one could examine the effect of ε-viniferin against COVID-19.

### References
Adrian, M., Jeandet, P., Veneau, J., Weston, L.A. and Bessis, R., 1997. Biological activity of resveratrol, a stilbenic compound from grapevines, against *Botrytis cinerea*, the causal agent for gray mold. *Journal of Chemical Ecology*, 23(7), pp. 1689–1702.

Arora, J., Roat, C., Goyal, S. and Ramawat, K.G., 2009. High stilbenes accumulation in root cultures of *Cayratia trifolia* (L.) domin grown in shake flasks. *Acta Physiologiae Plantarum, 31*(6), p. 1307.

Basri, D.F., Xian, L.W., Abdul S.N.I., et al., 2014. Bacteriostatic antimicrobial combination: antagonistic interaction between epsilon-viniferin and vancomycin against methicillin-resistant *Staphylococcus aureus. BioMed Research International, 2014.*

Breuil, A.C., Jeandet, P., Adrian, M., Chopin, F., Pirio, N., Meunier, P. and Bessis, R., 1999. Characterization of a pterostilbene dehydrodimer produced by laccase of *Botrytis cinerea. Phytopathology, 89*(4), pp. 298–302.

Han, Y., 2007. Synergic effect of grape seed extract with amphotericin B against disseminated candidiasis due to *Candida albicans. Phytomedicine, 14*(11), pp. 733–738.

Hsu, H.W., Wang, J.H., Liu, C.L., Chen, C.S. and Wang, Y.C., 2011. Anti-enterovirus 71 activity screening of Taiwanese folk medicinal plants and immune modulation of *Ampelopsis brevipedunculata* (Maxim.) Trautv against the virus. *African Journal of Microbiology Research, 5*(17), pp. 2500–2511.

Lee, S., Mailar, K., Kim, M.I., Park, M., Kim, J., Min, D.H., Heo, T.H., Bae, S.K., Choi, W. and Lee, C., 2019. Plant-derived purification, chemical synthesis, and in vitro/in vivo evaluation of a resveratrol dimer, viniferin, as an HCV replication inhibitor. *Viruses, 11*(10), p. 890.

Liu, D.Y., Ye, J.T., Yang, W.H., Yan, J., Zeng, C.H. and Zeng, S., 2004. Ampelopsin, a small molecule inhibitor of HIV infection targeting HIV-1 entry. *Biomedical and Environmental Sciences, 17*(2), pp. 153–164.

Mattio, L.M., Dallavalle, S., Musso, L., Filardi, R., Franzetti, L., Pellegrino, L., D'Incecco, P., Mora, D., Pinto, A. and Arioli, S., 2019. Antimicrobial activity of resveratrol-derived monomers and dimers against foodborne pathogens. *Scientific Reports, 9*(1), pp. 1–13.

Nguyen, T.N.A., Dao, T.T., Tung, B.T., Choi, H., Kim, E., Park, J., Lim, S.I. and Oh, W.K., 2011. Influenza A (H1N1) neuraminidase inhibitors from *Vitis amurensis. Food Chemistry, 124*(2), pp. 437–443.

Sahidin, I., Wahyuni, W., Malaka, M.H., Imran, I. (2017) Antibacterial and cytotoxic potencies of stilbene oligomers from stem barks of baoti (*Dryobalanops lanceolata*) growing in Kendari, Indonesia. *Asian Journal of Pharmaceutical and Clinical Research, 10*(8), pp. 139–143.

Sari, R.S.E., Soegianto, L. and Liliek, S., 2018. Uji aktivitas antimikroba ekstrak etanol daun cayratia trifolia terhadap *Staphylococcus aureus* dan *candida albicans. Jurnal Farmasi Sains dan Terapan, 5*(1), pp. 23–29.

Yim, N., Trung, T.N., Kim, J.P., et al., 2010. The antimicrobial activity of compounds from the leaf and stem of *Vitis amurensis* against two oral pathogens. *Bioorganic & medicinal chemistry letters, 20*(3), pp. 1165–1168.

### 6.1.2.3 *Cissus quadrangularis* L.

Synonyms: *Cissus edulis* Dalziel; *Vitis quadrangularis* (L.) Wall. ex W. & Arn.

Common names: Climbing cactus, edible-stemmed vine; harjora, vajravalli (India); tikel balung (Indonesia); patah tulang (Malaysia); sugpon-sugpon (the Philippines); khankho (Thailand); hồ dằng bốn cạnh (Vietnam)

Habitat: Cultivated

Distribution: Pakistan, India, Sri Lanka, Bangladesh, Myanmar, Cambodia, Laos, Vietnam, Thailand, Malaysia, Indonesia, and the Philippines

Botanical observation: This climber presents fleshy, edible, quadrangular, winged stems constricted at nodes, looking like some sort of bones, hence the Malay name *patah tulang*=broken bones. The stems develop slender and simple tendrils, which are leaf opposed. The leaves are simple, alternate, stipulate, and grow from the internodes. The petiole is 0.6–1.2 cm long.

The blades are fleshy, somewhat triangular cordate at base, round at apex, wavy-crenate, and 3–5 cm×3–5 cm. The inflorescence is a cyme on a peduncle, which grows up to 2.5 cm long. The calyx is cupular. The 4 petals are minute. A 4-lobed disc is present. The drupes are red, globose, glossy, succulent, very acidic, and 6–10 mm in diameter. The seeds are obovoid smooth, and 4–8 mm across.

Medicinal uses: Burns, fever, dysentery, otitis, and wounds (India)

*Broad-spectrum antibacterial halo developed by mid-polar extract:* Ethyl acetate extract of fresh stem (50 µg/mL of a 1 mg/mL solution/well) inhibited the growth of *Bacillus subtilis, Bacillus cereus, Staphylococcus aureus, Pseudomonas aeruginosa,* and *Streptococcus* species with inhibition zone diameters of 22, 15, 16, 7, and 16 mm, respectively (Murthy et al., 2003).

*Very weak broad-spectrum antibacterial mid-polar extract:* Ethyl acetate extract of stems inhibited the growth of *Bacillus subtilis* (ATCC No. 6633), *Pseudomonas aeruginosa* (ATCC No. 27853), *Salmonella typhi, Staphylococcus aureus* (ATCC No. 25923), and *Streptococcus pyogenes* with MIC values of 930, 1870, 3750, 930, and 3750 µg/mL, respectively (Kashikar & George, 2006).

*Very weak broad-spectrum antibacterial polar extract:* Methanol extract inhibited the growth of *Staphylococcus aureus, Streptococcus pneumoniae, Escherichia coli,* and *Klebsiella pneumoniae* (Raj et al., 2010).

*Moderate broad-spectrum antibacterial amphiphilic stilbenes:* Piceatannol (Log D=1.7 at pH 7.4; molecular weight=244.2) inhibited the growth of *Staphylococcus aureus* (ATCC 25923) and *Escherichia coli* (ATCC 25922) with MIC/MBC values of 125/125 and 100/125 µg/mL, respectively (Kusumaningtyas et al., 2020). Resveratrol inhibited the growth of *Staphylococcus aureus* (ATCC 25923) and *Escherichia coli* (ATCC 25922) with MIC/MBC values of 125/125 and 50/125 µg/mL, respectively (Kusumaningtyas et al., 2020).

Piceatannol

*Very weak antifungal (filamenous) stilbene:* Piceatannol inhibited spore germination of *colletotrichum falcatum* with an MIC value of 900 µM and germ tube growth with an MIC value of 280 µM (Brinker & Seigler, 1991).

*Strong antiviral (enveloped linear monopartite double-stranded DNA) amphiphilic stilbene:* Piceatannol inhibited the replication of the Human Cytomegalovirus in WI-38 cells with an $IC_{50}$ value of about 5 µM (Wang et al., 2020).

*Very weak anticandidal polar extract:* Methanol extract inhibited the growth of *Candida albicans* and *Candida tropicalis* with MIC values of 2500 and 1250 µg/mL, respectively (Raj et al., 2010).

Commentary: The plant generates stilbenes, such as resveratrol, piceatannol, pallidol, quadrangularins A, B, and C, and parthenocissin A (Adesanya et al., 1999).

**References**

Adesanya, S.A., Nia, R., Martin, M.T., Boukamcha, N., Montagnac, A. and Païs, M., 1999. Stilbene derivatives from *Cissus quadrangularis*. *Journal of Natural Products, 62*(12), pp. 1694–1695.

Brinker, A.M. and Seigler, D.S., 1991. Isolation and identification of piceatannol as a phytoalexin from sugarcane. *Phytochemistry, 30*(10), pp. 3229–3232.

Kashikar, N.D. and George, I., 2006. Antibacterial activity of *Cissus quadrangularis* Linn. *Indian Journal of Pharmaceutical Sciences, 68*(2), 245–247.

Kusumaningtyas, V.A., Syah, Y.M. and Juliawaty, L.D., 2020. Two stilbenes from Indonesian *Cassia grandis* and their antibacterial activities. *Research Journal of Chemistry and Environment, 24*, p. 1–4.

Murthy, K.C., Vanitha, A., Swamy, M.M. and Ravishankar, G.A., 2003. Antioxidant and antimicrobial activity of *Cissus quadrangularis* L. *Journal of Medicinal Food, 6*(3), pp. 99–105.

Raj, A.J., Selvaraj, A., Gopalakrishnan, V.K. and Dorairaj, S., 2010. Antimicrobial profile of *Cissus quadrangularis*. *Journal of Herbal Medicine and Toxicology, 4*(2), pp. 177–180.

Wang, S.Y., Zhang, J., Xu, X.G., Su, H.L., Xing, W.M., Zhang, Z.S., Jin, W.H., Dai, J.H., Wang, Y.Z., He, X.Y. and Sun, C., 2020. Inhibitory effects of piceatannol on human cytomegalovirus (hCMV) in vitro. *Journal of Microbiology*, pp. 1–8.

# Bibliography

Abe, R. and Ohtani, K., 2013. An ethnobotanical study of medicinal plants and traditional therapies on Batan Island, the Philippines. *Journal of Ethnopharmacology, 145*, pp. 554–565.

Abu Bakar, F.I., Abu Bakar, M.F., Abdullah, N., Endrini, S. and Rahmat, A., 2018. A review of Malaysian medicinal plants with potential anti-inflammatory activity. *Hindawi*, 13 pages, Article ID 8603602.

Acharya, K.P. and Acharya, M., 2010. Traditional knowledge on medicinal plants used for the treatment of livestock diseases in Sardikhola VDC, Kaski, Nepal. *Journal of Medicinal Plants Research, 4*(2), pp. 235–239.

Acharya, E. (Siwakoti) and Pokhrel, B., 2006. Ethno-medicinal plants used by Bantar of Bhaudaha, Morang, Nepal. *Our Nature, 4*, pp. 96–103.

Adams, M., Alther, W., Kessler, M., Kluge, M. and Hamburger, M., 2011. Malaria in the renaissance: Remedies from European herbals from the 16th and 17th century. *Journal of Ethnopharmacology, 133*, pp. 278–288.

Ahmad, S.S., 2007. Medicinal wild plants from Lahore-Islamabad Motorway (M-2). *Pakistan Journal of Botany, 39*(2), pp. 355–375.

Ahmad, K., Weckerle, C.S. and Nazir, A. Ethnobotanical investigation of wild vegetables used among local communities in northwest Pakistan. *Acta Societatis Botanicorum Poloniae, 88*(1), p. 3616.

Ahmed, N., Mahmood, A., Mahmood, A., Sadeghi, Z. and Farman, M., 2015. Ethnopharmacological importance of medicinal flora from the district of Vehari, Punjab province, Pakistan. *Journal of Ethnopharmacology, 168*, pp. 66–78.

Ahmed, M.J. and Murtaza, G., 2015. A study of medicinal plants used as ethnoveterinary: Harnessing potential phytotherapy in Bheri, District Muzaffarabad (Pakistan). *Journal of Ethnopharmacology, 159*, pp. 209–214.

Ainslie, W., 1813. *Materia Medica of Hindoostan, and Artisan's and Agriculturist's Nomenclature.* Madras: Government Press.

Al-Adhroey, A.H., Nor, Z.M., Al-Mekhlafi, H.M. and Mahmud, R., 2010. Ethnobotanical study on some Malaysian anti-malarial plants: A community based survey. *Journal of Ethnopharmacology, 132*, pp. 362–364.

Altundag, E. and Ozturk, M., 2011. Ethnomedicinal studies on the plant resources of east Anatolia, Turkey. *Procedia Social and Behavioral Sciences, 19*, pp. 756–777.

Amber, R., Adnan, M., Tariq, A. and Mussarat, S., A review on antiviral activity of the Himalayan medicinal plants traditionally used to treat bronchitis and related symptoms. *Journal of Pharmacy and Pharmacology.* doi: 10.1111/jphp.12669

Amini, M.H. and Hamdam, S.M., 2017. Medicinal plants used traditionally in Guldara District of Kabul, Afghanistan. *International Journal of Pharmacognosy and Chinese Medicine, 1*(3), 000118.

Amiri, M.S., Joharchi, M.R. and TaghavizadehYazdi, M.E., 2014. Ethno-medicinal plants used to cure jaundice by traditional healers of Mashhad, Iran. *Iranian Journal of Pharmaceutical Research, 13*(1), pp. 157–162.

Asiatic Society of Bengal, 1812. Asiatic researches or transactions of the Society instituted in Bengal, for inquiring into the history and antiquities, the arts, sciences, and literature, of Asia. The Bavarian State Library.

Atigur Rahman, M., Uddin, S.B. and Wilcok, C.C., 2007. Medicinal plants used by Chakma tribe in Hill tracts Districts of Bangladesh. *Indian Journal of Traditional Knowledge, 6*(3), pp. 508–517.

Ayyanar, M. and Ignacimuthu, S., 2011. Ethnobotanical survey of medicinal plants commonly used by Kani tribals in Tirunelveli hills of Western Ghats, India. *Journal of Ethnopharmacology, 134*, pp. 851–864.

Azhar, M.F., Aziz, A., Haider, M.S., Nawaz, M.F. and Zulfiqar, M.A., 2015. Exploring the ethnobotany of *Haloxylon recurvum* (KHAR) and *Haloxylon salicornicum* (LANA) in Cholistan desert, Pakistan. *Pakistan Journal of Agricultural Sciences, 52*(4), pp. 1085–1090.

Balfour, E., 1857. *The Cyclopaedia of India and of Eastern and Southern Asia.* Madras: Scottish Press.

Bentley, R., 1872. *Dr. Pereira's Elements of Materia Medica and Therapeutics: Abridged and Adapted for the Use of Medical and Pharmaceutical Practitioners and Students.* London: Longmans, Green, and Co.

Bhagya, B. and Sridhar, K.R., 2009. Ethnobiology of coastal sand dune legumes of Southwest coast of India. *Indian Journal of Traditional Knowledge, 8*(4), pp. 611–620.

Bhandary, M.J. and Chandrashekar, K.R., 2011. Herbal therapy for herpes in the ethno-medicine of Coastal Karnataka. *Indian Journal of Traditional Knowledge, 10*(3), pp. 528–532.

Bhatia, H., Sharma, Y.P., Manhas, R.K. and Kumar, K., 2014. Ethnomedicinal plants used by the villagers of district Udhampur, J&K, India. *Journal of Ethnopharmacology, 151*, pp. 1005–1018.

Bose, D., Roy, J.G., Mahapatra, S.D. (Sarkar), Datta, T., Mahapatra, S.D. and Biswas, H., 2015. Medicinal plants used by tribals in Jalpaiguri district, West Bengal, India. *Journal of Medicinal Plants Studies, 3*(3), pp. 15–21.

Bulut, G., Zeki Haznedaroğlu, M., Doğan, A., Koyu, H. and Tuzlacı, E., 2017. An ethnobotanical study of medicinal plants in Acipayam (Denizli-Turkey). *Journal of Herbal Medicine, 10*, pp. 64–81.

Chakraborty, M.K. and Bhattacharjee, A., 2006. Some common ethnomedicinal uses for various diseases in Purulia District, West Bengal. *Indian Journal of Traditional Knowledge, 5*(4), pp. 554–558.

Chang, C.-S., Kim, H. and Chang, K.S., 2014. *Provisional Checklist of Vascular Plants for the Korea Peninsula flora (KPF)*. 660 pp.

Chowdhury, M. and Mukherjee, R., 2010. Ethno-medicinal survey of Santal tribe of Malda District of West Bengal, India. *Journal of Economic and Taxonomic Botany, 34*(3).

Chowdhury, M.S.H., Koike, M., Muhammed, N., Halim, M.A., Saha, N. and Kobayashi H., 2009. Use of plants in healthcare: A traditional ethno-medicinal practice in southeastern rural areas of Bangladesh. *International Journal of Biodiversity Science and Management, United Kingdom, 5*(1), pp. 41–51.

Chowlu, K., Mahar, K.S. and Das, A.K., 2017. Ethnobotanical studies on orchids among the *Khamti* Community of Arunachal Pradesh, India. *Indian Journal of Natural Products and Resources, 8*(1), pp. 89–93.

Colonel Heber Drury, 1873. *The Useful Plants of India*. London: William H. Allen & Co.

Cox, P.A., 1993. Saving the ethnopharmacological heritage of Samoa. *Journal of Ethnopharmacology, 38*, pp. 181–188.

Das, D. and P. Ghosh., 2017. Some important medicinal plants used widely in Southwest Bengal, India. *International Journal of Engineering Science Invention, 6*(6), pp. 28–50.

Das, T., Mishra, S.B., Saha, D. and Agarwal, S., 2012. Ethnobotanical survey of medicinal plants used by ethnic and rural people in Eastern Sikkim Himalayan Region. *African Journal of Basic & Applied Sciences, 4*(1), pp. 16–20.

Das, A.K. and Tongbram, Y., 2014. Study on medicinal plants used by Meitei community of Bishnupur district, Manipur. *International Journal of Current Research, 6*(2), pp. 5211–5219.

Debala Devi, A., Ibeton Devi, O., Chand Singh, T. and Singh, E.J., 2014. A study of aromatic plant species especially in Thoubal District, Manipur, North East India. *International Journal of Scientific and Research Publications, 4*(6).

Defilipps, R.A. and Krupnick, G.A., 2018. The medicinal plants of Myanmar. *PhytoKeys, 102*, pp. 1–341.

Defilipps, R.A., Maina, S.L. and Pray, L.A., 1988. The palauan and yap medicinal plant studies of masay-oshi okabe, 1941–1943. Atoll Research Bulletin. National Museum of Natural History Smithsonian Institution, Washington, D.C., October 1988.

Dey, A. and De, J.N., 2010. A survey of ethnomedicinal plants used by the tribals of Ajoydha Hill region, Purulia District, India. *American-Eurasian Journal of Sustainable Agriculture, 4*(3), pp. 280–290.

Dey, A. and De, J.N., 2012. Ethnobotanical survey of Purulia district, West Bengal, India for medicinal plants used against gastrointestinal disorders. *Journal of Ethnopharmacology, 143*, pp. 68–80.

Durmuşkahya, C. and Öztürk, M., 2013. Ethnobotanical survey of medicinal plants used for the treatment of diabetes in Manisa, Turkey. *Sains Malaysiana, 42*(10), pp. 1431–1438.

Dutta, S.K., Vanlalhmangaiha, Akoijam, R.S., Lungmuana, Boopathi, T. and Saha, S., 2018. Bioactivity and traditional uses of 26 underutilized ethno-medicinal fruit species of North-East Himalaya, India. *Journal of Food Measurement and Characterization, 12*, pp. 2503–2514.

Elliott, E., Chassagne, F., Aubouy, A., Deharo, E., Souvanasy, O., Sythamala, P., Sydara, K., Lamxay, V., Manithip, C., Torres, J.A. and Bourdy, G., 2020. Forest Fevers: Traditional treatment of malaria in the southern lowlands of Laos. *Journal of Ethnopharmacology, 247*.

Flora of China Editorial Committee, 2018. *Flora of China*. Missouri Botanical Garden and Harvard University, Herbaria, USA.

Fluchiger, F.A. and Hangury, D., 1874. *A History of the Principal Drugs of Vegetable Origin, Met with in Great Britain and British India*. London: Macmillan and Co.

Frederick Porter Smith, 1871. *Contributions towards the Materia Medica and Natural History of China*. Shanghai: American Presbyterian Mission Press.

Gagnepain, F., 1934. Iridacées, Amaryllidacées et Liliacées nouvelles d'Asie, *Bulletin de la Société Botanique de France, 81*(1), pp. 66–74. doi: 10.1080/00378941.1934.10833934

Gao, X-M., Ji, B-K., Li, Y-K., Ye, Y-Q., Jiang, Z-Y., Yang, H-Y., Du, G., Zhou, M., Pan, X-X., Liu, W-X. and Hu, Q-F., 2016. New biphenyls from *Garcinia multiflora*. *The Journal of the Brazilian Chemical Society, 27*(1), pp. 10–14.

Gautam, R., Saklani, A. and Jachak, S.M., 2007. Indian medicinal plants as a source of antimycobacterial agents. *Journal of Ethnopharmacology, 110*, pp. 200–234.

Ghosh, C., 2017. Ethnobotanical survey in the Bamangola Block of Malda District, West Bengal (India): II. Medicinal and Aromatic plants. *Pleione, 11*(2), pp. 249–267.

Giesen, W., Wulffraat, S., Zieren, M. and Scholten, L., 2007. *Mangrove Guidebook for Southeast Asia.* Dharmasarn Co., Ltd., Bangkok, Thailand.

Girach, R.D., Brahmam, M., Misra, M.K. and Ahmed, M., 1998. Indigenous phytotherapy for filariasis from Orissa. *Ancient Science of Life, 17*(3), pp. 224–227.

Goswami, D., Mukherjee, P.K. and Kar, A., 2016. Screening of ethnomedicinal plants of diverse culture for antiviral potentials. *Indian Journal of Traditional Knowledge, 15*(3), pp. 474–481.

Grosvenor, P.W., Gothard, P.K., McWilliam, N.C., Suprlono, A. and Gray, D.O., 1995. Medicinal plants from Riau Province, Sumatra, Indonesia. Part 1: Uses. *Journal of Ethnopharmacology, 45*, pp. 75–95.

Gruyal, G.A. del Roasario, R. and Palmes, N.D., 2014. Ethnomedicinal plants used by residents in Northern Surigao del Sur, Philippines. *Natural Products Chemistry & Research, 2*, p. 4.

Gunawardana, S.L.A. and Jayasuriya, W.J.A.B.N. 2019. Medicinally important herbal flowers in Sri Lanka. *Hindawi*, 18 pages. Article ID 2321961.

Hanbury, D., 1862. Notes on Chinese Materia Medica. In: John E. Taylor (ed.), *The Pharmaceutical Journal and Transactions.*

Hasan, Md. M., Hossain, Sk. A., Ali, Md. A. and Alamgir, A.N.M., 2014. Medicinal plant diversity in Chittagong, Bangladesh: A database of 100 medicinal plants. *Journal of Scientific and Innovative Research, 3*(5), pp. 500–514.

Hooper, D., 1937. *Useful Plants and Drugs of Iran and Iraq.* The Library of the University of Illinois.

Hope, J. 1970. *Lectures on the Material Medica: Containing the Natural History of Drugs, Their Virtues and Doses: Also Directions for the Study of the Materia Medica; and An Appendix on the Method of Prescribing.* London: Edward and Charles Dilly.

Ijaz, F., Iqbal, Z., Rahman, I.U., Alam, J., Khan, S.M., Shah, G.M., Khan, K. and Afzal, A., 2016. Investigation of traditional medicinal floral knowledge of Sarban Hills, Abbottabad, KP, Pakistan. *Journal of Ethnopharmacology, 179*, pp. 208–233.

Jahan, I., Rahman, M.A. and Hossain, M.A., 2019. Medicinal species of Fabaceae occurring in Bangladesh and their conservation status. *Journal of Medicinal Plants Studies, 7*(4), pp. 189–195.

Jain, S.K. and Srivastava, S., 2005. Traditional uses of some Indian plants among islanders of the Indian Ocean. *Indian Journal of Traditional Knowledge, 4*(4), pp. 345–357.

Jan, G., Khan, M.A. and Gul, F., 2008. Ethnomedicinal plants used against diarrhea and dysentery in Dir Kohistan Valley (NWFP), Pakistan. *Ethnobotanical Leaflets, 12*, pp. 620–637.

Jazani, A.M., Maleki, R.F., Kazemi, A.h., Matankolaei, L.g., Targhi, S.T., kordi, S., Rahimi-Esboei, B. and Azgomi, R.N.D., 2018. Intestinal Helminths from the viewpoint of traditional Persian medicine versus modern medicine. *African Journal of Traditional, Complementary and Alternative Medicines, 15*(2), pp. 58–67.

Ji, H., Long, C. and Pei, S., 2004. An ethnobotanical study of medicinal plants used by the Lisu people in Nujiang, Northwest Yunnan, China. *Economic Botany, 58*, pp. S253–264. doi: 10.1663/0013-000

Kadir, M.F., Sayeed, M.S.B., Setu, N.I., Mostaf, A. and Mia, M.M.K., 2014. Ethnopharmacological survey of medicinal plants used by traditional health practitioners in Thanchi, Bandarban Hill Tracts, Bangladesh. *Journal of Ethnopharmacology, 155*, pp. 495–508.

Kar, A. and Borthakur, S.K., 2008. Medicinal plants used against dysentery, diarrhoea and cholera by the tribes of erstwhile Kameng District of Arunachal Pradesh. *Natural Product Radiance, 7*(2), pp. 176–181.

Kar, T., Mandal, K.K., Reddy, C.S. and Biswal, A.K., 2013. Ethnomedicinal plants used to cure diarrhoea, dysentery and cholera by some tribes of Mayurbhanj District, Odisha, India. *Life Sciences Leaflets, 2*, pp. 18–28.

Karki, M., Hill, R., Xue, D., Alangui, W., Ichikawa, K. and Bridgewater, P., 2017. Knowing our Lands and Resources: Indigenous and Local Knowledge and Practices related to Biodiversity and Ecosystem Services in Asia. Published in 2017 by the United Nations Educational, Scientific and Cultural Organization, 7, place de Fontenoy, 75352 Paris 07 SP, France.

Kayani, S., Ahmad, M., Sultana, S., Shinwari, Z.K., Zafar, M., Yaseen, G., Hussain, M. and Bibi, T., 2015. Ethnobotany of medicinal plants among the communities of Alpineand Sub-alpine regions of Pakistan. *Journal of Ethnopharmacology, 164*, pp. 186–202.

Ketpanyapong, W. and Itharat, A., 2016. Antibacterial activity of Thai medicinal plant extracts against microorganism isolated from post-weaning diarrhea in piglets. *Journal of the Medical Association of Thailand, 99*(Suppl. 4), pp. S203–S210.

Khadka, D., Dhamala, M.K., Li, F., Aryal, P.C., Magar, P.R., Bhatta, S., Bhatta, S., Basnet, A., Shi, S. and Cui, D., 2005. The use of medicinal plant to prevent COVID-19 in Nepal. Research Square. doi:10.21203/rs.3.rs–88908/v1

Khan, S.U., Khan, R.U., Mehmood, S., Khan, A., Ullah, I. and Bokhari, T.Z. Medicinal plants used to cure diarrhea and dysentery by the local inhabitants of District Bannu, Khyber PakhtoonKhwa, Pakistan. *Advances in Pharmaceutical and Ethnomedicines, 1*(1), pp. 15–18.

Khan, S.W. and Khatoon, S., 2008. Ethnobotanical studies on some useful herbs of Haramosh and Bugrote valleys in Gilgit, Northern areas of Pakistan. *Pakistan Journal of Botany, 40*(1), pp. 43–58.

Khare, C.P. *Indian Medicinal Plants: An Illustrated Dictionary.* New York: Springer.

Khisha, T., Karim, R., Chowdhury, S.R. and Banoo, R., 2012. Ethnomedical studies of Chakma communities of Chittagong Hill tracts, Bangladesh. *Bangladesh Pharmaceutical Journal, 15*(1), pp. 59–67.

Kirtikar, K.R., Basu, B.D., An, I.C.S. *Indian Medicinal Plants.* Delhi: Jayyed Press.

Koch, M., Kehop, D.A., Kinminja, B., Sabak, M., Wavimbukie, G., Barrows, K.M., Matainaho, T.K., Barrows, L.R. and Rai, P.P., 2015. An ethnobotanical survey of medicinal plants used in the East Sepik province of Papua New Guinea. *Journal of Ethnobiology and Ethnomedicine, 11*, p. 79.

Korkmaz, M. and Karakuş, S., 2015. Traditional uses of medicinal plants of Üzümlü district, Erzincan, Turkey. *The Pakistan Journal of Botany, 47*(1), pp. 125–134.

Koteswara Rao, J., Suneetha, J., Seetharami Reddi, T.V.V. and Aniel Kumar, O., 2011. Ethnomedicine of the Gadabas, a primitive tribe of Visakhapatnam district, Andhra Pradesh. *International Multidisciplinary Research Journal, 1/2*, pp. 10–14.

Kültür, Ş., 2007. Medicinal plants used in Kırklareli Province (Turkey). *Journal of Ethnopharmacology, 111*, pp. 341–364.

Lemmens, R.H.M.J. and Bunyapraphatsara, N., 2003. *Plant Resources of South-East Asia: Medicinal and Poisonous Plants.* Leiden: Backhuys Publishers.

Lemmens, R.H.M.J., Soerianegara, I. and Wong, W.C., 1995. *Plant Resources of South-East Asia: Timber Trees: Minor Commercial Timbers.* Leiden: Backhuys Publishers.

Li, D-l. and Xing, F-w., 2016. Ethnobotanical study on medicinal plants used by local Hoklos people on Hainan Island, China. *Journal of Ethnopharmacology, 194*, pp. 358–368.

Li, D.-l., Zheng, X.-l., Duan, L., Deng, S.-w., Ye, W., Wang, A.-h. and Xinga, F.-w., 2017. Ethnobotanical survey of herbal tea plants from the traditional markets in Chaoshan, China. *Journal of Ethnopharmacology, 205*, pp. 195–206.

Liu, B., Zhang, X., Bussmaan, R.W., Hart, R.H., Li, P., Bai, Y. and Long, C., 2017. *Garcinia* in Southern China: Ethnobotany, management, and niche modeling. *Economic Botany, 70*(4), pp. 416–430.

Lokho, K. and Narasimhan, D., 2013. Ethnobotany of Mao-Naga Tribe of Manipur, India. *Pleione, 7*(2), pp. 314–324.

Lone, P.A. and Bhardwaj, A.K., 2013. Traditional herbal based disease treatment in some rural areas of Bandipora district of Jammu and Kashmir, India. *Asian Journal of Pharmaceutical and Clinical Research, 6*(4).

Long, C-l. and Li, R., 2004. Ethnobotanical studies on medicinal plants used by the Red-headed Yao People in Jinping, Yunnan Province, China. *Journal of Ethnopharmacology, 90*, pp. 389–395.

Mahbubur Rahman, A.H.M. and Akter, M., 2013. Taxonomy and medicinal uses of Euphorbiaceae (Spurge) family of Rajshahi, Bangladesh. *Research in Plant Sciences, 1*(3), pp. 74–80.

Mahmood, A., Mahmood, A., Shaheen, H., Qureshi, R.A., Sangi, Y. and Gilani, S.A., 2011. Ethno medicinal survey of plants from district Bhimber Azad Jammu and Kashmir, Pakistan. *Journal of Medicinal Plants Research, 5*(11), pp. 2348–2360.

Malla, B., 2015. Ethnobotanical study on medicinal plants in Parbat district of western Nepal. A Dissertation submitted for the Partial Fulfilment of the Requirement for the Degree of Doctor of Philosophy in Environmental Science.

Manilal, K.S. and Sabu, T., 1985. Cyclea Barbata Miers (Menispermaceae): A new record of a medicinal plant from south India. *Ancient Science of Life, IV*(4), pp. 229–231.

Merrill, E.D., 1903. *A Dictionary of the Plant Names of the Philippine Islands.* Manila: Bureau of Public Printing.

Mikaili, P., Shayegh, J. and Asghari, M.H., 2012. Review on the indigenous use and ethnopharmacology of hot and cold natures of phytomedicines in the Iranian traditional medicine. *Asian Pacific Journal of Tropical Biomedicine*, pp. S1189–S1193.

Minh, V.V., Yen, N.T.K. and Thoa, P.T.K., 2014. Medicinal plants used by the Hre community in the Ba to district of central Vietnam. *Journal of Medicinal Plants Studies, 2*(3), pp. 64–71.

Mishra, D.N., 2009. Medicinal plants for the treatment of fever (*Jvaracikitsā*) in the *Madhavacikitsā* tradition of India. *Indian Journal of Traditional Knowledge, 8*(3), pp. 352–361.

Mitra, S. and Mukherjee, S.Kr., 2010. Ethnomedicinal usages of some wild plants of North Bengal plain for gastro-intestinal problems. *Indian Journal of Traditional Knowledge, 9*(4), pp. 705–712.

Mohammad, N.S., Milow, P. and Ong, H.C., 2012. Traditional medicinal plants used by the Kensiu Tribe of Lubuk Ulu Legong, Kedah, Malaysia. *Studies on Ethno Medicine, 6*(3), pp. 149–153.

Morteza-Semnani, K., 2015. A Review on Chenopodium botrys L.: Traditional uses, chemical composition and biological activities. *Pharmaceutical and Biomedical Research, 1*(2), pp. 1–9.

Moungsrimuangdee, B., 2016. A survery of Riparian species in the Bodhivijjalaya college's forest, Srinakharinwirot University, Sa Kaeo. *Thai Journal of Forestry, 35*(3), 15–29.

Movaliya, V. and Zaveri, M., 2014. A review on the Pashanbheda plant "Aerva javanica". *International Journal of Pharmaceutical Sciences Review and Research, 25*(2), pp. 268–275. Article No. 51.

Nazan Çömlekçđoğlu and Sengül Karaman, 2008. Kahramanmaras Sehir Merkezindeki Aktar'larda Bulunan Tıbbi Bitkiler. *KSU Journal of Science and Engineering, 11*(1).

Neamsuvan, O. and Ruangrit, T., 2017. A survey of herbal weeds that are used to treat gastrointestinal disorders from southern Thailand: Krabi and Songkhla provinces. *Journal of Ethnopharmacology, 196*, pp. 84–93.

Neamsuvan, O., Sengnon, N., Seemaphrik, N., Chouychoo, M., Rungrat, R. and Bunrasri, S., 2015. A survey of medicinal plants around upper Songkhla lake, Thailand. *African Journal of Traditional, Complementary and Alternative Medicines, 12*(2), pp. 133–143.

Neamsuvan, O., Singdam, P., Yingcharoen, K. and Sengnon, N., 2012. A survey of medicinal plants in mangrove and beach forests from sating Phra Peninsula, Songkhla Province, Thailand. *Journal of Medicinal Plants Research, 6*(12), pp. 2421–2437.

Nguyen-Pouplin, J., Tran, H., Tran, H., Phan, T.A., Dolecek, C., Farrar, J., Tran, T.H., Caron, P., Bodo, B. and Grellier, P., 2007. Antimalarial and cytotoxic activities of ethnopharmacologically selected medicinal plants from South Vietnam. *Journal of Ethnopharmacology, 109*, pp. 417–427.

Noor Hassan Sajib, S. Uddin, B. and Islam, Md. M., 2014. Angiospermic plant diversity of Subarnachar Upazila in Noakhali, Bangladesh. *Journal of the Asiatic Society Bangladesh, Science, 40*(1), pp. 39–60.

O'shaughnessy, W.B., 1842. *Bengal Dispensatory.* Calcutta: W. Thacker and Co., St. Andrew's Library.

Ong, H.G., Ling, S.M., Win, T.T.M., Kang, D-H., Lee, J-H. and Kim, Y-D., 2018. Ethnomedicinal plants and traditional knowledge among three Chin indigenous groups in Natma Taung National Park (Myanmar). *Journal of Ethnopharmacology, 225*, pp. 136–158.

Panda, S.K., Das, R., Leyssen, P., Neyts, J. and Luyten, W., 2018. Assessing medicinal plants traditionally used in the Chirang Reserve Forest, Northeast India for antimicrobial activity. *Journal of Ethnopharmacology, 225*, pp. 220–233.

Pandit, P.K. Inventory of Ethno Veterinary Medicinal Plants of Jhargram Division, West Bengal, India.

Patra, J.K. and Thatoi, H.N., 2011. Metabolic diversity and bioactivity screening of mangrove plants: A review. *Acta Physiol Plant, 33*, pp. 1051–1061.

Pereira, J., 1854. *The Elements of Materia Medica and Therapeutics.* Philadelphia: Blanchard and Lea.

Pereira, J., 1855. *Materia Medica and Therapeutics.* London: Longman, Brown, Green, and Longmans.

Perry, L.M. and Metzger J., 1980. *Medicinal Plants of Asia and Southeast Asia.* MIT Press, USA.

Pomet, P., Histoire générale des drogues. A Paris : Chez Jean-Baptiste Loyson, & Augustin Pillon, sur le Pont au Change, à la Prudence. Et au Palais, chez Estienne Ducastin, dans la Gallerie des Prisonniers, au bon Pasteur 1664.

Pradhan, B.K. and Badola, H.K., 2008. Ethnomedicinal plant use by Lepcha tribe of Dzongu valley, bordering Khangchendzonga Biosphere Reserve, in North Sikkim, India. *Journal of Ethnobiology and Ethnomedicine, 4*, p. 22.

Prakasa Rao, J. and Satish, K.V., 2016. *Persicaria perfoliata* (L.) H. Gross (Polygonaceae): A species new to Eastern Ghats of India. *Tropical Plant Research, 3*(2), pp. 249–252.

Qasim, M., Gulzar, S. and Khan, M.A., 2011. Halophytes as medicinal plants. *Conference Paper.*

Qureshi, R. and Raza Bhatti, G., 2008. Ethnobotany of plants used by the Thari people of Nara Desert, Pakistan. *Fitoterapia, 79*, pp. 468–473.

Rahaman, C.H. and Karmakar, S., 2014. Ethnomedicine of Santal tribe living around Susunia hill of Bankura District, West Bengal, India: The quantitative approach. *Journal of Applied Pharmaceutical Science, 5*(02), pp. 127–136.

Rahman, M.A., 2010. Indigenous knowledge of herbal medicines in Bangladesh. 3. Treatment of skin diseases by tribal communities of the hill tracts districts. *Bangladesh Journal of Botany, 39*(2), pp. 169–177.

Rahmatullah, M., Azam, Md. N.K., Malek, I., Nasrin, D., Jamal, F., Rahman, Md. A., Khatun, Z., Jahan, S., Seraj, S. and Jahan, R., 2012. An ethnomedicinal survey among the Marakh Sect of the Garo Tribe of Mymensingh District, Bangladesh. *International Journal of PharmTech Research, 4*(1), pp. 141–149.

Rahmatullah, M., Haque, M.E., Kabir Mondol, M.R., Hasan, M., Aziz, T., Jahan, R. and Seraj, S., 2014. Medicinal formulations of the Kuch Tribe of Bangladesh. *The Journal of Alternative and Complementary Medicine, 20*(6), pp. 428–440.

Rahmatullah, M., Mollik, Md. A.H., Ahmed, Md. N., Bhuiyan, Md. Z.A., Hossain, Md. M., Azam, Md. N.K., Seraj, S., Chowdhury, M.H., Jamal, F., Ahsan, S. and Jahan, R., 2010. A survey of medicinal plants used by folk medicinal practitioners in two villages of Tangail district, Bangladesh. *American-Eurasian Journal of Sustainable Agriculture, 4*(3), pp. 357–362.

Rajendran, A., Ravikumar, K. and Henry, A.N., 2000. Plant genetic resources and knowledge of traditional medicine in Tamil Nadu. *Ancient Science of Life, XX*(1&2), pp. 25–28.

Rama Chandra Prasad, P., Sudhakar Reddy, C., Raza, S.H. and Dutt, C.B.S., 2008. Folklore medicinal plants of North Andaman Islands, India. *Fitoterapia, 79*, pp. 458–464.

Rao, P.K., Hasan, S.S., Bhellum, B.L. and Manhas, R.K., 2015. Ethnomedicinal plants of Kathua district, J&K, India. *Journal of Ethnopharmacology, 171*, pp. 12–27.

Ray, T., 2014. Customary use of mangrove tree as a folk medicine among the sundarban resource collectors. *International Journal of Research in Humanities, Arts and Literature (IMPACT: IJRHAL), 2*(4), pp. 43–48.

Reed, C.F., 1977. Economically Important Foreign Weeds. Potential Problems in the United States. United States Department of Agriculture.

Rehman, H., Begum, W., Anjum, F. and Tabasum, H., 2014. Rheum emodi (Rhubarb): A Fascinating Herb. *Journal of Pharmacognosy and Phytochemistry, 3*(2), pp. 89–94.

Romulo, A., Zuhud, E.A.M., Rondevaldova, J. and Kokoska, L., 2018. Screening of in vitro antimicrobial activity of plants used in traditional Indonesian medicine. *Pharmaceutical Biology, 56*(1), pp. 287–293.

Roy, S., 2015. An ethnobotanical study on the medicinal plants used by Rajbanshis of Coochbehar district, West Bengal, India. *Journal of Medicinal Plants Studies, 3*(5), pp. 46–49.

Safa, O., Soltanipoor, M.A., Rastegar, S., Kazemi, M., Dehkordi, K.N. and Ghannadi, A., 2012. An ethnobotanical survey on hormozgan province, Iran. *Avicenna Journal of Phytomedicine, 3*(1), Winter 2013, pp. 64–81.

Saha, M.R., De Sarker, D. and Sen, A., 2014. Ethnoveterinary practices among the tribal community of Malda district of West Bengal, India. *Indian Journal of Traditional Knowledge, 13*(2), pp. 359–367.

Saikia, A.P., Ryakala, V.K., Sharma, P., Goswami, P. and Bora, U., 2006. Ethnobotany of medicinal plants used by Assamese people for various skin ailments and cosmetics. *Journal of Ethnopharmacology, 106*, pp. 149–157.

Sandakan, S., 2005. Preferred check-list of Sabah trees.

Sankaranarayanan, S., Bama, P., Ramachandran, J., Kalaichelvan, P.T., Deccaraman, M., Vijayalakshimi, M., Dhamotharan, R., Dananjeyan, B. and Sathya Bama, S., 2010. Ethnobotanical study of medicinal plants used by traditional users in Villupuram district of Tamil Nadu, India. *Journal of Medicinal Plants Research, 4*(12), pp. 1089–1101.

Sapma, S., 2017. Indigenous plant. *ECHO Asia Conference.* October 2–6, 2017. Chiang Mai.

Savithramma, N., Yugandhar, P. and Linga Rao, M., 2014. Ethnobotanical studies on Japali Hanuman Theertham-A sacred grove of Tirumala hills, Andhra Pradesh, India. *Journal of Pharmaceutical Sciences and Research, 6*(2), pp. 83–88.

Shah, A., Ahmad, M., Marwat, S.K. and Zafar, M., 2013. Ethnobotanical study of medicinal plants of semi-tribal area of Makerwal & Gulla Khel (lying between Khyber Pakhtunkhwa and Punjab Provinces). *American Journal of Plant Sciences, 4*, pp. 98–116.

Shang, X., Tao, C., Miao, X., Wang, D., Tangmuke, Dawa, Wang, Y., Yang, Y. and Pan, H., 2012. Ethno-veterinary survey of medicinal plants in Ruoergai region, Sichuan province, China. *Journal of Ethnopharmacology, 142*, pp. 390–400.

Sharma, J., Gairola, S., Sharma, Y.P. and Gaur, R.D., 2014. Ethnomedicinal plants used to treat skin diseases by Tharu community of District Udham Singh Nagar, Uttarakhand, India. *Journal of Ethnopharmacology, 158*, pp. 140–206.

Simpson, D.A. and Inglis, C.A., 2001. Cyperaceae of economic, ethnobotanical and horticultural importance: A checklist. *Kew Bulletin, 56*(2), pp. 257–360.

Singh, J.P., Rathore, V.S. and Roy, M.M., 2015. Notes about Haloxylon salicornicum (Moq.) Bunge ex Boiss., a promising shrub for arid regions. *An International Journal, 62*, pp. 451–463.

Singh, B., Sultan, P., Hassan, Q.P., Gairola, S. and Bedi, Y.S. Ethnobotany, traditional knowledge, and diversity of wild edible plants and fungi: A case study in the Bandipora District of Kashmir Himalaya, India. *Journal of Herbs, Spices & Medicinal Plants.* DOI: 10.1080/10496475.2016.1193833

Song, M-J. and Kim, H., 2011. Ethnomedicinal application of plants in the western plain region of North Jeolla Province in Korea. *Journal of Ethnopharmacology, 137*, pp. 167–175.

Song, P., Sekhon, H.S., Lu, A., et al., 2007. M3 Muscarinic receptor antagonists inhibit small cell lung carcinoma growth and mitogen-activated protein kinase phosphorylation induced by acetylcholine secretion. *Cancer Research, 67*, pp. 3936–3944.

Srivastava, R.C., 2014. Family polygonaceae in India. *Indian Journal of Plant Sciences, 3*(2), pp. 112–150.

Sur, P.R., Sen, R., Halper, A.C. and Bandyopadhyay, S., 1987. Observation on the ethnobotany of malda-west Dinajpur Districts West Bengal-I. *Journal of Economic and Taxonomic Botany, 10*(2).

Tabata, M., Sezik, E., Yeşilada, E., et al. Traditional medicine in Turkey III. Folk medicine in East Anatolia, Van and Bitlis Provinces. *Pharmaceutical Biology.* doi: 10.3109/13880209409082966

Thapa, L.B., Dhakal, T.M., Chaudhary, R. and Thapa, H., 2013. Medicinal plants used by Raji ethnic tribe of Nepal in treatment of gastrointestinal disorders. *Our Nature, 11*(2), pp. 177–186.

Tolken, H.R. The species of Arthrocnemum and Salicornia (Chenopodiaceae) in Southern Africa. *Bothalia, 9*(2), pp. 255–307.

Tribedi, G.N., Mudgal, V. and Pal, D.C., 1993. Some less known ethnomedicinal uses of plants in sunderbans, India. *Bulletin of the Botanical Survey of India, 35*(1–4), pp. 6–10.

Tripathi, Y.C., Prabhu, V.V., Pal, R.S. and Mishra, R.N., 1996. Medicinal plants of Rajasthan in Indian system of medicine. *Ancient Science of Life, XV*, pp. 190–212.

Tumpa, S.I., Hossain, Md. I. and Ishika, T., 2014. Ethnomedicinal uses of herbs by indigenous medicine practitioners of Jhenaidah District, Bangladesh. *Journal of Pharmacognosy and Phytochemistry, 3*(2), pp. 23–33.

Uddin, S.N. and Hassan, Md. A., 2012. Angiosperm flora of Rampahar reserve forest under Rangamati district in Bangladesh. I. Liliopsida (Monocots). *Bangladesh Journal of Plant Taxonomy, 19*(1), pp. 37–44.

Uddin, M.Z., Hassan, Md. A. and Sultana, M., 2006. Ethnobotanical survey of medicinal plants in Phulbari Upazila of Dinajpur district, Bangladesh. *Bangladesh Journal of Plant Taxonomy, 13*(1), pp. 63–68.

Uddin, K., Mahbubur Rahman, A.H.M. and Rafiul Islam, A.K.M., 2014. Taxonomy and traditional medicine practices of Polygonaceae (Smartweed) Family at Rajshahi, Bangladesh. *International Journal of Advanced Research, 2*(11), pp. 459–469.

Ullah, M., Khan, M.U., Mahmood, A., Malik, R.N., Hussain, M., Wazir, S.M., Daud, M. and Shinwari, Z.K., 2013. An ethnobotanical survey of indigenous medicinal plants in Wana district south Waziristan agency, Pakistan. *Journal of Ethnopharmacology, 150*, pp. 918–924.

Ullah, S., Khan, M.R., Shah, N.A., Shah, S.A., Majid, M. and Farooq, M.A., 2010. Ethnomedicinal plant use value in the Lakki Marwat District of Pakistan. *Journal of Ethnopharmacology.*

Upadhyay, O.P., Kumar, K. and Tiwari, R.K., 2008. Ethnobotanical study of skin treatment uses of medicinal plants of Bihar. *Pharmaceutical Biology, 36*(3), pp. 167–172.

Vaidya, B.N. and Joshee, N., Report on Nepalese Orchid Species with Medicinal Properties. https://www.researchgate.net/publication/312888202.

Van Sam, H., Nanthavong, K. and Kessler, P.J.A., 2004. Trees of Laos and Vietnam: A field guide to 100 economically or ecologically important species. *Blumea, 49*, pp. 201–349.

Veldkamp, J.F. and Flipphi, R.C.H., 1987. A revision of Leptonychia (Sterculiaceae) in Southeast Asia. *BLUMEA, 32*, pp. 443–457.

Wagh, V.V. and Jain, A.K. Ethnopharmacological survey of plants used by the *Bhil* and *Bhilala* ethnic community in dermatological disorders in Western Madhya Pradesh, India. *Journal of Herbal Medicine.*

Wangchuk, P., Keller, P.A. Pyne, S.G., Taweechotipatr, M., Tonsomboon, A., Rattanajak, R. and Kamchonwongpaisan, S., 2011. Evaluation of an ethnopharmacologically selected Bhutanese medicinal plants for their major classes of phytochemicals and biological activities. *Journal of Ethnopharmacology, 137*, pp. 730–742.

Warrier, K.C.S. and Warrier, R.R., 2019. *Sacred Groves: Repositories of Medicinal Plants.* Coimbatore: Institute of Forest Genetics and Tree Breeding.

Watt, G., 1889. *A Dictionary of the Economic Products of India.* Calcutta: Government Printing.

Wazir, A.R., Shah, S.M. and Razzaq, A. Ethno botanical studies on plant resources of Razmak, North Waziristan, Pakistan.

WHO, 2009. *Medicinal Plants in Papua New Guinea*. Western Pacific Region:WHO.

Yaseen, G., Ahmad, M., Shinwari, S., Potter, D., Zafar, M., Zhang, G., Shinwari, Z.K. and Sultana, S., 2019. Medicinal plant diversity used for livelihood of Public health in deserts and arid regions of Sindh-Pakistan. *The Pakistan Journal of Botany, 51*(2), pp. 657–679.

Yazan, L.S. and Armania, N., 2014. *Dillenia* species: A review of the traditional uses, active constituents and pharmacological properties from pre-clinical studies. *Pharmaceutical Biology, 52*(7), pp. 890–897.

Zheng, X-l., Wei, J-h., Sun, W., Li, R-t., Liu, S-b. and Dai, H-f., 2013. Ethnobotanical study on medicinal plants around Limu Mountains of Hainan Island, China. *Journal of Ethnopharmacology, 148*, pp. 964–974.

# Index

## A

(E)-ajoene 166
1′-acetoxychavicol acetate 257, 258
1′S-1′-acetoxyeugenol acetate 256, 259
1-allyl-2,6-dimethoxy-3,4-methylenedioxybnzene 51
4-allylphenyl acetate 255
6,8-di-C-arabinopyranosyl luteolin 249
8-acetylheterophyllisine 337
a qia, jingangteng (China) 158
aba-aba (Philippines) 128
abacavir 20
abang asuh (Malaysia) 316
aceriphyllic acid 370
*Aceriphyllum rossii* (Oliv.) Engl. 369–371
acetogenins 89
acetylcorynoline 296
achatsarmum (India) 308
achheni (Nepal) 119
*Aconitum carmichaeli* Debeaux 330
*Acrotrema costatum* Jack. 349
actein 333
adakut (Myanmar) 250
adlay (Philippines) 237
adlumidine 303, 304
aegop (Papua New Guinea) 182
aerial yam 140
afing gash (India) 312
afyum (India) 312
afzelin 86, 357
agya ghas (India) 238
ahiphena (India) 312
ahos (Philippines) 164
aidu (India) 260
air plant 356
air potato 140
ajacisine E 337
aka kai (Thailand) 93
akanadi (Bangladesh) 321
akanbindi 313
akar chenana (Malaysia) 93
akar julong (Malaysia) 35
akar ketola hutan (Malaysia) 39
akar kuning (Indonesia) 316
akar kusin (Malaysia) 326
akar mempelas (Malaysia) 351
akar patawali (Malaysia) 329
akar siak (Malaysia) 217
akar surai (Malaysia) 35
akar wangi (Indonesia, Malaysia) 248
*Akebia quinata* (Houtt.) Decne 284–285
alach (India) 260
alang-alang (Indonesia) 244
aliphatic hydroxyketones 252
alkaloid aristolactam 38
alkaloid protopine 282
alkaloids columbamine 279

alky sulfur compounds 163
Allicin 165, 166
allocryptopine 307, 310
Aloe leaf Cymbidium 203
aloe-emodin 214–216
aloin A 214, 215
*Alpinia conchigera* Griff. 255–257
*Alpinia galanga* (L.) Willd. 257–258
*Alpinia nigra* (Gaertn.) B. L. Burtt. 259–260
altholactone 96–97
*Altingia excelsa* Noronha 352–353
Altingiaceae Horaninow (1841) 352–354
amamali (Philippines) 373
amar poi (India) 356
amblyone 130
*Amborella trichopoda* Baill 1–2
ambuda (India) 232
American Aloe 175
American eleutherine 195
aminopeptidase N 54, 55
amla bela (Bangladesh) 129
amla lata (Bangladesh) 375
amlavetash (India) 375
*Amomum dealbatum* Roxb. 260–261
amora (Philippines) 248
amorseko (Philippines) 236
amphiphilic alkyl phenols 113
amphiphilic phenylpropanoid 255
ampicillin 338
amprenavir 20
an luo (China) 102
andong (Indonesia; Malaysia) 182
*Anemone biflora* DC. 331
*Anemone obtusiloba* D. Don 331–332
angustific acid A 20
anigorufone 228
*Annona reticulata* L. 88–90
*Annona squamosa* L. 90–92
Annonaceae 88
annonaine 110
annotemoyin 91
anthocyanins 197
anthraquinones 219
antikinari (Bangladesh) 162
ao chun jiang (China) 262
*Apama corymbosa* (Bl.) O. Ktze 35
apeli (Fiji) 90
apigenin 54
apiole 53
aporphine 90
aporphine alkaloids 87, 104, 105
*Aquilegia vulgaris* L. 332
ardika (India) 266
areca palm 210
arecoline 212
*Argemone mexicana* L. 287–288
arid-land peperomia 54

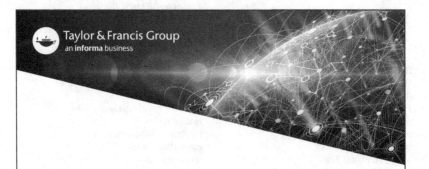

Printed in the United States
by Baker & Taylor Publisher Services

Printed in the United States
by Baker & Taylor Publisher Services